Animal Welfare

Volume 25

The Animal Welfare series has been designed to help contribute towards a culture of respect for animals and their welfare by producing academic texts addressing how best to provide for the welfare of the animal species that are managed and cared for by humans. Books in the series do not provide a detailed blue-print for the management of each species, but they do describe and discuss the major welfare concerns, often in relation to the wild progenitors of the managed animals. Welfare has been considered in relation to animals' needs, concentrating on nutrition, behaviour, reproduction and the physical and social environment. Economic effects of animal welfare provision were also considered where relevant, as were key areas where further research is required.

Silvana Mattiello • Monica Battini
Editors

The Welfare of Goats

Co-ordinating Editors: Annabelle Beaver and Clive Phillips

Springer

Editors
Silvana Mattiello
Department of Agricultural and Environmental Sciences - Production Landscape, Agroenergy
University of Milan
Milan, Italy

Monica Battini
Department of Agricultural and Environmental Sciences - Production Landscape, Agroenergy
University of Milan
Milan, Italy

This work contains media enhancements, which are displayed with a "play" icon. Material in the print book can be viewed on a mobile device by downloading the Springer Nature "More Media" app available in the major app stores. The media enhancements in the online version of the work can be accessed directly by authorized users.

ISSN 1572-7408
Animal Welfare
ISBN 978-3-031-62181-9   ISBN 978-3-031-62182-6 (eBook)
https://doi.org/10.1007/978-3-031-62182-6

This Springer imprint is published by the registered company Springer Nature Switzerland AG
The registered company address is: Gewerbestrasse 11, 6330 Cham, Switzerland

Springer Nature More Media App

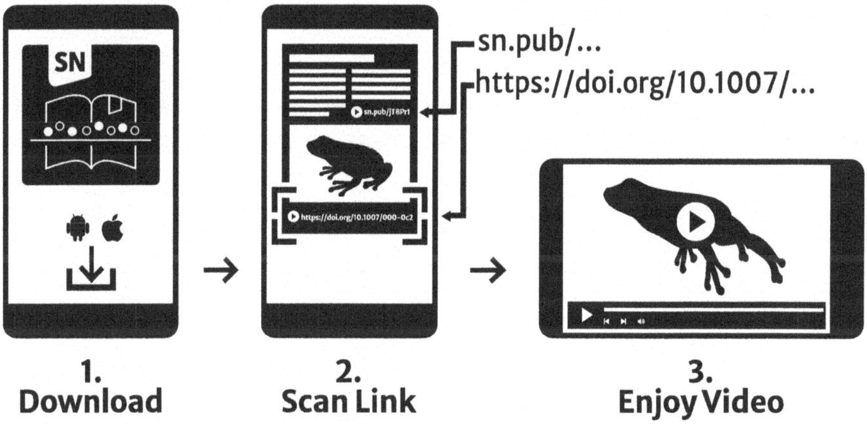

Support: customerservice@springernature.com

Animal Welfare Series Preface

Animal welfare is attracting increasing interest worldwide, and the knowledge and resources are available to, at least potentially, provide better management systems for farm animals, as well as companion, zoo, laboratory, and performance animals. The key requirements for adequate food, water, a suitable environment, companionship, and health are important for animals kept for all of these purposes.

The attention given to animal welfare in recent years derives largely from the fact that the relentless pursuit of financial reward and efficiency, to satisfy market demands, has led to the development of intensive animal management systems that challenge the conscience of many consumers, particularly in the farm and laboratory animal sectors. Livestock are the world's biggest land users, and the farmed animal population is increasing rapidly to meet the needs of an expanding human population. This results in a tendency to allocate fewer resources to each animal and to value individual animals less than the group. In these circumstances, the importance of each individual's welfare is diminished.

Increased attention to welfare issues is just as evident for zoo, companion, sport, and wild animals. Of growing importance is the ethical management of breeding programs since genetic manipulation is now technically advanced. There is less public tolerance of the breeding of extreme animals if it comes at the expense of animal welfare (e.g., brachycephalic dogs). The quest for producing novel genotypes has fascinated breeders for centuries. Dog and cat breeders have produced a variety of deformities that have adverse effects on their welfare, but nowadays the breeders are just as active in the laboratory, where the mouse is genetically manipulated with equally profound effects.

In developing countries, human survival is still a daily uncertainty for many, so that provision for animal welfare has to be balanced against human welfare. Animal welfare is usually a priority only if it supports the output of the animal, be it food, work, clothing, sport, or companionship. However, in many situations the welfare of animals is synonymous with the welfare of the humans that look after them, because happy, healthy animals will be able to assist humans best in their struggle for survival. In principle, the welfare needs of both humans and animals can be provided for, in both developing and developed countries, if resources are properly husbanded. In reality, the inequitable division of the world's riches creates physical and psychological poverty for humans and animals alike in many parts of the world.

The intimate connection between animals and humans that was once so essential for good animal welfare is rare nowadays, having been superseded by technologically efficient production systems where animals on farms and in laboratories are tended by increasingly few humans in the drive to enhance labor efficiency. With today's busy lifestyles, companion animals too may suffer from reduced contact with humans, although their value in providing companionship, particularly for certain groups such as the elderly, is beginning to be recognized. Animal consumers also rarely have any contact with the animals that are kept for their benefit.

In this estranged, efficient world, people struggle to find the moral imperatives to determine the level of welfare that they should afford to animals within their charge. A few people, and in particular many companion animal owners, strive for what they believe to be the highest levels of welfare provision, while others, deliberately or through ignorance, keep animals in impoverished conditions in which their health and wellbeing can be extremely poor. Today's multiple moral codes for animal care and use are derived from a broad range of cultural influences, including media reports of animal abuse, guidelines on ethical consumption, and campaigning and lobbying groups.

This series has been designed to contribute toward a culture of respect for animals and their welfare by producing learned treatises about the provision for the welfare of the animal species that are managed and cared for by humans. The early species-focused books were not detailed management blue-prints; rather they described and considered the major welfare concerns, often with reference to the behavior of the wild progenitors of the managed animals. Welfare was specifically focused on animals' needs, concentrating on nutrition, behavior, reproduction, and the physical and social environment. Economic effects of animal welfare provision were also considered where relevant, as were key areas where further research is required.

This volume addresses the welfare of goats, a species of farm animal whose welfare is neglected, for the first time in this Springer series. If the welfare of goats is considered, it is usually when they are kept in intensive housing systems of production, with concerns about the feeding, housing, health, behavior, reproduction, transport, and slaughter of the goats. However, most goats are kept in subsistence agriculture in developing regions of the world with very different concerns. This volume addresses this omission in some detail, including the impact that global climate change will have on these relatively unprotected extensive systems of production. It also considers the growing role of goats in urban and peri-urban settings, vital at a time when there is mass migration to the cities. The editors conclude that the welfare of goats is relatively understudied compared to more common livestock species. The volume addresses current gaps in knowledge and outlines recommendations for consideration for goat management. It is sure to be a key reference for those working with or interested in goats.

Perth, WA, Ausralia — Clive Phillips
Edgemont, UK — Annabelle Beaver

Contents

Where Do Goats Come From?

Paola Crepaldi, Arianna Bionda, and Licia Colli

Abstract

This chapter describes the history and biodiversity of the goat species, whose worldwide distribution demonstrates its great adaptability to a wide range of environmental conditions. We explain the genomic evidence clarifying the contribution of wild ancestors, the dynamics of the domestication process, and the post-domestication evolutionary history of this species, also reviewing the signals that both natural and human-mediated selection have left in goats' genomes. Animal welfare is related to the mental and physical state of a single subject, whereas genomic tools deal with populations. Hence, in the context of genetics, the concept of "fitness" depends on the interaction between genotype/phenotype and environments and is estimated as the individual's contribution of descendants to the next generation. During the domestication process, goats adapted to a new environment created by human activities and that differed substantially from the natural conditions in which they evolved. Similar to what happens in the natural context, also under anthropogenic conditions an increase in fitness comes at a cost, i.e., the morphological and physiological changes induced by the adaptation to human management and to the new environmental and nutritional conditions, which likely generated stress in the animals, particularly at the beginning of the process.

P. Crepaldi (✉) · A. Bionda
Department of Agricultural and Environmental Sciences - Production, Landscape, Agroenergy, Università degli Studi di Milano, Milano (MI), Italy
e-mail: paola.crepaldi@unimi.it

L. Colli
Department of Animal Science, Nutrition and Food, BioDNA Research Centre on Biodiversity and Ancient DNA, Università Cattolica del Sacro Cuore, Piacenza (PC), Italy

S. Mattiello, M. Battini (eds.), *The Welfare of Goats*, Animal Welfare 25,
https://doi.org/10.1007/978-3-031-62182-6_1

Keywords

Domestication · Genomics · Origins · Wild Goats · Selection · Adaptation

1 The Origin of Domestic Goats

The bezoar, *Capra aegagrus*, is the progenitor of domestic goats, *Capra hircus*. Available evidence indicates that, after a demographic expansion in the Late Pleistocene (ca. 250,000 years ago), wild goat populations inhabiting south-western Asia underwent strong reductions in census size during the last two glacial periods, i.e. 130–160,000 and 12–71,000 years ago (Colli et al. 2015; Nomura et al. 2013). During these glacial periods, the nuclei of individuals which survived in refugial areas in south-western Asia remained separate for thousands of years and became genetically differentiated. Thus, the wild goat populations that were involved in the domestication process were already highly differentiated at the level of both nuclear and uniparental (i.e. mitochondrial DNA, Y-chromosome) markers (Colli et al. 2015; Daly et al. 2018; Nijman et al. 2020).

Today the bezoar, which is classified as a near-threatened species by IUCN (International Union for Conservation of Nature), lives in the shrubland and mountainous areas of central-southern Asia, over a range spanning from the south-east of Turkey to Pakistan (Weinberg and Ambarli 2020) (Fig. 1). The bezoars are bigger than the domestic goats and present a coat with variable colouration, usually grey or brown, with darker areas on the muzzle, chest, back, and limbs, and a lighter abdomen. Both sexes have beard and horns, which in males are particularly long, curved like a scimitar, and with a sharp inside edge. Does and kids live in packs of about 50 individuals, separated from bucks, which stay alone or form packs of males only (Weaver 2020).

According to zooarchaeological and molecular evidence, goat domestication began during the Neolithic around 11,000 years ago in south-western Asia, over an area located north of the Fertile Crescent and spanning across the Zagros Mountains of north-western Iran, the Caucasus, and south-eastern Anatolia (Colli et al. 2015; Daly et al. 2018; Naderi et al. 2008; Vigne et al. 2005; Zeder et al. 2006). Palaeoclimatic studies report that, starting from the peak of the last glacial period (ca. 14,000 years Before the Common Era, BCE) until the beginning of the Holocene (ca. 9750 BCE), this area was characterized by an arid and cold steppe climate, followed by a slow transition to warmer and wetter conditions (Bosomworth et al. 2020). The climate change occurring in this area at the end of the Pleistocene is widely accepted as one of the main drivers of the transition from the human hunting-gathering culture to agriculture and herding: the harsh environmental conditions, paired with an excessive hunting pressure, led to a reduction in the inhabitable territories as well as in the number of wild ungulates. This made it necessary for humans to develop new ways to obtain and preserve food (Clutton-Brock 1999).

Therefore, the goat domestication process can be considered as the outcome of a gradual passage from active hunting to captivity and management of wild animals (Larson and Burger 2013; Vigne et al. 2005; Zeder 2008, 2012; Zeder and Hesse

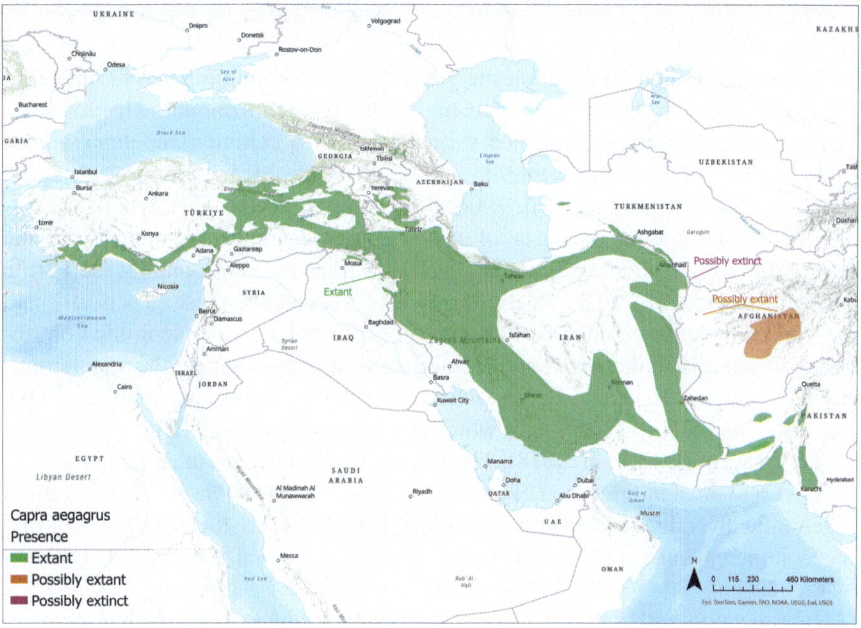

Fig. 1 Present-day distribution area of the bezoar, *C. aegagrus*, the wild ancestor of domestic goats. Data from NatureServe and IUCN (2020)

2000). Assemblages of bone remains deriving from hunting activities are, in fact, mainly attributable to large adult animals, especially males, which maximize the return in terms of meat. On the other hand, the archaeological assemblages found in Iraq and Iran and dating back to 9900 BCE mainly consisted of young males and adult females, indicating a transition to herding where females are slaughtered after finishing their productive career and excess males are culled at a young age. At that point, goats were still very similar to their wild counterparts, but some domestication-induced morphological changes, such as a reduction in body size and the alteration of horn shape and size, appeared about 500–1000 years later when the herds were moved to Iran lowlands, which were characterized by a hotter and dryer climate (Zeder 2008, 2012).

There are different reasons why goats and sheep were among the first domesticated species, whereas other ungulates sharing the same territories, such as deer and gazelles, did not undergo the same process. Among these reasons, the natural social structure of sheep and goats plays a role of primary importance: these animals follow a single dominant leader and, despite having a home range, namely a restricted area where they live and search for food, they are not territorial. This made them suitable to be led by herdsmen and to be gathered in compact and large groups. Certainly, goats' great resilience and varied diet also contributed to their success as domesticated animals; indeed, it has been suggested that Neolithic farmers might have taken advantage of goats' activity of clearing land after the forest was burnt or cut down (Clutton-Brock 1999).

As previously mentioned, domestic goats derive from multiple bezoar nuclei, as first suggested by the occurrence of six highly divergent mitochondrial haplogroups (i.e. groups of haplotypic sequence variants sharing a common ancestor), whose origin clearly predated the Neolithic age (Colli et al. 2015; Luikart et al. 2001; Naderi et al. 2008; Sardina et al. 2006; Sultana et al. 2003). This hypothesis was later confirmed by the analysis of ancient goat genomes spanning from the Palaeolithic to the Middle Ages and covering a large area surrounding the Fertile Crescent (Daly et al. 2018). According to this latter study, these genetically and geographically differentiated early domestic goat nuclei of the Neolithic contributed differentially to the gene pools of modern goat populations of the major continents, i.e., Africa, Asia, and Europe when, after domestication, goats followed the humans along the migration routes which spread agriculture and farming out of the Fertile Crescent (Daly et al. 2018). The traces of this pre-domestication population structure and ancient migration events are still clearly detectable in the distribution of genomic diversity of present-day goat populations, where the occurrence of distinct continental gene pools is still evident (Colli et al. 2018; Daly et al. 2018; Zheng et al. 2020).

Hybridization with wild relatives is another evolutionary force which seems to have impacted profoundly on the genetic makeup of domestic goat populations (Daly et al. 2022; Zheng et al. 2020). Nine wild goat species of the genus *Capra* are currently extant, with limited distribution areas scattered across Eurasia and North Africa (Amills et al. 2017) (Fig. 2). Being interfertile with *C. hircus*, many of these wild species can cross with domestic goats (Grossen et al. 2014; Hammer et al. 2008; Pidancier et al. 2006) and seem to have contributed differentially to pre- and post-domestication evolution through the so-called adaptive introgression phenomenon. This occurs when, after a hybridization event, natural selection leads to the retainment of genetic variants derived from a different species or source population and conferring a selective advantage and increased fitness (Hedrick 2013).

Evidence exists of hybridization between bezoars and a species sister to the present-day Caucasian turs (*Capra caucasica* and *Capra cylindricornis*) having occurred in the Caucasus region before the onset of the domestication process (Daly et al. 2022; Zheng et al. 2020). Later crossing of domestic goats and ibexes in the Alps (*C. hircus* x *Capra ibex*) and the Pyrenees (*C. hircus* x *Capra pyrenaica*) occurred when the early domestic populations reached Southern Europe (Alasaad et al. 2011; Grossen et al. 2014). This interspecific gene flow seems to have been bidirectional and may account for the discrepancies found between phylogenies based on different loci (Pidancier et al. 2006; Ropiquet and Hassanin 2006).

The high rusticity and capability to adapt even to harsh environments made goats a key resource during both the agricultural transition of the Neolithic and the diffusion of agriculture and farming in the subsequent millennia (Joshi et al. 2004). Local populations of this species became adapted to a large variety of different environmental conditions, often contrasting in terms of extremes of temperature, humidity, water and feed availability, and diseases (Bertolini et al. 2018). Early after

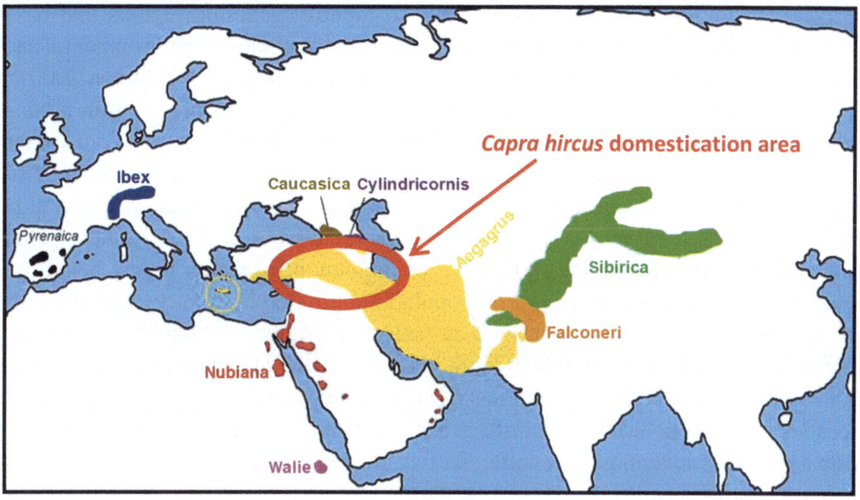

Fig. 2 Present-day distribution areas of the nine extant wild goat species and indication of the *C. hircus* domestication site. *Capra pyrenaica* (black) is present in the Iberian Peninsula; *Capra ibex* (blue) in the Alpine region; *Capra caucasica* (brown) and *Capra cylindricornis* (purple) on Great Caucasus Mountains; *C. aegagrus* (yellow) across Southwestern Asia and the Caucasus region as far as southwestern Turkey; *Capra sibirica* (green) in Central Asia, as far north as southern Siberia; *Capra falconeri* (orange) in Afghanistan, Pakistan, Tajikistan, Uzbekistan, and India; *Capra nubiana* (red) in Egypt, Ethiopia, and Arabian Peninsula; and *Capra walie* (pink) in Ethiopia. Modified from Amills et al. (2017)

domestication (i.e. 11,000 years ago), human-mediated selection began and started affecting morphological and production traits (Daly et al. 2018). In time, this led to the formation of phenotypically distinct breeds with clearly different production purposes and adapted to a variety of different environments.

2 The Effects of Domestication and Selection

2.1 Understanding the Origin and Diversity of Goats: DNA-Based Tools

Recent advances in genomic techniques have provided useful tools for understanding the origin and genetic diversity of domestic goats. The international scientific community has focused widely on studying the genome of several goat breeds around the world, to understand the molecular mechanisms underlying their ability to adapt to the different climatic and environmental contexts in which local goat populations are bred since the post-Neolithic expansion.

The evolution of techniques now offers molecular markers, such as single nucleotide polymorphisms (SNPs), or complete genome sequencing, that permit evaluation of similarities and differences between and within breeds. This allows us to reconstruct phylogenetic relationships and to identify genomic regions that have been affected by natural or anthropogenic selection in relation to environmental and climatic conditions, or to selection goals established by humans (Kreitman 2000).

Selection, indeed, induces specific changes in the patterns of variation among selected loci and in neutral loci linked to them. These genomic footprints are known as selection signatures and can be used to identify loci subjected to natural and artificial selection. A reduction in genetic variation in the genomic regions adjacent to the selected beneficial mutation is caused by the rapid fixation of that mutation after some generations (Kreitman 2000). The phenomenon by which a new beneficial mutation increases in frequency in a population and thereby reduces variability in the associated neutral sites is termed as "selective sweep" or "hitchhiking effect" (e.g. Maynard Smith and Haigh 1974; Stella et al. 2010). The identification of selection signatures has been a key interest of animal geneticists and breeders in the last decades as it can be used to identify genes and beneficial mutations that may provide a selective advantage to a particular livestock population (Zhao et al. 2015).

2.2 Domestication Syndrome: Selection for Behaviour?

Domesticated species share many morphological, behavioural, and physiological traits, collectively referred to as the "domestication syndrome". In animals, selection for tameness is believed to have been fundamental to the establishment of the human–animal relationship (Belyaev 1979). It has led to changes in the levels and patterns of animal development with the maintenance of neotenic features and the appearance of common phenotypic traits such as white patches, drooping ears, and curled tails. The expression of genes with pleiotropic effects, i.e., affecting several traits at the same time, related to the development of the nervous, immune, and endocrine systems (particularly adrenocortical hormones secretion), or involved in the fear/flight reaction and pigmentation, is well known in many domesticated animals (Wilkins et al. 2014).

Selective pressure during the initial stages of domestication has targeted behaviour, specifically the reduction of fearful or aggressive responses to humans. The decrease of acute fear and long-term stress is a prerequisite for assuring high welfare levels, as well as successful reproduction in captivity. The diminution in the size and function of the adrenal glands, which leads to a decrease of the physiological reaction to fear and stress, is also considered a fundamental step that occurred during domestication. Moreover, this goes along with a slower maturation of the adrenal gland and the consequent extension of the socialization window. During this period, kids are exposed to human presence and interaction and, thus, get familiar with them (Wilkins et al. 2014).

Adrenal gland development is closely linked to neural crest cells (NCCs), a specific class of stem cells that appear during early embryogenesis at the level of the dorsal "crest" of the neural tube and then migrate ventrally through the body, both in the head and trunk. NCC-derived tissues include much of the head, sympathetic ganglia, adrenal medulla, pigment-related melanoblasts in both the head and trunk, and tooth precursors. Traits associated with the domestication syndrome, such as behavioural and morphological neotenic characters (e.g. reduction in skull and body size, drooping ears, disappearance of homochromia typical of the wild species, and appearance of white patches) are all variations associated with migration and developmental patterns of NCCs (Wilkins et al. 2014).

Signatures have been detected in the genomes of several domesticated species pointing at selective pressures acting on genes with pleiotropic effects involved in the pathways described previously. In the following section, these signatures of selection identified in goats will be described.

2.3 Early Changes in the Goat Genome

In addition to these direct effects of domestication, further changes in the genome, and thus in the phenotype, of goats can be traced back to the earliest stages of the herding activity, and to the strong selective pressure exerted by the abrupt changes in lifestyle and feeding regimes. The comparison between genomes of Neolithic goats and modern bezoar allowed for the identification of some loci that showed high divergence and reduced variability. In particular, two pigmentation loci (*KIT* and *KITLG*), related to piebaldism, i.e., the absence of melanocytes producing pigment in certain areas of skin and hair, might have been favoured either indirectly within the context of the domestication syndrome or directly by humans as a way to better distinguish animals belonging to different herds. Other identified loci were related to a reduction of height at withers (*SIRT1*) and kidding interval (*EPGN*), as well as enhanced mammary gland development (*STAT1*). Moreover, domestic goats seem to have developed higher resistance to some xenobiotics (external substances), including the ones produced by fungi contaminating the cereals that were probably used to feed livestock, thus showing an adaptation to their new diet (Daly et al. 2018). Further molecular evidence pointed at selection signals in genes related to behaviour and resistance to gastrointestinal pathogens and parasites. Particularly, a favourable mutation in the *MUC6* gene conferring higher tolerance towards nematodes might have resulted from adaptive introgression from a wild species related to the West Caucasian tur followed by natural selection (Zheng et al. 2020).

While domestication has triggered positive selection for many traits in livestock species, it has also led to a relaxation of selection over traits of lesser importance in anthropogenic conditions, such as camouflage colouration, sexual selection, twinning, and the ability to escape or avoid predators (Alberto et al. 2018).

2.4 Back to the Wild

As reported before, the core of the bezoar's current distribution area is central-southern Asia; however, at present, some goats displaying bezoar-like phenotypic features are found also on different Mediterranean islands, like Crete, Youra, Montecristo, and Antimilos. The current opinion is that these populations are not native of these regions, but derive from domestic goats or early-domesticates introduced by humans in ancient times, from the Neolithic period onwards (Horwitz and Bar-Gal 2006; Masseti 2009; Somenzi et al. 2022).

Due to the lack of *ad hoc* genomic investigations, it has not yet been clarified to what extent the bezoar-like populations of the different islands derive from already domesticated goats or from wild taxa released on the islands as "living larders", or from the interbreeding of the starting wild stock with domesticated animals introduced in later times (Horwitz and Bar-Gal 2006; Masseti 2009). It is evident that these insular wild goats share phenotypical similarities with both the bezoars (e.g. coat pattern and horn shape) and domestic goats (e.g. smaller size, even though this could also be related to the so-called insular dwarfism).

Crete island hosted the largest population of free-living goats (*C. a. cretica*, often referred to as Agrimi). These animals used to live in all the mountainous regions of the Isle, but their range has since contracted (Horwitz and Bar-Gal 2006; Masseti 2009). Recent scientific literature about their current distribution is lacking, but they are reported to be found only on the three small islands of Dia, Thodorou, and Agii Pontes just off the shore of Crete (Colli, personal communication).

The island of Montecristo, located in the Tuscan Archipelago, is home to the homonymous goat. The origin of this population has not yet been clarified: some authors suggest it was introduced during the Neolithic, while others date this event between the sixth and thirteenth centuries and attribute it to the activities of the monk community that was present on the island at that time. During the last two centuries and until the establishment of a natural reserve in 1971, Montecristo island has been extensively used as a game reserve, which led to the necessity to restock the goat population with domesticated animals imported from the mainland, as indicated by a shared common ancestry highlighted by molecular evidence (Somenzi et al. 2022).

2.5 Adaptation and Selection

The earliest type of human-mediated selection, i.e., for tameness, was likely an involuntary process which was then followed by an active but empirically performed selection carried out for millennia to improve production traits (Clutton-Brock 1999). Later, phenotypic traits such as coat colours, body shape, and size started to be selected, when the standardization began in England during the Victorian era and subsequently spread mainly to Europe and the developed countries. To maintain a standardized appearance, between-breed crossbreeding was prevented which made it possible to push selection over performance characteristics further, such as meat, milk, or fibre production. Consequently, highly standardized breeds often became specialized for only one attitude, although many local breeds, especially in low-income countries, are still bred for multiple purposes (FAO 2015).

In the case of goats, differently from most livestock species, the extensive and nomadic or semi-nomadic breeding system has maintained the animals exposed to the pressures from the natural environment over time. Goat breeds have evolved to survive, breed, and produce in a variety of different and often harsh environments, and their genomes are likely to have segregated unique allele variants underlying adaptation to such diverse production regimes and natural conditions. At the same time, over the centuries, goat breeds have been selected for different economically important traits, such as growth, milk and meat production, and coat/wool features and quality. Environmental- and production-driven selective pressures are strongly interconnected: animals well adapted to their living conditions will perform better with respect to production, and therefore will be more likely selected as breeders by the farmer. This is especially true in extensive management systems, in which animals must cope largely with natural environmental conditions (Bertolini et al. 2018; Colli et al. 2018). Even though most local goat populations are still bred this way, it should also be noted that intensive breeding farms are growing in number in most developed countries (Bouwman et al. 2005). The industrial and highly specialized breeds reared in these facilities, such as the Saanen, are little subjected to the external environment, and their selection is aimed at maximizing their production potential.

2.5.1 Environmental Adaptation

Environmental factors are one of the key forces influencing agricultural and livestock productivity, exerting a strong selective pressure over domestic animal populations. The responses to this type of pressure include not only coping with extreme climates but also with poor forage and scarce water availability. Goat breeds and populations show great phenotypic variability, with peculiarities that can be related to the climatic condition of their breeding range. Recently, several genomic studies were carried out to detect selection signatures related to environmental adaptation by performing genome scans between groups of breeds with a limited geographic distribution and/or landscape genomic analyses in local goat breeds with a small census size. Several genes linked to circadian clock rhythms, coat colour, high-altitude adaptation, growth, immunity, and reproduction traits have been highlighted as under environmental selection (Bertolini et al. 2018; Cortellari et al. 2021a, b; Guo et al. 2018).

The goat is recognized as the domestic ruminant species with the greatest ability to adapt to hot and arid environments (Silanikove 2000). From a morphological point of view, goat populations that had to cope with these conditions developed mechanisms to regulate body temperature and water loss, goals that have been achieved in different ways by different populations. Dwarf goats are widely spread in the African continent, especially in wet and hot tropical regions, and are bred in very low-input systems. Small size is generally associated with a greater surface-to-volume ratio, which contributes to increase heat dissipation. This evidence is supported by the general tendency of animal populations living in hotter climates to be smaller compared with those in colder areas (Bergmann 1847). Dwarf goats are also reported as particularly resistant to pathogens typically spread in humid environments, such as *Trypanosoma* and *Haemonchus*, and to gastrointestinal parasites (Oseni et al. 2014). Thus, it is likely that different features within the same environment differentially exerted selective pressures on both size and pathogen resistance.

On the other hand, goat populations that developed in hot but arid areas, such as Sudanese and Egyptian desert goats, are often large-sized but characterized by long and thin appendages which increase surface available for heat dissipation. Moreover, larger animals tend to have lower mass-specific metabolic rates, which also implies that they produce less metabolic heat. A large body size might also have been favoured in goats mainly intended for meat production. Coat colour and texture are also important for thermoregulation: a light-coloured coat reflects higher amounts of solar radiation and is therefore recognized as being advantageous in hot climates, especially when paired with highly pigmented skin that blocks UV radiation (Sejian et al. 2018). Also, high-producing animals, particularly dairy ruminants, are generally considered as more susceptible to heat stress, given the positive correlation between metabolic heat load and milk production (Bernabucci et al. 2010). Adaptation to harsh and arid environments has been extensively studied also from the genetic perspective (e.g. Bertolini et al. 2018; Brito et al. 2017; Kim et al. 2016; Tarekegn et al. 2021; Waineina et al. 2022). The identification of regions undergoing selection consistently pinpointed loci with effects on thermo-tolerance (melanogenesis), body size and development, energy and digestive metabolism, and autoimmune response. Candidate loci differentiated in response to heat stress and altitude are linked to specific metabolic pathways, such as the insulin and glucose signalling pathway, glycogen metabolism, lipid biosynthetic processes, oxidative stress, and regulation of vasodilation (Bertolini et al. 2018).

Tibetan animals are ideal case studies to investigate adaptation to high-altitude environments characterized by low oxygen levels, low temperatures, and intense solar radiation. Several species, including Tibetan cashmere goats, showed selection signatures in genes related to the response to hypoxia; in particular, the *EPAS1* gene was identified as under selection in several studies on different species living in the Tibetan plateau (Guo et al. 2018; Song et al. 2016), while a haplotype variant at the *PAPSS2* gene significantly associated to high-altitude adaptability in Tibetan goats likely derived from a recent hybridization between goats and a wild caprid species, the markhor, *C. falconeri* (Li et al. 2022). Cashmere goats present a double coat with strata of long and wavy hair. This coat likely evolved to help animals cope with cold climates, but was later actively selected by humans, becoming one of the main products of this breed. Consistently, some genes related to both high-altitude adaptation and fibre production, such as *FGF5*, were found to harbour signatures of selection in Cashmere goats (Wang et al. 2016).

2.5.2 Production

When we speak about livestock selection for production traits, we usually think of selection programmes based on knowledge of phenotypes and genotypes. However, we should not forget that herders have been selecting animals since the beginning of domestication, by choosing as breeders the ones they preferred, either for personal reasons like aesthetic preference, or for real advantages in terms of management and production, or both. For most local breeds, this still holds true today. For example, archaeologic findings indicate that twisted horns have been preferred over other shapes since the Bronze Age, but the reasons are unclear: perhaps herdsmen preferred them for aesthetic reasons, or because they caused fewer damages during fights between males, or maybe they were somewhat related to other phenotypical characteristics. However, there were, and there are, some populations with exaggeratedly long horns that surely were difficult to manage, but that were appreciated for their peculiarity as a sign of prestige (Clutton-Brock 1999). The "modern" process of selection started to emerge only after the rediscovery of Mendel's laws in the twentieth century, thus covering only a minimal, but surely fundamental, span of the 11,000-year-long history of goats' selection and improvement.

The prolonged selection for performance traits left deep signatures in the goat genomes, which are clearly detectable in several populations around the world. In a large-scale investigation, a comprehensive screening of signatures of selection associated with external phenotypes (coat colour) and production traits (meat, milk, and wool) was carried out using different approaches in local and transboundary breeds sampled over five continents (Bertolini et al. 2018). Several genes associated with milk production in dairy breeds like Alpine and Saanen have been identified as under selection. In particular, selective sweeps were found in both the genomic region containing the cluster of casein genes, which are known to play a major role in milk composition with respect to cheese production, and close to genes, such as *ABCG2*, previously associated with milk yield and composition in cattle and sheep (Árnyasi et al. 2013; Cohen-Zinder et al. 2005). Genes involved in fibre production in the Angora breed included *CUX1* and *PLOD3*, associated with hair development and texture. Interestingly, signatures on these genes were found only in Angora goats, which have a single coat made of curly hairs like sheep, and not in the Cashmere breed, which instead presents a completely different hair conformation, as previously explained. When breeds were contrasted based on meat production, some of the regions under selection contained genes associated with muscle formation, for example *AMPD1* and *NRAS* on chromosome 3. Surprisingly, the Boer goat, which is the main breed for meat production worldwide, showed no evidence of strong selection for this trait. A possible explanation is that variation in meat production traits in this breed could be highly polygenic (i.e. many alleles with small effects determine the trait) (Bertolini et al. 2018).

3 Present Distribution

According to FAO data, the goat is one of the most diffused livestock species worldwide (http://www.fao.org/3/a-i4787e/index.html. Accessed 15 August 2022). The small size and lower management needs, if compared to other ruminants, allowed goats to follow humans during their migrations and movements. Thanks to their capability of withstanding harsh conditions in different types of ecosystems, goat populations spread and adapted to a large variety of different environments, in a process that is still underway and that will likely be pushed further by the ongoing climate changes: goats are, indeed, the ruminant species that can better adapt to the foreseen hotter and drier environmental conditions.

Collecting data about goat populations' consistency and diffusion might prove difficult, especially in countries in which they are bred in a context of subsistence, and sometimes nomadic, farming. The FAO database data indicate an increasing number of goats worldwide over the past 20 years (Fig. 3. FAOSTAT data, https://www.fao.org/faostat. Accessed 15 August 2022). In fact, even if goats are still less numerous than the two main farmed ruminant species, i.e., cattle and sheep, it is worth noticing that they showed the greatest growth in terms of number of heads, from ca. 760 million in 2000 to ca. 1.13 billion in 2020, with a net increase of about 370 million heads compared to the 200 million of cattle and sheep.

When we consider that in the same period the human population has increased from 6.14 billion to 7.79 billion (data taken from: https://www.worldometers.info/world-population), we can estimate that, while in 2000 there was 1 goat for every 8 people, in 2020 the ratio became 1 for less than 7 people. However, if we consider that the rural population, which includes the majority of goat breeders, decreased by 9.5%, at present there is 1 goat for every 3 people in rural areas.

Considering worldwide data (Fig. 4), goats are mainly distributed in poor or developing countries, where they constitute one of the main resources to fight poverty. Asia accounts for 51.4% of the total goat population, with a particularly high contribution from India and China. Recently, the goat population underwent a sharp increase in Africa, and at present the continent comprises 43.3% of the global population, with the highest number in Nigeria and Ethiopia. The remaining 5.3% of goats live in the Americas (especially in central-southern America), Europe (mainly in the Mediterranean countries), and Oceania.

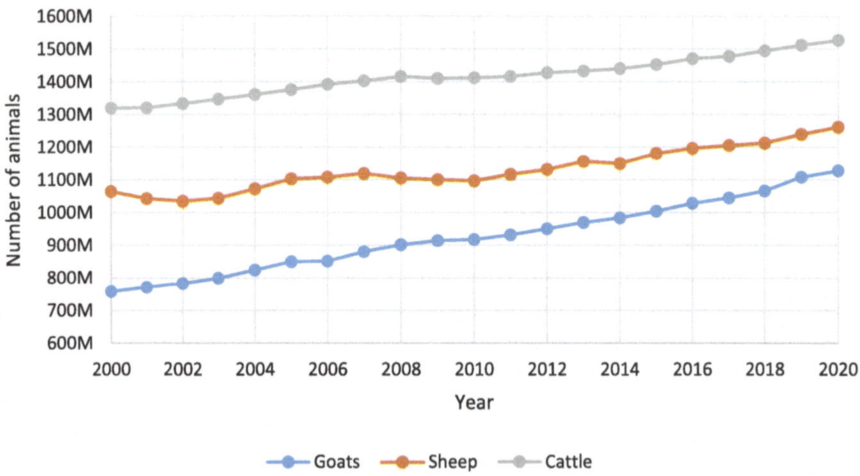

Fig. 3 Trend in the evolution of the number of the worldwide goat, sheep, and cattle populations during the last twenty years (FAOSTAT-world; the Authors thank Hassan R. Ghulami for contributing to the analysis of these data)

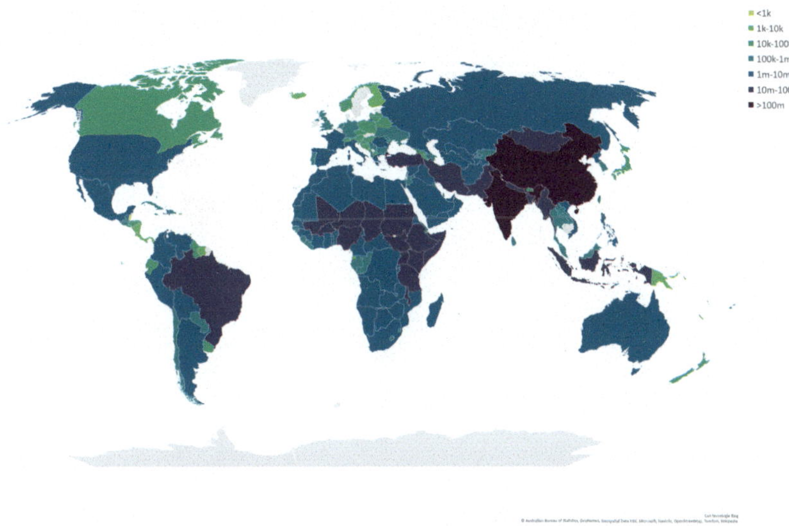

Fig. 4 Goat worldwide distribution and population consistency by country (https://www.fao.org/faostat for the year 2020. Accessed 15 August 2022). The darkest colours correspond to the highest numbers, whereas light grey indicates countries with missing data

3.1 Goat Biodiversity: Breeds and Populations

According to FAO statistics (FAO 2015), local goat breeds account for 21% of a total of 576 local ruminant breeds raised in the world.

From the list of breeds reported in the Domestic Animal Diversity Information System (DAD-IS) database, we reconstructed the distribution of the overall number of breeds reared in different countries of the world (Fig. 5), finding the highest numbers in China, followed by India and Italy. However, FAO database information can be incomplete, as it is managed independently in each country and is provided and updated on a voluntary basis: for example, only some countries include wild species, thus not providing a complete picture of their distribution.

Data about wild species can be found on the IUCN Red List website (www.iucnredlist.org). The bezoar (*C. aegagrus*) at present inhabits a discontinuous territory comprised between Turkey and Pakistan (Figs. 1 and 2) and, with an estimated census size of about 70,000 mature goats, it is considered “near threatened” (Weinberg and Ambarli 2020). The markhor (*C. falconeri*) is classified as “near threatened” as well, numbering less than 6000 adults, all living in the mountainous regions of central Asia (Michel and Rosen Michel 2015) (Fig. 2). The Alpine ibex (*C. ibex*) is listed as “least concern” species after risking extinction in the early nineteenth century. At present, it counts approximately 53,000 mature subjects and is endemic to Europe (Fig. 2), but it has also been introduced in Bulgaria and Argentina (Toïgo et al. 2020). Other wild goat species include the Siberian ibex

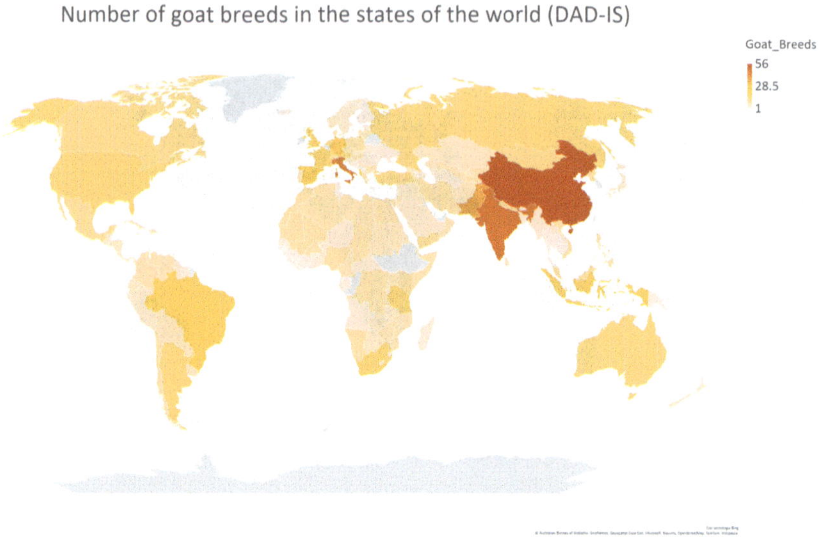

Fig. 5 Number of goat breeds recorded per country (data from DAD-IS)

(*C. sibirica*), the Ethiopian Walia ibex (*C. walie*), the Eastern tur (*C. cylindricornis*) living in the Great Caucasus, and the Iberian wild goat (*C. pyrenaica*) (Fig. 2).

Among the local goat breeds (Fig. 5), 50 are bred in different countries and are therefore referred to as "transboundary". The largest number of transboundary goat breeds is found in Africa, followed by Europe and Asia. In addition, a small number of breeds can be considered "cosmopolitan", being bred on all continents. Among these there are two highly specialized dairy breeds, Saanen and Alpine (Fig. 6, panels a and b), both native to Switzerland, as well as another important dairy breed, the Toggenburg (Fig. 6, panel c), which is now bred in most of the world. The Boer goat, instead, is the most widespread and productive meat breed. The intensive selection for this production goal was carried out during the twentieth century by the Dutch breeders living in South Africa, but formerly the breed was raised for milk production and it is recorded to have been crossed with Indian and European breeds during the eighteenth or nineteenth century to improve yields (Casey and Van Niekerk 1988; Porter 1996) (Fig. 6, panel d). Anglo-Nubian and Nubian breeds are also widespread in more than 50 countries worldwide (Fig. 6, panel e). The Anglo-Nubian, as its name implies, originated from a cross between Nubian and British breeds dating back to the mid-nineteenth century. Today it is recognized as a dual-purpose breed (meat-milk) particularly adapted to Australia's hot climates. In Africa, dwarf goats present a wide distribution (Fig. 6, panel f) and are used for multiple purposes, such as milk, meat, and leather. Regarding goats specialized for fibre production, Angora is the most widespread breed, followed by Cashmere and Mohair goats (Fig. 6, panels g and h). Fibre production, and in particular Cashmere, is still a characteristic of nomadic cultures in many countries (Ansari-Renani 2015).

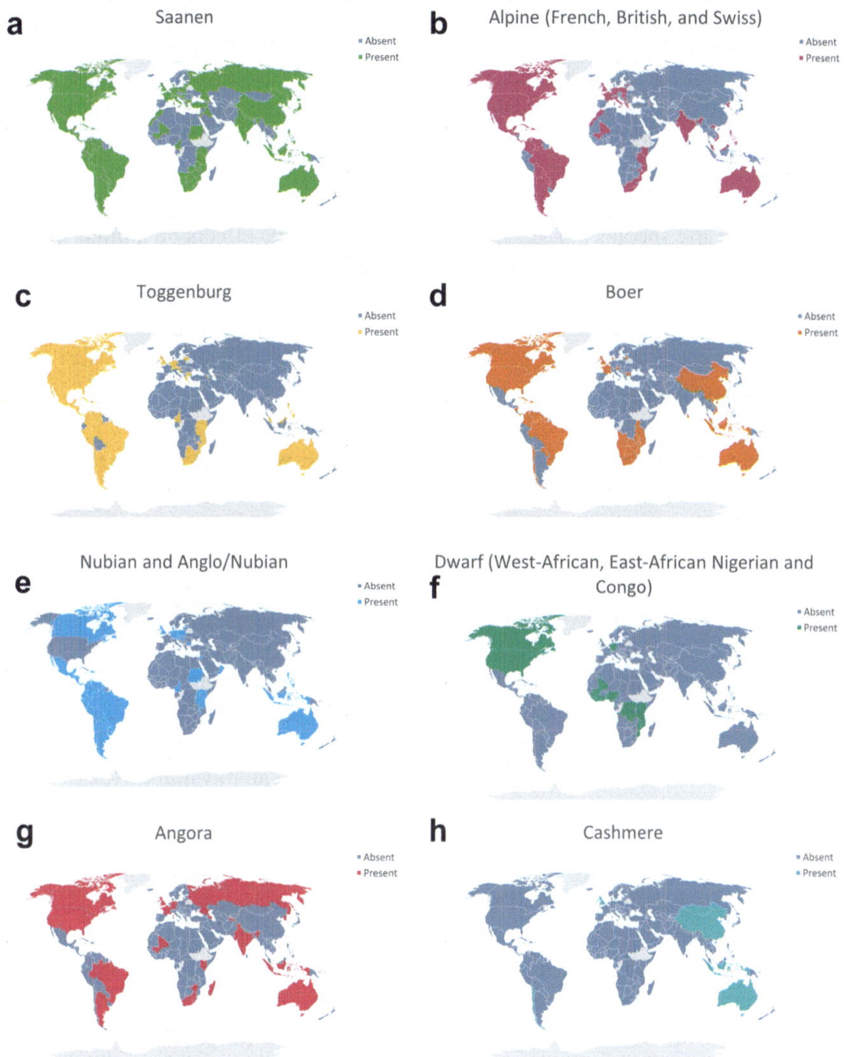

Fig. 6 Maps showing the worldwide distribution and occurrence of cosmopolitan and transboundary breeds. Coloured areas correspond to the countries where the breed is reported according to the DAD-IS database of FAO

The FAO database also includes information on the risk status of livestock breeds which adopted the following ranking categories similar to IUCN's ones: Unknown, Not at risk, Vulnerable, Endangered-Maintained, Endangered, Critical-maintained, Critical, Cryo-conserved only, and Extinct. Based on FAO report (FAO 2015), the risk status is unknown for about 34% of the recorded goat breeds, 50% are not endangered, while 9% are endangered, and 5% are in critical status. For 1% of the breeds in critical status and 2% of those at risk, conservation programmes have been put in place. The growing awareness of the importance of biodiversity and its preservation recently led to the development of projects for the conservation and enhancement of native animal resources in many countries and to the establishment of biobanks to store biological material to prevent extinction. Due to climate changes and growing socio-political uncertainty in many areas of the world, the survival of many local breeds is strongly threatened by the increasing spread of extreme weather phenomena (floods, landslides), economic crises, and war. This places the genetic variability of the species at risk, and might have negative effects on animal welfare, due to the loss of specific traits that have evolved over time, improving goats' adaptation and ability to cope with different environments.

4 Conclusions

Overall, the history and spread of goats around the world, together with the currently available genomic information, show how natural and artificial selection often act in synergy and how it is sometimes difficult to disentangle the effects of the two selective pressures in livestock populations.

As with other domestic species, the domestication process in goats has brought about major genomic, physiological, and behavioural changes that, together with anthropogenic selection, have determined the success of this species in the world. Pastoralism, a common practice in small ruminant management, has probably contributed to the complexity of forces that have shaped and are shaping the goat genome by exposing the animals to a variety of conditions in terms of environmental features, disease occurrence, feeding regimes, and water availability.

The biological architecture that regulates the expression of genes involved in complex traits has started to be dissected more thoroughly in recent years due to the availability of increasingly powerful molecular and bioinformatic tools, which provided a genome-wide picture of the genes and genomic regions which were subjected to selection along the history of goats and their relationship with humans. Although the intricate network of interactions between genes directly or indirectly involved in the mechanisms of adaptation is far from being fully understood, several studies have helped pinpoint genomic regions that are important putative targets of anthropogenic and/or natural selection for traits related to domestication, animal behaviour, and welfare (e.g. animal hardiness and resilience).

It is also worth remembering that selection goals in livestock are in constant evolution and are strictly related to socio-cultural and economic needs, which change over time. For example, recently increasing attention is being paid to livestock environmental adaptation, given the threats that animals and breeders will likely have to face in the near future due to climate change. Moreover, in countries in which food availability and supply are ensured, animal welfare, health, and fitness are becoming new objectives of selection. Indeed, on the one hand, these elements are advantageous for the breeders, contributing to the reduction of breeding costs and the production of healthier and more natural products, and on the other hand they respond to the increasing consumers' awareness about livestock well-being. For these reasons, the development of breeding goals is now oriented towards increasing the welfare, resilience, and longevity of animals, in the face of a decreased emphasis on production-related traits.

References

Alasaad S, Biebach I, Grossen C, Soriguer RC, Pérez JM, Keller LF (2011) Microsatellite-based genotyping of MHC class II DRB1 gene in Iberian and Alpine ibex. Eur J Wildl Res 58(4):743–748

Alberto FJ, Boyer F, Orozco-terWengel P, Streeter I, Servin B, de Villemereuil P, Benjelloun B, Librado P, Biscarini F, Colli L, Barbato M, Zamani W, Alberti A, Engelen S, Stella A, Joost S, Ajmone-Marsan P, Negrini R, Orlando L et al (2018) Convergent genomic signatures of domestication in sheep and goats. Nat Commun 9(1):813

Amills M, Capote J, Tosser-Klopp G (2017) Goat domestication and breeding: a jigsaw of historical, biological and molecular data with missing pieces. Anim Genet 48(6):631–644

Ansari-Renani HR (2015) Cashmere production, harvesting, marketing and processing by nomads of Iran – a review. Pastoralism 5(1):18

Árnyasi M, Komlósi I, Kent MP, Czeglédi L, Gulyás G, Jávor A (2013) Investigation of polymorphisms and association of the ABCG2 gene with milk production traits in sheep. Livest Sci 154(1–3):64–68

Belyaev DK (1979) Destabilizing selection as a factor in domestication. J Hered 70(5):301–308

Bergmann C (1847) Uber Die Verhaltnisse Der Warmekonomie Der Tiere Zu Ihrer Grosse. Gottinger Studien 3:595–708

Bernabucci U, Lacetera N, Baumgard LH, Rhoads RP, Ronchi B, Nardone A (2010) Metabolic and hormonal acclimation to heat stress in domesticated ruminants. Animal 4(7):1167–1183

Bertolini F, Servin B, Talenti A, Rochat E, Kim ES, Oget C, Palhière I, Crisà A, Catillo G, Steri R, Amills M, Colli L, Marras G, Milanesi M, Nicolazzi EL, Rosen BD, Van Tassell CP, Guldbrandtsen B, Sonstegard TS et al (2018) Signatures of selection and environmental adaptation across the goat genome post-domestication. Genet Sel Evol 50:57

Bosomworth M, Fleitman D, Rabbani M (2020) Palaeoclimate and environment of the Iraqi central Zagros. In: Matthews R, Matthews W, Raheem KR, Richardson A (eds) The early neolithic of the eastern fertile crescent. Oxbow Books, pp 35–42

Bouwman AF, Van der Hoek KW, Eickhout B, Soenario I (2005) Exploring changes in world ruminant production systems. Agric Syst 84(2):121–153

Brito LF, Kijas JW, Ventura RV, Sargolzaei M, Porto-Neto LR, Cánovas A, Feng Z, Jafarikia M, Schenkel FS (2017) Genetic diversity and signatures of selection in various goat breeds revealed by genome-wide SNP markers. BMC Genomics 18(1):229

Casey NH, Van Niekerk WA (1988) The boer goat. I. Origin, adaptability, performance testing, reproduction and milk production. Small Rumin Res 1(3):291–302

Clutton-Brock J (1999) A natural history of domesticated mammals, 2nd edn. Cambridge University Press

Cohen-Zinder M, Seroussi E, Larkin DM, Loor JJ, Everts-van der Wind A, Lee J-H, Drackley JK, Band MR, Hernandez AG, Shani M, Lewin HA, Weller JI, Ron M (2005) Identification of a missense mutation in the bovine ABCG2 gene with a major effect on the QTL on chromosome 6 affecting milk yield and composition in Holstein cattle. Genome Res 15(7):936–944

Colli L, Lancioni H, Cardinali I, Olivieri A, Capodiferro MR, Pellecchia M, Rzepus M, Zamani W, Naderi S, Gandini F, Vahidi SMF, Agha S, Randi E, Battaglia V, Sardina MT, Portolano B, Rezaei HR, Lymberakis P, Boyer F et al (2015) Whole mitochondrial genomes unveil the impact of domestication on goat matrilineal variability. BMC Genomics 16:1115

Colli L, Milanesi M, Talenti A, Bertolini F, Chen M, Crisà A, Daly KG, Del Corvo M, Guldbrandtsen B, Lenstra JA, Rosen BD, Vajana E, Catillo G, Joost S, Nicolazzi EL, Rochat E, Rothschild MF, Servin B, Sonstegard TS et al (2018) Genome-wide SNP profiling of worldwide goat populations reveals strong partitioning of diversity and highlights post-domestication migration routes. Genet Sel Evol 50(1):1–20

Cortellari M, Barbato M, Talenti A, Bionda A, Carta A, Ciampolini R, Ciani E, Crisà A, Frattini S, Lasagna E, Marletta D, Mastrangelo S, Negro A, Randi E, Sarti FM, Sartore S, Soglia D, Liotta L, Stella A et al (2021a) The climatic and genetic heritage of Italian goat breeds with genomic SNP data. Sci Rep 11:10986

Cortellari M, Bionda A, Negro A, Frattini S, Mastrangelo S, Somenzi E, Lasagna E, Sarti FM, Ciani E, Ciampolini R, Marletta D, Liotta L, Ajmone Marsan P, Pilla F, Colli L, Talenti A, Crepaldi P (2021b) Runs of homozygosity in the Italian goat breeds: impact of management practices in low-input systems. Genet Sel Evol 53:92

Daly KG, Maisano Delser P, Mullin VE, Scheu A, Mattiangeli V, Teasdale MD, Hare AJ, Burger J, Verdugo MP, Collins MJ, Kehati R, Erek CM, Bar-Oz G, Pompanon F, Cumer T, Çakırlar C, Mohaseb AF, Decruyenaere D, Davoudi H, … Bradley DG (2018) Ancient goat genomes reveal mosaic domestication in the Fertile Crescent. Science (New York, N.Y.) 361(6397):85–88. https://doi.org/10.1126/science.aas9411

Daly KG, Arbuckle BS, Rossi C, Mattiangeli V, Lawlor PA, Mashkour M, Sauer E, Lesur J, Atici L, Erek CM, Bradley DG (2022) A novel lineage of the *Capra* genus discovered in the Taurus Mountains of Turkey using ancient genomics. elife 11

FAO (2015) The second report on the state of the world's animal genetic resources for food and agriculture

Grossen C, Keller L, Biebach I, International Goat Genome Consortium, C, Croll D (2014) Introgression from domestic goat generated variation at the major histocompatibility complex of Alpine ibex. PLoS Genet 10(6):e1004438

Guo J, Tao H, Li P, Li L, Zhong T, Wang L, Ma J, Chen X, Song T, Zhang H (2018) Whole-genome sequencing reveals selection signatures associated with important traits in six goat breeds. Sci Rep 8(1):10405

Hammer SE, Schwammer HM, Suchentrunk F (2008) Evidence for introgressive hybridization of captive markhor (*Capra falconeri*) with domestic goat: cautions for reintroduction. Biochem Genet 46(3–4):216–226

Hedrick PW (2013) Adaptive introgression in animals: examples and comparison to new mutation and standing variation as sources of adaptive variation. Mol Ecol 22(18):4606–4618

Horwitz LK, Bar-Gal GK (2006) The origin and genetic status of insular caprines in the eastern mediterranean: a case study of free-ranging goats (*Capra aegagrus cretica*) on crete. Hum Evol 21(2):123–138

Joshi MB, Rout PK, Mandal AK, Tyler-Smith C, Singh L, Thangaraj K (2004) Phylogeography and origin of Indian domestic goats. Mol Biol Evol 21(3):454–462

Kim E-S, Elbeltagy AR, Aboul-Naga AM, Rischkowsky B, Sayre B, Mwacharo JM, Rothschild MF (2016) Multiple genomic signatures of selection in goats and sheep indigenous to a hot arid environment. Heredity 116(3):255–264

Kreitman M (2000) Methods to detect selection in populations with applications to the human. Annu Rev Genomics Hum Genet 1:539–559

Larson G, Burger J (2013) A population genetics view of animal domestication. Trends Genet TIG 29(4):197–205

Li C, Wu Y, Chen B, Cai Y, Guo J, Leonard AS, Kalds P, Zhou S, Zhang J, Zhou P, Gan S, Jia T, Pu T, Suo L, Li Y, Zhang K, Li L, Purevdorj M, Wang X, Li M, Wang Y, Liu Y, Huang S, Sonstegard T, Wang MS, Kemp S, Pausch H, Chen Y, Han JL, Jiang Y, Wang X (2022) Markhor-derived introgression of a genomic region encompassing PAPSS2 confers high-altitude adaptability in Tibetan goats. Mol Biol Evol 39(12):msac253

Luikart G, Gielly L, Excoffier L, Vigne JD, Bouvet J, Taberlet P (2001) Multiple maternal origins and weak phylogeographic structure in domestic goats. Proc Natl Acad Sci USA 98(10):5927–5932

Masseti M (2009) The wild goats *Capra aegagrus* Erxleben, 1777 of the Mediterranean Sea and the Eastern Atlantic Ocean islands. Mammal Rev 39(2):141–157
Maynard Smith J, Haigh J (1974) The hitch-hiking effect of a favourable gene. Genet Res 23(1):23–35
Michel S, Rosen Michel T (2015) *Capra falconeri* (errata version published in 2016). The IUCN Red List of Threatened Species 2015, e.T3787A97218336
Naderi S, Rezaei H-R, Pompanon F, Blum MGB, Negrini R, Naghash H-R, Balkiz O, Mashkour M, Gaggiotti OE, Ajmone-Marsan P, Kence A, Vigne J-D, Taberlet P (2008) The goat domestication process inferred from large-scale mitochondrial DNA analysis of wild and domestic individuals. Proc Natl Acad Sci USA 105(46):17659–17664
NatureServe, & IUCN (2020) The IUCN red list of threatened species. Version 2022-2. www.iucnredlist.org
Nijman IJ, Rosen BD, Zheng Z, Jiang Y, Cumer T, Daly KG, Bâlteanu VA, Berger B, Blichfeldt T, Boink G, Carolan S, Cubric-Curik V, Kantanen J, Martínez A, Mazza R, Khayatzadeh N, Kim N, Ouchene-Khelifi N-A, Pereira F et al (2020) Phylogeny and distribution of Y-chromosomal haplotypes in domestic, ancient and wild goats. BioRxiv
Nomura K, Yonezawa T, Mano S, Kawakami S, Shedlock AM, Hasegawa M, Amano T (2013) Domestication process of the goat revealed by an analysis of the nearly complete mitochondrial protein-encoding genes. PLoS One 8(8):e67775
Oseni SO, Ajayi BA, Oseni SO, Ajayi BA (2014) Phenotypic characterization and strategies for genetic improvement of WAD goats under backyard systems. Open Journal of Animal Sciences 04(05):253–262
Pidancier N, Jordan S, Luikart G, Taberlet P (2006) Evolutionary history of the genus *Capra* (Mammalia, Artiodactyla): discordance between mitochondrial DNA and Y-chromosome phylogenies. Mol Phylogenet Evol 40(3):739–749
Porter V (1996) Goats of the world. Farming Press
Ropiquet A, Hassanin A (2006) Hybrid origin of the Pliocene ancestor of wild goats. Mol Phylogenet Evol 41(2):395–404
Sardina MT, Ballester M, Marmi J, Finocchiaro R, van Kaam JBCHM, Portolano B, Folch JM (2006) Phylogenetic analysis of Sicilian goats reveals a new mtDNA lineage. Anim Genet 37(4):376–378
Sejian V, Bhatta R, Gaughan JB, Dunshea FR, Lacetera N (2018) Review: adaptation of animals to heat stress. Animal 12(S2):s431–s444
Silanikove N (2000) The physiological basis of adaptation in goats to harsh environments. Small Rumin Res 35(3):181–193
Somenzi E, Senczuk G, Ciampolini R, Cortellari M, Vajana E, Tosser-Klopp G, Pilla F, Ajmone-Marsan P, Crepaldi P, Colli L (2022) The SNP-based profiling of montecristo feral goat populations reveals a history of isolation, bottlenecks, and the effects of management. Genes 13(2)
Song S, Yao N, Yang M, Liu X, Dong K, Zhao Q, Pu Y, He X, Guan W, Yang N, Ma Y, Jiang L (2016) Exome sequencing reveals genetic differentiation due to high-altitude adaptation in the Tibetan cashmere goat (*Capra hircus*). BMC Genomics 17:122
Stella A, Ajmone-Marsan P, Lazzari B, Boettcher P (2010) Identification of selection signatures in cattle breeds selected for dairy production. Genetics 185(4):1451–1461
Sultana S, Mannen H, Tsuji S (2003) Mitochondrial DNA diversity of Pakistani goats. Anim Genet 34(6):417–421
Tarekegn GM, Khayatzadeh N, Liu B, Osama S, Haile A, Rischkowsky B, Zhang W, Tesfaye K, Dessie T, Mwai OA, Djikeng A, Mwacharo JM (2021) Ethiopian indigenous goats offer insights into past and recent demographic dynamics and local adaptation in sub-Saharan African goats. Evol Appl 14(7):1716–1731
Toïgo C, Brambilla A, Grignolio S, Pedrotti L (2020) Capra ibex. The IUCN Red List of Threatened Species 2020, e.T42397A161916377
Vigne J-D, Peters J, Helmer D (2005) The first steps of animal domestication: new archaeozoological approaches. Oxbow Books

Waineina RW, Okeno TO, Ilatsia ED, Ngeno K (2022) Selection signature analyses revealed genes associated with adaptation, production, and reproduction in selected goat breeds in Kenya. Front Genet 13:858923

Wang X, Liu J, Zhou G, Guo J, Yan H, Niu Y, Li Y, Yuan C, Geng R, Lan X, An X, Tian X, Zhou H, Song J, Jiang Y, Chen Y (2016) Whole-genome sequencing of eight goat populations for the detection of selection signatures underlying production and adaptive traits. Sci Rep 6:38932

Weaver S (2020) The goat: a natural and cultural history. Princeton University Press

Weinberg P, Ambarli H (2020) Capra aegagrus. The IUCN Red List of Threatened Species 2020, e.T3786A22145942

Wilkins AS, Wrangham RW, Fitch WT (2014) The "domestication syndrome" in mammals: a unified explanation based on neural crest cell behavior and genetics. Genetics 197(3):795–808

Zeder MA (2008) Domestication and early agriculture in the Mediterranean Basin: origins, diffusion, and impact. Proc Natl Acad Sci USA 105(33)

Zeder MA (2012) Pathways to domestication. In: Gepts P, Famula TR, Bettinger RL, Brush SB, Damania AB, McGuire PE, Qualset CO (eds) Biodiversity in agriculture: domestication, evolution, and sustainability. Cambridge University Press, pp 227–259

Zeder MA, Hesse B (2000) The initial domestication of goats (*Capra hircus*) in the Zagros mountains 10,000 years ago. Science (New York, NY) 287(5461):2254–2257

Zeder MA, Emshwiller E, Smith BD, Bradley DG (2006) Documenting domestication: the intersection of genetics and archaeology. Trends Genet TIG 22(3):139–155

Zhao F, McParland S, Kearney F, Du L, Berry DP (2015) Detection of selection signatures in dairy and beef cattle using high-density genomic information. Genet Sel Evol 47(1):49

Zheng Z, Wang XX, Li M, Li Y, Yang Z, Wang XX, Pan X, Gong M, Zhang Y, Guo Y, Wang Y, Liu J, Cai Y, Chen Q, Okpeku M, Colli L, Cai D, Wang K, Huang S et al (2020) The origin of domestication genes in goats. Science Advances 6(21):eaaz5216

Goats' Natural Behaviour

Gosia Zobel, Heather Whittaker Neave, and Christian Nawroth

Abstract

Understanding natural behaviour is vital for promoting good welfare of goats in human care. Observations of feral and wild populations, as well as goats in research settings, provide insight into natural behaviours. An overview is presented of the natural environments that goats thrive in, and how they interact with these environments. Goats are known to climb, hide, travel great distances, and be flexible in their foraging and social interactions; they use well-developed cognitive capacities throughout these activities. Yet, in human care, often these abilities are not promoted, potentially compromising welfare. These welfare concerns are typically system specific. For instance, indoor housed animals may face a lack of exercise opportunities, barren housing that does not promote climbing or hiding, lack of outdoor access, compromised doe-buck and doe-kid relationships, and large group sizes. Conversely, animals managed outdoors or extensively face exposure, predation, parasitic infections, teeth and toxicity issues, and potentially social isolation in the case of companion animal settings. This chapter reflects on how promoting natural behaviours can reduce these negative welfare outcomes, and importantly, how these behaviours themselves are positive welfare indicators. The discussion captures numerous systems, including

G. Zobel (✉)
EthicoNZ, Hamilton, New Zealand
e-mail: gosia@ethiconz.com

H. W. Neave
Department of Animal Science, Purdue University, West Lafayette, IN, USA

C. Nawroth
Research Institute for Farm Animal Biology, Institute of Behavioral Physiology, Dummerstorf, Germany

S. Mattiello, M. Battini (eds.), *The Welfare of Goats*, Animal Welfare 25,
https://doi.org/10.1007/978-3-031-62182-6_2

those producing dairy, meat, and fibre; captured ferals, companions and working animals are also considered.

Keywords

climbing · hiding · cognitive capacity · travelling · dairy · meat · fibre · companion · working

1 Introduction

A fundamental tenet of ensuring good goat welfare is achieving a balance between physical and biological health, affective state, and naturalness (Fraser 2008). Therefore, understanding goats' natural behaviour repertoire is vital towards establishing how to meet goats' needs. Garnering this knowledge is possible from various sources. First, although farmed goats are estimated to exceed 1.1 billion animals globally (FAOSTAT 2020), the vast majority are extensively managed; these goats provide insight into typical behaviours in relatively natural settings. Second, their resilience and adaptability mean that there are feral goat populations in numerous locations around the world that can be observed (e.g. New Zealand—Alley et al. 1995; Scotland—Shi et al. 2005; Northern Mariana Islands—Kessler 2002; Australia—Moseby et al. 2020; Majorca—Leidy et al. 2015; Seychelles—Bunbury et al. 2018; USA—Spatz and Mueller-Dombois 1973; Canada—Geist 1960). Furthermore, a great deal can be gleaned from the observation of goats in research settings, where their natural cognitive capacities can be assessed in a highly controlled manner that would not be possible in either commercial farm or feral populations. Lastly, while often threatened by human encroachment and environmental degradation, there are still several wild populations whose behaviour is conceivably the best indicator of truly natural behaviour. This chapter summarises the literature reporting on the behavioural repertoire of goats as context for understanding what is missing, and what should be promoted, when goats are in the care of humans. It is pertinent to note that many topics have not been greatly or robustly explored. We have done our best to present the most relevant literature available (sometimes using other species as proxies); additionally, we have discussed the robustness of observational literature that has merits, but that would benefit from further exploration.

Table 1 introduces the natural behaviours that are typical of goats and discusses them in terms of the benefits they serve, how their omission from management may compromise welfare, and the typical systems in which these omissions may have the most negative welfare impact. It is important to note that due to the sheer diversity in scale of goat production systems around the world, it is difficult to attribute one system type being better than another; for instance, dairy goat production ranges from smallholder nomadic herds of a few animals to intensively managed indoor herds made up of thousands of individuals. It is therefore helpful to think of size, production, and specific management when interpreting how natural behaviours can be promoted in each kind of system. Equally important to consider is that negative outcomes from thwarted natural behaviours rarely occur on their own; for example,

Table 1 Natural behaviours typical of goats

Natural Behaviour	What is promoted/ provided?	Welfare implications when behaviour not promoted	System in which welfare issues predominantly occur
Climbing and travel	Exercise	Compromised physical fitness	All
	Hoof maintenance	Hoof overgrowth and conformation issues	Predominantly intensive systems (indoor dairy, meat, and fibre)
	Choice and cognitive capabilities (sight, memory, navigation)	Boredom and lack of agency	All
Hiding	Shelter	Thermal exposure and predation	Fibre, meat, companion, working
	Security	Stress (resulting from exposure to human handling)	Feral (mustered for meat)
	Choice and cognitive capabilities (sight, memory)	Boredom and lack of agency	All
Foraging	Flexibility and diversity in feed	Toxicity and nutritional deficiencies	Fibre, meat, companion, working
		Teeth issues	All, but predominantly long living animals (companion, show)
	Flexibility in how feed is consumed	Parasite exposure	Fibre, meat, companion, working, outdoor dairy
		Competition and aggression	Predominantly intensive systems (indoor dairy, meat, and fibre)
	Choice and cognitive capabilities (sight, memory)	Boredom and lack of agency	All
Social	Choice and cognitive capabilities (sight, communication, memory) used to interact with preferred peers and groups	Stress due to large groups	All, excluding companion
		Stress due to regrouping	Predominantly dairy
		Isolation	Companion

(continued)

Table 1 (continued)

Natural Behaviour	What is promoted/ provided?	Welfare implications when behaviour not promoted	System in which welfare issues predominantly occur
Dam and kid relationships	Strong, long-lasting bonds formed early (olfactory, vocal, physical)	Stress of kid loss on doe when removed early (mental and physical)	Dairy, companion, show, working
		Stress of weaning (mental) on doe and kid when done later in life	Fibre, meat, working
Horn possession	Social signalling	Delayed social hierarchy development	Dairy, companion, show
	Thermal buffering	Heat stress susceptibility	Dairy, companion, show
	Scratching abilities	Chronic itchiness	Dairy, companion, show
Genetic considerations *(examples)*	Natural moulting of undercoat (down)	Thermal stress and death due to poorly managed fibre removal	Fibre
	Myotonia congenita* *(not a beneficial trait to the goat)*	Research lacking on how goats experience the chronic state of acute delays in muscle relaxation in response to stressors (including susceptibility to predation)	Meat, companion

goats managed in large outdoor groups with little forage choice may experience social aggression, lack of mental stimulation, and nutritional deficiencies and toxicity issues. Thus, to better understand the complex interaction between goats, their environment, and how management systems impact their welfare, it is appropriate to provide a brief overview of the various systems we place them in.

The humble goat conceivably takes the award for most flexible ruminant in terms of its contribution to human life. Goats are efficient milk, meat, and fibre producers, often thriving in environments that are not conducive for other species. This makes them a favourable option for subsistence and extensive management (reviewed by Silanikove 2000). Lohani and Bhandari (2021) report that over 90% of the world's goats are in Asia and Africa; these authors and Utaaker et al. (2021) present various reasons why many of these goats are extensively managed (e.g. cultural, social, economic, religious). It is tempting to assume that such systems will not compromise goat welfare; indeed, many extensively managed meat, fibre, and smallerholder dairy goats are afforded flexibility to perform natural behaviours (e.g. browsing opportunities, using cognitive capabilities to find forage, flexible social and shelter options). Nonetheless, these goats do face other burdens (e.g. high parasite loads, thermal challenges, predatory burdens) that they may not face if they were in intensive systems. Conversely, the burdens placed on dairy goats in intensive systems (e.g. boredom, poor hoof health, social stress due to large groups) may seem justified to industry when considering that these goats are afforded freedom from predation, for example.

An interesting offshoot in the meat goat realm is that of feral goat mustering; these goats are entirely unaccustomed to humans until the moment they are wrangled and held for feeding and then transported for slaughter. For most of their lives, they lead an entirely 'natural life', and while the remaining time in human care is relatively short, their welfare nonetheless should be considered.

Finally, an often-overlooked use of goats, at least from a scientific perspective, is in the working and companion animal realm. Their natural tendency to browse woody and shrub matter has made them an instrumental worker for reducing fire susceptibility of land (e.g. Lovreglio et al. 2014), for reducing herbicide use (Green et al. 1978), and even for restoring endangered grasslands (Elias and Tischew 2016). Feral goats can be attracted to areas in need of brush clearing (Pareja et al. 2020), or managed goat herds can be brought in temporarily (e.g. Free 2022; Nolden 2020). These goats will have similar welfare concerns as any extensively managed goats; if densely managed, their natural behaviour to segregate into preferred groups or to hide and isolate from predators may be challenged. These working animals may even face a unique burden of frequent exposure to new hazards, especially if they are being moved frequently to unknown territory (Divittorio 2022; Rodriguez 2019). Furthermore, there is a dearth of literature examining show and companion goat welfare, but their environment may present unique challenges (e.g. nutritional deficiencies stemming from lack of foraging diversity). Pet goats especially may face a welfare challenge that most other goats would not—that of potential lifelong solitude, often at the end of a tether.

Goats are known to have remarkable perceptual capacities. Their eyes are positioned on the sides of their heads, giving them a wide field of vision. Being able to see 320–340° (Hughes and Whitteridge 1973), coupled with horizontally elongated pupils, means that goats efficiently scan large areas of their environment, helping detect predators from afar (Banks et al. 2015). Furthermore, goats' large pupils and reflective retinas result in them possessing excellent night and low-light vision (Ollivier et al. 2004). Like other mammalian species, goats have dichromatic colour vision (Jacobs et al. 1998). Their sense of hearing is also highly developed and superior in range to humans (goats: 78 Hz to 37 kHz versus humans: 31 Hz to 17 kHz; Heffner and Heffner 1990). Additionally, goats have an exceptional sense of smell, which is important for identifying food sources and peers (Baldwin 1977; Scherer et al. 2019). Thus, for a species that has such impressive perceptual and cognitive capabilities, and diverse social and foraging preferences, and which has generally evolved to thrive in complex environments, it is no wonder that goat welfare may be compromised when some, or all, of their natural capacities and behaviours are not nurtured. However, interest in naturalness continues to grow, even at the regulatory level; for example, the Canadian National Farm Animal Care Council identified naturalness as a priority issue for inclusion in the most recent version of the Code of Practice for the Care and Handling of Goats (Menzies et al. 2020, https://www.nfacc.ca/codes-of-practice/goats). We hope that this chapter invigorates the ongoing discussion of how best naturalness can be promoted for goats in human care. And lastly, we posit a final consideration: there is a tendency in the scientific literature to group goats with sheep, presumably because they are small ruminants. We hope the reader might reflect on the applicability (or lack thereof) of the information presented in the context of sheep; we suggest that doing so may spur a newfound appreciation for the unique welfare requirements of goats that may often be dramatically different than sheep.

2 What Does an Ideal Goat Environment Look Like?

Wild and feral goats (*Capra* genus) are known for their ability to live in extreme, rocky, and elevated terrain. Eastern tur (*Capra cylindricornis*) gravitates towards steep, grassy terrain near cliff faces (Gavashelishvili et al. 2018; Weinberg et al. 2010). Himalayan ibex *(Capra ibex sibirica)* naturally resides above the tree line, up to 5000 m in elevation (Fedosenko and Blank 2001), moving in response to seasonal weather challenges. Wild goats in Iran (*Capra aegagrus*) favour heavily sloped, lower elevations (1400 m to 2800 m) (Sarhangzadeh et al. 2013), and Markhor (*Capra falconeri*) are typically seen below 3600 m, but may move higher if tree cover is still available (Michel and Rosen Michel 2015). Importantly, in response to increasing human settlement and climate change pressures, there has been evidence that some goat populations are using their affinity for higher elevations to re-establish territory when threatened; for instance, Gebremedhin et al. (2021) reported Walia ibex *(Capra walie)* at elevations over 2700 m and as high as 4600 m, a shift up of at least 400 m from lower elevations reported in previous decades (Nievergelt 1981). Such shifts further demonstrate goats' resilience. Their influence on the structure and composition of vegetation and on nutrient cycling is instrumental for their natural environment (Bagchi and Ritchie 2010). Additionally, their flexibility has led to the existence of feral goat populations around the world. Many of these animals, regardless of their global location, favour rocky and difficult-to-access terrain.

The culling literature that documents feral goat culling initiatives (e.g., organised hunting of goats, typically in environments in which they are not native) around the world is useful for its contribution to behavioural observations; it also highlights that goats use habitats that few other ungulates could (Fig. 1).

A frequently mentioned term in the literature examining goat habitats is 'escape'—indeed, goats have been reported to use the terrain as a means of maintaining a clear vantage point of observation, thus avoiding predation, including culling by humans (Iribarren and Kotler 2012; Kessler 2002; Parkes 1984). In addition to simply being less accessible, rocky terrains offer the opportunity to hide in caves and other crevasses. These provide shelter from the elements and likely a sense of security. Extensively managed dairy goats have been observed to use caves to avoid the heat on summer afternoons (Zobel et al. 2018) (Fig. 2), and feral goats spent their nights in cliff side caves (Boyd 1981; Shi et al. 2003, 2005).

Fig. 1 Extensively managed milking goats' browsing slopes which would be inaccessible for other ungulates (Location: Hut Nimi, Gordevio, Switzerland; Owner: Pietro Zanoli)

Fig. 2 Extensively managed Saanen milking goat hiding in a rocky outcrop (Location: Hut Nimi, Gordevio, Switzerland; Owner: Pietro Zanoli)

The desire to hide also appears in intensively managed goats. In a small pilot study, dairy goats sourced from a barren indoor commercial environment actively used a 3-sided platform to 'hide'. While only some immediately climbed the platform, all goats went under it within minutes of exposure; additionally, several

Fig. 3 Milking Saanens previously naive to a platform using it for resting and hiding (AgResearch Ltd)

individuals became frequent users, spending multiple hours per day 'hiding' (Zobel et al. 2017) (Fig. 3).

In addition to populating inaccessible terrain, goats are adept at utilising resource-poor land. Their success in such areas can be attributed to several behavioural adaptations. First, goats continue to climb even if the terrain is not rocky or mountainous. This results in feed sources that would be inaccessible to most other ungulates being well within the goats' reach. For instance, Delibes et al. (2017) report active tree-climbing in extensively managed Moroccan goats seeking the fruit of the Argan tree. It is suggested that a few goats naturally climbed the trees themselves to access browse when other sources of fodder were limited (el Aich et al. 2007); however, we do caution that this particular behaviour has been encouraged by their keepers, and the goats' welfare (as well as that of the trees) has been the subject of discussion (Hobart 2022). The lack of literature documenting tree-climbing in other regions suggests that this behaviour is possible, but only utilised by goats when necessary to access a sought-after resource. Indeed, they have put their climbing abilities to use in other situations. For instance, the remarkable dam-scaling feats of Alpine ibex (*Capra ibex*) in Italy have captured the attention of biomechanical scientific literature (Biancardi and Minetti 2017) and mainstream media (Dell'Amore 2010); their ability to climb near-vertical rock faces has afforded them access to a mineral source with no competition.

A second contributor to goats succeeding in harsh environments is their ability to maintain large ranges when necessary. One Australian study found that feral goats will move over 30 km between home ranges, with some animals having ranges of over 23,000 ha (Lethbridge 2016). In another large, multi-year Australian feral goat study, home ranges averaged from over 11,000 to nearly 16,000 ha (males and females, respectively), and while dependent on season and rainfall, the monitored animals travelled over 3 km/d on average (Moseby et al. 2020). Similar travel distances have been reported in extensively managed goats as well (dairy goats, Switzerland: Zobel et al. 2018; Angora goats, USA: Askins and Turner 1972; Mardi goats, West Africa: Ouédraogo-Koné et al. 2006). Zobel et al. (2019) postulate that because of the natural behaviour of the *Capra* genus to travel great distances on rough and rocky terrain, their hooves have evolved to grow continuously to combat the constant wear. This in turn has created one of the biggest welfare issues of intensively managed and companion goats: removing rocky terrain (Geist 1960) and preventing large range travel (Deiss et al. 2010) appear to create systematic hoof overgrowth and deformity (e.g. Anzuino et al. 2010; Sailer et al. 2021).

2.1 Welfare Issues Related to the Environment in Managed Goats and Potential Solutions

Complex natural environments provide goats with opportunities to perform behaviours such as climbing, hiding, and long-distance travel; all these activities involve goats making choices. The ability to choose is suggested to be rewarding and a potentially pleasurable experience in itself (Kurtycz 2015; Webster 2016). Providing complex environments that cognitively challenge and promote decision-making and the performance of different natural behaviours, is a notion frequently undertaken in zoo environments (e.g. Clark 2017); however, this concept is not readily discussed in commercial goat literature.

The type of system in which the goat is kept will have a major impact on what parts of its environment may compromise welfare. For instance, many dairy goat farms use open, yet relatively barren indoor housing, and provision of an outdoor space varies (e.g. UK survey: 17% of 46 farms, Anzuino et al. 2019; USA survey: 77% of 30 farms, Hempstead et al. 2021). Many of these facilities use straw or wood shavings as bedding (e.g. Hempstead et al. 2021; Sailer et al. 2021; Toussaint 1997), yet it is also not uncommon for dairy goats to be housed on grating with no bedding (e.g. Liang and Paengkoum 2019; Simensen et al. 2010). Regardless, none of the flooring options promote climbing or increased activity, thus limiting hoof wear and likely contributing to the prevalence of hoof overgrowth reported in dairy systems globally, even when some level of manual hoof trimming is occurring (e.g. Anzuino et al. 2010; Chakrabarti and Chandran 2016; Christodoulopoulos 2009; Hempstead et al. 2021; Sailer et al. 2021). Positively, goats are known to be attracted to different indoor (e.g. Bøe et al. 2007; Sutherland et al. 2017) and outdoor surfaces (e.g. Zobel et al. 2018), including hard and/or rocky ones, depending on the activity they are performing. Thus, when designing indoor facilities, a significant opportunity exists to facilitate goats wearing down their own hooves, either by providing indoor enrichments that promote climbing, or by providing additional outdoor space that provides enrichment and/or promotes exercise. In temperate climates, the housing may consist of open sided sheds with natural ventilation and opportunity for goats to have visual access to the outdoors (e.g. New Zealand: Prosser and Stafford 2017;

Scholtens et al. 2017), while in colder climates, barns are often completely enclosed (e.g. UK: Anzuino et al. 2019). From a naturalness perspective, having an opportunity to physically go outside is a major component to promoting good welfare for goats.

2.1.1 Outdoor Access

A common argument for keeping goats indoors without outdoor access is to limit parasite exposure, which in turn reduces use of anthelmintics for parasite control (Torres-Acosta and Hoste 2008). Goats are especially susceptible to parasite infection under grazing conditions compared to sheep, due to a reduced immune response (Hoste et al. 2007; Torres-Acosta and Hoste 2008), as well as growing anthelmintic resistance of many parasites. Thus, it is important to consider 'outdoor access' separate from 'grazing/pasture access'. Even if it is not possible to mirror goats' natural habitat, arguably the outdoors will always provide more variability than an indoor environment (Bøe et al. 2012). The very act of being able to go outside can be considered an enrichment, and once outside, even a non-grazing/pasture environment will provide goats with sights, sounds, and smells that they would not encounter indoors. Oesterwind et al. (2016) suggest that indoor environments may lead to boredom and frustration as they do not allow goats to investigate and explore. Also, given that some indoor facilities have been shown to have poor air quality, with high levels of ammonia, CO_2 and low airflow (e.g. Važić et al. 2019), providing goats with the opportunity to go outside would not only promote environmental choice but it may also have positive health effects.

Various observational studies examining goats' use of outdoor space exist; these studies have some confounding factors (i.e. single herd, limited observation times), they do nonetheless suggest outdoor space is valuable for goats. For instance, in a single herd of Norwegian goats, Bøe et al. (2012) reported that approximately 40–50% of the herd was outside during a springtime observation period. In a multi-farm study, Stachowicz et al. (2019) found that goats were sensitive to rain, wind, and cold. The general usage of the outdoor spaces across these same farms ranged between 17 and 38% (Stachowicz et al. 2018); it is important to note that the proportion was calculated based on different total times per farm, as each farm provided outdoor access differently. To our knowledge, just two controlled studies, with replicates and 24-hour observation periods, have been conducted, yet they too have some confounding factors. Bøe and Ehrlenbruch (2013) reported that 13–34% of their observations had goats using the outdoor space provided; however, the outdoor space was very small. Freeman (2018) found a diurnal effect, with approximately 30 and 40% of goats being outside at dawn and dusk, respectively. During the day, when groups had access to outdoor enrichment (climbing, hiding, scratching, and hard surfaces), the proportion of goats outside was 28% versus 21% when the outdoor space was empty; this pattern did not apply to the night-time. Although this was a strong study with a crossover design of 8 large groups, this work has yet to be peer-reviewed.

The literature exploring outdoor space provision and its use in goats is largely observational and may have limited observational times, as well as farm, time, and seasonal confounders; nonetheless, all the studies demonstrate that goats use outdoor space when it is available, provided that the space is suitable. For instance, if it is raining, goats will seek shelter; what this shelter looks like is not well documented. There is a need for further science-based, yet practical, examples of what features goats value most in an outdoor space to provide farmers with guidance. Importantly, goats may not choose to use an outdoor space as often as we might like (e.g. anecdotally farmers have expressed concern about spending money on providing the space if the goats are not using it frequently), however, from a welfare perspective, the key factor is that goats have a choice where they spend their time. Furthermore, while it has not been tested directly in goats, Freeman (2018) argued that goats may find exposure to sunlight pleasurable. It may be that this is due to the thermoregulatory properties provided by sunlight, but there is evidence that animals value warmth and light separately (chickens: Duncan et al. 1998). Goats may value the opportunity to interact with sunlight that outdoor access would afford to them; therefore, the benefits that goats get from outdoor space, including what sunlight means to them, is a much-needed area of research.

2.1.2 Climbing and Hiding

As described, wild, feral, and extensively managed goats climb, as well as seek shelter in caves or other covered areas, if given the opportunity. It is very likely that these types of environmental components tap into a hard-wired predator avoidance instinct, and that this remains even when the risk of predation is eliminated in human care. Indeed, it appears that providing vertical spaces as an alternative to the traditional flat pen results in goats experiencing a more positive social environment amongst themselves (Aschwanden et al. 2009a), as well as better learning performance in a computerised discrimination task (Oesterwind et al. 2016); the latter suggests that goats with a more enriching environment might be better able to cope with management changes.

Goats in a companion animal setting are arguably the most likely to be given the opportunity to express natural behaviours such as climbing and hiding; articles available for the public frequently discuss this topic (e.g. Cooper 2020). While not specifically for the promotion of hiding, some regions have regulations in place indicating that companion goats must be provided with shelter (e.g. New Zealand Code of Welfare: Goats 2018). In larger commercial systems, cleaning and bedding management may be impeded by additions of such environmental improvements, thus some of the outdoor space literature has suggested that climbing and hiding opportunities be included there. While they did not evaluate the value of resources, Freeman (2018) suggested provision of climbing outside may be more rewarding compared to indoors, due to the greater lines-of-sight available with increased elevation. Both Stachowicz et al. (2018) and Freeman (2018) suggested that climbing opportunities improve goats' use of the provided space, with the latter finding a 30% increase in proportion of goats outdoors in 'enriched' environments. Flint and Murray (2001) employed climbing opportunities in an outdoor feedlot setting and suggested it (in combination with other interactive objects) contributed to improved weight gain in growing meat goats; unfortunately, these authors did not reflect on the mechanism behind this finding. The provision of climbing and hiding opportunities specifically for decreasing predation risk in extensively managed goats (e.g., meat, fibre and working animals), has received little scientific attention.

Although not systematically documented, incorporation of climbing and hiding opportunities has been observed in commercial systems, both indoor and outdoor (Figs. 4, 5, 6, 7).

We suggest that a crucial area of needed research combines the development of 'farm friendly' (e.g. easy to use and manage, safe for goats) climbing and hiding opportunities, with behaviour measures that evaluate motivation and individual differences in preference, as well as animal health measures, such as cleanliness and hoof wear.

2.1.3 Travel and Ranges

Considering the distances goats can cover when in a natural environment (e.g. 3 km or more per day; ranges of thousands of ha), it is very likely that most housed commercial systems, particularly in the dairy sector where space allowances between 1 and 3 m^2 are common (e.g., Andersen and Bøe 2007; NZ Code of Welfare: Goats 2018), and even intensively managed outdoor systems, do not promote enough exercise. The implications of this are clear in their potential impact on hoof wear. Hoof overgrowth was reported as one of the main causes of euthanasia in wild goats enclosed in a Qatarian wildlife sanctuary (Deiss et al. 2010), and Lindgren (2016) reported that 3 out of 6 surveyed Swedish zoos had goats with hoof issues. Conversely, extensively managed dairy goats in the Swiss Alps were observed to have immaculate hoof structure (Zobel et al. 2019); although these goats had not been trimmed in at least 5 months (Fig. 8), they were recorded travelling over 3 km

Fig. 4 Milking goat in a smallholder setting with access to an outdoor area with climbing opportunities (Photo: Gosia Zobel)

Fig. 5 Young Deutsche Edelziege goats using an elevated platform incorporated into their indoor housing (© FBN/Nordlicht)

Fig. 6 Winter housing (for Nera Verzasca goats observed at summer location in Hut Nimi; Zobel et al., 2018) provides climbing and hiding opportunities (Location: Gordevio, Switzerland; Owner: Stefanie and Dario Martinoni)

Fig. 7 Extensively managed Saanen cross dairy goats using a platform designed for easy movement on farm (© AgResearch Ltd)

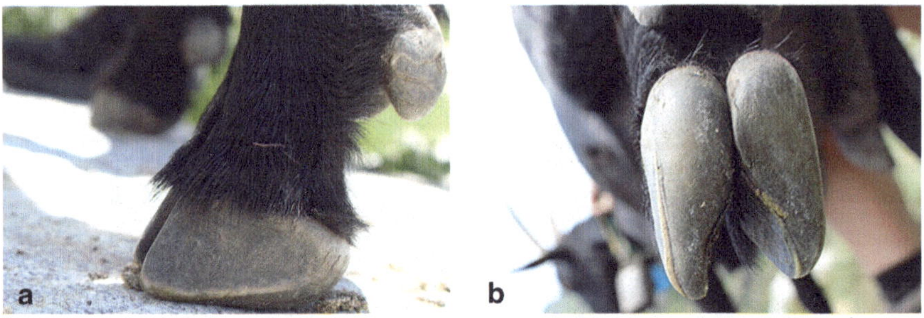

Fig. 8 (**a**) Rear right hoof of a Nera Verzasca extensively managed milking doe; (**b**) previous hoof trimming was approximately 5 months prior to photo being taken (Location: Hut Nimi, Switzerland; Owner: Stefanie and Dario Martinoni)

per day. This was a personal observation and only relates to these goats in the summertime (as they spend their winters enclosed in a barn); nonetheless, it shows the scale of the problem that has been created as goats no longer have large ranges when they are managed intensively, or even as companions on small blocks, or in backyards (Grant 2012).

On a similar note, a second physical welfare concern that has also evolved as goats have been shifted from large ranges to small space allowances is the need to manipulate horns. Chapter "Painful Procedures, Transport, and End-of-Life Management" discusses horns (and their removal) in more detail, but here we mention them as their removal in dairy systems is closely linked to the lack of space afforded. Horns play an important role in social communication and are used in a variety of behavioural repertoires; they provide offence and defence during competitive interactions, and their size is an indicator of social status (Shank 1972; Shi and Dunbar 2006). Use of horns during 'vegetation thrashing' (e.g. running horns back and forth through brush during non-consummatory events) is suggested to be a comfort rather than agonistic behaviour when performed simultaneously in groups of males (Shank 1972). Zobel et al. (2019) illustrate goats using horns for self-grooming. Finally, while not recently examined, Taylor (1966) suggested horns serve a thermoregulatory function, helping to keep goats cool in extreme environments. Many goats (e.g. some meat, fibre, working, companion) are still managed with horns; it is important to note that horns do serve important functions in the natural behaviour of goats, and thus our intensive restriction of some goats' range (e.g. dairy, some companion) and associated horn removal is a welfare concern beyond the shorter-term effects of pain.

Of all the natural behaviours, goats' capability (and physical need) to travel is likely the hardest to promote in most human-managed systems. From a hoof perspective, there is room to improve the opportunities we provide goats to help us manage their hooves (as described earlier), yet it is undeniable that good hoof trimming management must accompany these improvements. However, from an exercise and navigational needs perspective, this is where we must ask difficult questions and perhaps some concessions are needed. If goats are to be kept in our care, are we able to provide them with enough exercise opportunities? Perhaps some extensively managed farms can aim to move their goats around more, and intensive farms can build exercise yards, however there is a dearth of science-based information on the specifics of these solutions. There remains a massive gap in our understanding of how much goats value the opportunity to travel, the underlying benefits travel brings, and ultimately how to provide for these.

2.1.4 Sheltering

Due to their management, fibre producing goats (either mohair or cashmere) may be perceived to avoid many of the welfare concerns that are typical of intensively farmed goats. For instance, cashmere goats in hot and temperate climates are typically raised extensively and in relatively small herds. Shamsaddini-Bafti et al. (2012) describe the Iranian Raeini breed being reared in small extensive nomadic systems; while Wang et al. (2014) reported a mean herd size of 200 when surveying 12 extensively managed farms of Arbas Cashmere breed in Mongolia. Despite these management benefits, the major practice of fibre harvesting can dramatically impact welfare when not considered in balance with goats' natural behaviours. Most goat breeds have secondary hair follicles which produce fine fibres; genetic selection has resulted in some breeds being preferred for fibre production, resulting in a collection of 'cashmere' breeds. One exception is the Angora breed, which do not have secondary hair follicles but are also prized for their fibre. The production of high-quality fibre has resulted from a combination of traditional knowledge, and research about timing of natural moulting and sheering, manipulating photoperiod, genetics, and nutrition management (e.g. Bigham et al. 1993; Ivey et al. 2000; McGregor 2009; Merchant and Riach 2002; Qiu et al. 2018). However, there is little work exploring how the harvesting of fibre from goats impacts their welfare. McGregor et al. (1991) discussed the essential protectiveness of the primary guard hair against weather extremes in cashmere breeds. The author expressed concern for a shift from traditional harvesting methods (which work around natural moulting of the fine fibres linked to photoperiod and season; e.g. Rhind and McMillen 1995) to a complete shearing approach that removes both hair layers. Anecdotal reports in Angora goats (which produce mohair and do not have the benefit of two sets of hair follicles) have linked shearing, a lack of shelter and exposure to rain, to 'offshear deaths' (Shelton 1995). Goats actively seek shelter when it is available; preventing this natural behaviour for fibre goats has even more dire consequences than unpleasantness of being wet and cold. Fibre animals should be provided with shelter when shorn, and ideally the method used should not deplete the animal of all of its fibre if harvesting occurs in poor weather (NZ Code of Welfare: Goats 2018).

2.1.5 Environmental Consideration for Captured Feral Goats

In many places around the world, goats are a 'successful invader', persisting in difficult environments if sufficient numbers and range are available (e.g. Clout and Russell 2008). Literature considering their culling is plentiful (e.g. Bunbury et al. 2018; Campbell and Donlan 2005; Forsyth and Parkes 1995; Moseby et al. 2020), yet little discussion focuses on their welfare when they go from being in a 'wild state' to being in human care. In some places (e.g. Australia), mustering, feeding, and subsequent sale into the human food chain is a common occurrence (e.g. Chapman 2003; Thompson et al. 2002). A key natural behaviour these goats lose opportunity to perform at mustering is the motivation to seek shelter and hide. Leidy et al. (2015) suggested that the feral goats of Majorca (Spain) are more gregarious than truly wild ones, attributing this to their domestic background; however, their study was purely observational and did not involve handling of the goats. The assessment of stress levels in mustered feral goats lacks quantification. Underwood (2002) discusses opportunities for 'self-mustering'; these systems work by habituating animals to enter and exit a 'trap yard' installed around a water source; when animals are to be mustered, the exit gates are closed. This approach can reduce welfare concerns stemming from intensive human or air-based mustering. Yet the report does not describe indicators for animal welfare once the animals are in human care. There is a clear need for research in this area; if we can set aside the overarching ethical dilemma of mustering and placing feral animals into commercial food chains, there is certainly room for exploring how best to house these animals that have gone from a 'natural state' to a very unnatural one. Miller et al. (2018) reported reduced behavioural reactivity when captured feral goats were exposed to calm human interactions in their first few weeks of captivity; however, they did not record activity levels. As suggested earlier, environments with some level of complexity might be beneficial for meat animals; the potential of providing shelter points, as well as exercise, would be an excellent research area to explore.

2.1.6 Environmental Quality and a Potential Link to Stereotypic Behaviours

Anecdotal evidence suggests stereotypic (repetitive and apparently non-functional) behaviour is a common problem, particularly in indoor housed goats. For instance, during field visits with dairy goat farmers, they have commented that their goats will 'chew anything they can get their mouths on'. Furthermore, in our own research facility, we found that goats that came to us from a commercial housing environment stripped paint from metal doors in the milking area which our (extensively housed) goats never touched. We postulate these stereotypic behaviours may relate to goats' excellent cognitive capabilities that are not stimulated in barren, flat, and/or small housing environments. Surprisingly, there is little robust research examining how the environment influences stereotypic behaviour in goats. In an observation of two pens of goats (one enriched and one not), Kakarash et al. (2021) suggested that the provision of various 'enrichments' into goat housing reduced stereotypies; their work could be followed up with larger groups and provision of different biologically relevant enrichments. Although not typically considered a 'stereotypic behaviour', we suggest that the act of 'continuous escaping' (as is often mentioned by companion goat owners) is a manifestation of a behavioural need not being met. It should be considered an early alarm that the environment in which the goat is housed is not sufficient.

2.1.7 Environmental Enrichment Through Cognitive Stimulation

Provision of environmental enrichment can also take the form of cognitive stimulation, an approach that is often taken in captive zoo animals (reviewed by Clark 2017) but rarely in commercially managed livestock. Studies exploring cognitive enrichment examine whether animals will invest cognitive effort into a task to receive a reward, and if such cognitive challenge can improve animal welfare (Meehan and Mench 2007). For many animals, successful coping with a challenge has intrinsic reinforcing properties beyond the food reward that is usually gained from completing the task itself (Hagen and Broom 2004). Animals will even work for a reward that is freely available, a phenomenon called contrafreeloading, suggesting that the 'challenge of challenge' is important for animals under human care (Franks 2019; Osborne 1977). To the authors' knowledge, contrafreeloading has not been observed in the wild, however.

Goats have impressive skills at solving complex discrimination, operant, and categorisation tasks. They are quick to differentiate objects of different colours when the choice of these objects is associated with a food reward (Nawroth et al. 2017, 2022). They can learn a complex 2-step task of pulling and then lifting a lever to deliver a reward and recall their response for months (Briefer et al. 2014). Goats can also learn arbitrary 2D symbols in an automated choice task (Langbein et al. 2004), successfully put similar symbols into categories (Meyer et al. 2012), and identify symbols that did not belong to a group (Roitberg and Franz 2004). Once learned, goats are also able to reverse learned contingencies, both using real objects (Nawroth et al. 2022) and 2D images on screens (Langbein et al. 2008). Given their intrinsic curiosity and considerable problem-solving skills, it is not surprising that

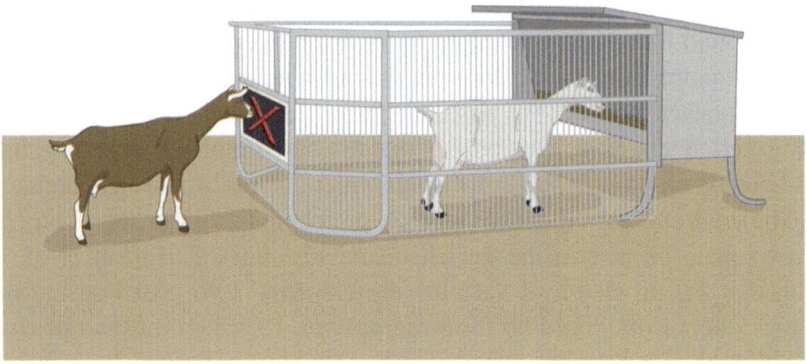

Fig. 9 Proposed cognitive enrichment providing access to creep feeder via RFID tags and reader (Zobel and Nawroth 2020)

goats are also motivated to engage in cognitive tasks simply for participation itself. For instance, dwarf goats housed in a research setting engaged in a computer-based discrimination task even when an identical reward was freely available outside of the test apparatus (Langbein et al. 2009). When confronted with a manipulation task, in which dwarf and dairy goats could either move a small sliding door to receive a reward or receive a similar reward for free, goats opted for the contrafreeloading option in almost half the trials (Rosenberger et al. 2020). Dairy, but not dwarf goats, were also faster in approaching the contrafreeloading option, compared to the free one. These results show that cognitive challenges may be rewarding in themselves, and thus an important form of stimulation for goats. The lack of cognitive challenges in the managed environments of goats under human care may deprive them of an important rewarding aspect of their natural behaviour, thus resulting in negative affective states. The provision of cognitive stimulation in their housing environment (e.g. Zobel and Nawroth 2020; Fig. 9) should therefore be regarded as an option for improving welfare in all systems, intensive or extensive, and small or large.

3 Foraging

'Flexibility' is a common theme that arises in all aspects of a goat's feeding and foraging behaviour. Goats are not only highly flexible in what they eat, but also in how they eat and where they find food. Even the timing of their feeding patterns can be flexible, depending on feeding conditions. Observations of modern day extensively managed and feral goats suggest that goats are opportunistic, feeding on what is preferred as well as available. Phytolith (silica particle) analysis of goat faecal pellets indicates that management of how goats eat began relatively early in the goat–human relationship; evidence suggest even in the Late Neolithic period (e.g. 7000 BCE), goats were being encouraged to graze, rather than browse, at certain times of the year (Landau et al. 2020). Such discoveries add a level of complexity when determining what is truly 'natural feeding behaviour', and what has evolved with human intervention. Regardless of the origin, certain behaviours have been consistently observed in goats over millennia, and as such have been actively recorded even before modern times. For instance, in The Rochester Bestiary, a rare and extravagantly illustrated Mediaeval guide of animals, goats are the only animals presented engaging in multiple feeding behaviours; goats are depicted grazing on elevated slopes, and bipedally browsing trees (Fig. 10). In this section, we describe the natural feeding and foraging behaviour of goats in their natural environment, and the animal welfare concerns that arise in managed goat systems. We propose solutions that incorporate elements of natural feeding behaviour, some of which have been tested in practice.

3.1 Finding Forage in the Natural Environment

Goats are skilled spatial navigators, possessing many adaptations that help them cope with challenges as they search for food and remember solutions over time. Brain scans of goats revealed modifications to the hippocampus, particularly elongations of the distal CA1 (Watson and Binks 2019), which enhances the ability to

Fig. 10 'Wild goats grazing' from The Rochester Bestiary circa 1230–1400

integrate objects in space. This specialisation appears to be especially apparent in migrating ungulates, allowing them to integrate topographical features while navigating their spatial environment. This evolutionary specialisation is seen in one reported eradication effort of feral goats: translocated individuals were able to navigate up to 8 km to reunite with other goats (Bunbury et al. 2018).

There are few reports of how goats navigate from field observations, but several controlled studies aid our understanding of these skills. Recent work has experimentally demonstrated the behavioural flexibility of goats (compared to sheep) when faced with a navigation task. Both species were tasked with navigating around a straight barrier through a gap at its edge to find a food reward; when this gap was moved to the opposite end of the barrier, goats were generally faster than sheep at reorientating to the new gap position (Raoult et al. 2021). This work demonstrates the navigational flexibility of goats when situations change. The authors proposed this may arise from the differences in foraging strategies between goat and sheep species, where goats tend to rely more on patchily distributed food sources. Another experimental demonstration of the navigational foraging strategies of goats was in a T-maze task that examined how goats make decisions about where to initiate a foraging bout. Initially, Hosoi et al. (1995) presented goats and sheep with high-quality food in both arms of the maze, and both species tended to return to the same arm they previously chose (demonstrating a 'win-stay' strategy). In a follow-up experiment, one arm now contained a lower quality food (representing a 'losing' choice compared to a 'winning' choice in the other arm). Here, goats more often changed to the other arm if they chose the 'losing' arm (demonstrating a 'lose-change' strategy), while sheep did not. This study demonstrates that foraging choices in goats may be more strongly driven by negative or 'losing' experiences that are later recalled when navigating and making decisions about where to forage. Several other studies using maze tests have demonstrated that goats are skilled navigators (Langbein 2012; Liddell 1927; Qiu et al. 2018), and that they can retain this spatial information over at least 3 months (Langbein 2012).

Evidence from other types of tasks suggest that goats retain different types of information well over long periods. For instance, goats observed how a reward was hidden in full sight in one of two opaque containers that then switched positions; most goats chose the correct container containing the non-visible reward after following its movement trajectory (Nawroth et al. 2015). Furthermore, goats that were trained to pull and lift a lever to obtain a food reward remembered the task 10 months later (Briefer et al. 2014). This evidence together points to the possibility that goats also have good memory of spatial navigation routes in their natural habitat, which would be highly advantageous when it comes to foraging for high-quality feed in variable environments.

3.2 What to Forage in the Natural Environment

Goats are both browsers and grazers (reviewed by: Goetsch et al. 2010); this means they forage on grasses, herbs, and shrubbery, including foliage, buds, flowers, and woody stems (Askins and Turner 1972). They are selective foragers, using their flexible lips and tongues to select for the best quality buds, leaves, and fruits (Ngwa et al. 2000; Ouédraogo-Koné et al. 2006). They also have a narrower mouth and shorter jaws compared to other ruminant species, which contributes to their selection abilities and permits faster chewing (Mellado et al. 2007; Perez-Barberia and Gordon 1999); this anatomy permits successive bites to sever the tender small leaves of shrub species (Mellado et al. 2007). Consequently, the diet of goats is often of greater nutritional value than if the animal ingested the whole plant (Lu 1988). Selection of particular plants to ingest appears to be based on familiarity, and goats readily learn where preferred plants are located (Provenza et al. 1994). One study found that goats spent 50% more time foraging for thistle types that were familiar compared to unfamiliar, and this feeding pattern was also seen in their offspring (Arviv et al. 2016). Overall, goats are well known for being selective, flexible foragers, although domestic goats appear to be less selective than feral goats (Aldezabal and Garin 2000).

When goats are faced with changing conditions or need to forage in different locations, they can readily shift their diet and increase the amount of lower quality feed as needed. For example, when Argentinian Creole goats (*Capra aegagrus hircus*) were given a choice of 9 shrubs and trees and 5 grass types, grasses were the predominant choice when they were available in the summer (13% of their diet was just one type of grass), but in the winter when grasses dwindled, the goats shifted their diet to 98% shrubs and trees (Egea et al. 2014). Similarly, goats in the southern Mediterranean region included shrubs and trees in 60–90% of their diet and continued to seek preferred shrubbery despite low availability (Chebli et al. 2020). In the USA, surveyed farmers reported goats' forage preferences as one of the top 4 reasons for why they began managing meat goats; their flexibility made goats a good option for browsing land already grazed by other species such as cattle (Gillespie et al. 2016). Indeed, goats are very efficient at digesting poor quality roughage (Silanikove 1997; Ventura-Cordero et al. 2018) and can tolerate bitter flavours (Bell 1959). They even tolerate potentially toxic compounds like hydrolysable tannins (Jansen et al. 2007), which further expands their diet composition, and may even have anthelmintic benefits (Kahiya et al. 2003). Their mobile lips even allow goats to consume parts of cactus species with large spikes when in semi-arid environments (Nair et al. 2021).

Unfortunately, a consequence of this ability to shift to different vegetation has resulted in the disruption of native tree populations on many island ecosystems (e.g. Campbell and Donlan 2005; Chynoweth et al. 2013), where feral goats will freely debark and kill trees (Spatz and Mueller-Dombois 1973). However, such disruptions to landscapes may be related to the habitat and density. A study of domestic free-ranging goats on Brazilian Caatinga rangeland (semi-arid desert where 95% of goats in Brazil reside) monitored animal movement with GPS and their dietary preference; the authors found that goats preferred to stay in open, rather than forested, areas and consumed a diet reflecting the relative abundance of plant species in the area (Jamelli et al. 2021), demonstrating their generalist feeding behaviour. The authors concluded that these free-ranging goats were not a threat to damaging the local vegetation, so long as densities remained low. Studies of other goat populations consistently find that when goat density is high, they are damaging to vegetation structure (e.g. Canary Islands: Arevalo et al. 2011; USA: Severson and Debano 1991; South Africa: Scogings and Macanda 2005). Overall, studies consistently find that goats can exploit a range of habitat types and heterogeneous environments, but dietary selection depends on a range of factors including plant diversity, season, livestock density, and size of the home range.

The type of feed that goats consume greatly impacts another natural behaviour—water intake. Genetics, lactation status, feed composition, and environmental factors play significant roles in goats' water requirements (Giger-Reverdin and Gihad 1991). While not a well-studied topic, there is arguably a discrepancy between what goats need and what they want in terms of their water allocation. For instance, goats are naturally sensitive to the quality of their water; Runa et al. (2019) demonstrated that even minute concentrations of salt were avoided by Boer goats. Older literature demonstrates that some desert breeds can go up to four days without water (e.g. Bramer breed—(Khan et al. 1979), and that dehydration plays a dramatic role in the milk production of desert Bedouin goats (Maltz and Shkolnik 1984). Indeed, goats' ability to survive on minimal water (particularly those animals residing in arid, desert areas) has been discussed (e.g. Silanikove 1994). Nonetheless, it is generally accepted that animals managed under human care should be provided with *ad libitum* access to water. The latter is particularly important in the case of animals that are expected to be highly productive. For instance, it has been reported that lactating goats (Swedish breed) drink approximately 8 L/d when not heat stressed, and their voluntary water intake increases to approximately 10–12 L/d, with increasing temperature challenge (Hartmann et al. 2021). As mentioned earlier, depending on the situation, goats may even value water as enough of a treat that they will work for it when water is freely available (Langbein et al. 2009). Thus, we suggest that water, its quality, and how it is allocated, is an important welfare consideration.

3.3 How to Forage in the Natural Environment

Flexibility is exhibited in how goats forage. Goats are selective foragers of 'browse' and have adopted a range of ways to feed. Browsing behaviour may include some low-level feeding (e.g., grazing), but most often requires a range of higher-level feeding to access foliage, buds, flowers, and shrubbery at varying heights. Foraging at or above head level therefore substantially expands feeding possibilities. Goats can often be seen feeding at eye-level, and in bipedal positions where they rear onto their hind legs (Tölü et al. 2012). Their ability to bipedally forage above head level exceeds that of cattle (at 1.65 m) and may be up to 2.1 m in West African goats (Sanon et al. 2007). Although there are limited reports specifically about feeding heights, the few available suggest goats can and will forage at substantial heights (Delibes et al. 2017), demonstrating their remarkable ability to exploit forages in areas that most other ruminants cannot reach.

When, and how often, goats adopt these different feeding positions depends on several factors. Feeding duration changes with time of day (e.g. feral goats, *Capra hircus*, Shi et al. 2003), temperature (e.g. marked ibex, *Capra ibex*, Aublet et al. 2009), season (e.g. Nubian ibex, *Capra Nubiana*, Tadesse and Kotler 2010, 2011), and social competition or facilitation (extensively managed milking goats, *Capra hircus*, Shrader et al. 2007). When managed on rangeland, Angora goats spent a third of their time grazing and the remaining time browsing (Askins and Turner 1972). Sanon et al. (2007) reported that goats browsed for more than 60% of their feeding time during the rainy season and for more than 90% of their feeding time during the dry season when the herbaceous components of plants began to decline. In a mixed herd of non-pregnant and non-lactating dairy goats (Granadina and Criollo breeds) extensively managed in an arid northern Mexican environment, browse composed the majority of reported dietary intake (Mellado et al. 2005), especially for taller goats (Mellado et al. 2004). The majority of browsing time appears to be spent in the quadrupedal position, primarily because adopting a bipedal position may not justify the energy requirements if the browse at this height is not worthwhile (Pfister et al. 1988). However, foraging in bipedal positions becomes much more common when other herbaceous vegetation is lacking. Even when there is plenty of forage at lower levels, goats have been reported spending up to 8% of time foraging in the bipedal position (Pfister et al. 1988). However, time spent foraging at these different heights can differ by breed (e.g. at least twice as much bipedal feeding in Turkish Saanens compared to other breeds; Tölü et al. 2012).

In addition to expanding dietary selection, an evolutionary advantage of 'eye-level' feeding may allow for quicker identification of predators (because the head is elevated above ground; Beauchamp 2015) and may reduce exposure to noxious insects and parasites whose eggs are usually at ground-level (Lu 1988). Indeed, Hoste et al. (2001) reported that individuals that more often browsed, rather than grazed, had lower nematode infection. This tendency to avoid grazing likely contributes to Forsyth and Parkes (1995) conclusion that baiting feral goats (for culling purposes) was not effective because the animals did not like eating off the ground. However, when goats do feed from plants close to the ground, they use their narrow mouth and short jaw to select leaves without insects, or to shake individual leaves when they detect insects by repeatedly touching leaves with their muzzle (Berman et al. 2017). This ability to detect and avoid noxious insect ingestion was shown to be innate by testing in both naïve goat kids and adults (Berman et al. 2019).

Overall, reports from a range of regions and breeds suggest that goats will adopt a variety of feeding positions to access forage at ground-level, eye-level, and above the head. These feeding positions allow goats to access a much wider range of forages than other ruminants and almost certainly contribute to their ability to thrive in harsh environments.

3.4 Welfare Issues Related to Feeding Behaviour in Managed Goats and Potential Solutions

Most human-managed systems offer a markedly different feeding environment for goats than is typical in a natural environment. Providing opportunities for goats to express their natural feeding behaviour in such systems is not necessarily easy, especially when many commercial systems manage goats indoors or in extensive environments where foraging diversity is not a key consideration. However, there is growing interest and scrutiny from the public in how food animals are reared and managed; this attention has mainly focused on poultry, swine, and dairy cattle industries, but the concerns expressed extend to goats in commercial production (including production for meat, dairy, and fibre). The most common objection from society is the unnaturalness of management practices (e.g. Clark et al. 2016; Placzek et al. 2021). In this regard, there are several practices in goat industries that could be viewed as unnatural, which have been reviewed by Zobel et al. (2019). Here we specifically focus on aspects of the feeding environment in management systems that could be improved to better meet the natural needs of individuals.

3.4.1 Presentation of Feed

The way that feed is presented currently in commercial housing systems is typically uniform; the diet is presented along a feed rail, in a trough or within a rack, and usually in the same place; Zobel et al. (2019) show several photos of such feeding locations. Many housing systems for goats provide feed at ground-level, where the tractor drives along the front of a feed bunk to dispense feed, and goats access this feed from inside the pen. This leaves little opportunity for goats to feed from elevated heights that mimic browsing behaviour. We suggest commercial farms can implement opportunities to present feed at higher levels, such as offering elevated feed racks, or raising the drive alley to allow goats to feed from eye-level (Fig. 11). The idea behind providing these raised surfaces is to promote browsing positions. There are only a few studies that have examined the feeding behaviour of goats when offered elevated feeders, with some positive results. In an experimental study, Neave et al. (2018b) demonstrated that non-lactating dairy goats eat more from a feed bunk that is elevated above the ground (head angled upwards above the neck in a browsing position, with front feet on a step), compared to a feed bunk at ground-level (where the head is lowered in a grazing position). A follow-up study expanded on this concept and suggested that presentation is influenced by feed type, with goats trading elevated feeding of grass when leaves were available at ground-level (Cellier et al., 2022). These findings are in line with Van et al. (2005) who showed that goats consume more dry matter when foliage is hanging versus in a trough.

In the Neave et al. (2018b) study, dairy goats competed for access to the elevated feeder by more often displacing individuals from the feed trough, suggesting they were motivated to obtain feed from higher positions. However, the authors acknowledged that all goats likely could not feed comfortably together at a single feeder, so this may have contributed to more displacements while feeding. This highlights the

Fig. 11 Elevated drive alley providing feed to a commercial dairy herd at eye-level. Illustration modelled after dairy goat farm in Ontario Canada (Henry and Anja van der Vlies and family)

importance of offering sufficient feeding space for all goats so that they can feed where, and when, they like. This is also necessary to manage aggression among goats, especially for those that are horned (Loretz et al. 2004), or when goats are mixed with unfamiliar animals (Aschwanden et al. 2008). Aschwanden et al. (2009b) found lower agonistic behaviour and increased feeding time when goats were offered a vertical, elongated hayrack with a narrow gap, and goats could stand on a platform to reach the upper portions of the rack. Providing elevated platforms from which goats can eat promotes longer feeding bouts and fewer uninterrupted resting bouts (Aschwanden et al. 2009a). A small study of two treatment groups compared hexagonal and linear feeders that provided feed at eye-level (Kaur et al. 2021). The authors observed more 'comfortable and normal' feeding postures and feeding time at the linear raised feeder (but the definition of these postures was not defined); the hexagonal feeder had more competitive social interactions, especially from dominant goats, possibly due to greater visual access to other goats. When re-designing feed bunks, goats must comfortably reach feed. According to Keil et al. (2017), the feeding surface should be raised at least 10 cm to allow goats to feed in a relaxed posture. It has been noted that goats may eat while kneeling in commercial systems (with no reports in feral herds), suggesting an uncomfortable feeding environment. It is also possible that these goats are facing hoof health or other chronic inflammatory challenges (without necessarily being overtly lame; Zobel et al. 2023); while several welfare assessments mention kneeling as an issue (e.g. Anzuino et al. 2010; Battini et al. 2016a), the causes are not discussed, and thus, we suggest that kneeling-based feeding behaviour warrants further investigation.

In addition to improvement of feed presentation in commercial systems, it is important to consider animal origin. As mentioned by Forsyth and Parkes (1995), the feral goats in their observations were unwilling to eat bait off the ground; therefore, we suggest that opportunities exist to improve how feral goats mustered for feedlotting, and eventual slaughter, are fed. Considering Neave et al. (2018b) showed that goats ate more from elevated feeders, a consideration for how these mustered animals are fed may not only promote their natural behaviour, but could have production benefits too.

3.4.2 Type of Feed

It is common for intensive goat management systems to provide mixed rations comprised heavily of concentrates and hay, straw, and/or silages (e.g. Gürsoy et al. 2011; Zobel et al. 2015); in some regions, diets may also largely be made up of fresh cut grass (Murney et al. 2019). These diets are homogenous and provided consistently each day in the same form and usually at the same time, which limits the foraging choices of goats. In fact, this is often precisely why commercial farms provide a mixed ration, to prevent selection of particular components since the ration is mixed to specifically meet daily nutrient requirements and to maximise growth or yield (Maltz et al. 1991). We suggest that diets can still be balanced for good yield while providing opportunities for natural foraging behaviour and feed choice. Without these opportunities, goats are at risk of boredom and lack of agency. This is not to suggest that commercial goats should be turned out on rangeland to obtain their dietary needs; indeed, it has been shown that goats left to forage entirely on woody rangeland had reduced milk yield when compared to goats managed on cultivated pasture or fed hay indoors (Steinshamn et al. 2014). Thus, a supplementary approach may be a good compromise for maintaining high yield of commercially managed goats, while providing opportunities for natural behaviour. Outdoor foraging areas connected to standard indoor housing are an option if they are well managed and maintained with fast growing trees or shrubs (e.g. willow, poplar, or other region-dependent browse option). This outdoor space access would allow goats to supplement their diet with browsing materials and promote their natural selective foraging behaviour. Furthermore, if such areas were varied in their planting, goats could employ some spatial skills to re-locate preferred forage patches. Controlled studies of goats fed supplementary browse material show potential for this kind of supplementary feeding to promote greater feed intake and growth compared to control diets (Mabeza et al. 2018; Okoruwa and Ikhimioya 2020); similar effects were observed when browse leaves were fed *ad libitum* (Sanon et al. 2008). To the authors' knowledge, there are no studies that have examined feed intake or production outcomes of commercially managed indoor goats with outdoor access to browse.

Farms may not have the outdoor space to provide foraging areas and, depending on region, seasonal constraints may limit the availability of forage outdoors. We suggest there are still ways to provide feed diversity to indoor managed goats, for instance by cutting shrub and tree material and presenting it in creative ways in barns. This could include hanging cut foliage (Van et al. 2005), attaching to a 'foraging board' fixed to a wall at different heights (Dziba et al. 2003), presenting in elevated feed bunks (Neave et al. 2018b) or on elevated platforms (Aschwanden et al. 2009b). Although a drawback of this approach is the added labour to source, cut, and carry foliage into the barn, these different presentation methods will stimulate goats' foraging skills, reduce boredom, and provide agency in obtaining feed. Indeed, each of these studies that tested different feed presentation methods found largely positive results (e.g. greater feed intake, feeding time, or reduced aggression); Neave et al. (2019) mentioned increased displacements when goats were provided with an option to feed with an elevated head position. While this is a negative outcome, the authors suggested that this was an indicator of the goats' natural motivation to perform this feeding stance.

Another way to provide diverse feeds is through supplementary automated feeders. These systems have been used for many years in dairy cattle (Wierenga and Hopster 1991) and swine industries (Young and Lawrence 1994), and as suggested by Zobel and Nawroth (2020) such feeders could also be implemented in indoor housing systems for goats as an addition to their regular feed provision. Such automated feeders could be stocked with different types of feeds and programmed with unique signals corresponding to specific goats. Research has shown that goats will regularly interact with an automated learning device, where they were required to discriminate between a variety of visual shapes to obtain a reward (Langbein et al. 2009). Thus, incorporation of automated feeding stations with diverse feeds could also serve as a form of cognitive enrichment requiring goats to learn a visual discrimination (e.g. Oesterwind et al. 2016) or categorisation task (e.g. Meyer et al. 2012) to access the feeder. It is important to note that this system should not withhold feed but rather supplement readily available feed through a cognitively beneficial task. In addition, attention to the nutritional requirements and diet balancing of goats is still necessary, especially for high producing dairy goats, and fibre producing goats (which divert energy to hair production; e.g. Angora breeds: Cronje 1992; Shahjalal et al. 1992).

3.4.3 Health Issues Related to Feeding Practices

Goats are susceptible to parasite infection when intensively grazed; this is, at least partly, likely why browsing is part of a goat's natural behavioural repertoire, because feeding above the ground limits exposure to parasitic eggs that are in grasses (Hoste et al. 2001; Lu 1988). Rout et al. (2002) propose a hypothesis that some breeds possess anatomical features that may have served to limit parasite exposure by promoting browsing; for instance, the Jamunapari breed feature long ears (about 20–30 cm in length) that touch the ground when the goat grazes, and a protruded lower jaw (prognathism) that limits biting ability when the lower lip touches the ground first. Conversely, Geiger et al. (2021) discuss hypotheses that suggest the protruding jaw

Fig. 12 Damascus breed with distinctive head anatomy (Photo: Serge-Yan Landau)

may in fact be maladaptive when foraging opportunities are limited. Anecdotal reports suggest that this latter feature (not the ears) compromise production. Given that numerous breeds possess some level of prognathism (e.g. Damascus; Fig. 12), it is surprising that little scientific inquiry exists around the potential welfare implications of this facial structure, or its implications for reducing worm burden.

Evidence from naturally parasitised goats and sheep on rangeland indicate that goats are less often infected when they can exploit browsing opportunities (Hoste et al. 2007). Additional benefits of browsing include possible anthelmintic properties (such as tannins) of certain plant species (Hammond et al. 1997; Hoste et al. 2001, 2006), which contribute to reducing the negative effects of parasitism. Therefore, in commercial systems, we suggest providing outdoor areas with plentiful browsing (eye-level or above the head foraging) and limiting ground-level feeding, particularly in areas that cannot be frequently left dormant for extended periods; if space is limited, access could be provided periodically. These approaches would promote feed choice and natural foraging behaviour while lowering the risk of parasite exposure. Consideration should be given for tannin content of the plants; browse high in tannins has been shown to reduce milk somatic cell count (Min et al. 2005), promote ruminating time (Nascimento et al. 2021), and improve the quality and weight of goat meat (Pimentel et al. 2021).

Poor grazing management can lead to toxicity issues. A case report of tansy ragwort (*Jacobaea vulgaris*) poisoning in two pet pygmy goats described how ingestion of high amounts of this plant can lead to toxicosis (Anholt and Britton 2017), which was never previously reported in North America. The goats were managed on poor pasture and without any active control of the ragwort, resulting in large, concentrated areas of seeding ragwort on which goats browsed frequently due to a dearth of other feed options. The ingestion of toxic plants occurs more often when goat herds are stocked on rangeland at high densities (reviewed by Holechek 2002). Mellado et al. (2003) showed that reduced shrubbery and grass cover resulted in goats altering their diet selection by consuming more toxic plant species; this feeding approach had negative consequences for body condition, weight gain, and fertility compared to goats managed at lower stocking rates on the same rangeland. Proposed solutions include rotational grazing, conditioned food aversion (e.g. training animals not to eat toxic plants; Ralphs 1992), and co-grazing with multiple livestock species (e.g. sheep and cattle, with goats; (Animut and Goetsch 2008; Taylor and Ralphs 1992). It is important to note that co-grazing with sheep may introduce other issues however, such as a higher worm burden for the goats (Jallow et al. 1994). Together these reports demonstrate the importance of good pasture and rangeland management to prevent depletion of preferred plants and over-ingestion of toxic forages. In production industries, this is especially relevant for meat and fibre goats that are typically managed outdoors. Importantly, 'pasture management' is often not sufficiently considered in companion goat settings, especially if these goats find themselves in urban backyards.

Goats can experience excessive teeth wear, a condition that generally affects herbivorous mammals. One of the primary causes of teeth wear is from an abrasive silicate component of plant matter called phytoliths, which grazing animals regularly ingest from grasses (Hummel et al. 2011). However, there is considerable argument that abrasions from ingestion of dust and grit are the more likely cause of teeth wear (Sanson et al. 2007). Animals that primarily graze are more likely to ingest soils close to the ground, increasing their likelihood of teeth wear (Damuth and Janis 2011). A review of the literature indicated that teeth wear rates are similar across ungulates regardless of body mass, but interestingly it is not well correlated with percentage of grass in the diet; instead, teeth wear seems to be more related to habitat characteristics and feeding behaviours that lead to ingestion of soil and grit (Damuth and Janis 2014). Browsers had lower rates of teeth wear, likely because trees, shrubs, and herbs have lower phytolith content (Hummel et al. 2011) and because feeding above ground-level reduces incidental soil ingestion (Damuth and Janis 2014). Teeth wear clearly has implications for the animal as it affects grazing, ingestion, chewing, and rumination of forage, especially if there is eventual teeth loss (McGregor 2011). A study of the feeding behaviour of commercial dairy and meat goats with or without worn teeth showed that goats with eroded teeth consumed more browse species and generally avoided grasses; the authors suggested this was due to the teeth wear limiting their ability to bite and masticate the thick cell walls of grass (Mellado et al. 2005). Indeed, when given the choice, goats will avoid diets that are high in course abrasive components (e.g., phytolith particles in the plant, dust, grit) while sheep do not (Ackermans et al. 2019). Surprisingly there are very few reports of implications of teeth wear on productivity of goats, despite tooth condition being a common reason for culling on commercial sheep farms (McGregor 2011). A study of Angora goats did find that increased teeth wear of the permanent first incisors reduced mohair fleece production by up to 30% (McGregor and Butler 2011); the authors attributed this production loss to reduced feed intake during grazing. Together these reports illustrate that goats managed outdoors should have browsing opportunities to reduce teeth wear. Browsing will limit ingestion of soil and phytolith-containing plants that accelerate teeth wear, and access to tender-leaved plants will place less pressure on already worn teeth. While we were not able to find literature in the companion and working goat space about teeth wear, we suggest that due to the potential of these animals being long living (compared to their production animal counterparts), attention to tooth health may be an especially important consideration for these animals.

4 Social Behaviour

4.1 Structure, Social Recognition and Bonds, and Agonistic and Affiliative Behaviour

Feral goats form a complex social structure with constant formation and dissolution of the group (the so-called fission-fusion dynamics), and a stable dominance hierarchy maintained by agonistic and affiliative behaviours (Miranda-de la Lama and Mattiello 2010; Shank 1972; Stanley and Dunbar 2013). Although the structure of goats' social groups can persist for several years (Barroso et al. 2000; Rout et al. 2002), the fission–fusion component (e.g. individuals drifting away as they forage and compete for resources) can make this structure look quite flexible (Dunbar and Shi 2008).

Reports of 'typical' feral goat group sizes range from 1 to 10 subjects but can vary dramatically depending on the population and its environmental context (e.g. Canada: 1–100 individuals, Shank 1972; Scotland: 1–3 individuals, Shi et al. 2005; Hawaii: 2–25 individuals, Yocum 1967). Using Social Network Analysis, Stanley and Dunbar (2013) identified that the feral goat population on the Isle of Rum formed 'cliques' of between 12 and 13 animals, which was also suggested to be the maximum stable group size. Others have also reasoned that group sizes larger than this are more susceptible to dissolution (Calhim et al. 2006; Dunbar and Shi 2008). Groups consist of adult females and younger individuals of both sexes. After adolescence, males leave the herd and join bachelor groups; older bucks live in solitude and only join herds during breeding season (O'Brien 1988).

Goats possess sophisticated socio-cognitive capacities when it comes to differentiating and recognising conspecifics (Chapter "Goat Emotions, Cognition, and Personality"); these are crucial traits in the cohesion of groups. In parent–offspring recognition, scent (Bordi et al. 1994) and vocalisations are key factors in individual recognition (Briefer et al. 2012; Terrazas et al. 2003). Visual cues also play a role in peer-peer recognition, while facial cues (such as colour and/or shape) might be less relevant. Keil et al. (2012) reported that goats could tell familiar goats apart when the heads of the other animals were not visible. They also appear to form cross-modal representations, combining visual and auditory cues, to distinguish between familiar and unfamiliar individuals (Pitcher et al. 2017). Goats form non-random bonds with other individuals in their group, preferring some individuals over others. They often seek out contact and remain close to preferred herd mates; however, these preferences can vary in strength and number of preferred goats (Stanley and Dunbar 2013).

Agonistic behaviour can be found in all social species and is needed to establish and maintain social hierarchies (Blanchard et al. 1993); this behaviour secures an individual's access to crucial resources, such as food, shelter, and reproductive partners. Agonistic behaviour in goats can be categorised into behaviours involving physical contact, such as biting and head bumping, and behaviour without physical contact, such as threat displays and chases (e.g. Alvarez et al. 2003; Miranda-de la Lama and Mattiello 2010). Other aggressive behaviours that have been observed specific to foraging include locking of horns, butting flanks, and ear biting (Syme and Syme 1979; Tölü and Savaş 2007). Rearing and striking one another's heads has also been observed during foraging conflict; interestingly, this same behaviour is also a common play behaviour seen at non-conflict times.

Socio-positive relationships or affiliative behaviours facilitate the establishment of bonds between individuals. In goats, common affiliative behaviours include grooming, sniffing, and resting within close proximity. Affiliative behaviours increase group cohesion and reduce aggression in herds of goats (Andersen et al. 2011).

4.2 Welfare Issues Related to Social Behaviour in Managed Goats and Potential Solutions

Goats naturally live in small, dynamic groups whereas in commercial systems, particularly dairy, herds can be comprised of hundreds, and even thousands, of individuals (e.g. described by Zobel et al. 2019). Stocking density may be as important as group size to reduce agonistic behaviour (Miranda-de la Lama and Mattiello 2010); however, the literature on feral goat behaviour demonstrates that goats are clearly adapted to live in smaller groups (Stanley and Dunbar 2013). The cognitive capabilities of remembering and recognising individual peers are certainly challenged when kept in large groups (e.g. more than 100 individuals in a single pen; Battini et al. 2016a; Muri et al. 2013), likely resulting in goats experiencing frustration and becoming aggressive (e.g. Miranda-de la Lama and Mattiello 2010). These outcomes can be indirectly seen when goats are regrouped (Andersen et al. 2011; Miranda-de la Lama and Mattiello 2010; Patt et al. 2013a). When introduced into a new herd, goats are confronted with unfamiliar animals which will act aggressively towards the 'intruder' (Patt et al. 2013b). In addition to increasing the risk of injuries, group restructuring negatively impacts milk production in lactating goats (Fernández et al. 2007).

There are several practical solutions to create a socio-positive environment. Reducing herd sizes to a number that reliably allows recognition of individual goats by every group member (Stanley and Dunbar 2013) will lead to a significant drop in aggression (Flint and Murray 2001; Miranda-de la Lama and Mattiello 2010).

Environmental enrichment may reduce fighting and competitive behaviours, but we caution that this conclusion was based on observations of just 4 groups of 3 animals (Gomes et al., 2018), and only descriptive statistics are provided making inferences on the effectiveness of the enrichment premature. Given the other positive benefits of enrichment, an opportunity exists for the design of robust studies exploring how the provision of opportunities for hiding, climbing, and object play may reduce aggression.

When a goat needs to be introduced into an existing but unfamiliar group, it can be advantageous to have it accompanied by a peer; adding more than one new goat to the group decreases aggression directed at newly introduced individuals and lowers cortisol levels compared to single animal introductions (Patt et al. 2013a). The use of a neighbouring pen to provide a visual introduction prior to animals coming into physical contact minimises aggression. Based on previous literature, Zobel and Nawroth (2020) make a series of suggested environmental enrichment suggestions for improving social choice, but these suggestions still require scientific evaluation.

When goats require specialist care or veterinary attention, the temporary isolation of individuals is often necessary (Price and Thos 1980), but this is known to compromise welfare. Isolated individuals show behavioural and physiological indicators associated with a heightened stress response, such as low activity and increased freezing behaviour, as well as higher cortisol levels (e.g. Kannan et al. 2002; Siebert et al. 2011). While these responses stem from a single animal being removed from their home group, companion goats may face a more chronic isolation stress; indeed, many pet goats are kept as single pets. The long-term implications of lone-goat housing have not been studied, yet given goats' social nature, it likely negatively impacts their welfare. The ideal situation is for goats to never be isolated; in New Zealand, the best practice recommendations are that goats should be socially housed (NZ Code of Welfare: Goats 2018). If temporary isolation is necessary (e.g. for quarantine, veterinary treatment) maintaining a sense of perceived social support, by allowing olfactory, vocal, and/or visual contact with other goats may decrease the stress response. The needs of the individual must be balanced with the length of time the animal is to be isolated; Siebert et al. (2011) suggested that partial isolation methods resulted in increased escape attempts (e.g. rearing, jumping) and vocalisations. It is therefore possible that a high motivation to restore social contact might be counterproductive for the animal (particularly if the animal is injured and should not be moving). Finally, as mentioned earlier, reintegration should be done carefully to minimise the effect on the isolated and remaining group of goats.

5 Mating Behaviour

In feral goats, males join female herds only during the breeding season (O'Brien 1988). While younger bucks live in bachelor groups, older bucks often become solitary. During rut, bucks emit a strong, characteristic odour and display behaviours such as 'tongue flapping' and urinating on their beard and mouth (enurination; Fritz

2017). As in many other group-living animals, dominant feral males monopolise mating during the breeding season (Kilgour and Ross 1980). At breeding season, it is not uncommon for does' ovulation cycles to become synchronised, especially if the buck is kept separated and suddenly introduced; this is sometimes referred to as the 'buck effect'.

Does may form harems around specific bucks, and high-ranking does may displace low-ranking does; this hierarchy can also impact oestrous, with lower-ranking animals ovulating later (Alvarez et al. 2003). Anecdotal reports suggest that an adult buck can inseminate up to 50 does.

5.1 Welfare Issues Related to Mating Behaviour in Managed Goats and Potential Solutions

The ratio of bucks to does has a potential to impact welfare. On the one hand, it has been observed that lower doe-to-buck ratios increase mating success in production systems (Zarazaga et al. 2018); on the other hand, there may actually be merit to higher ratios due to their potential to lower stress levels of individual animals. For example, it is possible that less dominant males will have more mating success if there are more does, simply because social interactions between the same individuals are reduced in larger groups (Andersen et al. 2011). Furthermore, having fewer bucks may reduce the stress experienced by does when in oestrous; anecdotally, farmers have commented on being uncertain regarding how many times a doe should be bred, particularly by multiple males. However, the ideal doe-to-buck ratio, that considers both mating success and doe and buck welfare, has not been established in the scientific literature.

For management purposes, the separation of bucks from does outside the breeding season is advisable. In smallholder applications, this may create the issue of isolation for the buck, particularly for younger males which naturally would form bachelor groups. Most larger farms have multiple bucks, eliminating the isolation concern, but potentially creating the opposite effect (as older males naturally seek solitude). If a buck has been left intact (i.e. not neutered) for any reason, consideration should be given to ensure it is not left alone if it seeks social contact. We have observed some smallholders that only keep one intact buck, housing it with a wether (neutered male) for company.

Artificial insemination is increasingly being used, particularly where high value progeny is being sought in dairy, companion, or pedigree/show animals. In such situations, buck contact is omitted. This is not without its issues, however. These include the necessity of artificial synchronisation (due to lack of buck effect), lower conception rates (compared to natural mating), handling stress (if personnel are not well trained in artificial insemination techniques), and challenges in collecting off-season semen from bucks. In the case of the latter, contact with an oestrus doe, as well as another buck, may stimulate sexual activity in bucks (Ahmad and Noakes 1996). Importantly, thwarting buck and doe contact in itself may be a welfare concern; Mellor et al. (2020) suggest that sexual activity is part of a suite of affiliative interactions that all contribute to positive welfare. Therefore, if promoting states such as comfort, pleasure and attachment are important, natural mating may be the preferred breeding method.

6 Maternal Behaviour

Gestation in goats lasts approximately 150 days (Sambraus and Wittmann 1989). Depending on the population, multiple offspring can be common (e.g. New Zealand feral goats: 44% twins/triplets; Alley et al. 1995), however this appears to be linked to location, genetics, or nutrition (e.g. on the Scottish Isle of Rum, just over 3% of births were twins; Shi et al. 2003). Before birth (kidding), the pregnant doe becomes restless and leaves the herd, but usually stays in visual contact. After the kid is born, she licks it clean and encourages it to stand up. The licking is important for olfactory recognition; however, there is evidence that a variety of factors contribute to recognition, including visual and acoustic cues (e.g. reviewed by Poindron et al. 2007). Most kids stand up within 15 minutes after birth (Lickliter 1985). After an initial high frequency of suckling cycles right after birth, suckling usually takes place 1 to 2 times per hour for single kids, while twins suckle more often (Sambraus and Wittmann 1989). Although cross-sucking (e.g. an abnormal behaviour where kids sucks on one another) exists, it is rare; Sambraus and Wittmann (1989) suggested occurrence might be approximately 1%. After just 4 hours postpartum, most does are bonded with their kids and recognise them if removed and reunited, specifically due to vocal and olfactory cues (e.g. Bordi et al. 1994; Poindron et al. 2003; Terrazas et al. 2003). Once bonded, strong ties exist; Briefer et al. (2012) showed dams remember vocal cues from kids after 1 year of separation.

Perhaps due to the environment in which they would naturally be born, goat kids typically begin life as 'hiders'. Initially, they will remain separated from the herd, typically hidden, and the dam will return to nurse them. O'Brien (1984) suggested that there are individual differences in how dams handle this period; some stay close by, while others leave the kid alone. This period can last up to 7 days (Lickliter 1985). Gradually, the kids will follow their mother, and eventually kids form kid groups within the herd. Bungo et al. (1998) found that goat kids become more independent after approximately 5 weeks of age, and weaning occurs gradually over time as the mother reduces her ongoing investment into her young through less frequent and shorter suckling events. Feral goat observations on the Isle of Rum suggest that while kids are feeding on their own at 3 or 4 months of age, they maintain some suckling of their dam, and natural weaning appears to occur somewhere between 4 and 6 months of age (Shi et al. 2003) (Fig. 13).

Fig. 13 Kid (approximately 7 months old) and dam resting together (Location: Hut Nimi, Switzerland; Owner: Pietro Zanoli)

6.1 Welfare Issues Related to Maternal Behaviour in Managed Goats and Potential Solutions

6.1.1 Commercial Systems that Practise Natural Suckling (Doe and Kid Remain Together Until Weaning)

In commercial systems, some industries (e.g. meat, fibre, working) leave kids to be reared by their mothers; this practice affords opportunities to express some natural maternal behaviours described above. In such systems, it is important that a strong mother–offspring bond develops, by avoiding disturbances to the animals during the first few days. Does naturally segregate, at least partially, from the rest of the herd at parturition (Lickliter 1985). Therefore, whenever possible, does should not be isolated, and instead be kept in groups with other pre-parturient dams, while also having the opportunity to self-segregate (Fig. 14).

For young kids, segregation, either alone as described above, or in groups of similar-aged kids, is common. This behaviour could be promoted with the use of a separate kid-only area. For example, the use of creep feeders is common in beef and sheep production systems; young animals can access a separate area that excludes adult animals, typically because of a modified entrance limiting the size of the animals that can enter. These creep feeders are strategically used to provide supplement to young animals (Foster et al. 2013); they are often promoted to meat goat producers however there is only minimal supporting research. Evidence in meat goat kids managed on pasture with their dams found creep feeders provisioned with supplementary feed improved pre-weaning average daily gain and weaning weights compared to kids that could not access the creep feeder; there was no improvement in dam traits (i.e. postpartum weight change, parasite load) or financial returns (Hayes

Fig. 14 Provision of segregation for preferred social partners and during kidding (Zobel and Nawroth 2020)

Fig. 15 Group of kids in an elevated 'creep' environment, where they have access to additional browse that the adult goats cannot access (Photo: Gosia Zobel)

et al. 2019). In addition to providing supplement for weaning kids, the creep area concept could also serve as a segregation opportunity from the rest of the herd, providing a hiding, shelter, or climbing area (Fig. 15).

The value of these aspects of creep feed areas has never been researched.

A similar concept could be expanded to allow dams to access segregated areas, both during prepartum to accommodate their natural behaviour to isolate from the herd before kidding, and during postpartum when dams may also desire time away from the herd. As suggested by Zobel and Nawroth (2020), goats could be equipped with RFID tags that, after some initial training, would allow goats' automated access to a separate area of the pen or paddock away from other goats. This area could have different feeds, or just allow for segregation into smaller groups. Such a system could allow adult goats (as well as kids) a choice about their environment.

A clear advantage for the kids managed with their dams is that it fulfils the natural behaviour of suckling milk from the dam, and permits *ad libitum* intake of milk, which is known to improve growth of kids during pre-weaning (compared to kids on restricted milk feeding programmes, Goetsch et al. 2001). There are conflicting reports of whether *ad libitum* milk intake from natural rearing improves growth of kids compared to artificial rearing (Argüello et al. 2004; Delgado-Pertíñez et al. 2009); however, a few studies have explored behavioural and cognitive benefits of natural rearing systems (e.g. Toinon et al. 2021, 2022). We suggest that natural rearing systems allow both dam and kid to express their natural behaviours and this should be valued considering evidence that natural suckling systems do not negatively impact growth of kids. Self-reported farmer survey results suggest such systems may improve health outcomes as well (e.g. reduction in respiratory illness; Bélanger-Naud et al. 2021); however, this requires verification in controlled research.

Kids reared in natural suckling systems eventually must be weaned from milk and from the dam, which can be stressful given that high milk intakes before weaning are known to suppress solid feed intake and can result in hunger during and immediately after the milk weaning process (Zobel et al. 2020); these are common issues also discussed in dairy calf rearing (Khan et al. 2011). Despite weaning being a crucial step in the growth and development of kids, there is minimal robust research examining the best methods for how to wean suckling kids from their dams. In an observational study of three pens, Ugur et al. (2007) reported that kids that could suckle from dams that were not milked had greater weight gains, while kids with dams that were milked out daily had greater solid feed intake, likely because the partially restricted milk intake promoted an increase in solid feed. Another limited study of kids with full-time access to suckle their dams versus restricted suckling of their dams (3 times per day for 15 min) showed no difference in heart rate or heart rate variability during baseline and on the day of weaning, suggesting that both methods were equally stressful (Sporkmann et al. 2012). In the beef cow industry, calves are often weaned from the dam using a fenceline system that allows visual and restricted physical contact over a fence, while preventing suckling. Positively, when compared to abrupt separation, reductions in negative behavioural responses and improved growth have been reported for this approach in both beef (Taylor et al. 2020) and dairy cattle systems (Johnsen et al. 2015); in lambs, reduced behavioural responses were reported, but there were no positive growth effects of two-stage weaning. This type of fenceline weaning system has not been scientifically explored for use in natural suckling systems for goats, but these other species' results suggest a potential exists to reduce behavioural stress responses in kids as well.

Research typically focuses on the impact of weaning practices on the young; however, the mother can also experience substantial stress due to removal of her offspring. The impacts of weaning and separation practices on the mother have recently been discussed for sheep (Freitas-de-Melo et al. 2022), and these topics are also highly relevant for does in systems that practise natural suckling of kids. We suggest research should explore alternative weaning approaches to abrupt weaning that may reduce behavioural responses of both doe and kid (e.g. progressive, two-step, and fenceline weaning methods) or even hormonal treatments such as progesterone with anxiolytic effects that could reduce stress of the mother at weaning (Freitas-de-Melo et al. 2013).

There is some evidence that rearing kids with their dams positively influences social behaviours and stress coping abilities. Goats that were reared without their dam (in groups of kids of the same age) showed different social behaviours compared to kids reared with their dam, including more play-fighting, racing, stepping on and standing in contact with other kids, and less allogrooming (Toinon et al. 2022). It was speculated that these differences in social behaviours arose due to a lack of maternal presence and maternal care, possibly resulting in redirection of behaviours towards conspecifics in the absence of the mother. Furthermore, Toinon et al. (2021) found that artificial rearing of kids affected their behavioural responses to stressful situations, such as increased vocalisations, running and jumping when exposed to a novel environment and novel object. However, the artificially reared kids showed less avoidance of a familiar human after weaning, which would be expected due to the closer contact and reliance on humans for feeding. Together, these studies suggest that the rearing environment can have important effects on how kids interact with conspecifics and respond to stressful novel situations. While dam-rearing of kids provides benefits of a natural suckling opportunity and maternal care, there should be consideration for preparing kids for a post-weaning environment without maternal presence and with greater human contact.

6.1.2 Commercial Systems that Practise Artificial Rearing (Doe and Kid Separated at Birth)

On commercial dairy goat farms, as well as when kids are selected to be companion animals, the mother–infant bond is often interrupted early. Separation from the dam is accomplished soon after birth; this approach diverts the dam's milk production for commercial purposes (e.g. Lu and Potchoiba 1988). Depending on the region and disease prevalence, the timing of kid removal from dams will vary (e.g. in an on-farm monitoring study, 4/16 farms removed kids immediately, 11/16 farms removed kids within 24 h, and 1 farm left kids with does for longer; Zobel et al. 2016). Immediate removal is often done to prevent the transmission of caprine arthritis encephalitis (CAE) and Johne's disease. CAE can be transmitted from doe to kid through colostrum, milk, and saliva of infected does (Reina et al. 2009), so many guidelines currently recommend complete suckling prevention (see review: Bélanger-Naud and Vasseur 2021). However, this recommendation may be easier to make than to put into practice; for instance, a disconnect appeared to exist between actual and perceived prevention of suckling of colostrum by dairy goat farmers that were aiming to remove kids immediately after birth (Zobel et al. 2016). It is important to note that the actual effect of kid removal on preventing CAE and Johne's transmission has not been thoroughly investigated. The continued prevalence of CAE on dairy goat farms means that immediate separation of the kid and doe at birth will remain recommended practice for the near future. However, it is important to not ignore societal expectations; increasingly, consumers have expressed concern about the separation of mothers and their offspring, particularly in dairy cattle systems in which this practice is prevalent (Placzek et al. 2021). By extension, we can expect similar concerns regarding doe–kid separation in the dairy goat industry. A recent survey of farmer's willingness to adopt practice change was focused on artificially reared kids, and therefore unfortunately excluded responses from farmers which practised dam-rearing of kids (Vickery et al. 2022b); follow-up work including such farmers (as has been done for dairy cow farmers rearing calves with cows; Neave et al. 2022) would be valuable for goats.

The separation of doe and kid at birth means that kids receive milk artificially, usually through a nipple, which enables the natural behaviour to suckle from a teat. Depending on farm practice, kids may be fed *ad libitum*, or only have limited access to milk and at restricted quantities. The latter goes against the natural behaviour of kids that freely suckle from their dams. There is surprisingly little literature examining the effects of different milk feeding quantities for artificially reared kids. For instance, even the amount that dam fed kids drink per day has not been quantified, as it has in calves (de Passillé and Rushen 2006). Recommendations of the amount of milk to be fed reviewed by Bélanger-Naud and Vasseur (2021) suggest 0.5 to 1.5 L per day in the first two weeks, then 1.8 to 2 L per day thereafter up to weaning; however, these recommendations do not appear to be science-based. As expected, the available research suggests that feeding higher versus lower amounts of milk to artificially reared kids improves growth (Argüello et al. 2004; Goetsch et al. 2001; Yan et al. 1993), but no research has examined how restricted milk feeding affects the behaviour of goat kids. There is plenty of evidence in dairy calves that limiting milk intake has negative impacts on growth (reviewed by Khan et al. 2011) and behaviour, including increased behavioural indicators of hunger (e.g. (Rosenberger et al. 2017) and reduced frequency of play (e.g. Jensen et al. 2015). Such findings could also be evident in dairy goat kids on restricted feeding programmes, which pose welfare concerns that have yet to be examined.

As with kids in natural suckling systems, there are issues related to weaning from milk in artificially reared kids; the stress related to weaning from milk can be mitigated to some extent with appropriate weaning practices. The timing of weaning is a major factor because kids must be developmentally ready to transition from milk onto solid feed. Weaning criteria are highly variable across farms and countries. A multi-country survey of kid rearing practices found that the majority of farmers reported weaning based on age (Vickery et al. 2022a); the median weaning age they reported (84 days) was similar to that recorded on multiple farms in New Zealand (86 days; Todd et al. 2019), however reports exist of kids being weaned much earlier (e.g. 61 days, Tölü et al. 2016; 45 to 56 days, Ugur et al. 2004, 2007), and indeed Vickery et al. (2022a) reported a range of 42 and 168 days, depending on country. The majority of reported weaning ages are earlier than what has been observed in nature (e.g. approximately 4 to 6 months; Shi et al. 2003). Evidence supports weaning at later rather than earlier ages to promote growth in dairy goat kids (Panzuti et al. 2018) and to promote testicular growth, sexual behaviour and reduce age at puberty in male goat kids (Amjad et al. 2021).

Other common weaning criteria reported by Bélanger-Naud and Vasseur (2021) include weaning based on weight or solid feed intake. Again, weaning at later, heavier weights (15 kg) resulted in faster growth to reach reproductive weight, compared to weaning at earlier, lighter weights (10 kg) (Palma and Galina 1995). In nature, weaning would be initiated by the dam, and when kids are reliably eating solid feed to promote rumen development. Thus, artificial weaning based on solid feed intake would achieve some level of naturalness and would also accommodate individual variability in how quickly some kids may begin eating solid feed. For instance, in dairy calves, there is high individual variability in solid feed intake (Neave et al. 2018a, 2019), indicating that weaning based on other criteria may result in some individuals being weaned from milk when they are not yet prepared to rely on solid feed. Weaning based on solid feed consumption therefore provides opportunity to tailor weaning to the needs of calves who need more time on milk (Benetton et al. 2019; Welk et al. 2022). However, these approaches to weaning have not been experimentally explored in goats, and no work indicates the optimal amount of solid feed for a smooth transition through weaning.

In addition to the timing of weaning, the method of milk removal contributes to weaning stress. It is still reasonably common for artificially reared kids to be abruptly weaned (30% of farms in a multi-country survey, Vickery et al. 2021; 37% in Canada; Bélanger-Naud 2020). Gradual weaning methods are also used, and these include reducing milk volume, milk concentration by diluting with water, or skipping a milk feeding (Vickery et al. 2022a; Zobel et al. 2020). When weaned at approximately 3 months of age, Zobel et al. (2020) found minimal effects of abrupt versus gradual weaning methods on kid growth, solid feed intake and behaviour, apart from an initial reduction in weight gain in kids that were weaned by diluting the milk. Magistrelli et al. (2013) also found no differences in growth, abnormal behaviours, or physiological stress indicators in kids that were progressively weaned (by slowly reducing milk quantity over 17 days) compared to abruptly weaned. Both authors (Magistrelli et al. 2013; Zobel et al. 2020) concluded either weaning strategy were suitable, providing the goat kids are of a suitable age. It is well documented in dairy calves that abrupt weaning can be especially stressful for individuals fed high amounts of milk, since high milk intakes suppress consumption of solid feed (see review by Khan et al. 2011). Goat farms feeding *ad libitum* milk to goat kids were more likely to practise abrupt weaning (Vickery et al. 2022a) suggesting that these kids may experience weaning stress. A reduction in milk in the latter phase of the milk-feeding period successfully increased solid feed intake in goat kids (Goetsch et al. 2001), so this approach may be useful for *ad libitum* fed kids. There is a need for goat-specific research on weaning methods to better understand which weaning strategies affect welfare; we predict that methods mimicking natural weaning (e.g. gradual reduction of milk intake, timing weaning with kids already eating some solid feed) are most likely to reduce stress and promote growth.

7 Final Thoughts

While the primary aim of this chapter was to highlight how promoting naturalness and natural behaviours is positive to goat welfare, it is pertinent to point out where genetic selection of a natural trait has arguably been detrimental for goat welfare. Myotonia congenita is a hereditary condition characterised by delayed muscle relaxation following abrupt contraction, as well as general muscle hypertrophy (Beck et al. 1996). While relatively rare (e.g. estimated 10,000 individuals in the USA; Sponenberg 2019), myotonic goats nonetheless deserve mention. Some promising (yet unreplicated) data suggesting better muscle development has meant that they are touted for their meat production potential (Dzakuma and Risch 1996). They are also said to be easier to manage and enclose than some more active breeds (Sponenberg 2019). Indeed, the same author postulated that these goats were perpetuated by some herd managers as sacrificial animals (falling over when startled by predators, thus saving others in the herd); regardless of the validity of this claim, it does highlight that their condition will likely increase their susceptibility to predation. Furthermore, their propensity to become immobile and fall over when startled has made them a popular companion animal option; this behavioural novelty, coupled with an increasing 'urban goat' movement (Grant 2012) will likely hasten their breeding and availability. While it is beyond the scope of this chapter to present an in-depth, balanced argument for and against maintaining such a breed, it is important to at least be aware that while a natural trait, the condition may compromise goat welfare. No research exists that documents how these goats experience the chronic and sudden onset of muscle stiffness and falling. Together, these points might be reasons to consider the ethics of perpetuating this trait (and some other natural, yet maladaptive traits), and at the very least to spur some balanced scientific exploration.

Goats have impressive cognitive capabilities and are naturally flexible in their behaviour; this flexibility extends to the environments in which they can thrive, their social behaviours and their feeding and foraging behaviour. They adjust to most environments and are considered one of the world's great 'invaders', as can be attested to the large amount of literature focusing on the eradication of feral goats. They are highly social, have preferred social companions, and will travel large distances to maintain their social groupings. They readily shift their diet to cope with changing conditions and adopt a range of feeding positions to access forage, ranging from grazing on the ground to bipedal standing to reach far above their heads. A key feeding feature of goats that distinguishes them from other ruminants is their highly selective foraging and browsing behaviour, resulting in a diet comprised of shrubbery, buds, flowers, woody stems, and herbs, in addition to grasses. Their narrow mouth, short jaws, and mobile lips allow goats to access parts of plants that are difficult to reach and permit careful selection to avoid noxious insects. These abilities, combined with their foraging flexibility, ability to climb near-vertical rock faces, and their capacity to navigate their environment both socially and spatially, have certainly contributed to their success in harsh environments.

Our understanding of the natural behaviours of goats can contribute to improving their management in all their various uses, including meat, dairy, fibre, as well as the working and companion realms. Identifying the cognitive capacities that underlie natural behavioural expressions will help to better target species-specific needs in goat management and handling. When housing goats, environmental complexity should be balanced for the level of stimulation best suited for goats' capacities. Offering monotonous feeding regimes may thus result in boredom. On the other hand, many of our typical management systems might be providing too much social stimulation; feral goats live in small to medium-sized groups, so housing domestic goats in large groups may result in issues with recognition of individuals, leading to aggression and injuries.

A general uniformity of the environments in which we keep goats can reduce their welfare, but there are plenty of opportunities to incorporate environmental diversity and promote natural social and foraging behaviours. Whether they are intensive indoor facilities, extensive outdoor ranges, or somewhere in between, we suggest that most systems can include climbing, hiding, and sheltering opportunities. Does and their kids should have opportunities to self-segregate. Housing, feeding, and rearing systems can be made more complex to promote cognitive capabilities and alleviate boredom. Elevated feed bunks, known to promote natural browsing postures, could be included with the added benefit of increasing feed intake. Feed could be presented in different locations in different ways; in large-scale systems feed could be delivered on elevated platforms, while in smaller holdings, additional vertical hay racks at different levels could suffice. If space and season allows, indoor systems could provide outdoor access to promote exercise, climbing, hiding, and/or browsing. We also strongly encourage diversifying the surfaces to which goats have access to promote hoof wear and cleanliness. Furthermore, provision of a variety of feed types will promote natural selective foraging behaviours. The latter could include providing supplementary browse (e.g. woody shrubs, buds, and leaves) in automated feeders or in outdoor areas planted with fast growing species. When goats are managed outdoors, good pasture management is critical to prevent health issues related to foraging. Parasite exposure, toxicosis, and teeth wear can all be controlled by ensuring pasture is not grazed too low and by permitting access to browsing areas. Care must be taken to ensure goats are able to avoid predation and that management practices (e.g. sheering of fibre animals) are matched with sheltering opportunities.

Overall, our knowledge of the natural behaviours of goats can aid in improving how we care for commercially managed, and companion and working, goats. Incorporating flexibility and diversity in their environment, in how and what goats are eating, and in how they are able to choose to interact socially, will promote cognitive and foraging skills, prevent boredom, and provide agency in how they spend their time.

References

Ackermans NL, Martin LF, Hummel J, Müller DWH, Clauss M, Hatt JM (2019) Feeding selectivity for diet abrasiveness in sheep and goats. Small Rumin Res 175(May):160–164

Ahmad N, Noakes DE (1996) Seasonal variations in the semen quality of young British goats. Br Vet J 152(2):225–236

el Aich A, el Assouli N, Fathi A, Morand-Fehr P, Bourbouze A (2007) Ingestive behavior of goats grazing in the Southwestern Argan (Argania spinosa) forest of Morocco. Small Rumin Res 70(2–3):248–256

Aldezabal A, Garin I (2000) Browsing preference of feral goats (Capra hircus L.) in a Mediterranean mountain scrubland. J Arid Environ 44(1):133–142

Alley JC, Fordham RA, Minot EO (1995) Mother-offspring interactions in feral goats—a behavioural perspective of maternal investment. NZ J Zool 22(1):17–23

Alvarez L, Martin GB, Galindo F, Zarco LA (2003) Social dominance of female goats affects their response to the male effect. Appl Anim Behav Sci 84(2):119–126

Amjad M, Hameed N, Khan MI-R, Farid Ullah AS (2021) Effect of weaning method on body weight, testicular growth, sexual behavior, age of puberty, and testosterone concentration of Beetal bucks. Anim Sci J 92(1):e13520

Andersen IL, Bøe KE (2007) Resting pattern and social interactions in goats—the impact of size and organisation of lying space. Appl Anim Behav Sci

Andersen IL, Tønnesen H, Estevez I, Cronin GM, Bøe KE (2011) The relevance of group size on goats' social dynamics in a production environment. Appl Anim Behav Sci 134(3–4):136–143

Anholt H, Britton A (2017) Presumptive chronic pyrrolizidine alkaloid poisoning in 2 pygmy goats due to ingestion of tansy ragwort (Jacobaea vulgaris) in southwestern British Columbia. Can Vet J 58(11):1171–1175

Animut G, Goetsch AL (2008) Co-grazing of sheep and goats: benefits and constraints. Small Rumin Res 77(2–3):127–145

Anzuino K, Bell NJ, Bazeley KJ, Nicol CJ (2010) Assessment of welfare on 24 commercial UK dairy goat farms based on direct observations. Vet Rec 167(20):774–780

Anzuino K, Knowles TG, Lee MRF, Grogono-Thomas R (2019) Survey of husbandry and health on UK commercial dairy goat farms. Vet Record 185(9):267

Arevalo JR, de Nascimento L, Fernandez-Lugo S, Saro I, Camacho A, Mata J, Bermejo L (2011) Effects of abandoning long-term goat grazing on species composition and species richness of pastures at La Gomera, Canary Islands. Span J Agric Res 9(1):113

Argüello A, Castro N, Capote J (2004) Growth of milk replacer kids fed under three different managements. J Appl Anim Res 25(1):37–40

Arviv A, Muklada H, Kigel J, Voet H, Glasser T, Dvash L, Ungar ED, Landau SY (2016) Targeted grazing of milk thistle (Silybum marianum) and Syrian thistle (Notobasis syriaca) by goats: preference following preconditioning, generational transfer, and toxicity. Appl Anim Behav Sci 179:53–59

Aschwanden J, Gygax L, Wechsler B, Keil NM (2008) Social distances of goats at the feeding rack: influence of the quality of social bonds, rank differences, grouping age and presence of horns. Appl Anim Behav Sci 114(1–2):116–131

Aschwanden J, Gygax L, Wechsler B, Keil NM (2009a) Loose housing of small goat groups: influence of visual cover and elevated levels on feeding, resting and agonistic behaviour. Appl Anim Behav Sci 119(3–4):171–179

Aschwanden J, Gygax L, Wechsler B, Keil NM (2009b) Structural modifications at the feeding place: effects of partitions and platforms on feeding and social behaviour of goats. Appl Anim Behav Sci 119(3–4):180–192

Askins GD, Turner EE (1972) A behavioral study of Angora goats on West Texas range. J Range Manag 25:82–87

Aublet JF, Festa-Bianchet M, Bergero D, Bassano B (2009) Temperature constraints on foraging behaviour of male Alpine ibex (Capra ibex) in summer. Oecologia 159(1):237–247

Bagchi S, Ritchie ME (2010) Herbivore effects on above- and below ground plant production and soil nitrogen availability in the Trans-Himalayan shrub-steppes. Oecologia 164(4):1075–1082
Baldwin BA (1977) Ability of goats and calves to distinguish between conspecific urine samples using olfaction. Appl Anim Ethol 3(2):145–150
Banks MS, Sprague WW, Schmoll J, Parnell JA, Love GD (2015) Why do animal eyes have pupils of different shapes? Sci Adv 1:e1500391
Barroso FG, Alados CL, Boza J (2000) Social hierarchy in the domestic goat: effect on food habits and production. Appl Anim Behav Sci 69(1):35–53
Battini M, Barbieri S, Vieira A, Stilwell G, Mattiello S (2016a) Results of testing the prototype of the awin welfare assessment protocol for dairy goats in 30 intensive farms in northern Italy. Ital J Anim Sci 15(2):283–293
Battini M, Barbieri S, Waiblinger S, Mattiello S (2016b) Validity and feasibility of human-animal relationship tests for on-farm welfare assessment in dairy goats. Appl Anim Behav Sci 178:32–39
Beauchamp G (2015) Animal vigilance: monitoring predators and competitors. Academic Press
Beck CL, Fahlke C, George AL (1996) Molecular basis for decreased muscle chloride conductance in the myotonic goat. Proc Natl Acad Sci USA 93(20):11248–11252
Bélanger-Naud S (2020) Raising healthy and productive kids: a survey of dairy goat kid rearing practices on Canadian farms and their associations with farm performance. A Thesis Submitted to McGill University in Partial Fulfillment of the Requirements of the Degree of Master of Science in Animal Science, February
Bélanger-Naud S, Vasseur E (2021) Graduate student literature review: current recommendations and scientific knowledge on dairy goat kid rearing practices in intensive production systems in Canada, the United States, and France. J Dairy Sci 104(6):7323–7333
Bélanger-Naud S, Cinq-Mars D, Julien C, Arsenault J, Buczinski S, Lévesque J, Vasseur E (2021) A survey of dairy goat kid-rearing practices on Canadian farms and their associations with self-reported farm performance. J Dairy Sci 104(9):9999–10009
Bell FR (1959) Preference thresholds for taste discrimination in goats. J Agric Sci 52:125–129
Benetton JB, Neave HW, Costa JHC, von Keyserlingk MAG, Weary DM (2019) Automatic weaning based on individual solid feed intake: effects on behavior and performance of dairy calves. J Dairy Sci 102(6):5475–5491
Berman TS, Ben-Ari M, Glasser TA, Gish M, Inbar M (2017) How goats avoid ingesting noxious insects while feeding. Sci Rep 7(1):1–10
Berman TS, Messeri N, Glasser TA, Inbar M (2019) Innate ability of goats to sense and avoid ingestion of noxious insects while feeding. Royal Soc Open Sci 6(2)
Biancardi CM, Minetti AE (2017) Gradient limits and safety factor of Alpine ibex (Capra ibex) locomotion. Hystrix Ital J Mammal 28(1):50–60
Bigham ML, Morris CA, Southey BR, Baker RL (1993) Heritabilities and genetic correlations for live weight and fibre traits in New Zealand Cashmere goats. Livest Prod Sci 33(1–2):91–104
Blanchard DC, Sakai RR, McEwen BS, Weiss SM, Blanchard RJ (1993) Subordination stress: behavioral, brain, and neuroendocrine correlates. Behav Brain Res 58(1–2):113–121
Bøe KE, Ehrlenbruch R (2013) Thermoregulatory behavior of dairy goats at low temperatures and the use of outdoor yards. Can J Anim Sci 93(1):35–41
Bøe KE, Andersen IL, Buisson L, Simensen E, Jeksrud WK (2007) Flooring preferences in dairy goats at moderate and low ambient temperature. Appl Anim Behav Sci 108(1–2):45–57
Bøe KE, Ehrlenbruch R, Andersen IL (2012) Outside enclosure and additional enrichment for dairy goats—a preliminary study. Acta Vet Scand 54(August 2014):68
Bordi A, de Rosa G, Napolitano F, Litterio M, Marino V, Rubino R (1994) Postpartum development of the mother-young relationship in goats. Appl Anim Behav Sci 42(2):145–152
Boyd IL (1981) Population changes and the distribution of a herd of feral goats (Capra Sp.) on Rhum, Inner Hebrides, 1960–1978. J Zool Lond 193:287–304
Briefer EF, de la Torre MP, McElligott AG (2012) Mother goats do not forget their kids' calls. Proc R Soc B Biol Sci 279:3749–3755

Briefer EF, Haque S, Baciadonna L, McElligott AG (2014) Goats excel at learning and remembering a highly novel cognitive task. Front Zool 11(1):20
Bunbury N, Von Brandis RV, Currie JC, Van De Crommenacker JV, Accouche W, Birch D, Seng LC, Doak N, Haupt P, Haverson P, Baptiste MJ, Fleischer-Dogley F (2018) Late stage dynamics of a successful feral goat eradication from the unesco world heritage site of aldabra atoll, seychelles. Biol Invasions 20(7):1735–1747
Bungo T, Shimojo M, Nakano Y, Okano K, Masuda Y, Goto I (1998) Relationship between nursing and suckling behaviour in Tokara native goats. Appl Anim Behav Sci 59(4):357–362
Calhim S, Shi J, Dunbar RIM (2006) Sexual segregation among feral goats: testing between alternative hypotheses. Anim Behav 72(1):31–41
Campbell K, Donlan CJ (2005) Feral goat eradications on islands. Conserv Biol 19(5):1362–1374
Chakrabarti A, Chandran PC (2016) Foot Disorders of Goats (Capra Hircus) in Bihar, India. Int J Environ Ecol Family Urban Stud (IJEEFUS) 6(1):171–176
Chapman M (2003) Kangaroos and feral goats as economic resources for graziers: some views from south-west Queensland. Rangel J 25(1):20–36
Chebli Y, el Otmani S, Chentouf M, Hornick JL, Bindelle J, Cabaraux JF (2020) Foraging behavior of goats browsing in southern Mediterranean forest rangeland. Animals 10(2):1–19
Christodoulopoulos G (2009) Foot lameness in dairy goats. Res Vet Sci 86(2):281–284
Chynoweth MW, Litton CM, Lepczyk CA, Hess SC, Cordell S (2013) Biology and impacts of pacific island invasive species. 9. Capra hircus, the feral goat (Mammalia: Bovidae). Pac Sci 67(2):141–156
Clark FE (2017) Cognitive enrichment and welfare: current approaches and future directions. Anim Behav Cognit 4(1):52–71
Clark B, Stewart GB, Panzone LA, Kyriazakis I, Frewer LJ (2016) A systematic review of public attitudes, perceptions and behaviours towards production diseases associated with farm animal welfare. J Agric Environ Ethics 29(3):455–478
Clout MN, Russell JC (2008) The invasion ecology of mammals: a global perspective. Wildl Res 35(3):180–184
Cooper T (2020) 7 Steps to goat-friendly housing. Goat J 98(5):48–51
Cronje PB (1992) Differences in nitrogen and urea metabolism between goats bred for fibre production (Angora goat) or meat production (Boer goat). South Afr J Anim Sci 22(5):143–148
Damuth J, Janis CM (2011) On the relationship between hypsodonty and feeding ecology in ungulate mammals, and its utility in palaeoecology. Biol Rev 86(3):733–758
Damuth J, Janis CM (2014) A comparison of observed molar wear rates in extant herbivorous mammals. Ann Zool Fenn 51(1–2):188–200
Deiss R, Hammer C, Muller D, Deb A, Clauss M, Hammer S (2010) Mortality patterns in non-domestic hoofstock (Ovis orientalis laristanica, Capra aegagrus, Capra ibex nubiana) indicate species-specific differences in disease susceptibility in small ruminants. Proc Int Conf Dis Zoo Wild Anim May:59–66
Delgado-Pertíñez M, Guzmán-Guerrero JL, Caravaca FP, Castel JM, Ruiz FA, González-Redondo P, Alcalde MJ (2009) Effect of artificial vs. natural rearing on milk yield, kid growth and cost in Payoya autochthonous dairy goats. Small Rumin Res 84(1–3):108–115
Delibes M, Castañeda I, Fedriani JM (2017) Tree-climbing goats disperse seeds during rumination. Front Ecol Environ 15(4):222–223
Dell'Amore C (2010) "Mind-Boggling" pictures: goats scale dam in Italy. Natl Geogr. https://www.nationalgeographic.com/science/article/101101-ibex-goats-dam-italy-bighorn-sheep-wyoming-science
Divittorio M (2022, July 12) Police: investigation underway into death of goat working in Aspinwall park. Trib Total Media. https://triblive.com/local/valley-news-dispatch/police-investigation-underway-into-the-death-of-a-goat-working-in-an-aspinwall-park/
Dunbar RIM, Shi J (2008) Sex differences in feeding activity results in sexual segregation of feral goats. Ethology 114(5):444–451
Duncan IJH, Widowski TM, Malleau AE, Lindberg AC, Petherick JC (1998) External factors and causation of dustbathing in domestic hens. Behav Process 43(2):219–228

Dzakuma JM, Risch E (1996) The potential of Tennessee Stiff-legged goats (TS) in meat production. VI international conference on goats, 136–139
Dziba LE, Scogings PF, Gordon IJ, Raats JG (2003) Effects of season and breed on browse species intake rates and diet selection by goats in the False Thornveld of the Eastern Cape, South Africa. Small Ruminant Res 47(1):17–30
Egea AV, Allegretti L, Paez Lama S, Grilli D, Sartor C, Fucili M, Guevara JC, Passera C (2014) Selective behavior of Creole goats in response to the functional heterogeneity of native forage species in the central Monte desert, Argentina. Small Ruminant Res 120(1):90–99
Elias D, Tischew S (2016) Goat pasturing—a biological solution to counteract shrub encroachment on abandoned dry grasslands in Central Europe? Agric Ecosyst Environ 234:98–106
FAOSTAT (2020) Live animals – goats (world total). https://www.fao.org/faostat/en
Fedosenko AK, Blank DA (2001) Capra sibirica. Mammalian Species 675(675):1–13
Fernández MA, Alvarez L, Zarco L (2007) Regrouping in lactating goats increases aggression and decreases milk production. Small Rumin Res 70(2–3):228–232
Flint M, Murray PJ (2001) Lot-fed goats – the advantages of using an enriched environment. Aust J Exp Agric 41(4):473–476
Forsyth DM, Parkes JP (1995) Suitability of aerially sown artificial baits as a technique for poisoning feral goats. N Z J Ecol 19(1):73–76
Foster JL, Carter JN, Lamb GC, Sollenberger LE, Blount AR, Myer RO, Maddox MK, Adesogan AT (2013) Performance of beef cattle creep fed concentrate or creep grazed on warm-season legumes. Crop Sci 53(4):1818–1825
Franks B (2019) What do animals want? Anim Welf 28(1):1–10
Fraser D (2008) Understanding animal welfare. Acta Vet Scand 50(Suppl 1):1–7
Free C (2022, July 21) Goats with big appetites are front-line workers in fire prevention. Washington Post. https://www.washingtonpost.com/lifestyle/2022/06/21/goats-hire-wildfire-prevention-california/
Freeman H (2018) Providing outdoor space for indoor-housed dairy goats
Freitas-de-Melo A, Banchero G, Hötzel MJ, Damián JP, Ungerfeld R (2013) Progesterone administration reduces the behavioural and physiological responses of ewes to abrupt weaning of lambs. Animal 7(8):1367–1373
Freitas-de-Melo A, Orihuela A, Hötzel MJ, Ungerfeld R (2022) What do we know and need to know about weaning in sheep? An overview of weaning practises, stress and welfare. Front Anim Sci 3(February)
Fritz Jr WF (2017) Self-enurination in the domesticated male goat (Capra hircus). ProQuest Dissertations and Theses, May, 168. http://ucelinks.cdlib.org:8888/sfx_local?url_ver=Z39.88-2004&rft_val_fmt=info:ofi/fmt:kev:mtx:dissertation&genre=dissertations+%26+theses&sid=ProQ:ProQuest+Dissertations+%26+Theses+Global&atitle=&title=Self-Enurination+in+the+Domesticated+Male+Goat+%28Cap
Gavashelishvili A, Yarovenko YA, Babayev EA, Mikeladze G, Gurielidze Z, Dekanoidze D, Kerdikoshvili N, Levan N, Nika P (2018) Modeling the distribution and abundance of eastern tur (Capra cylindricornis) in the Caucasus. J Mammal 99(4):885–897
Gebremedhin B, Chala D, Flagstad Ø, Bekele A, Bakkestuen V, van Moorter B, Ficetola GF, Zimmermann NE, Brochmann C, Stenseth NC (2021) Quest for new space for restricted range mammals: the case of the endangered Walia Ibex. Front Ecol Evol 9(July):1–12
Geiger M, Schoenebeck JJ, Schneider RA, Schmidt MJ, Fischer MS, Sánchez-Villagra MR (2021) Exceptional changes in skeletal anatomy under domestication: the case of brachycephaly. Integr Organis Biol, 1–31
Geist V (1960) Feral goats in British Columbia. Murrelet 41(3):34–40
Giger-Reverdin S, Gihad EA (1991) Water metabolism and intake in goats. In P. Morand-Fehr (ed) Goat nutrition (pp 37–45). Pudoc Wageningen
Gillespie J, Nyaupane N, Dunn B, McMillin K (2016) Why do farmers decide to produce meat goats? Evidence from the United States. Agric Hum Values 33(4):911–927
Goetsch AL, Detweiler G, Sahlu T, Dawson LJ (2001) Effects of different management practices on preweaning and early postweaning growth of Alpine kids. Small Rumin Res 41(2):109–116

Goetsch AL, Gipson TA, Askar AR, Puchala R (2010) Invited review: feeding behavior of goats. J Anim Sci 88(1):361–373
Gomes KAR, Valentim JK, Lemke SSR, Dallago GM, Vargas RC, Paiva AL d C (2018) Behavior of Saanen dairy goats in an enriched environment. Anim Prod 40:e42454
Grant J (2012) City goats: the goat justice league's guide to backyard goat keeping. Skipstone Books
Green LR, Hughes CL, Graves WL (1978) Goat control of brush regrowth on Southern California fuelbreaks. Rangelands 1(3):451–455. http://catalog2.nmsu.edu:2048/login?url=http://search.ebscohost.com/login.aspx?direct=true&db=lah&AN=19792326751&site=ehost-live&scope=site
Gürsoy O, Şentut T, Çankaya S (2011) Feedlot performance and carcass characteristics of Kilis goat breed. Macedonian J Anim Sci 1(1):39–51
Hagen K, Broom DM (2004) Emotional reactions to learning in cattle. Appl Anim Behav Sci 85(3–4):203–213
Hammond JA, Fielding D, Bishop SC (1997) Prospects for plant anthelmintics in tropical veterinary medicine. Vet Res Commun 21(3):213–228
Hartmann E, Högberg M, Olsson K, Dahlborn K (2021) Physiological and behavioural responses of Swedish domestic goats and their kids (Capra hircus) to 15 days of heat exposure. Acta Agric Scand A Anim Sci 70(1):41–49
Hayes EG, Lourençon RV, Browning R (2019) Effects of creep feeding and its interactions with other factors on the performance of meat goat kids and dams when managed on pasture. Transl Anim Sci 3(4):1466–1474
Heffner RS, Heffner HE (1990) Hearing in domestic pigs (Sus scrofa) and goats (Capra hircus). Hear Res 48(3):231–240
Hempstead MN, Lindquist TM, Shearer JK, Shearer LC, Cave VM, Plummer PJ (2021) Welfare Assessment of 30 Dairy Goat Farms in the Midwestern United States. Front Vet Sci 8(April):1–12
Hobart E (2022) The real story behind Morocco's tree-climbing goats. Natl Geogr. https://www.nationalgeographic.co.uk/animals/2022/06/the-real-story-behind-moroccos-tree-climbing-goats
Holechek JL (2002) Do most livestock losses to poisonous plants result from "poor" range management? J Range Manag 55(3):270–276
Hosoi E, Swift DM, Rittenhouse LR, Richards RW (1995) Comparative foraging strategies of sheep and goats in a T-maze apparatus. Appl Anim Behav Sci 44(1):37–45
Hoste H, Leveque H, Dorchies P (2001) Comparison of nematode infections of thc gastrointestinal tract in Angora and dairy goats in a rangeland environment: relations with the feeding behaviour. Vet Parasitol 101(2):127–135
Hoste H, Jackson F, Athanasiadou S, Thamsborg SM, Hoskin SO (2006) The effects of tannin-rich plants on parasitic nematodes in ruminants. Trends Parasitol 22(6):253–261
Hoste H, Torres-acosta JFJ, Aguilar-caballero AJ (2007) Nutrition–parasite interactions in goats: is immunoregulation involved in the control of gastrointestinal nematodes? Parasite Immunol 30(2):79–88
Hughes A, Whitteridge D (1973) The receptive fields and topographical organization of goat retinal ganglion cells. Vis Res 13:1101–1114
Hummel J, Findeisen E, Südekum KH, Ruf I, Kaiser TM, Bucher M, Clauss M, Codron D (2011) Another one bites the dust: Faecal silica levels in large herbivores correlate with high-crowned teeth. Proc R Soc B Biol Sci 278(1712):1742–1747
Iribarren C, Kotler BP (2012) Foraging patterns of habitat use reveal landscape of fear of Nubian ibex Capra nubiana. Wildl Biol 18(2):194–201
Ivey DS, Owens FN, Sahlu T, Teh TH, Claypool PL, Goetsch AL (2000) Growth and cashmere production by Spanish goats consuming ad libitum diets differing in protein and energy levels. Small Rumin Res 35(2):133–139
Jacobs GH, Deegan JF, Neitz J (1998) Photopigment basis for dichromatic color vision in cows, goats, and sheep. Vis Neurosci 15(3):581–584

Jallow O, McGregor B, Anderson N, Holmes J (1994) Intake of trichostrongylid larvae by goats and sheep grazing together. Aust Vet J 77(11):361–364

Jamelli D, Bernard E, Melo FPL (2021) Habitat use and feeding behavior of domestic free-ranging goats in a seasonal tropical dry forest. J Arid Environ 190(April):104532

Jansen DAWAM, van Langevelde F, de Boer WF, Kirkman KP (2007) Optimisation or satiation, testing diet selection rules in goats. Small Rumin Res 73(1–3):160–168

Jensen MB, Duve LR, Weary DM (2015) Pair housing and enhanced milk allowance increase play behavior and improve performance in dairy calves. J Dairy Sci 98(4):2568–2575

Johnsen JF, Ellingsen K, Grøndahl AM, Bøe KE, Lidfors L, Mejdell CM (2015) The effect of physical contact between dairy cows and calves during separation on their post-separation behavioural response. Appl Anim Behav Sci 166(1):11–19

Kahiya C, Mukaratirwa S, Thamsborg SM (2003) Effects of Acacia nilotica and Acacia karoo diets on Haemonchus contortus infection in goats. Vet Parasitol 115(3):265–274

Kakarash NA, Ramzi DO, Ismaeel DO, Marif HF (2021) Effects of environmental enrichment on behaviours and welfare of meriz goat. Assiut Vet Med J 67(170):11–18

Kannan G, Terrill TH, Kouakou B, Gelaye S, Amoah EA (2002) Simulated preslaughter holding and isolation effects on stress responses and live weight shrinkage in meat goats. J Anim Sci 80(7):1771–1780

Kaur G, Kaswan S, Singla M, Sharma A, Lamba JS (2021) Behaviour of Beetal does and bucks at linear vs. hexagonal feeder with special reference to homologous regrouping. Appl Anim Behav Sci 234:105210

Keil NM, Imfeld-Mueller S, Aschwanden J, Wechsler B (2012) Are head cues necessary for goats (Capra hircus) in recognising group members? Anim Cogn 15(5):913–921

Keil NM, Pommereau M, Patt A, Wechsler B, Gygax L (2017) Determining suitable dimensions for dairy goat feeding places by evaluating body posture and feeding reach. J Dairy Sci 100(2):1353–1362

Kessler CC (2002) Eradication of feral goats and pigs and consequences for other biota on Sarigan Island, Commonwealth of the Northern Mariana Islands. In CR Veitch, MN Clout (eds), Turning the tide: the eradication of invasive species (pp 132–140)

Khan MS, Sasidharan TO, Ghosh PK (1979) Water regulation in Barmer goat of the Rajasthan desert. Experientia 35(9):1185–1186

Khan MA, Weary DM, von Keyserlingk MAG (2011) Invited review: effects of milk ration on solid feed intake, weaning, and performance in dairy heifers. J Dairy Sci 94(3):1071–1081

Kilgour R, Ross DJ (1980) Feral goat behaviour—a management guide. N Z J Agric 141(5):20

Kurtycz LM (2015) Choice and control for animals in captivity. The Psychologist 28:892–894

Landau SY, Dvash L, Ryan P, Saltz D, Deutch T, Rosen SA (2020) Faecal pellets, rock shelters, and seasonality: the chemistry of stabling in the Negev of Israel in late prehistory. J Arid Environ 181:104219

Langbein J (2012) Investigations on training, recall and reversal learning of a Y-maze by dwarf goats (Capra hircus): the impact of lateralisation. Behav Process 89(3):304–310

Langbein J, Nürnberg G, Manteuffel G (2004) Visual discrimination learning in dwarf goats and associated changes in heart rate and heart rate variability. In Physiology and behavior (Vol 82, Issue 4, pp 601–609)

Langbein J, Siebert K, Nuernberg G (2008) Concurrent recall of serially learned visual discrimination problems in dwarf goats (Capra hircus). Behav Process 79(3):156–164

Langbein J, Siebert K, Nürnberg G (2009) On the use of an automated learning device by group-housed dwarf goats: do goats seek cognitive challenges? Appl Anim Behav Sci 120(3–4):150–158

Leidy RS, Jorge CR, Elena BR, Jordi BF (2015) Comparative study of trophic behaviour and herd structure in wild and feral goats living in a mediterranean island: management implications. Appl Anim Behav Sci 165:81–87

Lethbridge M (2016) Insights into feral goat movement in Australia using dynamic Brownian Bridges for movement analysis. Rangel J 38(4):343–359

Liang JB, Paengkoum P (2019) Current status, challenges and the way forward for dairy goat production in Asia – conference summary of dairy goats in Asia. Asian Australas J Anim Sci 32(8):1233–1243

Lickliter RE (1985) Behavior associated with parturition in the domestic goat. Appl Anim Behav Sci 13(4):335–345

Liddell HS (1927) Higher nervous activity in the thyroidectomised sheep and goat. Q J Exp Physiol 17(1):41–51

Lindgren FK (2016) Evaluation of enclosure design, with focus on wild and domesticated herbivores in Swedish zoos (Issue 639). www.slu.se/husdjurmiljohalsa

Lohani M, Bhandari D (2021) The importance of goats in the world. Profession Agric Workers J 6(2):14

Loretz C, Wechsler B, Hauser R, Rüsch P (2004) A comparison of space requirements of horned and hornless goats at the feed barrier and in the lying area. Appl Anim Behav Sci 87(3–4):275–283

Lovreglio R, Meddour-Sahar O, Leone V (2014) Goat grazing as a wildfire prevention tool: a basic review. IForest – Biogeosciences and Forestry 7(4):260–268

Lu CD (1988) Grazing behavior and diet selection of goats. Small Rumin Res 1(3):205–216

Lu CD, Potchoiba MJ (1988) Milk feeding and weaning of goat kids – a review. Small Rumin Res 1(2):105–112

Mabeza G, Masama E, Mpofu IDT (2018) Evaluation of browse legume diets (Acacia angustissima, Leucaena trichandra and Calliandra calothyrsus) on feed intake and growth of goats. Univ J Agric Res 6(1):18–22

Magistrelli D, Aufy AA, Pinotti L, Rosi F (2013) Analysis of weaning-induced stress in Saanen goat kids. J Anim Physiol Anim Nutr 97(4):732–739

Maltz E, Shkolnik A (1984) Milk composition and yield of the black Bedouin goat during dehydration and rehydration. J Dairy Res 51(1):23–27

McGregor BA (2009) A review of cashmere nutrition experiments with suggestions for improving their design and conduct. Small Rumin Res 82(2–3):71–83

McGregor BA (2011) Incisor development, wear and loss in sheep and their impact on ewe production, longevity and economics: a review. Small Rumin Res 95(2–3):79–87

McGregor BA, Butler KL (2011) The relationship between permanent incisor wear and mohair production and attributes in grazing adult Angora goats. Small Rumin Res 100(1):37–43

McGregor BA, An M, Jiang Y (1991) Fleece metrology of Liaoning cashmere goats. Small Rumin Res 4(1):61–71

Meehan CL, Mench JA (2007) The challenge of challenge: can problem solving opportunities enhance animal welfare? Appl Anim Behav Sci 102:246–261

Mellado M, Valdez R, Lara LM, Lopez R (2003) Stocking rate effects on goats: a research observation. J Range Manag 56(2):167–173

Mellado M, Rodríguez A, Villarreal JA, Lopez R (2004) Height to withers and abdominal circumference effects on diets of grazing goats. Appl Anim Behav Sci 88(3–4):263–274

Mellado M, Rodríguez A, Villarreal JA, Rodríguez R, Salinas J, López R (2005) Gender and tooth wear effects on diets of grazing goats. Small Rumin Res 57(2–3):105–114

Mellado M, Olivares L, Pittroff W, Díaz H, López R, Villarreal JA (2007) Oral morphology and dietary choices of goats on rangeland. Small Rumin Res 71(1–3):194–199

Mellor DJ, Beausoleil NJ, Littlewood KE, McLean AN, McGreevy PD, Jones B, Wilkins C (2020) The 2020 five domains model: Including human–animal interactions in assessments of animal welfare. Animals 10(10):1–24

Menzies P, Bauman C, Uehlinger F, Zobel G, Moggy M, Kerkvliet P (2020) *Code of practice for the care and handling of goats: Review of scientific research on priority issues* (Issue August). http://www.nfacc.ca/resources/codes-of-practice/equine/Equine_SCReport_Aug23.pdf

Merchant M, Riach DJ (2002) The effect of plane of nutrition and shearing on the pattern of the moult in Scottish Cashmere goats. Anim Sci 74(1):177–188

Meyer S, Nürnberg G, Puppe B, Langbein J (2012) The cognitive capabilities of farm animals: categorisation learning in dwarf goats (Capra hircus). Anim Cogn 15(4):567–576

Michel S, Rosen Michel T (2015) Capra falconeri. The IUCN Red List of Threatened Species, 8235, e.T3787A97218336

Miller DW, Fleming PA, Barnes AL, Wickham SL, Collins T, Stockman CA (2018) Behavioural assessment of the habituation of feral rangeland goats to an intensive farming system. Appl Anim Behav Sci 199(November 2017):1–8

Min BR, Hart SP, Miller D, Tomita GM, Loetz E, Sahlu T (2005) The effect of grazing forage containing condensed tannins on gastro-intestinal parasite infection and milk composition in Angora does. Vet Parasitol 130(1–2):105–113

Miranda-de la Lama GC, Mattiello S (2010) The importance of social behaviour for goat welfare in livestock farming. Small Rumin Res 90(1–3):1–10

Maltz E, Silanikove N, Karaso Y, Shefet G, Meltzer A, Barak M (1991) A note on the effects of feeding total mixed ration on performance of dairy goats in late lactation. https://www.sciencedirect.com/science/article/abs/pii/037784019190095A

Moseby KE, Read JL, Andersen GE (2020) Goat movement patterns inform management of feral goat populations in semiarid rangelands. Wildl Res 48(1):44–54

Muri K, Stubsjøen SM, Valle PS (2013) Development and testing of an on-farm welfare assessment protocol for dairy goats. Anim Welf 22(3):385–400

Murney R, Burggraaf V, Mapp N, Ganche E, King W (2019)The effect of cultivated mixed-species green fodder on intake, milk production and milk composition of housed dairy goats. Animal 13 (12):2802–2810

Nair MRR, Sejian V, Silpa MV, Fonsêca VFC, de Melo Costa CC, Devaraj C, Krishnan G, Bagath M, Nameer PO, Bhatta R (2021) Goat as the ideal climate-resilient animal model in tropical environment: revisiting advantages over other livestock species. Int J Biometeorol 65(12):2229–2240

Nascimento TVC, Oliveira RL, Menezes DR, de Lucena ARF, Queiroz MAÁ, Lima AGVO, Ribeiro RDX, Bezerra LR (2021) Effects of condensed tannin-amended cassava silage blend diets on feeding behavior, digestibility, nitrogen balance, milk yield and milk composition in dairy goats. Animal 15(1):100015

Nawroth C, von Borell E, Langbein J (2015) Object permanence in the dwarf goat (Capra aegagrus hircus): perseveration errors and the tracking of complex movements of hidden objects. Appl Anim Behav Sci 167:20–26

Nawroth C, Prentice PM, Mcelligott AG (2017) Individual personality differences in goats predict their performance in visual learning and non-associative cognitive tasks. Behav Process 134:43–53

Nawroth C, Rosenberger K, Keil NM, Langbein J (2022) Goats (Capra hircus) from different selection lines differ in their behavioural flexibility. Front Psychol 12(February):1–7

Neave HW, Costa JHC, Weary DM, von Keyserlingk MAG (2018a) Personality is associated with feeding behavior and performance in dairy calves. J Dairy Sci 101(8):7437–7449

Neave HW, von Keyserlingk MAG, Weary DM, Zobel G (2018b) Feed intake and behavior of dairy goats when offered an elevated feed bunk. J Dairy Sci 101(4):3303–3310

Neave HW, Costa JHC, Benetton JB, Weary DM, von Keyserlingk MAG (2019) Individual characteristics in early life relate to variability in weaning age, feeding behavior, and weight gain of dairy calves automatically weaned based on solid feed intake. J Dairy Sci 102(11):10250–10265

Neave HW, Sumner CL, Henwood RJT, Zobel G, Saunders K, Thoday H, Watson T, Webster JR (2022) Dairy farmers' perspectives on providing cow-calf contact in the pasture-based systems of New Zealand. J Dairy Sci 105(1):453–467

Ngwa AT, Pone DK, Mafeni JM (2000) Feed selection and dietary preferences of forage by small ruminants grazing natural pastures in the Sahelian zone of Cameroon 88:253–266

Nievergelt B (1981) Ibexes in an African environment: ecology and social system of the Walia ibex in the Simen Mountains. Springer, Ethiopia

Nolden C (2020) Goat dietary selections, performance and browsing effects on a brush-invaded oak savanna in southwest Wisconsin [University of Wisconsin-Madison]. https://www.sfa-mn.org/wp-content/uploads/2020/03/2020-Nolden-Goat-Browsing-MS-Thesis.pdf

NZ Code of Welfare: Goats, 51 (2018). http://www.mpi.govt.nz/protection-and-response/animal-welfare/codes-of-welfare/
O'Brien PH (1984) Leavers and stayers: maternal post-partum strategies in feral goats. Appl Anim Behav Sci 12(3):233–243
O'Brien PH (1988) Feral goat social organization: a review and comparative analysis. Appl Anim Behav Sci 21(3):209–221
Oesterwind S, Nürnberg G, Puppe B, Langbein J (2016) Impact of structural and cognitive enrichment on the learning performance, behavior and physiology of dwarf goats (Capra aegagrus hircus). Appl Anim Behav Sci 177:34–41
Okoruwa MI, Ikhimioya I (2020) Influence of browse-tree leaves supplementation on digestibility, rumen fermentation and performance of goats fed mixed grass hay. Livest Res Rural Dev 32(6) http://www.lrrd.org/lrrd32/6/odion32093.html
Ollivier FJ, Samuelson DA, Brooks DE, Lewis PA, Kallberg ME, Komáromy AM (2004) Comparative morphology of the tapetum lucidum (among selected species). Vet Ophthalmol 7:11–22
Osborne SR (1977) The free food (contrafreeloading) phenomenon: a review and analysis. Anim Learn Behav 5(3):221–235
Ouédraogo-Koné S, Kaboré-Zoungrana CY, Ledin I (2006) Behaviour of goats, sheep and cattle on natural pasture in the sub-humid zone of West Africa. Livest Sci 105(1–3):244–252
Palma JM, Galina MA (1995) Effect of early and late weaning on the growth of female kids. Small Rumin Res 18(1):33–38
Panzuti C, Mandrile G, Duvaux-Ponter C, Dessauge F (2018) Early weaning and high feeding level in post-weaning period did not impact milk production in Alpine dairy goats. J Dairy Res 85(3):277–280
Pareja J, Baraza E, Ibáñez M, Domenech O, Bartolomé J (2020) The role of feral goats in maintaining firebreaks by using attractants. Sustain For 12:7144
Parkes JP (1984) Feral goats on Raoul Island. I. Effect of control methods on their density, distribution, and productivity. N Z J Ecol 7(May):85–93
de Passillé AMB, Rushen J (2006) Calves' behaviour during nursing is affected by feeding motivation and milk availability. Appl Anim Behav Sci 101(3–4):264–275
Patt A, Gygax L, Wechsler B, Hillmann E, Palme R, Keil NM (2013a) Behavioural and physiological reactions of goats confronted with an unfamiliar group either when alone or with two peers. Appl Anim Behav Sci 146(1–4):56–65
Patt A, Gygax L, Wechsler B, Hillmann E, Palme R, Keil NM (2013b) Factors influencing the welfare of goats in small established groups during the separation and reintegration of individuals. Appl Anim Behav Sci 144(1–2):63–72
Perez-Barberia FJ, Gordon IJ (1999) The functional relationship between feeding type and jaw and cranial morphology in ungulates. Oecologia 118:157–165
Pfister JA, Malechek JC, Balph DF (1988) Foraging behaviour of goats and sheep in the Cattinga of Brazil. J Appl Ecol 25(2):379–388
Pimentel PRS, Pellegrini CB, Lanna DPD, Brant LMS, Ribeiro CVDM, Silva TM, Barbosa AM, da Silva Júnior JM, Bezerra LR, Oliveira RL (2021) Effects of Acacia mearnsii extract as a condensed-tannin source on animal performance, carcass yield and meat quality in goats. Anim Feed Sci Technol 271(September 2020):114733
Pitcher BJ, Briefer EF, Baciadonna L, McElligott AG (2017) Cross-modal recognition of familiar conspecifics in goats. r. Soc. Open Sci 4:160346
Placzek M, Christoph-Schulz I, Barth K (2021) Public attitude towards cow-calf separation and other common practices of calf rearing in dairy farming—a review. Org Agric 11(1):41–50
Poindron P, Gilling G, Hernandez H, Serafin N, Terrazas A (2003) Early recognition of newborn goat kids by their mother: I. Nonolfactory discrimination. Dev Psychobiol 43(2):82–89
Poindron P, Terrazas A, de la Luz Navarro Montes de Oca M, Serafín N, Hernández H (2007) Sensory and physiological determinants of maternal behavior in the goat (Capra hircus). Horm Behav 52:99–105

Price EO, Thos J (1980) Behavioral responses to short-term social isolation in sheep and goats. Appl Anim Ethol 6(4):331–339

Prosser C, Stafford K (2017) Goat production. In: Stafford K (ed) Livestock production in New Zealand. Massey University Press, pp 147–169

Provenza FD, Lynch JJ, Burritt EA, Scott CB (1994) How goats learn to distinguish between novel foods that differ in postingestive consequences. J Chem Ecol 20(3):609–624

Qiu X, Chen Y, Xiao X, Ling W, Wang M, Li Y (2018) The difference between White Cashmere goats and Merino sheep in cognitive abilities for visual and spatial discriminations. Acta Veterinaria et Zootechnica Sinica 49(11):2384–2393. http://www.xml-data.org/XMSYXB/html/2018/11/20181109.htm http://ovidsp.ovid.com/ovidweb.cgi?T=JS&CSC=Y&NEWS=N&PAGE=fulltext&D=caba6&AN=20183374581 http://va6hg2ev9v.search.serialssolutions.com/?V=1.0&N=25&L=VA6HG2EV9V&S=AC_T_B&C=Acta+Veterinaria+et+Zootec

Ralphs MH (1992) Continued food aversion: training livestock to avoid eating poisonous plants. J Range Manag 45(1):46–51

Raoult CMC, Osthaus B, Hildebrand ACG, McElligott AG, Nawroth C (2021) Goats show higher behavioural flexibility than sheep in a spatial detour task. Royal Soc Open Sci 8(3)

Reina R, Berriatua E, Luján L, Juste R, Sánchez A, de Andrés D, Amorena B (2009) Prevention strategies against small ruminant lentiviruses: an update. Vet J 182(1):31–37

Rhind SM, McMillen SR (1995) Seasonal patterns of secondary fibre growth, moulting and hair follicle activity in Siberian and Icelandic X Scottish feral goats offered high and low levels of dietary protein. Small Rumin Res 16(1):69–76

Rodriguez KC (2019, July 28) Owner: Leesburg's Weed Goats May Have Been Bitten, Poisoned. LoudounNow. https://www.loudounnow.com/archives/owner-leesburgs-weed-goats-may-have-been-bitten-poisoned/article_160cbe26-a814-57d1-94bd-4971063e5104.html

Roitberg E, Franz H (2004) Oddity learning by African dwarf goats (Capra hircus). Anim Cogn 7(1):61–67

Rosenberger K, Costa JHC, Neave HW, von Keyserlingk MAG, Weary DM (2017) The effect of milk allowance on behavior and weight gains in dairy calves. J Dairy Sci 100(1):504–512

Rosenberger K, Simmler M, Nawroth C, Langbein J, Keil N (2020) Goats work for food in a contrafreeloading task. Sci Rep 10(1):1–12

Rout PK, Mandal A, Singh LB, Roy R (2002) Studies on behavioral patterns in Jamunapari goats. Small Rumin Res 43(2):185–188

Runa RA, Brinkmann L, Riek A, Hummel J, Gerken M (2019) Reactions to saline drinking water in Boer goats in a free-choice system. Animal 13(1):98–105

Sailer LM, Holinger M, Burla JB, Wechsler B, Zanolari P, Friedli K (2021) Influence of housing and management on claw health in Swiss dairy goats. Animals 11(7):1–20

Sambraus H, Wittmann M (1989) Observations of the birth and suckling behavior of goats. Tierarztl Prax 17(4):359–365

Sanon HO, Kaboré-Zoungrana C, Ledin I (2007) Behaviour of goats, sheep and cattle and their selection of browse species on natural pasture in a Sahelian area. Small Rumin Res 67(1):64–74

Sanon HO, Kaboré-Zoungrana C, Ledin I (2008) Growth and carcass characteristics of male Sahelian goats fed leaves or pods of Pterocarpus lucens or Acacia senegal. Livest Sci 117(2–3):192–202

Sanson GD, Kerr SA, Gross KA (2007) Do silica phytoliths really wear mammalian teeth? J Archaeol Sci 34(4):526–531

Sarhangzadeh J, Yavari AR, Hemami MR, Jafari HR, Shams-Esfandabad B (2013) Habitat suitability modeling for wild goat (Capra aegagrus) in a mountainous arid area, central Iran. Caspian J Environ Sci 11(1):41–51

Scherer R, Gerlach K, Südekum KH (2019) Decision-making of goats when exposed to choice feeding: triggered by taste or smell? Appl Anim Behav Sci 210(September 2018):46–51

Scholtens MR, Smith R, Lopez-Lozano SR, Lopez-Villalobos N, Burt D, Harper L, Tuohy M, Thomas DG, Carr AJ, Gray DI, Tozer P, Schreurs NM (2017) Brief communication: the current state of the New Zealand goat industry. Proc N Z Soc Anim Prod 77, 164–168. http://www.nzsap.org/system/files/proceedings/59 Scholtens.pdf

Scogings P, Macanda M (2005) Acacia karroo responses to early dormant season defoliation and debarking by goats in a semi-arid subtropical savanna. Plant Ecol 179(2):193–206
Severson KE, Debano LF (1991) Influence of Spanish Goats on Vegetation and Soils in Arizona Chaparral. J Range Manag 44(2):111–117
Shahjalal M, Galbraith H, Topps J (1992) The effect of changes in dietary protein and energy on growth, body composition and mohair fibre characteristics of British Angora goats. Anim Sci 54(3):405–412
Shamsaddini-Bafti M, Salehi M, Maghsoudi A, Tehrani AM, Mirzaei F, Momen SMS (2012) Effect of sex and rearing system on the quality and mineral content of fiber from raeini cashmere goats. J Anim Sci Biotechnol 3(1):2–7
Shank CC (1972) Some aspects of social behaviour in a population of feral goats (Capra hircus L.). Z Tierpsychol 30(5):488–528
Shelton M (1995) Angora goat and mohair production
Shi J, Dunbar RIM (2006) Feeding competition within a feral goat population on the Isle of Rum, NW Scotland. J Ethol 24(2):117–124
Shi J, Dunbar RIMM, Buckland D, Miller D (2003) Daytime activity budgets of feral goats (Capra hircus) on the Isle of Rum: Influence of season, age, and sex. Can J Zool 81(5):803–815
Shi J, Dunbar RIM, Buckland D, Miller D (2005) Dynamics of grouping patterns and social segregation in feral goats (Capra hircus) on the Isle of Rum, NW Scotland. Mammalia 69(2):185–199
Shrader AM, Kerley GIH, Kotler BP, Brown JS (2007) Social information, social feeding, and competition in group-living goats (Capra hircus). Behav Ecol 18(1):103–107
Siebert K, Langbein J, Schön PC, Tuchscherer A, Puppe B (2011) Degree of social isolation affects behavioural and vocal response patterns in dwarf goats (Capra hircus). Appl Anim Behav Sci 131(1–2):53–62
Silanikove N (1994) The struggle to maintain hydration and osmoregulation in animals experiencing severe dehydration and rapid rehydration: the story of ruminants. Exp Physiol 79(3):281–300
Silanikove N (1997) Why goats raised on harsh environment perform better than other domesticated animals. In: Lindberg JE, Gonda HL, Ledin I (eds) Recent advances in small ruminant nutrition. Ciheam – Options Mediterraneennes, pp 185–194
Silanikove N (2000) The physiological basis of adaptation in goats to harsh environments. Small Rumin Res 35(3):181–193
Simensen E, Hardeng F, Lunder T (2010) Housing of Norwegian goat herds and associations with milk yield and milk quality. Acta Agric Scand A Anim Sci 60:187–193
Spatz G, Mueller-Dombois D (1973) The influence of feral goats on Koa tree reproduction in Hawaii Volcanoes National Park. Ecology 54(4):870–876
Sponenberg DP (2019) Local goat breeds in the United States. In: Kukovics S (ed) Goats (Capra) – from ancient to modern. IntechOpen, pp 1–13
Sporkmann K, Georg H, Bender S, Ude G (2012) Heart rate variability of goat kids to evaluate stress in different weaning situations. Landtechnik 67:417–20
Stachowicz J, Gygax L, Hillmann E, Wechsler B, Keil NM (2018) Dairy goats use outdoor runs of high quality more regardless of the quality of indoor housing. Appl Anim Behav Sci 208(June):22–30
Stachowicz J, Lanter A, Gygax L, Hillmann E, Wechsler B, Keil NM (2019) Under temperate weather conditions, dairy goats use an outdoor run more with increasing warmth and avoid light wind or rain. J Dairy Sci 102(2):1508–1521
Stanley CR, Dunbar RIM (2013) Consistent social structure and optimal clique size revealed by social network analysis of feral goats, Capra hircus. Anim Behav 85(4):771–779
Steinshamn H, Inglingstad RA, Ekeberg D, Mølmann J, Jørgensen M, Steinshamn H, Inglingstad RA, Ekeberg D, Mølmann J, Jørgensen M, Inglingstad RA, Ekeberg D, Mølmanna J, Jørgensena M (2014) Effect of forage type and season on Norwegian dairy goat milk production and quality. Small Rumin Res 122(1–3):19–30

Sutherland MA, Lowe GL, Watson TJ, Ross CM, Rapp D, Zobel GA (2017) Dairy goats prefer to use different flooring types to perform different behaviours. Appl Anim Behav Sci 197(May):24–31
Syme GJ, Syme LA (1979) Social structure in farm animals. Elsevier
Tadesse SA, Kotler BP (2010) Habitat choices of Nubian ibex (Capra Nubiana) evaluated with a habitat suitability modeling and isodar analysis. Israel J Ecol Evol 56(1):55–74
Tadesse SA, Kotler BP (2011) Seasonal habitat use by Nubian ibex (Capra Nubiana) evaluated with behavioral indicators. Israel J Ecol Evol 57:223–246
Taylor CR (1966) The vascularity and possible thermoregulatory function of the horns in goats. Physiol Zool 39(2):127–139
Taylor CA, Ralphs MH (1992) Reducing livestock losses from poisonous plants through grazing management. J Range Manag 45(1):9–12
Taylor JD, Gilliam JN, Mourer G, Stansberry C (2020) Comparison of effects of four weaning methods on health and performance of beef calves. Animal 14(1):161–170
Terrazas A, Serafin N, Hernández H, Nowak R, Poindron P (2003) Early recognition of newborn goat kids by Their Mother: II. Auditory recognition and evidence of an individual acoustic signature in the neonate. Dev Psychobiol 43(4):311–320
Thompson J, Riethmuller J, Kelly D, Miller E, Scanlan JC (2002) Feral goats in South-Western Queensland: a permanent component of the Grazing Lands. Rangel J 24(2):268–287
Todd CG, Bruce B, Deeming L, Zobel G (2019) Short communication: survival of replacement kids from birth to mating on commercial dairy goat farms in New Zealand. J Dairy Sci 102(10):9382–9388
Toinon C, Waiblinger S, Rault JL (2021) Maternal deprivation affects goat kids' stress coping behaviour. Physiol Behav 239(June):113494
Toinon C, Waiblinger S, Rault JL (2022) Maternal deprivation affects goat kids' social behavior before and after weaning. Dev Psychobiol 64(4):1–13
Tölü C, Savaş T (2007) A brief report on intra-species aggressive biting in a goat herd. Appl Anim Behav Sci 102(1–2):124–129
Tölü C, Yurtman IY, Baytekin H, Ataşoğlu C, Savaş T (2012) Foraging strategies of goats in a pasture of wheat and shrubland. Anim Prod Sci 52(12):1069–1076
Tölü C, Göktürk S, Sava T (2016) Effects of weaning and spatial enrichment on behavior of Turkish Saanen Goat Kids. Asian-Australas J Anim Sci 29(6):879–886
Torres-Acosta JFJ, Hoste H (2008) Alternative or improved methods to limit gastro-intestinal parasitism in grazing sheep and goats. Small Rumin Res 77:159–173
Toussaint G (1997) The housing of milk goats. Livest Prod Sci 49(2):151–164
Ugur F, Savas T, Dosay M, Karabayir A, Atasoglu C (2004) Growth and behavioral traits of Turkish Saanen kids weaned at 45 and 60 days. Small Rumin Res 52(1–2):179–184
Ugur F, Atasoglu C, Tolu C, Diken F, Pala A (2007) Effects of different weaning programs on growth of Saanen kids. Anim Sci J 78(3):281–285
Underwood C (2002) Total grazing management field guide self-mustering systems for cattle, sheep and goats
Utaaker KS, Chaudhary S, Kifleyohannes T, Robertson LJ (2021) Global goat! Is the expanding goat population an important reservoir of Cryptosporidium? Front Vet Sci 8(March)
Van DTT, Mui NT, Ledin I (2005) Tropical foliages: effect of presentation method and species on intake by goats. Anim Feed Sci Technol 118(1–2):1–17
Važić B, Batinić V, Savić N, Erbez M (2019) Characteriscs of indoor microclimate status in goat barns in bosnia and herzegovina. J Anim Sci Bih 1(2):43–49
Ventura-Cordero J, González-Pech PG, Torres-Acosta JFJ, Sandoval-Castro CA, Tun-Garrido J (2018) Sheep and goat browsing a tropical deciduous forest during the rainy season: why does similar plant species consumption result in different nutrient intake? Anim Prod Sci 59:66–72
Vickery HM, Neal RA, Meagher RK (2022a) Rearing goat kids away from their dams 1. A survey to understand rearing methods. Animal 16(6):100547
Vickery HM, Neal RA, Meagher RK (2022b) Rearing goat kids away from their dams 2. Understanding farmers' views on changing management practices. Animal 16(6):100548

Wang Z, Wang R, Li J, Zhang W, Wuriliga, Su R, Liu Z, Zhou J, Wei Y, Meng R, Zhang Y (2014) Modeling genetic covariance structure across ages of fleece traits in an Inner Mongolia cashmere goat population using repeatability and multivariate analysis. Livest Sci 161(1):1–5

Watson C, Binks D (2019) Elongation of the CA1 field of the septal hippocampus in ungulates. J Comp Neurol 527(4):818–832

Webster J (2016) Animal welfare: freedoms, dominions and "A life worth living". Animals 6(6):2–7

Weinberg PJ, Akkiev MI, Buchukuri RG (2010) Clineal variation in caucasian tur and its taxonomic relevance. Galemys 22:375–394

Welk A, Neave HW, Spitzer HB, von Keyserlingk MAG, Weary DM (2022) Effects of intake-based weaning and forage type on feeding behavior and growth of dairy calves fed by automated feeders. J Dairy Sci 105(11):9119–9136

Wierenga HK, Hopster H (1991) Behaviour of dairy cows when fed concentrates with an automatic feeding system. Appl Anim Behav Sci 30(3–4):223–246

Yan T, Cook JE, Gibb MJ, Ivings WE, Treacher TT (1993) The effects of quantity and duration of milk feeding on the intake of concentrates and growth of castrated male Saanen kids to slaughter. Anim Sci 56(3):327–332

Yocum CF (1967) Ecology of feral goats in Haleakala National Park, Maui, Hawaii. Am Midl Nat 77:418–451

Young RJ, Lawrence AB (1994) Feeding behaviour of pigs in groups monitored by a computerized feeding system. Anim Prod 58(1):145–152

Zarazaga LÁ, Gatica MC, Gallego-Calvo ML, Guzmán JL (2018) When using photostimulated bucks to induce the male effect in female goats living at Mediterranean latitudes, a male: female ratio of 1:20 is optimum. J Appl Anim Res 46(1):883–887

Zobel G, Nawroth C (2020) Current state of knowledge on the cognitive capacities of goats and its potential to inform species-specific enrichment. Small Rumin Res 192(December 2019):106208

Zobel G, Leslie K, Weary DM, von Keyserlingk MAG (2015) Ketonemia in dairy goats: effect of dry period length and effect on lying behavior. J Dairy Sci 98(9):6128–6138

Zobel G, Tan B, Deeming L (2016) Brief communication: immediate removal of dairy goat kids from the doe as a strategy to manage colostrum intake. Proc N Z Soc Anim Prod 76:169–171

Zobel G, Neave H, Webster J (2017) The use of a climbing platform in mid-lactation dairy goats is related to production and behavioural responses to novelty. 7th International conference on the assessment of animal welfare at farm and group level

Zobel G, Freeman H, Schneider D, Henderson H, Johnstone P, Webster J (2018) Behaviour of dairy goats managed in a natural alpine environment. 52nd Congress of the international society for applied ethology

Zobel G, Neave HW, Webster J (2019) Understanding natural behavior to improve dairy goat (Capra hircus) management systems. Transl Anim Sci 3(1):212–224

Zobel G, Freeman H, Watson T, Cameron C, Sutherland M (2020) Effect of different milk-removal strategies at weaning on feed intake and behavior of goat kids. J Vet Behav 35

Zobel G, Barnaby E, Watson T (2023) Seeing beyond gait: a case study identifying chronic laminitis in 3 goats. Small Rumin Res 227:107068

Goat Emotions, Cognition, and Personality

Marianne A. Mason, Elodie F. Briefer, Stuart Semple, and Alan G. McElligott

Abstract

Positive welfare is a relatively modern concept based on the idea that on the whole, an animal's positive experiences should outweigh their negative ones. Intuitively, we therefore require indicators to measure when animals are experiencing positive and negative emotional states, as well as providing opportunities for animals to create positive experiences. Goats typically live rich social and emotional lives, are naturally inquisitive, and seek cognitive challenges, highlighting the importance of a complex and cognitively stimulating environment for their welfare. However, how goats respond to their environment is not the same for each individual, underlining the shortcomings of a 'one size fits all' approach to enhancing welfare. In sum, to achieve positive welfare in goats, we

This chapter is adapted from Mason (2023; doctoral thesis).

Supplementary Information The online version contains supplementary material available at https://doi.org/10.1007/978-3-031-62182-6_3. The videos can be accessed individually by clicking the DOI link in the accompanying figure caption or by scanning this link with the SN More Media App.

M. A. Mason (✉) · S. Semple
School of Life and Health Sciences, University of Roehampton, London, UK

E. F. Briefer
Behavioural Ecology Group, Section for Ecology and Evolution, Department of Biology, University of Copenhagen, Copenhagen, Denmark

A. G. McElligott (✉)
Department of Infectious Diseases and Public Health, Jockey Club College of Veterinary Medicine and Life Sciences, City University of Hong Kong, Hong Kong SAR, China

Centre for Animal Health and Welfare, Jockey Club College of Veterinary Medicine and Life Sciences, City University of Hong Kong, Hong Kong SAR, China
e-mail: alan.mcelligott@cityu.edu.hk

S. Mattiello, M. Battini (eds.), *The Welfare of Goats*, Animal Welfare 25,
https://doi.org/10.1007/978-3-031-62182-6_3

must possess a means to monitor their emotional experiences, understand their cognitive abilities and provide sufficient opportunities to apply these, while ensuring welfare measures are tailored at an individual level. This chapter will therefore focus on potential behavioural, physiological, and cognitive indicators used to measure goat emotional states in the short and long term; their cognitive abilities in the physical and social domains; and the effect of personality on how goats interact with their social environment and on cognitive abilities. It contains enriching material in the form of videos and goat call recordings to complement your reading experience.

Keywords

Emotional indicators · Human–animal interactions · Learning · Memory · Positive welfare · Temperament

1 Introduction

Goats (*Capra hircus*) were one of the first livestock species to be domesticated, around 10,500 years ago (Zeder and Hesse 2000; MacHugh et al. 2017). They have a global population of around 1.3 billion, with most farmed goats living in Africa (43.3%) and Asia (51.4%: FAO 2020). While a minority of goats live in Europe (1.4% of global farmed goats) and the USA (0.2% of goats), increasing intensification of production systems in these regions has moved more goats indoors, strengthening the mismatch between the domestic and natural environment (Zobel et al. 2019). Given the increasing pressure on other nations to compete, and with the growing interest in goat farming as a 'greener' alternative to other livestock production systems, this mismatch is projected to continue (Darcan and Silanikove 2018; Zobel et al. 2019). To ensure welfare standards are still met amidst growing public concern (Alonso et al. 2020), researchers continue to develop indicators for measuring multiple aspects of goat welfare.

A role for emotional states has long been acknowledged as being central to farm animal welfare, but only more recently have positive experiences been explicitly recognised as being important, rather than simply minimising negative ones (FAWC 1993). Modern welfare concepts generally promote 'positive welfare', and key to this is the idea that positive emotional experiences should on the whole outweigh negative ones (Green and Mellor 2011; Mellor and Beausoleil 2015). Therefore, to appraise and improve goat welfare, we clearly require an objective means to assess their positive and negative emotional states, in order to then be able to promote positive experiences.

Goats are naturally inquisitive, motivated to seek cognitive challenges and, as browsers, explore the world through their mouth, using their mobile lips to manipulate and investigate novel objects (Briefer et al. 2014; Berman et al. 2017; Mahdy et al. 2020; Rosenberger et al. 2020). Such tendencies, along with an impressive set of foraging skills evolved in the mountains navigated by their ancestors (Silanikove 2000; Amills et al. 2017; Zobel and Nawroth 2020). However, in the barren indoor environments commonly faced by many goats today, these behavioural tendencies may lead to boredom and frustration (Franks 2018; Zobel et al. 2019). Furthermore, the altered social conditions on farms (e.g. larger group sizes, higher densities, and regrouping of individuals) can pose a challenge for goats from a social cognitive perspective, as they attempt to form social organisations and stable relationships with group members; this being contingent on their ability to discriminate between individuals (Miranda-de la Lama and Mattiello 2010; Patt et al. 2012; Stanley and Dunbar 2013). An understanding of how goats perceive and process information from their physical and social environment is central to maintaining positive emotional states. Further, as learning can be inherently rewarding in itself, understanding the cognitive abilities of goats and other livestock is imperative to designing enrichment strategies to provide complex, cognitively stimulating environments while avoiding frustrating animals through overly difficult challenges (Jensen 1963; Inglis et al. 1997; Franks 2018). However, how individuals respond to their environment varies from goat-to-goat (e.g. Briefer et al. 2015a).

Individual differences in how goats perceive their environment result in individual-level variation in welfare outcomes (Richter and Hintze 2019). Consistent behavioural and physiological variation among goats can be captured through measuring and categorising individuals according to their personality. Accordingly, for the purposes of this chapter, we will discuss emotions in goats, while focussing on indicators of emotional responses, as well as their cognition (physical and social) and personality, including how personality feeds into emotional experiences and cognition.

2 Emotions

Emotional responses play a key role in tailoring an animal's behaviour towards environmental information, such as food, predators, and mates (Kremer et al. 2020). In psychology, such processes are principally defined either through breaking up emotions into discrete categories (for example, happiness, anger, and disgust; Ekman 1992), or as graded responses measured over multiple axes or dimensions. The most widely used of the latter, or 'dimensional approaches' splits emotion into two main axes: 'valence', which ranges from positive (e.g. happiness) to negative (e.g. anger/fear) states and 'arousal', which reflects the level of bodily activation (e.g. calm to excited; Russell 1980; Mendl et al. 2010; Fig. 1).

2.1 Emotional Indicators

We can assess emotional responses according to dimensional criteria, through measuring and combining constituents of an animal's overall emotional response (Mormède et al. 2007). Emotional experiences are the short-lived culmination of cognitive and neurophysiological changes, which may facilitate motor (behavioural) responses, and there may also be a subjective element that supposes a conscious 'feeling' of emotions experienced (Désiré et al. 2002; Špinka 2012). Indeed,

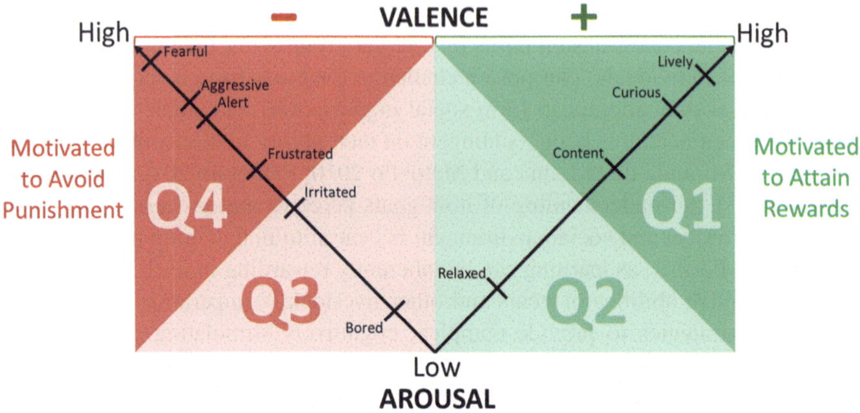

Fig. 1 Dimensional model splitting animal emotional state by valence (negative: motivated to avoid punishment and positive: motivated to attain rewards) and arousal (low to high; schematic adapted from Briefer 2020). Examples provided derive from AWIN (2015) recommended descriptors for goat behaviour. Quadrant 1 (Q1): high arousal, positive emotions; Quadrant 2 (Q2): low arousal, positive emotions; Quadrant 3 (Q3): low arousal, negative emotions; Quadrant 4 (Q4): high arousal, negative emotions

although neural structures implicated in emotional processes appear conserved across species, unlike humans who can provide verbal reports, definitive evidence of conscious feelings in non-human animals remains elusive (de Waal 2011; Le Doux 2021). By contrast, emotionally mediated changes in physiological responses and brain activity can be measured using specialised equipment. Behavioural outcomes are directly observable, and cognitive effects can be investigated via certain experimental paradigms (e.g. judgement bias tests: Briefer and McElligott 2013; Baciadonna et al. 2016). Measuring emotion provides insight into how animals appraise their environment; as care for goats and other livestock is provided by humans, these environmental conditions are under human control, making emotional indicators indispensable for informing welfare-motivated management decisions.

2.1.1 Neurophysiological Indicators

The brain processes environmental information and initiates cascading emotional responses (Machado and da Silva 2019). However, measuring patterns of brain activity linked with emotions often requires animals to be immobilised using anaesthesia or training (Hui et al. 2016; Thompkins et al. 2021). New techniques, such as functional near-InfraRed Spectroscopy (fNIRS), enable remote measurement of brain activity and have been employed in goats using wearable sensors, but these lack the precision of traditional approaches, such as functional Magnetic Resonance Imaging (fMRI; Gygax et al. 2013). Neural activity is, in turn, affected by feedback from all levels of the emotional response, which further modifies the animal's central emotional state (Anderson and Adolphs 2014; Zych and Gogolla 2021).

One prominent level of the emotional response involves the release of hormones. For example, cortisol levels can increase during stressful situations (particularly in more excitable goats: Kannan et al. 2022) and lower oxytocin and β-endorphin levels have been observed in individually tethered versus pair-housed goats (Hydbring-Sandberg et al. 2022). Assessing changes in hormonal profiles is often achieved through blood sampling, but this can exacerbate emotional responses (e.g. stress). Although less invasive methods do exist, such as sampling milk, saliva, hair, urine, or faeces, these are often less sensitive and come with their own disadvantages (Mormède et al. 2007). Hormone release can also be a relatively slow process, especially if changes are part of the hypothalamo–pituitary–adrenocortical (HPA) axis response (e.g. corticosteroid release; Mormède et al. 2007). This lag in release can limit its usefulness in assessing an animal's current emotional state, such as their responses to sudden or transient events. Therefore, we will focus below on two relatively rapid and non-invasive physiological measures: cardiac responses and skin temperature changes assessed using thermal imaging.

2.1.1.1 Cardiac Measures

Nervous system activation elicited during emotional responses leads to rapid and predictable changes in cardiac activity, namely in heart rate (HR) and heart rate variability (HRV; Wascher 2021). Relative level of activity between two of the branches of the autonomic nervous system is thought to vary according to emotional state, with the sympathetic nervous system dominating for negative emotions, and parasympathetic for positive ones (Sorinas et al. 2020). HR, primarily quantified as heartbeats per minute, is governed by the interaction between the parasympathetic (primarily through the vagal nerves) and sympathetic nervous systems. Specifically, increased parasympathetic nervous activity reduces HR and decreased activity increases HR, perhaps in concert with or independently of increases in sympathetic activity, which often plays a more prominent role in elevating HR (von Borell et al. 2007). HR increases form part of the initial physiological stress response enabling animals to cope with challenges imposed by internal and/or external stimuli, and is more generally associated with arousal in a variety of species and emotional contexts (for review, see Wascher 2021).

In goats, increases in HR have been observed in response to positive (feeding and food anticipation) and negative (food frustration) contexts of higher arousal compared to more neutral situations (with no external stimulus: Briefer et al. 2015b; Baciadonna et al. 2020). HR has also been shown to vary according to social context, with lower HRs observed in goats engaged in affiliative interactions with conspecifics compared to other behaviours, such as agonistic interactions and eating and/or drinking (Briefer et al. 2015a). Intriguingly, HR seems to increase during grooming by humans despite behavioural indications that such interactions were positively received (Baciadonna et al. 2016) and opposite patterns observed in other species experiencing similar interactions with humans (expected to have calming effect, Coulon et al. 2015; Lange et al. 2021). Although HR remains an accurate and widely used measure of arousal, it tells us little about the valence of emotions experienced (Wascher 2021). By contrast, HRV may be a promising tool for contextualising emotional responses according to their valence.

HRV is a measure of the variability in duration between successive heartbeat intervals. Unlike HR, where underlying sympathetic and parasympathetic influences are difficult to disentangle, many researchers consider HRV a reliable indicator of parasympathetic nervous system activity (particularly, e.g., root mean square of successive interbeat differences, RMSSD: de Geus et al. 2019; Thomas et al. 2019). Specifically, increases in vagal nerve activity are associated with increases in HRV (Eckberg 1983, 2003). This association, and by extension with positive emotions, has led some researchers to suggest HRV as a valence indicator, with some investigations finding corroborating increases or decreases in HRV in positive and negative situations, respectively (e.g. Reefmann et al. 2009; Kowalik et al. 2017), including in goats (Briefer et al. 2015a; Baciadonna et al. 2019). However, results are not always consistent (e.g. Baciadonna et al. 2020; Lange et al. 2021) and this inference is not universally accepted, particularly in respect to research failing to control for emotional arousal (e.g. Briefer et al. 2015a; Travain et al. 2016). Where control measures have been implemented (e.g. through comparing situations of opposite valence, but similar arousal), HRV, like HR, more closely reflects changes in arousal, rather than the valence of a situation (Briefer et al. 2015b). Indeed, a non-linear negative relationship exists between HR and HRV, leading some researchers to even propose HRV as a 'surrogate' measure for HR (Monfredi et al. 2014, 2015). Ultimately, it may be safer to preliminarily conclude that HRV, like HR, acts more as an indicator of emotional arousal, with higher HRV measures corresponding to situations where stress decreases (e.g. release from stress, animal relaxing). Further research is needed to examine its value as a valence indicator.

A common means of obtaining cardiac data in goats and other livestock is through belt monitors fitted around the animal's thorax (e.g. Briefer et al. 2015a, b; Baciadonna et al. 2019, 2020). These approaches, while useful, do have several drawbacks. For example, the belt needs to be fitted tightly around the animal, and as these belts are largely designed for humans, they can necessitate a lot of handling to attain an ECG trace before the test and again if it is lost later on (usually through animal movement). Animals, therefore, need to be well-habituated to wearing the belt, otherwise emotional responses can be elicited by the device itself, compromising the validity of emotional measures. Remote techniques for measuring cardiac activity do exist, including those based on video footage (e.g. through measuring subtle head motion associated with movement of blood from the heart to the head). However, the accuracy of these is affected by animal movement, and they are currently scarcely used in livestock (for review on cardiac monitoring approaches, see Nie et al. 2020) .

2.1.1.2 Thermal Imaging

Thermal imaging is a remote technology increasingly used in animal welfare research to measure the physiological component of emotional responses. Thermal imaging technology captures infrared radiation emitted from objects to generate a thermal profile, enabling users to visualise and measure subtle variations in peripheral temperature (Knízková et al. 2007). When the subject of interest is a living mammal, local variations in blood flow and metabolism create a complex, somewhat fluid topography of temperature zones occurring across the skin's surface

(Godyń et al. 2019; e.g. Fig. 2). Through focusing on certain 'regions of interest' (ROIs), typically relatively furless areas, such as the eyes and nose, researchers can track changes in skin temperature over time (Tattersall 2016). Importantly, changes in skin temperature are known to be influenced by emotional processes. When stressed, sympathetically-mediated vasoconstriction causes blood (and body heat) to be diverted away from peripheral blood vessels to the core, limiting blood loss from potential injuries and infusing vital organs and tissues important in the pending behavioural response (Travain and Valsecchi 2021).

In goats, current investigations of emotional responses using thermal imaging are limited to negative situations. In both adult and juvenile meat goats, maximum eye temperature, in conjunction with respiration rate, has been shown to increase following a stressful event, i.e. an experimenter guiding animals loudly with exaggerated arm movements to a nearby pen (Bartolomé et al. 2019). As eye temperatures tend to bear a stronger relationship with core temperature relative to other regions (George et al. 2014; de Ruediger et al. 2018; Bleul et al. 2021; Kim and Cho 2021), including in a preliminary study in goats (Marques et al. 2021), we may expect eye and core temperature to increase in tandem with decreases in temperature of other regions, such as the muzzle (e.g. Proctor and Carder 2016). Indeed, given this association with core temperature, as well as its proximity to emotional centres of the brain and rich blood supply, the eyes, and the region around the posterior border of the eyelid and lacrimal caruncle especially, have been proposed as sensitive regions to measure thermal responses to emotional events (Kim and Cho 2021; Travain and Valsecchi 2021; Fig. 2).

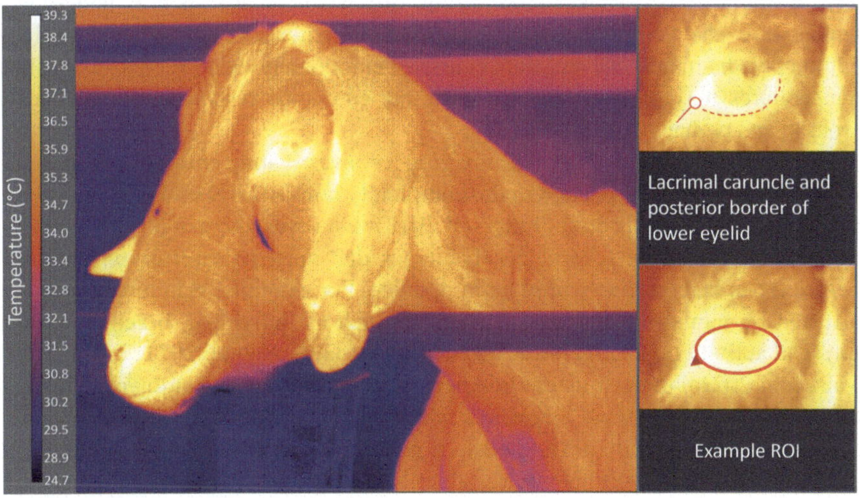

Fig. 2 Thermal image of a goat. Magnified version of key region of interest (ROI), the eye is shown in the panel to the right. Lacrimal caruncle is circled and the posterior border of the eyelid is indicated by a dashed line. An example ROI incorporating both of these sensitive areas is shown below (image credit: Marianne Mason)

Temperature changes in positive contexts have yet to be investigated in goats, but in cattle (*Bos taurus*), positive emotions of both low and high arousal can cause a drop in nasal temperatures (Proctor and Carder 2015, 2016). By contrast, nasal and withers temperatures increased in sheep (*Ovis aries*) following brushing, a low arousal, positive experience (Tamioso et al. 2017), while dogs (*Canis familiaris*) showed increases in eye temperature during feeding (high arousal, positive event: Travain et al. 2016). Confusingly, direction of temperature changes within a single ROI can be consistent between both positive and negative contexts (e.g. in cows: Proctor and Carder 2015, 2016), but go in opposite directions in situations of the same emotional valence (eye temperature: in goats, Bartolomé et al. 2019; Zebaria et al. 2021; sheep, Stubsjøen et al. 2009; Cannas et al. 2018; cows, Stewart et al. 2008a, b; Gómez et al. 2018).

Although the direction of emotionally mediated temperature changes is not always uniform and so can be hard to interpret (e.g. Bartolomé et al. 2019; Zebaria et al. 2021), to understand these changes in the context of the dimensional framework, we must consider the intensity of the response and its time course. More arousing emotional events may provoke larger changes in peripheral temperatures and prolonged persistence of temperature changes in the initial (acute) phase of emotional responses, and for very stressful experiences, initial decreases may be accompanied by a delayed temperature increase, exceeding baseline temperatures (chronic phase; Stewart et al. 2007, 2008a; Herborn et al. 2015). Onset of temperature changes has also been shown to vary with valence in cows, with a greater lag in response observed when responding to positive compared with negative stimuli, despite these contexts being of similar arousal (Proctor and Carder 2016). Since negative emotions are implicated in 'fight-or-flight' responses to threatening stimuli, we might expect quicker mobilisation of physiological responses underlying temperature changes to negative over positive contexts (Chotard et al. 2018). Ultimately, the magnitude and time course of emotionally induced changes in skin temperatures may offer a means to examine emotional responses in relation to arousal, and perhaps even valence.

The main advantage of thermal imaging for investigating emotional responses in goats and other animals lies in the completely non-invasive and remote way in which data can be collected (Tattersall 2016). However, the use of this technology also suffers from several disadvantages that must be carefully considered before it can be effectively applied in practice. For example, skin temperature is affected by numerous endogenous factors, such as individual, breed, sex, and physical activity (also affects cardiac measures; Jansson et al. 2021; Travain and Valsecchi 2021), as well as environmental variables, such as ambient temperature, humidity, and wind intensity, which influence peripheral temperatures directly and through internal homeostatic processes (Jansson et al. 2021; Marques et al. 2021). Further extraneous variables that affect the accuracy of peripheral temperature readings include distance or angle of the subject to the camera (Tattersall 2016; Ijichi et al. 2020), the ROI chosen, its size and the specific measure used (i.e. maximum, minimum, or average temperature; Uddin et al. 2020; Kim and Cho 2021).

2.1.2 Behavioural Indicators

The behavioural component can be considered the final stage and external manifestation of an emotional response (Anderson and Adolphs 2014). Given it is through behaviour that animals directly interact with their environment, it is this component upon which selection most directly operates (Kremer et al. 2020). In an applied setting, behaviour can provide a directly observable indication of an animal's current emotional state (Machado and da Silva 2020).

2.1.2.1 Body Posture, Behaviour, and Qualitative Behaviour Assessment

The level of general activity has been suggested as an indicator of emotional arousal, with goats in positive situations of relatively high arousal tending to move around more and show increased head movements, relative to those in less arousing, negative and neutral contexts (Baciadonna et al. 2020). However, to understand the link between activity level and arousal, we must also consider the specific situation and the goat's motivation. For example, although complete social isolation appears to be more stressful than partial isolation (where they could still hear and smell other goats), greater activity has been observed in partially isolated goats, as they attempted to restore contact with their separated group mates (Siebert et al. 2011). Directional responses such as movement towards or away from an object, differentiation in looking behaviours, and levels of physical interaction can be used to judge the level of emotional arousal elicited by a stimulus (e.g. Nawroth et al. 2018; Baciadonna et al. 2019). Particular postures in goats have also been linked to different emotions, with tails expected to be raised more in positive, relative to negative situations (Briefer et al. 2015b). Other valence indicators may include play, which reflects positive emotions (mostly in juveniles: Ahloy-Dallaire et al. 2018), and displacement behaviours, which are assumed to indicate anxiety. Displacement behaviours refer to actions performed that seem irrelevant to an animal's current behavioural context (Breed and Moore 2016) and goats have been shown to engage in displacement activities such as scratching and self-grooming following conflict. Interestingly, these behaviours decrease in receivers of aggression following reconciliation with their aggressor (Schino 1998). The behaviours and postures discussed in this section so far can be employed quantitatively to investigate emotions in goats through measuring, for example, the frequency and/or duration over which they occur. However, focus on individual indices can prove problematic, with some behaviours only happening infrequently (e.g. play) and although certain behaviours are relatively simple to interpret in terms of emotional valence (e.g. affiliative behaviours), others are more suggestive of arousal (e.g. activity level; Cooper and Wemelsfelder 2020).

As an alternative to quantitative approaches, qualitative techniques do exist and require observation of an animal as a 'whole' rather than focussing on specific aspects of its behaviour. Qualitative behaviour assessment (QBA) requires assessors to observe and form impressions of animal 'body language' according to a set of predefined descriptors (Battini et al. 2016). These descriptors define how goats or groups of goats interact with their environment (Cooper and Wemelsfelder 2020). Although there appears to be greater coupling of negative emotions with states of higher arousal, as descriptive categories vary in arousal for both positive and negative emotions, QBA can be contextualised using dimensional models (Mattiello et al. 2019; Fig. 1). For example, in terms of negative emotions, the AWIN welfare assessment protocol (part of the European Animal Welfare Indicators Project) employs words such as 'fearful' and 'suffering' to describe goats experiencing negative, high arousal states, and 'bored' for those in low arousal, negative states (AWIN 2015). Conversely, in terms of positive emotions, descriptors such as 'curious' and 'relaxed' are used to denote higher and lower arousal states, respectively. Descriptor use has been shown to vary predictably with husbandry and body condition, with Grosso et al. (2016) finding that goats with pasture access tended to be described as 'content/calm' and 'curious/attentive', whereas negative descriptors such as 'irritated' or 'suffering' were used more often for indoor-housed goats. Similarly, Battini et al. (2018) found that goats with normal coats were scored as being more relaxed and sociable than goats with coats in poorer condition. However, such methods are not without shortcomings, and indeed, the inherent subjectivity of QBA can bring into question its reliability for measuring emotion in domesticated species (e.g. Bokkers et al. 2012; Tuyttens et al. 2014).

When assessors observe animals, they will be unavoidably aware of the conditions in which animals are kept, and therefore may, to some extent, assign descriptors based on their own presumptions of how animals should be behaving (Spigarelli et al. 2020). Accordingly, Tuyttens et al. (2014) found that observers tended to favour descriptors linked to more positive emotions than negative ones, when told hens (*Gallus gallus domesticus*) were from an organic rather than a conventional farm, even though they were scoring modified versions of the same video. Strength of biases varied based on observer opinions regarding the relative welfare benefits for hens on organic farms, and additional biases can be introduced, for example, through differences in professional background and experience (e.g. Grosso et al. 2016) and perhaps even gender (Tallet et al. 2010). However, it seems that such observer biases can be partly overcome with training. Grosso et al. (2016) found that trained observers had a good level of broad agreement on goat emotional state (i.e. positive or negative), level of arousal and sociability. Finer-grained agreement on specific terms was only possible for some descriptors; however, given the ubiquity of dimensional approaches in modelling animal emotions, this additional refinement may be unnecessary. Furthermore, QBA scores in other livestock species have been shown to vary predictably with other behavioural measures (including in goats: e.g. Miller et al. 2018) and physiological markers of stress (for review see, Fleming et al. 2016). Ultimately, to overcome the subjectivity of qualitative methods, it may be best to employ QBA in tandem with quantitative methods, with QBA in turn complementing and contextualising quantitative approaches at the 'whole animal' level. It must be noted, however, that the majority of QBA research, including the principal AWIN protocol, has been designed around European dairy goats (AWIN 2015), with no similar developments for meat or fibre goats (Leite et al. 2017).

2.1.2.2 Facial Expressions

Humans convey their emotions using a rich array of facial expressions. However, competing demands on facial musculature to facilitate communication and emotional expression may diminish the link between facial expressions and underlying state (Fridlund 1991; Schmidt and Cohn 2001). Outside of humans and other primates, conscious control and complexity of facial musculature is relatively limited, potentially rendering facial expression a more 'honest' indicator of an animal's internal state (Descovich et al. 2017).

Goats adjust their behavioural responses towards photographs of other goat's faces taken in emotional contexts of opposite valence (Bellegarde et al. 2017), suggesting that emotional experiences create perceivable shifts in their facial expressions. However, what these shifts are, in terms of changes in various facial muscles have not been studied in detail, with the exception of ear position, for which movement is governed by facial muscles (Descovich et al. 2017). In goats, a greater proportion of time spent with their ears pointing forward, relative to the side, is indicative of heightened arousal, whereas time spent with the ears pointed backwards acts as a valence indicator, being more expected in negatively valenced situations (Briefer et al. 2015b; Bellegarde et al. 2017; Hussein and Hidayet 2019). Additionally, goats in a positive, low arousal emotional state induced through human grooming tend to exhibit ears positioned in a lowered, turned down position (Bellegarde et al. 2017). However, facial expressions are composite emotional responses requiring coordinated action of multiple facial muscles. This focus on a single component may not provide a comprehensive view of how goats may be expressing emotions facially. Techniques such as Facial Action Coding Systems (FACS) which classify facial expressions in relation to the shape and action of the underlying musculature may provide a promising means to analyse facial expressions as a whole (e.g. Wathan et al. 2015), but this method has yet to be applied in goats. Intriguingly, as animals produce distinctive facial expressions when vocalising, vocal analysis may prove complementary to the measurement of facial expressions and is an area, which has been studied to a greater extent in goats (Briefer et al. 2015b; Neethirajan et al. 2021).

2.1.2.3 Vocal Expressions

Goats are highly vocal animals, with calls encoding a variety of information including an individual's identity, motivation, age, sex, body size, social environment, and moreover, emotional state (for example, goat calls, see Figs. 3, 4, 5, 6, 7, 8, and 9; Briefer and McElligott 2011a, b, 2012; Briefer et al. 2015b). Similar to facial expressions, diminished conscious control over a vocalisation's structural elements (e.g. frequency) observed in farm animals, compared to humans, results in vocalisations being considered a more 'honest' manifestation of their internal state (Jürgens 2009; Briefer 2012, 2020). Indeed, vocal communication enables state information to be transmitted over longer distances, experiencing less attenuation from intervening objects than visual cues, and shifts in state can be conveyed more rapidly than through chemical communication (Johnstone and Scherer 2000; Laurijs et al. 2021). However, not all emotions elicit vocal responses (e.g. low arousal, negatively valenced depressive states) and like facial expressions, as vocalisations are often inherently communicative in nature, they are rooted in social context (Briefer 2020).

In goats, different calls are favoured in different social situations; for example, bucks produce gobbling vocalisations during courtship, does emit low frequency, affiliative 'murmuring' vocalisation shortly before and after parturition, and alarm snorts are used to alert herd members to predators (Figs. 3 and 4; Shank 1972;

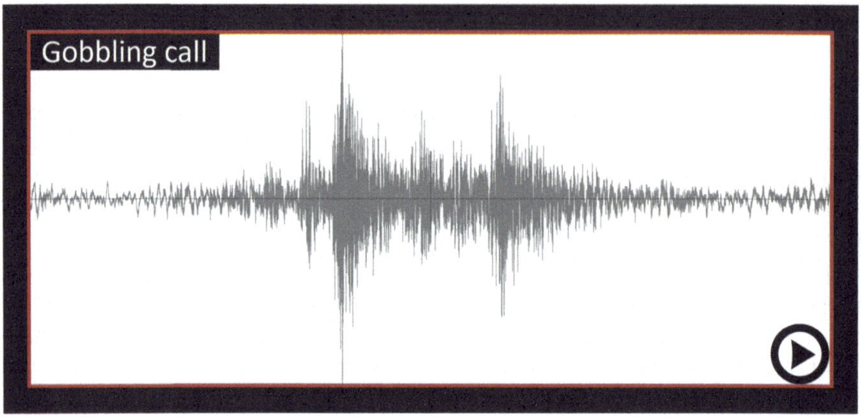

Fig. 3 Playback of goat gobbling call. This vocalisation is produced by bucks during courtship. Recordings collected by the VOCAPRA project (▶ https://doi.org/10.1007/000-c4j)

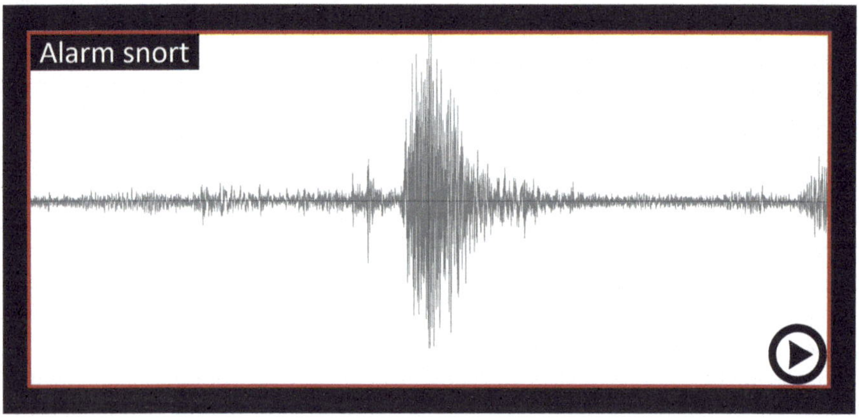

Fig. 4 Playback of alarm snort which goats produce to alert group members to potential threats, like predators. Recordings collected by the VOCAPRA project (▶ https://doi.org/10.1007/000-c4f)

Miranda-de la Lama and Mattiello 2010; Longpre et al. 2011; Weaver 2021). Social context in gregarious species such as goats often induces predictable shifts in valence and arousal making call-type a useful indicator of animal emotional state, albeit one requiring knowledge of a species-specific communication system (Briefer 2012, 2020). Nevertheless, vocalisations produced in multiple emotional situations are considered particularly useful for assessing animal emotional state as they enable better comparison of acoustic features across contexts (Briefer 2012, 2020). The principal form of vocalisation emitted by goats are contact calls, comprising closed- and open-mouthed bleats, which are generally thought to be produced to maintain social cohesion over shorter to longer distances, respectively (Briefer and McElligott 2011a; O'Bryan et al. 2019). Furthermore, one investigation found that closed-mouthed bleats were more likely to be produced by socially isolated goats

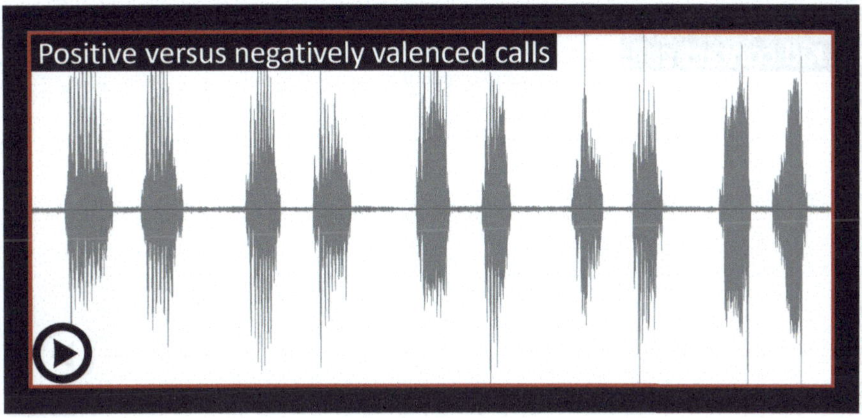

Fig. 5 Pairs of calls, one collected from a goat anticipating a preferred food item, and another from the same goat taken while it was socially isolated, repeated for five different individuals (Briefer et al. 2015b) (▶ https://doi.org/10.1007/000-c4g)

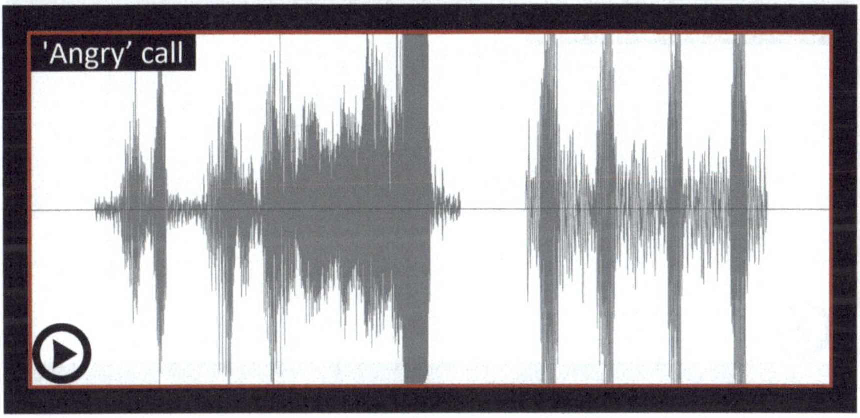

Fig. 6 Playbacks of closed-mouthed calls produced by goats after being displaced by a group member. Recordings collected by Elodie Briefer (unpublished data) (▶ https://doi.org/10.1007/000-c4h)

(and especially those isolated for longer periods) than those that could still hear and smell conspecifics (Siebert et al. 2011). Accordingly, this type of call has been speculated to have an additional role in 'auto-communication', produced to self-regulate an individual's emotional state. Moreover, both open- and closed-mouthed bleats vary predictably with emotional arousal and valence (Briefer et al. 2015b).

The source-filter theory posits that mammal vocalisations are initiated via vibrations of the vocal folds (the source, which determines fundamental frequency or pitch, F0), before being filtered through the vocal tract (Taylor and Reby 2010). This filtering process modulates the vocalisation's energy distribution, dampening certain frequencies and amplifying others, giving rise to spectral peaks, or

Fig. 7 Playbacks of a week-old kid responding to a playback of their mother's contact call (Briefer and McElligott 2011a) (▶ https://doi.org/10.1007/000-c4e)

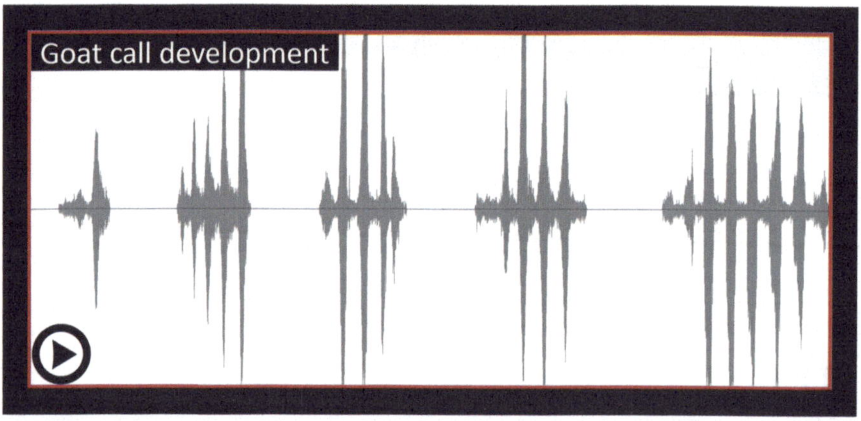

Fig. 8 The recordings illustrate how goat contact calls change over the developmental period. Calls were produced by the same kid, over different time points (1, 5, 9, 35, and 104 days old; Briefer and McElligott 2011b) (▶ https://doi.org/10.1007/000-c4k)

'formants'. In goats, like in other taxa, both source- and filter-related components of vocalisations have been shown to vary somewhat predictably with emotional arousal (Briefer 2012, 2020). Briefer et al. (2015b) found that goats in highly arousing situations called more, and the calls they produced had a higher overall frequency, as well as a higher F0 (Fig. 5). Furthermore, goats in positive situations produced calls with a lower and more stable F0 relative to negative situations. Thus, as well as which call-type is favoured (e.g. gobbling, snorting; Briefer 2012, 2020), some characteristics of F0 may act as a useful valence indicator (Briefer et al. 2015b).

Given consistent correlates of emotions present in goat vocalisations (Briefer et al. 2015b), there is obvious scope for automated bioacoustic techniques for monitoring their welfare, as have already been investigated in pigs (*Sus scrofa domesticus*), cows and chickens (*Gallus gallus domesticus*; for review, see Mcloughlin et al.

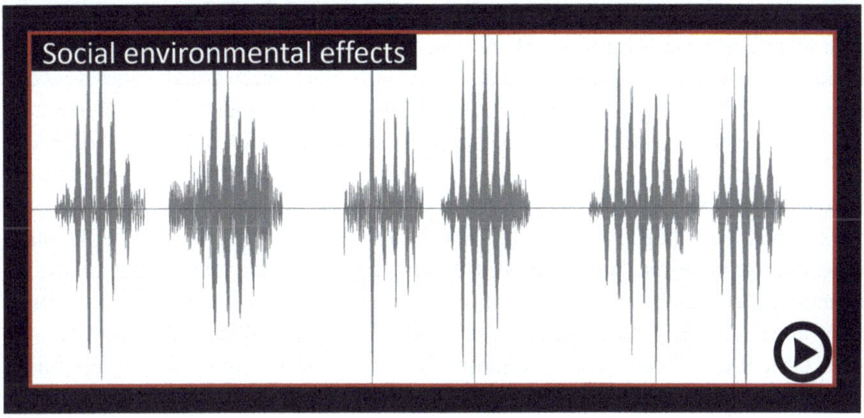

Fig. 9 These recordings illustrate the influence of the social environment experienced during development on kid contact calls. Pairs of recordings of goat kids from the same group (for three groups in total; Briefer and McElligott 2012) (▶ https://doi.org/10.1007/000-c4m)

2019). Indeed, vocalisations provide a long-range, instantaneous signal of animal emotional state, which in the context of precision livestock farming can provide a highly informative, remote means to assess welfare in real-time (Briefer 2020). A single microphone mounted in a goat's indoor environment would be sufficient to determine emotional state at a group level (e.g. Herborn et al. 2020). To achieve a finer-grained analysis, the use of specialised microphones, or several microphones in conjunction with sound triangulation techniques (Matthews et al. 2016; Laurijs et al. 2021) or even wearable audio-recorders fitted to collars would be necessary (O'Bryan et al. 2019).

Until now we have focused on methods to measure short-term emotional experiences in goats provoked by specific events. In the next section, we discuss how these methods link to longer-term emotional states and how they can be measured.

2.1.3 Cognitive Indicators

Although emotions can be considered as short-lived shifts in affective state prompted by a specific stimulus or event, they can have cumulative and lasting impacts on an animal's overall 'mood' (Russell 2003; Kremer et al. 2020). This, in turn, affects cognitive processes, influencing how goats and other animals attend to, learn, memorise, and appraise stimuli in general (Baciadonna and McElligott 2015). Indeed, animal affective states can cause attention biases, meaning individuals experiencing a negative emotional state, like anxiety, can become more attentive towards negative or novel stimuli (for review see, Crump et al. 2018). Similarly, long-term emotional states can induce memory biases, with stimuli or events in line with the animal's state being easier to recall; for example, depression in humans is linked to selective retention of negative memories (e.g. Dillon and Pizzagalli 2018). However, given the greater focus and support for the use of judgement bias tests compared to attention or memory biases in non-human animals (Lagisz et al. 2020), we will concentrate on this test for the current section.

To investigate the presence of judgement biases, animals are first trained, for example, to approach a rewarded cue, such as a particular location or colour container baited with food, and to suppress approach to an unrewarded or aversive cue (e.g. unpalatable food; go/no-go tasks; e.g. Bethell and Koyama 2015; Bučková et al. 2019). Alternatively, animals may be trained to associate different locations or colour cues with different value rewards (e.g. more or less food; preferred or non-preferred food). In these active choice (or go/go tasks) animals are expected to respond in some way to both cues (e.g. Brilot et al. 2010; Murphy et al. 2013). Once a set learning criterion has been reached, animals are then presented with intermediate cues, for example, a container at a new location positioned in between the two previously reinforced ones. The expectation is that animal decision-making will be in line with their current emotional state, with individuals in a positive state behaving optimistically, so will predict positive outcomes when appraising ambiguous cues, compared to those in negative states which are expected to behave pessimistically (Harding et al. 2004; Baciadonna and McElligott 2015). Changes in judgement biases have been shown in relation to, for example, housing and enrichment in hamsters (*Mesocricetus auratus*) and horses (*Equus caballus*; Bethell and Koyama 2015; Henry et al. 2017), pair-housing in cows (Bučková et al. 2019), and dominance rank and proactive-reactive personality traits in pigs (Asher et al. 2016; Horback and Parsons 2019).

In goats trained in a go/no-go task to approach a rewarded, but not an unrewarded location, contrary to predictions, short-term positive handling and prior poor welfare did not induce an optimistic or pessimistic bias, respectively, when exploring ambiguous locations (Briefer and McElligott 2013; Baciadonna et al. 2016). Despite evidence that human grooming was positively perceived, given goats at the study site received regular positive contact with humans and were generally well-cared for, short-term handling appeared to be insufficient for producing measurable shifts in judgement biases (Baciadonna et al. 2016). Indeed, another investigation carried out at the same site suggested that female goats subject to past mistreatment tended to inspect ambiguous locations faster, indicating that these individuals were more optimistic than control females, while previously mistreated males expressed similar levels of optimism to control males (Briefer and McElligott 2013). Judgement bias tasks and especially those involving training, such as the go/no-go task used for the above experiments, are not without criticism (Lagisz et al. 2020). Repeated presentations of ambiguous stimuli over multiple trials can cause a reduction in response towards these unrewarded locations based on learning, rather than a pessimistic tendency (Perdue 2017). Ultimately, however, although further refinement of the methodology may be warranted, the use of judgement bias tasks has so far provided a promising means to assess the positive impacts of good welfare on goat rehabilitation and long-term emotional state.

2.2 The Social Dimension of Emotion

For a social species, emotions do not happen in a vacuum, and as such, goats are well positioned to freely detect and interpret the emotional cues of group members

(Baciadonna et al. 2018). Indeed, irrespective of whether emotions are intentionally communicated, such expressions can be highly informative, enabling receivers to anticipate a signaller's changing motivations (Schmidt and Cohn 2001; Seyfarth and Cheney 2003). From here, receivers can use another's emotions referentially, to guide social interactions with that individual, remain informed of external stimuli or events (e.g. predators) and to respond more appropriately towards novel objects or social partners (e.g. Seyfarth and Cheney 2003; Merola et al. 2014, 2015; Proops et al. 2018).

Emotions can be communicated in a variety of sensory modalities (e.g. visual, auditory, olfactory, and gustatory), and sensitivity to emotional cues expressed by other goats has been demonstrated across two such channels. When presented with facial photographs of another goat taken under situations of opposing valence, goats spent more time with their ears pointed forward in response to conspecifics conveying negative (in discomfort) relative to positive emotions (being groomed), which was interpreted as them expressing greater interest in the former (Bellegarde et al. 2017). This difference suggests not only that goats register differences in valence of conspecific facial expressions, but that negatively valenced facial expressions may have greater perceived relevance, as they could indicate potential threats, so have clearer implications for survival and fitness. Similarly, having habituated to contact calls conveying one valence (e.g. negative), goats renewed behavioural responses (or dishabituated) when calls expressing the opposite valence were presented (e.g. positive; Baciadonna et al. 2019). This suggests that they can perceive differences in emotional content in the vocalisations of other goats. In addition, shifts in valence seemed to be accompanied by changes at physiological levels, with HRV being higher when experiencing positive calls in the first and last of three playback phases. These increases in HRV suggest greater parasympathetic influences on cardiac activity (implicated in rest and digest states), which could indicate that positive calls were less emotionally arousing to goats than negative ones. However, these changes occurred in the absence of shifts in HR, a (probably) more sensitive measure of emotional arousal (Baciadonna et al. 2020). In sum, it appears goats can discriminate between, and show perceptual biases towards, conspecific facial and vocal emotional cues based on valence.

For animals living closely with another species, as goats do with humans, emotional cues from members of another species may be almost as relevant as those of conspecifics. Although throughout their domestic history, goats generally have not possessed a close working relationship with humans in the same way as horses (but see pack goats, Sutliff 2019), nor are household companions like domestic cats (*Felis catus*) and dogs, they still rely on us for food, shelter, and protection from predators (Jardat and Lansade 2022). Given this shared history with us, goats might be expected to develop a suite of social cognitive abilities to discriminate between and interpret our cues as well as their own (e.g. Hare et al. 2005). As the outcome of interactions with humans and their predictability can have substantial consequences for animal welfare and productivity (Zulkifli 2013; Rault et al. 2020), human emotional cues may be salient for goats, which might use them to anticipate our behaviour. Indeed, goats discriminate human facial expressions of opposing valence and display inherent preferences for happy over angry facial expressions (Nawroth et al. 2018; also see, Mason et al. 2024). Given the likelihood of happy facial expressions being paired with positive human–goat interactions and angry with negative, such preferences towards human emotional cues are perhaps to be expected.

Ultimately, the ability to interpret the emotions of others, most likely from other goats, but perhaps even from humans as has been seen in dogs (e.g. Katayama et al. 2019), provides an avenue for emotions to be transmitted between individuals. This phenomenon is known as emotional contagion (yet to be clearly demonstrated in goats; Briefer 2018; Düpjan et al. 2020; Pérez-Manrique and Gomila 2022). Discriminating social cues, including emotional ones, can be defined under the umbrella of social cognition.

3 Cognition

Cognition can be defined as the process through which animals collect, process, retain, and respond to environmental information (e.g. perception, memory, learning, and decision-making: Shettleworth 2010). Cognition is often split into two broad domains, those concerning an animal's mental adaptations to their physical environment, i.e., physical cognition, and to their social one, i.e. social cognition.

3.1 Physical Cognition

Physical cognition concerns an animal's understanding of the physical properties of objects found in their environment and of an object's causal and spatial relationships with other objects and/or events (Auersperg et al. 2017; Nawroth et al. 2019). The simplest means by which animals can establish these causal links is through associative learning; for example, if an animal repeatedly experiences *a*-stimulus with *b*-event, it will learn *a* predicts *b* (Auersperg et al. 2017). Now say *c* and *d* share properties with *a* and also predict *b*, animals may be expected to generalise their responses to *a* to *c*, *d* and other similar objects adding a further level of sophistication. However, to establish existence of 'complex' cognitive abilities in goats and other species, we must move beyond these learnt contingencies and generalisations to evaluating their understanding of more abstract causal relationships, which apply across a greater range of situations (Auersperg et al. 2017). Therefore, in addition to examining object discrimination and categorisation in goats (generalisation), we will discuss evidence for 'higher order' reasoning, specifically inferential reasoning and object permanence, as well as long-term memory and behavioural flexibility (important for coping with environmental change: Coppens et al. 2010). Finally, we will discuss contrafreeloading, in which cognitive challenges can in themselves be inherently rewarding (Inglis et al. 1997).

3.1.1 Object Discrimination and Categorisation

Discrimination refers to the ability to differentiate between objects, while categorisation is the capacity to perceptually group objects based on their shared properties (Shettleworth 2010). Domestic goats, when given the opportunity, tend to choose a diet rich in browse (Landau and Provenza 2020). High browse consumption means exposure to plant secondary compounds such as tannins, which can reduce food intake and cause liver damage. Being able to discriminate between and categorise plant species is therefore a crucial skill for goats, enabling them to avoid plants containing higher concentrations of these compounds (e.g. Provenza et al. 1994), and/or consume a more optimal mix of plant species to enhance feed intake (Estell 2010). In an experimental setting, goat expertise in grouping foraging resources extends to categorising artificial symbols (e.g. Meyer et al. 2012).

Goats can be readily trained to discriminate between artificial symbols presented using an automated learning device. Use of this paradigm has shown that goats can generalise responses between symbols, forming open-ended categories based on shared visual properties (Meyer et al. 2012), and even transfer knowledge of the task itself to learn new problems, in other words 'learning to learn' (or developing a 'learning set'). Specifically, goats trained to select one out of four symbols showed increases in learning performance after being presented with further sets of symbols (Langbein et al. 2007). Similar investigations have shown that some goats can learn the 'oddity concept', selecting the odd symbol out of three other uniform symbols (Roitberg and Franz 2004). The ability of goats to discriminate between and categorise objects may partially explain their impressive ability to adapt and colonise novel environments (Chynoweth et al. 2013).

3.1.2 Inferential Reasoning

Inferential reasoning refers to the capacity to integrate indirect information regarding an event or stimulus to form inferences, enabling animals to better cope with novel challenges (Heimbauer et al. 2019). Solutions must be reached without explicit learning and are assumed to be particularly advantageous in changing physical or social environments, as long as changes are predictable (Völter and Call 2017; Nawroth et al. 2019). For example, if we have three goats competing for a resource, and *Goat A* loses to *Goat B*, but observes *Goat C* beat *B*, *A* may infer that should they challenge *C*, they would also be likely to lose. Nawroth et al. (2014) demonstrated that both goats and sheep could select a rewarded over an unrewarded cup when given direct information (sight of food), but most goats (8/11), and no sheep were able to infer the presence of food from indirect information (absence of food under the unrewarded cup). Support for inferential reasoning in goats, but not sheep, can be interpreted as reflecting differences in feeding ecology between these two related species; goats, as browsers, are more selective in what they eat, compared to sheep, which are grazers. As a result, goats but not sheep flexibly learn, for example, to avoid high fibre foods when preferred foods of lower fibre become available (Hosoi et al. 1995). For the wild ancestors of domestic goats inhabiting mountainous terrain, and contemporary free-ranging domestic and feral populations in harsh environments, foraging resources are often patchily distributed and can require substantial effort to navigate between (e.g. Negi et al. 1993; Shrader et al. 2012). Therefore, reaching informed decisions based on both direct and indirect information may be extremely valuable to locate and prioritise foraging patches (Nawroth et al. 2014).

3.1.3 Object Permanence

Object permanence refers to the understanding that an object continues to exist even when it is no longer visible to the observer (Piaget 1954). Such a capacity enables animals to predict the location and trajectory of a hidden physical or social object and can have key implications for an animal's fitness (Jaakkola 2014). For example, it can help animals to keep track of the relative locations of group members, allowing them to maintain group cohesion in dense vegetation or even follow the movements of a hidden predator (Noë and Laporte 2014). Goats have been shown to express a relatively sophisticated ability for object permanence, being capable of tracking the movement of a reward hidden under an opaque cup as it switched positions with an adjacent cup (Nawroth et al. 2015a). Especially when colour and shape of the cups differed, some goats were able to select the rewarded cup with a moderate to high level of success (3/9 goats performance exceeded chance levels). The ability to infer the location of out-of-view objects can substantially increase how predictable a goat perceives its environment to be.

3.1.4 Long-Term Memory

Goats have demonstrated a high propensity to discriminate and learn in physical cognitive challenges (e.g. Langbein et al. 2007), and to retain the solutions of these tasks in the long term. For example, goats have an excellent memory for spatial and visual cues, remembering the solutions to a Y-maze 3 months after it had been learned (Langbein 2012). Similarly, Langbein et al. (2008) trained goats to learn a rewarded symbol for each of ten different sets of four symbols presented consecutively. When presented with these tasks concurrently, goats were able to solve the task immediately in at least half of cases, even tasks learnt several weeks before, and others were relearned at a faster rate than it took originally. However, one of the best examples of long-term memory in goats has been demonstrated by their performance in a two-step foraging task (Briefer et al. 2014). Goats were first trained to

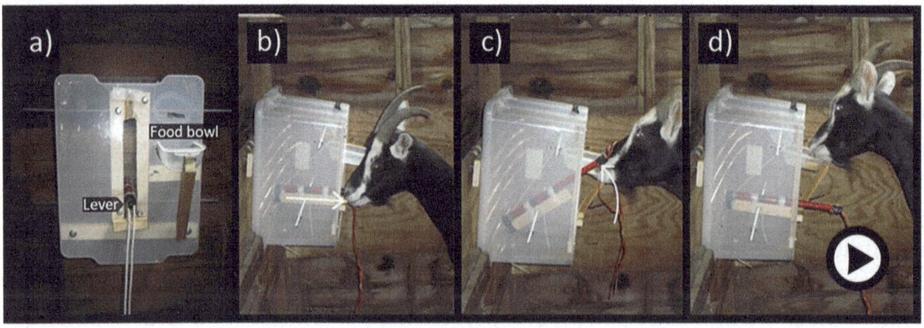

Fig. 10 Video of a goat solving the two-step foraging task (taken with permission from Briefer et al. 2014 supplementary materials). Goats had to first pull the lever from the box, before lifting it upwards. Goats were then able to retrieve the food reward from the bowl (▶ https://doi.org/10.1007/000-c4n)

extract food from a box through pulling out a lever and lifting it upwards (for video of a goat solving the task, see Fig. 10). Not only were nine out of 12 goats able to learn this highly novel foraging task in a relatively small number of trials (8–22), but they also retained the knowledge in the long term, exhibiting no decrease in performance even 10 months after the task had been learnt (and anecdotally over 7 years).

3.1.5 Behavioural Flexibility

Behavioural flexibility, i.e., the degree to which an animal's behaviour can be shaped by external parameters, is a key determinant in their ability to adapt and survive in changing environments (Coppens et al. 2010). We have already described goat proficiency for flexible object categorisation, learning, and decision-making (e.g. Langbein et al. 2007; Meyer et al. 2012; Nawroth et al. 2014). However, behavioural flexibility is more commonly inferred from how well animals can inhibit or control a behaviour of interest or through their performance in reversal learning tasks (Nawroth et al. 2016a; Raoult et al. 2021), both of which have been explored in a variety of contexts in goats.

Inhibitory control of behaviour can be defined as an animal's capacity to suppress a more direct, impulsive response in favour of an action more appropriate for a given situation (Diamond 2013). For example, after training goats to obtain a reward from an opaque cylinder, Langbein (2018) substituted the cylinder for a transparent one, requiring animals to avoid direct attempts to retrieve the reward through the cylinder, and instead detour around its side. Goats were shown to perform this feat at a comparable level or better than a variety of bird and mammal species (average accuracy = 63%). However, despite the wide use of this task, given that an individual's performance can be affected by their previous experience with, or understanding of artificial, transparent barriers, its ecological validity is questionable (van Horik et al. 2018). In an arguably more ecologically valid task, goats (but not all sheep) were shown to immediately detour around a physical barrier to retrieve food, despite momentarily increasing the distance between themselves and the reward (Nawroth et al. 2016a; Raoult et al. 2021).

In addition to testing detour performance, Raoult et al. (2021) investigated reversal learning, by moving the gap in the barrier that animals could use to reach the reward to the opposite side following training. Reversal learning paradigms first reinforce a particular behaviour, such as selecting a specific rewarded location, before the reinforced behaviour is switched with a previously unrewarded pattern of response. More flexible goats are expected to be quicker in adjusting behaviours in line with task parameters, and, therefore, show less persistence in performing the previously reinforced behaviour, otherwise known as perseveration, or A-not-B errors. Indeed, when the solution to the detour task was changed, Raoult et al. (2021) found that goats discovered the new gap in the barrier more quickly than sheep and were, therefore, faster to reach the target (also see, Nawroth et al. 2016a). Given the relatively close phylogenetic relatedness, as well as similarities in social structure and domestic history between goats and sheep, this difference in behavioural flexibility was again thought to be due to differences in their feeding ecology.

Goats as browsers, unlike sheep, exploit more patchily distributed, ephemeral food resources and thus probably benefit more from expressing a flexible foraging strategy. Accordingly, when presented with a T-maze in which both arms contained a food reward, goats and sheep tended to repeatedly return to the same arm over multiple trials (Hosoi et al. 1995). However, when the quality of the food patch in one of the arms was lowered, goats, but not sheep, responded by increasing the frequency of switching between the food patches. Intriguingly, differences in behavioural flexibility have also been found between goat breeds, in that although dairy and dwarf goats performed similarly in an initial visual discrimination task, dairy goats were slower to learn in a subsequent reversal task (Nawroth et al. 2022). This was interpreted as reflecting a trade-off between behavioural flexibility and production traits in dairy goats, and may mean they are slower to adapt to new environments and/or husbandry conditions.

3.1.6 Contrafreeloading

Goats appear to be motivated to use their impressive suite of physical cognitive abilities to solve tasks (Briefer et al. 2014; Zobel and Nawroth 2020). Contrafreeloading refers to the phenomenon where animals continue to work for a resource when an identical one is freely available (Jensen 1963; Inglis et al. 1997). For example, goats trained in a visual discrimination task to receive water, continued to direct a third of their interactions towards the learning device (and as many as 90%), despite having free access to a water dispenser (Langbein et al. 2009). Similarly, goats trained to slide open a door to access food, continued to do so even when an identical food resource was provided through a nearby open door (Rosenberger et al. 2020). Goat HR has been shown to decrease with increasing competence in a discrimination task; however, when the challenge was removed (all choices rewarded), the subjects HR increased once again (Langbein et al. 2004). Engagement in contrafreeloading coupled with shifts in response at autonomic levels, suggests that mastery of cognitive tasks induces positive emotions in goats, which extend beyond acquisition of the reward alone (Langbein et al. 2009). Indeed, cognitive tasks have been suggested to provide goats with an opportunity to manipulate their environment, with success increasing their perceived control over their surroundings, which, in turn, motivates them to continue engaging with the task (White 1959; Rosenberger et al. 2020). Ultimately, the indicators of advanced cognition and contrafreeloading in goats highlight the importance of cognitive challenges for their welfare.

3.2 Social Cognition

Goats navigate a complex social world featuring relatively stable, near-linear dominance hierarchies (Barroso et al. 2000; Saunders et al. 2005; Aschwanden et al. 2008), and aggregations of individuals that form based on sex, social bonds, and kinship, but fluctuate in composition hourly as groups fuse, and/or separate (fission; Calhim et al. 2006; Stanley and Dunbar 2013). Although goats incur the costs of group-living through increased aggression and resource competition (e.g. Shi and Dunbar 2006), the effects of negative interactions can be counteracted with positive ones, which help reduce stress and mend damaged social bonds (Schino 1998; Briefer et al. 2015a). Goats further show sensitivity to their wider social environment, intervening in conflicts between other goats (Keil and Sambraus 1998), and to the orientation and motion of fellow foragers. Motion of certain group members can initiate collective movement, allowing foraging parties to exploit new resources (Sankey et al. 2021). Indeed, by paying attention to where group members are and what they are interacting with, goats can gain valuable information regarding food sources and threats (e.g. Kaminski et al. 2005; Shrader et al. 2007).

Goats have developed a suite of cognitive abilities to cope with challenges imposed by the social environment, from discriminating between group members, to interpreting their attentional state and motivations, to using social learning to determine where foraging resources are, and which ones are best to exploit. However, in the domestic environment, it is not only conspecific social cues that are relevant. The interdependent relationship between humans and domesticated animals spanning thousands of generations has provided a unique setting for goats and other livestock to develop more sophisticated forms of interspecific communication and perception of human cues (Avarguès-Weber et al. 2013). Accordingly, we will discuss goat ability to recognise, attribute attention to and learn socially from both other goats and humans and their interpretation of human gestural cues.

3.2.1 Recognition

Through social recognition, animals can attribute outcomes of previous interactions to particular individuals (individual recognition), or groups of individuals (class-based recognition, e.g. sex, familiar versus unfamiliar) and use these to anticipate future behaviour (Yorzinski 2017). Discriminating among group members may, for example, enable goats to keep track of dominance relationships (Miranda-de la Lama and Mattiello 2010) and avoid costs of misdirected parental or affiliative behaviours (Yorzinski 2017). In the mother–offspring context, initial recognition and social bonding is facilitated through olfactory cues (Poindron et al. 2007). However, individualised calls develop quickly in kids (before 5–9 days old), and although differences in call characteristics between individuals become further emphasised with age, mothers and offspring appear capable of discriminating each other vocally from other familiar kids and females, respectively, by this stage (Figs. 7 and 8; Briefer and McElligott 2011a). This recognition persists at least to 13 months post-weaning in mothers (Briefer et al. 2012). Visual cues also play a role, and in colour dimorphic herds at least, pelage colour appears to be important for kids to identify their mother (Ruiz-Miranda 1993). However, head cues appear to be non-essential for social recognition, with goats able to discriminate group members from unfamiliar individuals even when the head is hidden (Keil et al. 2012). Goats can even combine cues in different sensory modalities, using visual and vocal cues to cross-modally recognise preferred conspecific social partners (Pitcher et al. 2017). Such a capacity suggests goats can develop cognitive templates for familiar individuals, which can be used to compare against available cues; with these templates being a prerequisite for individual recognition (Proops et al. 2009). However, despite the research exploring how goats recognise each other, curiously no current research has tested if and how they recognise their next most important social partners, humans.

3.2.2 Attributing Attention

Direction and level of attention conveyed by conspecifics and humans provide important cues for gregarious livestock such as goats, as it increases the predictability of social interactions (Nawroth et al. 2019). Goats are sensitive to human attentional cues, with body orientation (Nawroth et al. 2015b, 2016b; Nawroth and McElligott 2017), and to a lesser extent head orientation (observed in: Nawroth et al. 2015b, 2016b; but not: Nawroth and McElligott 2017), and whether the eyes are open or closed (in: Nawroth and McElligott 2017; but not: Nawroth et al. 2016b) allowing them to appraise the level of attention they receive from humans. Goats appear to use these attentional cues to adjust investment in human-directed communication. When faced with a task that they were unable to solve independently, goats looked faster and for longer, and moreover performed more gaze alternations between experimenter and the task, when an experimenter faced towards them rather than away (Nawroth et al. 2016c).

Referential communication refers to the transfer of signals that encode information about external stimuli or events (Evans 1997). Gaze alternations can be interpreted as a form of referential communication, with goats attempting to draw the experimenter's attention to an external stimulus, such as a task they cannot solve, in a bid to solicit help (Nawroth et al. 2016c). However, perhaps unsurprisingly, given the mixed evidence of the use of head orientation to attribute attention to humans, goats do not appear to follow human gaze (Kaminski et al. 2005; Nawroth et al. 2015b, 2020), but have been shown to use conspecific gaze referentially, to locate hidden food (Kaminski et al. 2005). This ability is not only important for locating foraging resources, but also to detect predators over distance, which is crucial in prey species like goats (Nawroth et al. 2015b; Kaminski 2017). Gaze following may be an example of an innate behaviour, or one achieved through learning (e.g. cue learning, stimulus enhancement). Kaminski et al. (2006) found that subordinate goats that had received aggression were more likely to consume a piece of food that was not visible to the dominant aggressor, whereas subjects that had not received aggression consumed visible pieces as well. This suggests that subordinate goats possessed an understanding of the perceptual experience of their dominant social partner, adjusting their own behaviour accordingly to reduce costs of receiving further aggression. Sensitivity to the perspective of others may underlie the presence of gaze-following behaviour in goats (Kaminski et al. 2005; Kaminski 2017).

3.2.3 Interpretation of Human Gestural Communication

Comprehension of human pointing gestures has been investigated in a variety of species, from our closest living relatives, the great apes, to domesticated (dogs especially) and non-domesticated species, such as seals and bats (for review, see Krause et al. 2018). For goats, pointing gestures are uniquely human signals and, although they were largely domesticated for products rather than behavioural cooperation (e.g. dogs and horses; MacHugh and Bradley 2001), their domestic background *per se* may have sensitised goats to human behaviour (Hare et al. 2005). In line with this argument, goats have been shown to use human pointing and tapping gestures to infer the location of a food reward hidden under one of two cups (Kaminski et al. 2005; Nawroth et al. 2015b). Nawroth et al. (2020) investigated the extent to which goats understand the referential intent of these cues by examining their ability to follow an array of different pointing gestures to locate a reward placed in one of two buckets. Goats successfully used proximal (nearest arm pointed towards bucket) and crossed pointing gestures (arm crossed across body), where the distance between the experimenter's index finger and rewarded bucket was smaller, but not asymmetrical pointing cues, where the experimenter's body was closer to the unrewarded bucket. This suggests that, although goats can use a variety of pointing gestures, they do not understand their referential motivation and rely on simpler mechanisms, such as local or stimulus enhancement. Given performance did not improve over time (eight trials per gesture), it appears that no learning took place.

3.2.4 Use of Social Cues and Social Learning

Animals need to make frequent decisions, such as *where should I forage?*, *with whom should I mate?* and *where should I live?* (Valone and Templeton 2002). To select a more optimal behaviour for any given situation, they must gather information about their environment, either through their own trial-and--error (hereafter personal information) or by observing the behaviour of others facing similar challenges (social information; Danchin et al. 2004). The value of information in decision-making will depend on the degree to which it reduces uncertainty about environmental conditions, offset against the cost of acquiring this information (e.g. time, energy, and predation risk: Dall et al. 2005; Aguiñaga et al. 2021). The net value will, in turn, vary according to sampling technique used, with personal information often assumed to be more costly but more accurate, and social learning a cheaper but error-prone alternative (Laland 2004). As a herd-living species, goats are well positioned to acquire social information (Shrader et al. 2007), but whether they should rely on it will depend on whether the costs of obtaining personal information are high and whether personal information is unreliable or outdated (van Bergen et al. 2004; Rieucau and Giraldeau 2011). In addition, to decide when to use social information, animals must also decide whom to acquire it from, with several strategies having been proposed, including adopting the majority's behaviour, learning from more successful (e.g. more dominant) individuals, from older, more experienced individuals, or from kin (Laland 2004). Indeed, the most prominent early life role model for many species, including goats, is the mother (e.g. Glasser et al. 2009).

3.2.4.1 Vertical Learning

Mothers provide a wealth of social information for their offspring, from shaping species-specific social behaviours, including mating preferences (i.e. sexual imprinting: Kendrick et al. 1998, 2001), to influencing what plants kids will forage on in later life (Landau and Provenza 2020). These feeding preferences can be initially formed through passive learning *in utero*, via exposure to flavour molecules in the amniotic fluid (Hai et al. 2014), and later through the mother's milk and microbiome transfer through saliva and social contact (Nolte and Provenza 1992; Landau and Provenza 2020). Later, passive learning mechanisms are combined pre-weaning and replaced post-weaning with active learning from mothers. For example, Mamber and Damascus breeds differ in their propensity to consume tannin-rich lentisk (*Pistacia lentiscus*), but when Mamber kids are fostered by Damascus mothers and vice versa, differences in kid preference for lentisk most closely reflect that of their foster mother than their own breed (Glasser et al. 2009). This suggests that breed-specific feeding preferences are largely actively and/or passively learned from mothers rather than being genetically predetermined.

Processes underlying vertical learning may include a mixture of 'local enhancement' and 'production imitation' (Landau and Provenza 2020). Specifically, kids may learn through local enhancement when the mother's presence or actions draws their attention to a particular location, such that they are more likely to visit or interact with stimuli at that location (Hoppitt and Laland 2013). This process is generally considered to be relatively simple and can occur with or without learning, although they can also mediate social learning indirectly. By contrast, production imitation requires performance of a novel action, or sequence of actions, after observing a demonstrator (the mother) performing the same behaviour. Feeding preferences in a Martinique population of criollo goats reflect matrilineal lineage, but also year of birth (Biquand and Biquand-Guyot 1992). This convergence in feeding preferences within age cohorts suggests a role both for independent environmental sampling and social transmission of feeding preferences among peers.

3.2.4.2 Horizontal Learning

After observing their peer's foraging decisions, goats have been shown to select patches with the highest quantity of food first, while those prevented from collecting such social information distributed themselves equally among available foraging patches regardless of their profitability (Shrader et al. 2007). Through observing others, goats probably found foraging patches through local enhancement caused by the presence and/or inadvertent actions of foragers, using 'public information' of their peer's foraging success to assess patch quality (Valone 1989; Hoppitt and Laland 2013). By contrast, Baciadonna et al. (2013) found that observers that had watched a conspecific demonstrator select one of two foraging patches, did not show preference for the patch demonstrators chose, relying more on personal information instead. Similarly, goats observing a demonstrator solve a two-step foraging task depended on trial-and-error, being no faster to learn than goats that had not received a demonstration (Briefer et al. 2014; Fig. 10). However, this may be due to the difficulty of the task. Ultimately, although goats use simpler social learning mechanisms, such as local enhancement, evidence for complex peer learning in adult goats is limited. Goats could thus lack sophisticated forms of social learning, as simple forms may be sufficient to effectively exploit the patchily distributed resources that wild and free-ranging goat populations often encounter (Shrader et al. 2012; Baciadonna et al. 2013; Briefer et al. 2014). Alternatively, it could be a by-product of domestication, caused by relaxed reliance on conspecific social information (Range and Virányi 2014), and increased sensitivity towards human cues (Udell et al. 2010).

Through monitoring human and conspecific behaviour, goats gain valuable information regarding foraging and other resources, but unlike other goats, humans provide but do not compete for resources and thus supply a unique source of social information in the domestic environment (Avarguès-Weber et al. 2013). Indeed, goats faced with a detour task where they had to navigate around a V-shaped hurdle, performed better when they could watch a human complete this task before they attempted it independently (Nawroth et al. 2016a). As the demonstrator's actions actively maintained goat attention throughout their demonstration (shaking the box containing food), goats could have improved task performance using local or stimulus enhancement, where the demonstrator's behaviour drew their attention to the box or its location. However, given most goats receiving the demonstration took the same route as the demonstrator on their first trial, there is a possibility that goats used imitation to replicate human behaviour.

4 Personality

Variation is the substrate of natural selection, and one that might be expected to erode over time as traits conferring the highest fitness become fixed in a population (Darwin 1859; Endler 1986; Wilson 1998). However, one continuous source of behavioural variation retained across a diverse range of taxa is personality (Wolf and McNamara 2012; Wilson et al. 2019). Personality refers to behavioural (and physiological) differences between individuals that are consistent over time and across contexts (Réale et al. 2010; Finkemeier et al. 2018). Resulting variation in behaviour is expected to lead to differences in costs and benefits accrued based on personality. For example, bolder individuals that control resources through aggression may be expected to have greater reproductive success, but also to bear costs (e.g. injury) reducing their longevity relative to shyer individuals (Smith and Blumstein 2008). Variation in payoff across contexts and in the short and long term, coupled with frequency-dependent selection (fitness of one personality depends on the proportion of the population with the same personality), is expected to allow personality differences to become entrenched in animal populations (Wolf and McNamara 2012; Lichtenstein and Pruitt 2015). In an applied setting, given that personality can predict how individual animals respond to their environment, the importance of personality research is being increasingly recognised from an animal welfare perspective (Richter and Hintze 2019; Vaz et al. 2022). With this in mind, here we discuss evidence for personality in goats, with focus on the effects of personality on how goats interact with their social environment and how far personality can explain differences in cognition among individuals.

4.1 Terminology

In goats, as for many species, several related terms have been used to define inter-individual differences in behaviour. For example, temperament can refer to heritable, early emerging behavioural tendencies that persist over time and is considered to serve as a foundation for personality (Lyons et al. 1988; Gosling 2001; Finkemeier et al. 2018; Kannan et al. 2022). By contrast, identity profiles describe consistent social strategies adopted by certain individuals (i.e. aggressive, affiliative, passive, and evasive), combined with the behavioural and morphological characteristics that often accompany their expression (Miranda-de la Lama et al. 2011; Pascual-Alonso et al. 2013). Finally, coping styles refer to how animals respond to and mitigate the effects of aversive stimuli (Wechsler 1995; Toinon et al. 2021). However, the terminology has been inconsistently defined and is often used interchangeably. For example, some authors prefer the term 'temperament' to avoid anthropomorphising non-human animals through ascribing them with a personality (MacKay and Haskell 2015). Therefore, for the sake of simplicity, we will employ the umbrella term 'personality' to incorporate all related terminology.

4.2 Personality and Social Behaviour

The social environment is often synonymous with conflict, as individuals compete with conspecifics to gain resources such as mates, food, and space (Bergmüller and Taborsky 2010). For goats, effects of social conflict can become further exacerbated on farms as space is more limited, opportunities for social assortment (based on, e.g., social preferences and rank) are reduced, and resources are often clumped (e.g. food troughs), enabling dominant individuals to monopolise their access (e.g. Jørgensen et al. 2007; Stanley and Dunbar 2013). In the face of such costs, divergences in behaviour that reduce the scope for conflict among competitors will be favoured (Bergmüller and Taborsky 2010). Given that consistency in inter-individual behavioural differences is expected to increase with rising costs of conflict, shifts in behaviour may manifest as differences in personality, which can translate into asocial contexts as well.

Neave and Zobel (2020) measured personality in female dairy goats based on their behaviour towards novel objects, a novel environment and startling (opening umbrella), and threatening stimuli (a dog). In a competitive feeding context, 'bolder' goats tended to aggressively control resource access, fearful goats were less aggressive and spent less time feeding, and exploratory goats fed more from elevated feeders than those at ground-level feeders (the latter being subject to higher levels of competition).

Although the proximate and ultimate explanations underlying the link between personality and physiology are poorly understood, changes in physiology underpin shifts in behaviour and vice versa (Finkemeier et al. 2018; Wilson et al. 2019; McMahon and Cavigelli 2021). More social goats have been found to have a more stable HR than less social goats, with the latter expressing greater HR reactivity according to behavioural context (e.g. greater decrease in HR when engaged in affiliative behaviours; more elevated HR during eat/drink: Briefer et al. 2015a). A concurrent investigation found that shyer and/or more active goats expressed a sympathetic bias in nervous system activity at rest (higher standard deviation of inter-beat intervals in normal sinus beats, SDNN, a measure of HRV: Finkemeier et al. 2019). This, coupled with greater HR reactivity, likely predisposes shy (and more active) goats to mobilise and adapt more quickly to physical and social challenges than those with other personality types.

Personality not only affects interactions with other goats, but can have implications for their responses towards humans and ultimately, the human–animal relationship. In dairy goats, consistent differences in 'timidity' (related to boldness/fearfulness) have been found between individuals which predicted their behavioural and physiological responses (serum corticosteroid levels) towards moving and stationary humans, as well as across contexts, including responses to an unfamiliar goat, novel objects and an opportunity to explore a novel environment (Lyons et al. 1988). These behavioural and physiological differences were also to an extent present at group levels, with artificially reared kids being less timid than those that had been left with their mothers, although these differences, including their responses towards humans, faded over time. Importantly, the literature discussed in this section so far has used a mix of breeds and ways to define and test goat personality, making comparison between investigations more difficult.

In addition to production traits, goat breeds are well known to be associated with certain behavioural dispositions. For example, Nubians are tame and affectionate, British Saanen are less aggressive to group members, and Old English goats tend to be more active and vigilant (Weaver 2021). These differences in behavioural tendencies may translate into differences in how personality is expressed across breeds. Through observing the behaviour of dairy and angora (fibre) goats, Miranda-de la Lama et al. (2011) and Pascual-Alonso et al. (2013), respectively, found four personality traits associated with behavioural and morphological characteristics. Aggressive individuals were larger, more dominant, and better able to monopolise feed, while affiliative goats were of middling rank and gained resource access through prosocial interactions, rather than aggression. Both passive and evasive goats were lower ranked, but passive individuals were less active and showed little avoidance of agonistic interactions, whereas evasive goats actively avoided conflict, which in dairy goats extended to avoiding affiliative interactions as well. Indeed, although there were subtle differences in expression of these personalities between goats of different breeds, it is interesting that the same personality types emerge. Furthermore, personality in angora goats appears to relate to cognitive ability (Pascual-Alonso et al. 2013).

4.3 Personality and Cognition

Like personality, cognitive ability can vary substantially among members of a population (e.g. Nawroth et al. 2015a; Finkemeier et al. 2022). The extent to which individual differences in cognitive abilities relate to differences in personality has become of increasing interest to empirical researchers over the last decade or so (Dougherty and Guillette 2018). The mechanisms driving such covariation have been suggested to, for example, result from a shared neuroendocrinological system acting on both personality and cognition over development (e.g. stress hormones), and/or from personality traits causing differences in an individual's opportunity to encounter and learn about stimuli present in their environment (Griffin et al. 2015; Wilson et al. 2019). For example, bolder, and/or more exploratory individuals may be expected to approach novel tasks faster, or be more attentive to changes in their environment, to retain information more quickly, and/or form and memorise associations between environmental stimuli more easily (Wilson et al. 2019). However, bolder, more proactive individuals are also often predicted to more readily form routines, so may be less flexible in adapting to changing environments. Indeed, although bolder goats were no faster to learn an initial visual discrimination task, they appeared to be slower to adjust their behaviour in line with task requirements, being outperformed by shyer, more reactive individuals in a subsequent reversal learning task (but not in two further reversal learning tasks: Finkemeier et al. 2022). Furthermore, more aggressive goats (associated with boldness: e.g. Neave and Zobel 2020) have also been shown to be slower to solve and learn the solution to a T-maze than those with other personality types (Pascual-Alonso et al. 2013). Less dominant goats, and especially those with a passive personality, were the quickest to solve the T-maze on both their first and second exposure to the task. As lower-ranking goats regularly receive aggression from higher ranking individuals, they may have acquired greater sensitivity to food-related cues, such as the feeding-related sounds used to direct goats to the rewarded location (e.g. rattling of feed pellets) to locate food before they can be displaced. Similarly, goats that were more or less social and exploratory were found to express distinct differences in performance of cognitive tasks accordingly (Nawroth et al. 2017).

During a transposition task (used to test object permanence), less exploratory goats were better at tracking the movements of a reward hidden under one cup as it switched places with an adjacent cup than more exploratory individuals (Nawroth et al. 2017). Rather than a learning impairment in more exploratory individuals, this was interpreted as reflecting less exploratory goats being less distracted by external cues, or more exploratory goats being less motivated by the food reward. Conversely, less social goats were quicker to learn a visual discrimination task than more social individuals, perhaps due to the greater levels of stress experienced by the latter at being separated from conspecifics or because they were more attentive to other cues (e.g. spatial ones). In sum, although personality seems to affect cognition, there is little evidence that goats with different personalities diverge in their 'general cognitive ability', with variation in performance appearing to be more task-specific, based on motivation and/or differences in how goats learn and process particular cues.

5 Linking Emotions, Cognition, Personality, and Welfare

As emotional experiences are a key component of goat welfare, approaches that enable us to objectively measure emotion have strong value as welfare indicators (Désiré et al. 2002). The indicators discussed over the course of this chapter have assessed multiple levels of the composite emotional response, from internal neurophysiological changes (e.g. thermal imaging and cardiac responses) to behaviour (e.g. postural and vocal expression) and moreover, cognitive processes (Briefer et al. 2015b; Bartolomé et al. 2019; Baciadonna et al. 2020). Indeed, how environmental information is processed, retained, and acted upon are all guided by an animal's emotional state; with the cumulative effect of emotional experience on creating cognitive biases, and on decision-making in particular (judgement biases) being the best investigated (Briefer and McElligott 2013; Baciadonna et al. 2016). Cognitive biases provide a further means to measure welfare and long-term 'mood' states in particular; however, the relationship between emotions and cognition is not unidirectional (Russell 2003; Kremer et al. 2020).

A goat's ability to discriminate, learn, and form expectations about its environment can have substantial implications for its emotional state and welfare. For example, interpretation of social or physical cues enables goats to anticipate positive and negative events (Kaminski et al. 2006; Baciadonna et al. 2020). Furthermore, in the case of positive predictors, emotional changes can extend beyond the rewarding properties of the event itself, and towards the process of building the association, i.e. learning. Learning seems to have inherent value to goats and drives them to seek cognitive challenges (Langbein et al. 2009; Rosenberger et al. 2020). When considered from a social perspective, the ability to, for example, learn the identities of group members and humans can serve as a foundation for social bonding, acting to promote prosocial behaviours in goats, which can alleviate stress and create opportunities to experience positive emotions (Schino 1998; Stanley and Dunbar 2013; Briefer et al. 2015a; Baciadonna et al. 2016; Pitcher et al. 2017). Indeed, goats have demonstrated a sophisticated suite of cognitive abilities both in the social (e.g. cross-modal recognition and perspective taking) and physical domains, including evidence for abstract thinking (e.g. inferential reasoning and object permanence; Kaminski et al. 2006; Nawroth et al. 2014, 2015a; Pitcher et al. 2017). The general problem-solving ability established in this species, and behavioural flexibility in particular may, for example, enable goats to adapt more quickly to changes in husbandry regime and their environment, but also predispose them towards boredom and anhedonia (Coppens et al. 2010; Briefer et al. 2014; Franks 2018; Raoult et al. 2021). Ultimately, accumulating evidence of the emotional lives and cognitive sophistication, or sentience of goats and other animals under our care, not only helps us assess how far their cognitive needs are being met, but often our perceived level of obligation in ensuring their welfare (Broom 2010; Franks 2018). Indeed, the recognised importance of learning and cognition on animal welfare is particularly evident in the widespread use of environmental and cognitive enrichment to create cognitively stimulating environments for captive animals (Oesterwind et al. 2016; Clark 2017; Zobel and Nawroth 2020). However, given that both cognitive ability and propensity to seek cognitive challenges appear to vary among individual goats (Langbein et al. 2009; Nawroth et al. 2015a), the importance of cognitive stimulation on welfare will not be the same for every individual.

Differences in personality can be used to predict consistent differences in how goats process and respond behaviourally and emotionally to environmental cues, underlining the fact that species- or group-level welfare interventions can be insufficient for ensuring welfare needs are met for every goat (Briefer et al. 2015a; Nawroth et al. 2017; Finkemeier et al. 2019; Richter and Hintze 2019; Vaz et al. 2022). In this chapter, we have focused on the top-down effects of personality on cognition and emotions. Although beyond the scope of our review, it is clear that early life emotional experiences such as maternal deprivation, hand-rearing (Lyons et al. 1988; Toinon et al. 2021, 2022), and problem-solving opportunities (Oesterwind et al. 2016) can create long-term differences in behaviour among goats. Notwithstanding genetic influences, if these experiences occur during 'sensitive windows' of heightened developmental plasticity (largely occurring during ontogeny), they can potentially create persistent or even irreversible changes in an adult animal's personality (Wilson et al. 2019). Ultimately, personality affects how goats emotionally respond to, process, and deal with environmental challenges. However, both emotional experiences (particularly those associated with the HPA axis) and problem-solving opportunities can interact in complex ways during development to affect personality.

6 Conclusions and Future Directions

Over the last decade or so, a growing body of research has illustrated the rich emotional lives goats lead, as well as the sophisticated cognitive adaptations and inter-individual differences they express when navigating their challenging physical and social environments. Indeed, a suite of indicators have been developed to assess goat emotional states, both in the short and long term (Briefer and McElligott 2013; AWIN 2015; Briefer et al. 2015b). Given the majority of these indicators more reliably relate to either arousal (particularly physiological variables: e.g. cardiac measures and thermal imaging) or valence (e.g. certain ear postures and judgement biases), assessing emotions according to dimensional criteria necessitates integration of multiple indices of a goat's overall response. There is a growing interest in monitoring farm animal emotions in precision livestock farming, especially in the field of bioacoustics (Mcloughlin et al. 2019). Given what is known about emotional signatures present in goat vocalisations (Siebert et al. 2011; Briefer et al. 2015b), there is obvious scope for implementation of automated emotional monitoring techniques which, for example, have been used to classify kid calls according to individual, age, and group membership (Favaro et al. 2014). Goats not only exhibit a sophisticated set of physical and social cognitive abilities but actively seek opportunities to apply them to solve tasks (e.g. Briefer et al. 2014; Pitcher et al. 2017; Rosenberger et al. 2020). These attributes underscore the significance of cognitive challenges for goat welfare (see Zobel and Nawroth 2020 for review on how this can inform species-specific enrichment). However, cognitive performance along with behaviour and emotional responses vary among individuals, and these to an extent can be predicted by a goat's personality (Briefer et al. 2015a; Nawroth et al. 2017; Neave and Zobel 2020). Ultimately, to improve welfare in goats (Green and Mellor 2011; Mellor and Beausoleil 2015), we must monitor their emotional as well as physical welfare, give opportunities for them to apply their impressive cognitive abilities and to recognise the importance of individual-level strategies.

References

Aguiñaga J, Gomulkiewicz R, Watts HE (2021) Effect of social information on an individual's assessment of its environment. Anim Behav 178:267–277

Ahloy-Dallaire J, Espinosa J, Mason G (2018) Play and optimal welfare: Does play indicate the presence of positive affective states? Behav Process 156:3–15

Alonso ME, González-Montaña JR, Lomillos JM (2020) Consumers' concerns and perceptions of farm animal welfare. Animals 10:385

Amills M, Capote J, Tosser-Klopp G (2017) Goat domestication and breeding: A jigsaw of historical, biological and molecular data with missing pieces. Anim Genet 48:631–644

Anderson DJ, Adolphs R (2014) A framework for studying emotions across species. Cell 157:187–200

Aschwanden J, Gygax L, Wechsler B, Keil NM (2008) Social distances of goats at the feeding rack: Influence of the quality of social bonds, rank differences, grouping age and presence of horns. Appl Anim Behav Sci 114:116–131

Asher L, Friel M, Griffin K, Collins LM (2016) Mood and personality interact to determine cognitive biases in pigs. Biol Lett 12:20160402

Auersperg A, Teschke I, Tebbich S (2017) Physical cognition and tool use in birds. In: ten Cate C, Healy SD (eds) Avian cognition. Cambridge University Press, Cambridge, pp 163–183

Avarguès-Weber A, Dawson EH, Chittka L (2013) Mechanisms of social learning across species boundaries. J Zool 290:1–11

AWIN (2015) AWIN welfare assessment protocol for goats. https://doi.org/10.13130/AWIN_GOATS_2015

Baciadonna L, McElligott AG (2015) The use of judgement bias to assess welfare in farm livestock. Anim Welf 24:81–91

Baciadonna L, McElligott AG, Briefer EF (2013) Goats favour personal over social information in an experimental foraging task. PeerJ 1:e172

Baciadonna L, Nawroth C, McElligott AG (2016) Judgement bias in goats (*Capra hircus*): Investigating the effects of human grooming. PeerJ 4:e2485

Baciadonna L, Duepjan S, Briefer EF, Padilla de la Torre M, Nawroth C (2018) Looking on the bright side of livestock emotions—The potential of their transmission to promote positive welfare. Front Vet Sci 5:218

Baciadonna L, Briefer EF, Favaro L, McElligott AG (2019) Goats distinguish between positive and negative emotion-linked vocalisations. Front Zool 16:25

Baciadonna L, Briefer EF, McElligott AG (2020) Investigation of reward quality-related behaviour as a tool to assess emotions. Appl Anim Behav Sci 225:104968

Barroso FG, Alados CL, Boza J (2000) Social hierarchy in the domestic goat: Effect on food habits and production. Appl Anim Behav Sci 69:35–53

Bartolomé E, Azcona F, Cañete-Aranda M, Perdomo-González DI, Ribes-Pons J, Terán EM (2019) Testing eye temperature assessed with infrared thermography to evaluate stress in meat goats raised in a semi-intensive farming system: A pilot study. Arch Anim Breed 62:199–204

Battini M, Barbieri S, Vieira A, Stilwell G, Mattiello S (2016) Results of testing the prototype of the AWIN welfare assessment protocol for dairy goats in 30 intensive farms in Northern Italy. Ital J Anim Sci 15:283–293

Battini M, Barbieri S, Vieira A, Can E, Stilwell G, Mattiello S (2018) The use of qualitative behaviour assessment for the on-farm welfare assessment of dairy goats. Animals 8:123

Bellegarde LG, Haskell MJ, Duvaux-Ponter C, Weiss A, Boissy A, Erhard HW (2017) Face-based perception of emotions in dairy goats. Appl Anim Behav Sci 193:51–59

Bergmüller R, Taborsky M (2010) Animal personality due to social niche specialisation. Trends Ecol Evol 25:504–511

Berman TS, Ben-Ari M, Glasser TA, Gish M, Inbar M (2017) How goats avoid ingesting noxious insects while feeding. Sci Rep 7:14835

Bethell EJ, Koyama NF (2015) Happy hamsters? Enrichment induces positive judgement bias for mildly (but not truly) ambiguous cues to reward and punishment in *Mesocricetus auratus*. R Soc Open Sci 2:140399

Biquand S, Biquand-Guyot V (1992) The influence of peers, lineage and environment on food selection of the criollo goat (*Capra hircus*). Appl Anim Behav Sci 34:231–245

Bleul U, Hässig M, Kluser F (2021) Anwendung einer kleinen tragbaren wärmebildkamera zum screening auf kühe mit fieber. Tierärztliche Praxis Ausgabe G: Großtiere/Nutztiere 49:12–20

Bokkers EAM, De Vries M, Antonissen ICMA, De Boer IJM (2012) Inter-and intra-observer reliability of experienced and inexperienced observers for the Qualitative Behaviour Assessment in dairy cattle. Anim Welf 21:307–318

Breed MD, Moore J (2016) Homeostasis and time budgets. In: Breed MD, Moore J (eds) Animal behaviour, 2nd edn. Academic, Oxford, pp 109–144

Briefer EF (2012) Vocal expression of emotions in mammals: Mechanisms of production and evidence. J Zool 288:1–20

Briefer EF (2018) Vocal contagion of emotions in non-human animals. Proc R Soc B 285:20172783

Briefer EF (2020) Coding for ‘dynamic’ information: Vocal expression of emotional arousal and valence in non-human animals. In: Aubin T, Mathevon N (eds) Coding strategies in vertebrate acoustic communication. Springer Nature, Cham, pp 137–162

Briefer EF, McElligott AG (2011a) Mutual mother-offspring vocal recognition in an ungulate hider species (*Capra hircus*). Anim Cogn 14:585–598

Briefer EF, McElligott AG (2011b) Indicators of age, body size and sex in goat kid calls revealed using the source-filter theory. Appl Anim Behav Sci 133:175–185

Briefer EF, McElligott AG (2012) Social effects on vocal ontogeny in an ungulate, the goat, *Capra hircus*. Anim Behav 83:991–1000

Briefer EF, McElligott AG (2013) Rescued goats at a sanctuary display positive mood after former neglect. Appl Anim Behav Sci 146:45–55

Briefer EF, Padilla de la Torre M, McElligott AG (2012) Mother goats do not forget their kids' calls. Proc R Soc B 279:3749–3755

Briefer EF, Haque S, Baciadonna L, McElligott AG (2014) Goats excel at learning and remembering a highly novel cognitive task. Front Zool 11:20

Briefer EF, Oxley JA, McElligott AG (2015a) Autonomic nervous system reactivity in a free-ranging mammal: Effects of dominance rank and personality. Anim Behav 110:121–132

Briefer EF, Tettamanti F, McElligott AG (2015b) Emotions in goats: Mapping physiological, behavioural and vocal profiles. Anim Behav 99:131–143

Brilot BO, Asher L, Bateson M (2010) Stereotyping starlings are more ‘pessimistic’. Anim Cogn 13:721–731

Broom DM (2010) Cognitive ability and awareness in domestic animals and decisions about obligations to animals. Appl Anim Behav Sci 126:1–11

Bučková K, Špinka M, Hintze S (2019) Pair housing makes calves more optimistic. Sci Rep 9:20246

Calhim S, Shi J, Dunbar RIM (2006) Sexual segregation among feral goats: Testing between alternative hypotheses. Anim Behav 72:31–41

Cannas S, Palestrini C, Canali E, Cozzi B, Ferri N, Heinzl E, Minero M, Chincarini M, Vignola G, Dalla Costa E (2018) Thermography as a non-invasive measure of stress and fear of humans in sheep. Animals 8:146

Chotard H, Ioannou S, Davila-Ross M (2018) Infrared thermal imaging: Positive and negative emotions modify the skin temperatures of monkey and ape faces. Am J Primatol 80:e22863

Chynoweth MW, Litton CM, Lepczyk CA, Hess SC, Cordell S (2013) Biology and impacts of Pacific Island invasive species. 9. *Capra hircus*, the feral goat (Mammalia: Bovidae). Pac Sci 67:141–156

Clark FE (2017) Cognitive enrichment and welfare: Current approaches and future directions. Anim Behav Cogn 4:52–71

Cooper R, Wemelsfelder F (2020) Qualitative behaviour assessment as an indicator of animal emotional welfare in farm assurance. Livestock 25:180–183

Coppens CM, de Boer SF, Koolhaas JM (2010) Coping styles and behavioural flexibility: Towards underlying mechanisms. Philos Trans R Soc B 365:4021–4028

Coulon M, Nowak R, Peyrat J, Chandèze H, Boissy A, Boivin X (2015) Do lambs perceive regular human stroking as pleasant? Behaviour and heart rate variability analyses. PLoS One 10:e0118617

Crump A, Arnott G, Bethell EJ (2018) Affect-driven attention biases as animal welfare indicators: Review and methods. Animals 8:136

Dall SR, Giraldeau LA, Olsson O, McNamara JM, Stephens DW (2005) Information and its use by animals in evolutionary ecology. Trends Ecol Evol 20:187–193

Danchin E, Giraldeau LA, Valone TJ, Wagner RH (2004) Public information: From nosy neighbours to cultural evolution. Science 305:487–491

Darcan NK, Silanikove N (2018) The advantages of goats for future adaptation to climate change: A conceptual overview. Small Rumin Res 163:34–38

Darwin C (1859) On the origin of species. John Murray, London

de Geus EJ, Gianaros PJ, Brindle RC, Jennings JR, Berntson GG (2019) Should heart rate variability be "corrected" for heart rate? Biological, quantitative, and interpretive considerations. Psychophysiology 56:e13287

de Ruediger FR, Yamada PH, Barbosa LGB, Chacur MGM, Ferreira JCP, de Carvalho NAT, Soriano GAM, Codognoto VM, Oba E (2018) Effect of oestrous cycle phase on vulvar, orbital area and muzzle surface temperatures as determined using digital infrared thermography in buffalo. Anim Reprod Sci 197:154–161

Descovich KA, Wathan J, Leach MC, Buchanan-Smith HM, Flecknell P, Framingham D, Vick SJ (2017) Facial expression: An under-utilised tool for the assessment of welfare in mammals. ALTEX 34:409–429

Désiré L, Boissy A, Veissier I (2002) Emotions in farm animals: A new approach to animal welfare in applied ethology. Behav Process 60:165–180

de Waal FBM (2011) What is an animal emotion? Ann N Y Acad Sci 1224:191–206

Diamond A (2013) Executive functions. Annu Rev Psychol 64:135–168

Dillon DG, Pizzagalli DA (2018) Mechanisms of memory disruption in depression. Trends Neurosci 41:137–149

Dougherty LR, Guillette LM (2018) Linking personality and cognition: A meta-analysis. Philos Trans R Soc B 373:20170282

Düpjan S, Krause A, Moscovice LR, Nawroth C (2020) Emotional contagion and its implications for animal welfare. CABI Rev 15:1–6

Eckberg DL (1983) Human sinus arrhythmia as an index of vagal cardiac outflow. J Appl Physiol 54:961–966

Eckberg DL (2003) Topical Review: The human respiratory gate. J Physiol 548:339–352

Ekman P (1992) An argument for basic emotions. Cognit Emot 6:169–200

Endler JA (1986) Natural selection in the wild. Princeton University Press, Princeton, NJ

Estell RE (2010) Coping with shrub secondary metabolites by ruminants. Small Rumin Res 94:1–9

Evans CS (1997) Referential signals. Perspect Ethol 12:99–143

FAO (2020) FAOSTAT—Crops and livestock products. https://www.fao.org/faostat/en/#data/QCL. Accessed 19 May 2022

Favaro L, Briefer EF, McElligott AG (2014) Artificial neural network approach for revealing individuality, group membership and age information in goat kid contact calls. Acta Acustica United Acustica 100:782–789

FAWC (1993) Second report on priorities for research and development in farm animal welfare. Farm Animal Welfare Council (FAWC), Department of Environment, Food and Rural Affairs, London

Finkemeier MA, Langbein J, Puppe B (2018) Personality research in mammalian farm animals: Concepts, measures, and relationship to welfare. Front Vet Sci 5:131

Finkemeier MA, Oesterwind S, Nürnberg G, Puppe B, Langbein J (2019) Assessment of personality types in Nigerian dwarf goats (*Capra hircus*) and cross-context correlations to behavioural and physiological responses. Appl Anim Behav Sci 217:28–35

Finkemeier MA, Krause A, Tuchscherer A, Puppe B, Langbein J (2022) Personality traits affect learning performance in dwarf goats (*Capra hircus*). Front Vet Sci 9:916459

Fleming PA, Clarke T, Wickham SL, Stockman CA, Barnes AL, Collins T, Miller DW (2016) The contribution of qualitative behavioural assessment to appraisal of livestock welfare. Anim Prod Sci 56:1569–1578

Franks B (2018) Cognition as a cause, consequence, and component of welfare. In: Mench J (ed) Advances in agricultural animal welfare: Science and practice. Woodhead, Oxford, pp 3–24

Fridlund AJ (1991) Evolution and facial action in reflex, social motive, and paralanguage. Biol Psychol 32:3–100

George WD, Godfrey RW, Ketring RC, Vinson MC, Willard ST (2014) Relationship among eye and muzzle temperatures measured using digital infrared thermal imaging and vaginal and rectal temperatures in hair sheep and cattle. J Anim Sci 92:4949–4955

Glasser TA, Ungar ED, Landau SY, Perevolotsky A, Muklada H, Walker JW (2009) Breed and maternal effects on the intake of tannin-rich browse by juvenile domestic goats (*Capra hircus*). Appl Anim Behav Sci 119:71–77

Godyń D, Herbut P, Angrecka S (2019) Measurements of peripheral and deep body temperature in cattle—A review. J Therm Biol 79:42–49

Gómez Y, Bieler R, Hankele AK, Zähner M, Savary P, Hillmann E (2018) Evaluation of visible eye white and maximum eye temperature as non-invasive indicators of stress in dairy cows. Appl Anim Behav Sci 198:1–8

Gosling SD (2001) From mice to men: What can we learn about personality from animal research? Psychol Bull 127:45–86

Green TC, Mellor DJ (2011) Extending ideas about animal welfare assessment to include 'quality of life' and related concepts. N Z Vet J 59:263–271

Griffin AS, Guillette LM, Healy SD (2015) Cognition and personality: An analysis of an emerging field. Trends Ecol Evol 30:207–214

Grosso L, Battini M, Wemelsfelder F, Barbieri S, Minero M, Dalla Costa E, Mattiello S (2016) On-farm Qualitative Behaviour Assessment of dairy goats in different housing conditions. Appl Anim Behav Sci 180:51–57

Gygax L, Reefmann N, Wolf M, Langbein J (2013) Prefrontal cortex activity, sympatho-vagal reaction and behaviour distinguish between situations of feed reward and frustration in dwarf goats. Behav Brain Res 239:104–114

Hai PV, Schonewille JT, Van Tien D, Everts H, Hendriks WH (2014) Improved acceptance of *Chromonaela odorata* by goat kids after weaning is caused by in utero exposure during late but not early pregnancy. Appl Anim Behav Sci 159:50–54

Harding EJ, Paul ES, Mendl M (2004) Cognitive bias and affective state. Nature 427:312

Hare B, Plyusnina I, Ignacio N, Schepina O, Stepika A, Wrangham R, Trut L (2005) Social cognitive evolution in captive foxes is a correlated by-product of experimental domestication. Curr Biol 15:226–230

Heimbauer LA, Johns TN, Weiss DJ (2019) Inferential reasoning in the visual and auditory modalities by cotton-top tamarins (*Saguinus oedipus*). J Comp Psychol 133:496–501

Henry S, Fureix C, Rowberry R, Bateson M, Hausberger M (2017) Do horses with poor welfare show 'pessimistic' cognitive biases? Sci Nat 104:8
Herborn KA, Graves JL, Jerem P, Evans NP, Nager R, McCafferty DJ, McKeegan DE (2015) Skin temperature reveals the intensity of acute stress. Physiol Behav 152:225–230
Herborn KA, McElligott AG, Mitchell MA, Sandilands V, Bradshaw B, Asher L (2020) Spectral entropy of early-life distress calls as an iceberg indicator of chicken welfare. J R Soc Interface 17:20200086
Hoppitt W, Laland KN (2013) Social learning mechanisms. In: Hoppitt W, Laland KN (eds) Social Learning: An introduction to mechanisms, methods, and models. Princeton University Press, Princeton, NJ, pp 62–104
Horback KM, Parsons TD (2019) Judgement bias testing in group-housed gestating sows. Behav Process 159:86–92
Hosoi E, Swift DM, Rittenhouse LR, Richards RW (1995) Comparative foraging strategies of sheep and goats in a T-maze apparatus. Appl Anim Behav Sci 44:37–45
Hui JJ, Xi GJ, Liu SS, Li XL, Geng LY, Teng GJ, Nie BB, Shan BC, Yan J, Dong L, Reynolds GP, Zhang ZJ (2016) Blood oxygen level-dependent signals via fMRI in the mood-regulating circuit using two animal models of depression are reversed by chronic escitalopram treatment. Behav Brain Res 311:210–218
Hussein NJ, Hidayet HM (2019) Changes in ear postures of kid goats in response to ear tagging. Basrah J Agric Sci 32:25–33
Hydbring-Sandberg E, von Walter LW, Forkman B (2022) Cortisol is not enough: A complex stress reaction in tethered goats. Anim Welf 31:91–98
Ijichi C, Evans L, Woods H, Yarnell K (2020) The Right Angle: Validating a standardised protocol for the use of infra-red thermography of eye temperature as a welfare indicator. Anim Welf 29:123–131
Inglis IR, Forkman B, Lazarus J (1997) Free food or earned food? A review and fuzzy model of contrafreeloading. Anim Behav 53:1171–1191
Jaakkola K (2014) Do animals understand invisible displacement? A critical review. J Comp Psychol 128:225–239
Jansson A, Lindgren G, Velie BD, Solé M (2021) An investigation into factors influencing basal eye temperature in the domestic horse (*Equus caballus*) when measured using infrared thermography in field conditions. Physiol Behav 228:113218
Jardat P, Lansade L (2022) Cognition and the human-animal relationship: A review of the socio-cognitive skills of domestic mammals toward humans. Anim Cogn 25:369–384
Jensen GD (1963) Preference for bar pressing over "freeloading" as a function of number of rewarded presses. J Exp Psychol 65:451–454
Johnstone T, Scherer KR (2000) Vocal communication of emotion. In: Lewis M, Haviland-Jones JM (eds) Handbook of emotions. Guildford, New York, pp 220–235
Jørgensen GHM, Andersen IL, Bøe KE (2007) Feed intake and social interactions in dairy goats—The effects of feeding space and type of roughage. Appl Anim Behav Sci 107:239–251
Jürgens U (2009) The neural control of vocalisation in mammals: A review. J Voice 23:1–10
Kaminski J (2017) Mind reading in animals? In: Call J, Burghardt GM, Pepperberg IM, Snowdon CT, Zentall T (eds) APA handbook of comparative psychology: Perception, learning, and cognition. American Psychological Association, Washington, DC, pp 723–744
Kaminski J, Riedel J, Call J, Tomasello M (2005) Domestic goats, *Capra hircus*, follow gaze direction and use social cues in an object choice task. Anim Behav 69:11–18
Kaminski J, Call J, Tomasello M (2006) Goats' behaviour in a competitive food paradigm: Evidence for perspective taking? Behaviour 143:1341–1356
Kannan G, Batchu P, Naldurtiker A, Dykes GS, Kouakou B, Terrill TH, Gurrapu P (2022) Influence of excitability rate on physiological responses to stress in goats. Animals 12:1023
Katayama M, Kubo T, Yamakawa T, Fujiwara K, Nomoto K, Ikeda K, Mogi K, Nagasawa M, Kikusui T (2019) Emotional contagion from humans to dogs is facilitated by duration of ownership. Front Psychol 10:1678

Keil NM, Sambraus HH (1998) "Intervenors" in agonistic interactions amongst domesticated goats. Zeitschrift für Säugetierkunde 63:266–272
Keil NM, Imfeld-Mueller S, Aschwanden J, Wechsler B (2012) Are head cues necessary for goats (*Capra hircus*) in recognising group members? Anim Cogn 15:913–921
Kendrick KM, Hinton MR, Atkins K, Haupt MA, Skinner JD (1998) Mothers determine sexual preferences. Nature 395:229–230
Kendrick KM, Haupt MA, Hinton MR, Broad KD, Skinner JD (2001) Sex differences in the influence of mothers on the sociosexual preferences of their offspring. Horm Behav 40:322–338
Kim SM, Cho GJ (2021) Validation of eye temperature assessed using infrared thermography as an indicator of welfare in horses. Appl Sci 11:7186
Knízková I, Kunc P, Gürdil G, Pinar Y, Selvi KÇ (2007) Applications of infrared thermography in animal production. Anadolu Tarım Bilimleri Dergisi 22:329–336
Kowalik S, Janczarek I, Kędzierski W, Stachurska A, Wilk I (2017) The effect of relaxing massage on heart rate and heart rate variability in purebred Arabian racehorses. Anim Sci J 88:669–677
Krause MA, Udell MA, Leavens DA, Skopos L (2018) Animal pointing: Changing trends and findings from 30 years of research. J Comp Psychol 132:326–345
Kremer L, Holkenborg SK, Reimert I, Bolhuis JE, Webb LE (2020) The nuts and bolts of animal emotion. Neurosci Biobehav Rev 113:273–286
Lagisz M, Zidar J, Nakagawa S, Neville V, Sorato E, Paul ES, Bateson M, Mendl M, Løvlie H (2020) Optimism, pessimism and judgement bias in animals: A systematic review and meta-analysis. Neurosci Biobehav Rev 118:3–17
Laland KN (2004) Social learning strategies. Anim Learn Behav 32:4–14
Landau SY, Provenza FD (2020) Of browse, goats, and men: Contribution to the debate on animal traditions and cultures. Appl Anim Behav Sci 232:105127
Langbein J (2012) Investigations on training, recall and reversal learning of a Y-maze by dwarf goats (*Capra hircus*): The impact of lateralisation. Behav Process 89:304–310
Langbein J (2018) Motor self-regulation in goats (*Capra aegagrus hircus*) in a detour-reaching task. PeerJ 6:e5139
Langbein J, Nürnberg G, Manteuffel G (2004) Visual discrimination learning in dwarf goats and associated changes in heart rate and heart rate variability. Physiol Behav 82:601–609
Langbein J, Siebert K, Nürnberg G, Manteuffel G (2007) Learning to learn during visual discrimination in group housed dwarf goats (*Capra hircus*). J Comp Psychol 121:447–456
Langbein J, Siebert K, Nuernberg G (2008) Concurrent recall of serially learned visual discrimination problems in dwarf goats (*Capra hircus*). Behav Process 79:156–164
Langbein J, Siebert K, Nürnberg G (2009) On the use of an automated learning device by group-housed dwarf goats: Do goats seek cognitive challenges? Appl Anim Behav Sci 120:150–158
Lange A, Waiblinger S, van Hasselt R, Mundry R, Futschik A, Lürzel S (2021) Effects of restraint on heifers during gentle human-animal interactions. Appl Anim Behav Sci 243:105445
Laurijs KA, Briefer EF, Reimert I, Webb LE (2021) Vocalisations in farm animals: A step towards positive welfare assessment. Appl Anim Behav Sci 236:105264
Le Doux JE (2021) What emotions might be like in other animals. Curr Biol 31:R824–R829
Leite LO, Stamm FDO, Garcia RDCM (2017) Indicators to assess goat welfare on-farm in the semiarid region of Brazilian Northeast. Ciência Rural 47:e20161073
Lichtenstein JLL, Pruitt JN (2015) Similar patterns of frequency-dependent selection on animal personalities emerge in three species of social spiders. J Evol Biol 28:1248–1256
Longpre KM, Koepfinger ME, Katz LS (2011) Female goats use courtship display as an honest indicator of male quality. Horm Behav 60:505–511
Lyons DM, Price EO, Moberg GP (1988) Individual differences in temperament of domestic dairy goats: Constancy and change. Anim Behav 36:1323–1333
Machado M, da Silva IJO (2019) Does farm animals experience emotions and feelings? J Anim Behav Biometeorol 7:170–175

Machado M, da Silva IJO (2020) Body expressions of emotions: Does animals have it? J Anim Behav Biometeorol 8:1–10
MacHugh DE, Bradley DG (2001) Livestock genetic origins: Goats buck the trend. Proc Natl Acad Sci USA 98:5382–5384
MacHugh DE, Larson G, Orlando L (2017) Taming the past: Ancient DNA and the study of animal domestication. Annu Rev Animl Biosci 5:329–351
MacKay JR, Haskell MJ (2015) Consistent individual behavioural variation: The difference between temperament, personality and behavioural syndromes. Animals 5:455–478
Mahdy MA, Mohamed SA, Abdalla KE (2020) Morphological investigations on the lips and cheeks of the goat (*Capra hircus*): A scanning electron and light microscopic study. Microsc Res Tech 83:1095–1102
Marques JI, Leite PG, Lopes Neto JP, Furtado DA, Lopes FFDM (2021) Estimation of rectal temperature of goats based on surface temperature. Engenharia Agrícola 41:591–598
Mason MA (2023) Goat perception of human cues & reliability of thermal imaging in welfare research. Doctoral thesis, University of Roehampton. Ethos. https://ethos.bl.uk/OrderDetails.do?uin=uk.bl.ethos.879491
Mason MA, Semple S, Marshall HH, McElligott AG (2024) Goats discriminate emotional valence in the human voice. Anim Behav 209:227–240
Matthews SG, Miller AL, Clapp J, Plötz T, Kyriazakis I (2016) Early detection of health and welfare compromises through automated detection of behavioural changes in pigs. Vet J 217:43–51
Mattiello S, Battini M, De Rosa G, Napolitano F, Dwyer C (2019) How can we assess positive welfare in ruminants? Animals 9:758
Mcloughlin MP, Stewart R, McElligott AG (2019) Automated bioacoustics: Methods in ecology and conservation and their potential for animal welfare monitoring. J R Soc Interface 16:20190225
McMahon EK, Cavigelli SA (2021) Gaps to address in ecological studies of temperament and physiology. Integr Comp Biol 61:1917–1932
Mellor DJ, Beausoleil NJ (2015) Extending the 'Five Domains' model for animal welfare assessment to incorporate positive welfare states. Anim Welf 24:241–253
Mendl M, Burman OH, Paul ES (2010) An integrative and functional framework for the study of animal emotion and mood. Proc R Soc B 277:2895–2904
Merola I, Prato-Previde E, Lazzaroni M, Marshall-Pescini S (2014) Dogs' comprehension of referential emotional expressions: Familiar people and familiar emotions are easier. Anim Cogn 17:373–385
Merola I, Lazzaroni M, Marshall-Pescini S, Prato-Previde E (2015) Social referencing and cat-human communication. Anim Cogn 18:639–648
Meyer S, Nürnberg G, Puppe B, Langbein J (2012) The cognitive capabilities of farm animals: Categorisation learning in dwarf goats (*Capra hircus*). Anim Cogn 15:567–576
Miller DW, Fleming PA, Barnes AL, Wickham SL, Collins T, Stockman CA (2018) Behavioural assessment of the habituation of feral rangeland goats to an intensive farming system. Appl Anim Behav Sci 199:1–8
Miranda-de la Lama GC, Mattiello S (2010) The importance of social behaviour for goat welfare in livestock farming. Small Rumin Res 90:1–10
Miranda-de la Lama GC, Sepúlveda WS, Montaldo HH, María GA, Galindo F (2011) Social strategies associated with identity profiles in dairy goats. Appl Anim Behav Sci 134:48–55
Monfredi O, Lyashkov AE, Johnsen AB, Inada S, Schneider H, Wang R, Nirmalan M, Wisloff U, Maltsev VA, Lakatta EG, Zhang H, Boyett MR (2014) Biophysical characterization of the underappreciated and important relationship between heart rate variability and heart rate. Hypertension 64:1334–1343
Monfredi O, Zhang H, Boyett MR (2015) Letter by Monfredi et al. regarding article, "Physical activity and heart rate variability in older adults: The cardiovascular health study". Circulation 131:e348

Mormède P, Andanson S, Aupérin B, Beerda B, Guémené D, Malmkvist J, Manteca X, Manteuffel G, Prunet P, van Reenen CG, Richard S, Veissier I (2007) Exploration of the hypothalamic-pituitary-adrenal function as a tool to evaluate animal welfare. Physiol Behav 92:317–339

Murphy E, Nordquist RE, van der Staay FJ (2013) Responses of conventional pigs and Göttingen miniature pigs in an active choice judgement bias task. Appl Anim Behav Sci 148:64–76

Nawroth C, McElligott AG (2017) Human head orientation and eye visibility as indicators of attention for goats (*Capra hircus*). PeerJ 5:e3073

Nawroth C, von Borell E, Langbein J (2014) Exclusion performance in dwarf goats (*Capra aegagrus hircus*) and sheep (*Ovis orientalis aries*). PLoS One 9:e93534

Nawroth C, von Borell E, Langbein J (2015a) Object permanence in the dwarf goat (*Capra aegagrus hircus*): Perseveration errors and the tracking of complex movements of hidden objects. Appl Anim Behav Sci 167:20–26

Nawroth C, von Borell E, Langbein J (2015b) 'Goats that stare at men': Dwarf goats alter their behaviour in response to human head orientation, but do not spontaneously use head direction as a cue in a food-related context. Anim Cogn 18:65–73

Nawroth C, Baciadonna L, McElligott AG (2016a) Goats learn socially from humans in a spatial problem-solving task. Anim Behav 121:123–129

Nawroth C, von Borell E, Langbein J (2016b) 'Goats that stare at men'—Revisited: Do dwarf goats alter their behaviour in response to eye visibility and head direction of a human? Anim Cogn 19:667–672

Nawroth C, Brett JM, McElligott AG (2016c) Goats display audience-dependent human-directed gazing behaviour in a problem-solving task. Biol Lett 12:20160283

Nawroth C, Prentice PM, McElligott AG (2017) Individual personality differences in goats predict their performance in visual learning and non-associative cognitive tasks. Behav Process 134:43–53

Nawroth C, Albuquerque N, Savalli C, Single MS, McElligott AG (2018) Goats prefer positive human emotional facial expressions. R Soc Open Sci 5:180491

Nawroth C, Langbein J, Coulon M, Gabor V, Oesterwind S, Benz-Schwarzburg J, von Borell E (2019) Farm animal cognition—Linking behaviour, welfare and ethics. Front Vet Sci 6:24

Nawroth C, Martin ZM, McElligott AG (2020) Goats follow human pointing gestures in an object choice task. Front Psychol 11:915

Nawroth C, Rosenberger K, Keil NM, Langbein J (2022) Goats (*Capra hircus*) from different selection lines differ in their behavioural flexibility. Front Psychol 12:796464

Neave HW, Zobel G (2020) Personality of dairy goats affects competitive feeding behaviour at different feeder heights. Small Rumin Res 192:106222

Neethirajan S, Reimert I, Kemp B (2021) Measuring farm animal emotions—Sensor-based approaches. Sensors 21:553

Negi GCS, Rikhari HC, Ram J, Singh SP (1993) Foraging niche characteristics of horses, sheep and goats in an alpine meadow of the Indian Central Himalaya. J Appl Ecol 30:383–394

Nie L, Berckmans D, Wang C, Li B (2020) Is continuous heart rate monitoring of livestock a dream or is it realistic? A review. Sensors 20:2291

Noë R, Laporte M (2014) Socio-spatial cognition in vervet monkeys. Anim Cogn 17:597–607

Nolte DL, Provenza FD (1992) Food preferences in lambs after exposure to flavours in milk. Appl Anim Behav Sci 32:381–389

O'Bryan LR, Abaid N, Nakayama S, Dey T, King AJ, Cowlishaw G, Rubenstein DI, Garnier S (2019) Contact calls facilitate group contraction in free-ranging goats (*Capra aegagrus hircus*). Front Ecol Evol 7:73

Oesterwind S, Nürnberg G, Puppe B, Langbein J (2016) Impact of structural and cognitive enrichment on the learning performance, behaviour and physiology of dwarf goats (*Capra aegagrus hircus*). Appl Anim Behav Sci 177:34–41

Pascual-Alonso M, María GA, Sepúlveda WS, Villarroel M, Aguayo-Ulloa L, Galindo F, Miranda-De La Lama GC (2013) Identity profiles based on social strategies, morphology, physiology, and cognitive abilities in goats. J Vet Behav 8:458–465

Patt A, Gygax L, Wechsler B, Hillmann E, Palme R, Keil NM (2012) The introduction of individual goats into small established groups has serious negative effects on the introduced goat but not on resident goats. Appl Anim Behav Sci 138:47–59

Perdue BM (2017) Mechanisms underlying cognitive bias in nonhuman primates. Anim Behav Cogn 4:105–118

Pérez-Manrique A, Gomila A (2022) Emotional contagion in nonhuman animals: A review. Wiley Interdiscip Rev Cogn Sci 13:e1560

Piaget J (1954) The construction of reality in the child. International Universities Press, New York

Pitcher BJ, Briefer EF, Baciadonna L, McElligott AG (2017) Cross-modal recognition of familiar conspecifics in goats. R Soc Open Sci 4:160346

Poindron P, Terrazas A, Montes MDLLN, Serafín N, Hernández H (2007) Sensory and physiological determinants of maternal behaviour in the goat (*Capra hircus*). Horm Behav 52:99–105

Proctor HS, Carder G (2015) Nasal temperatures in dairy cows are influenced by positive emotional state. Physiol Behav 138:340–344

Proctor HS, Carder G (2016) Can changes in nasal temperature be used as an indicator of emotional state in cows? Appl Anim Behav Sci 184:1–6

Proops L, McComb K, Reby D (2009) Cross-modal individual recognition in domestic horses (*Equus caballus*). Proc Natl Acad Sci USA 106:947–951

Proops L, Grounds K, Smith AV, McComb K (2018) Animals remember previous facial expressions that specific humans have exhibited. Curr Biol 28:1428–1432

Provenza FD, Lynch JJ, Burritt EA, Scott CB (1994) How goats learn to distinguish between novel foods that differ in postingestive consequences. J Chem Ecol 20:609–624

Range F, Virányi Z (2014) Wolves are better imitators of conspecifics than dogs. PLoS One 9:e86559

Raoult CMC, Osthaus B, Hildebrand ACG, McElligott AG, Nawroth C (2021) Goats show higher behavioural flexibility than sheep in a spatial detour task. R Soc Open Sci 8:201627

Rault JL, Waiblinger S, Boivin X, Hemsworth P (2020) The power of a positive human-animal relationship for animal welfare. Front Vet Sci 7:590867

Réale D, Garant D, Humphries MM, Bergeron P, Careau V, Montiglio PO (2010) Personality and the emergence of the pace-of-life syndrome concept at the population level. Philos Trans R Soc B 365:4051–4063

Reefmann N, Wechsler B, Gygax L (2009) Behavioural and physiological assessment of positive and negative emotion in sheep. Anim Behav 78:651–659

Richter SH, Hintze S (2019) From the individual to the population—And back again? Emphasising the role of the individual in animal welfare science. Appl Anim Behav Sci 212:1–8

Rieucau G, Giraldeau LA (2011) Exploring the costs and benefits of social information use: An appraisal of current experimental evidence. Philos Trans R Soc B 366:949–957

Roitberg E, Franz H (2004) Oddity learning by African dwarf goats (*Capra hircus*). Anim Cogn 7:61–67

Rosenberger K, Simmler M, Nawroth C, Langbein J, Keil N (2020) Goats work for food in a contrafreeloading task. Sci Rep 10:22336

Ruiz-Miranda CR (1993) Use of pelage pigmentation in the recognition of mothers in a group by 2-to 4-month-old domestic goat kids. Appl Anim Behav Sci 36:317–326

Russell JA (1980) A circumplex model of affect. J Pers Soc Psychol 39:1161–1178

Russell JA (2003) Core affect and the psychological construction of emotion. Psychol Rev 110:145–172

Sankey DWE, O'Bryan LR, Garnier S, Cowlishaw G, Hopkins P, Holton M, Fürtbauer I, King AJ (2021) Consensus of travel direction is achieved by simple copying, not voting, in free-ranging goats. R Soc Open Sci 8:201128

Saunders FC, McElligott AG, Safi K, Hayden TJ (2005) Mating tactics of male feral goats (*Capra hircus*): Risks and benefits. Acta Ethologica 8:103–110

Schino G (1998) Reconciliation in domestic goats. Behaviour 135:343–356

Schmidt KL, Cohn JF (2001) Human facial expressions as adaptations: Evolutionary questions in facial expression research. Am J Phys Anthropol 116:3–24

Seyfarth RM, Cheney DL (2003) Meaning and emotion in animal vocalisations. Ann N Y Acad Sci 1000:32–55

Shank CC (1972) Some aspects of social behaviour in a population of feral goats (*Capra hircus L.*). Zeitschrift fur Tierpsychologie 30:488–528

Shettleworth SJ (2010) Cognition, evolution, and behaviour. Oxford University Press, Oxford

Shi J, Dunbar RIM (2006) Feeding competition within a feral goat population on the Isle of Rum, NW Scotland. J Ethol 24:117–124

Shrader AM, Kerley GI, Kotler BP, Brown JS (2007) Social information, social feeding, and competition in group-living goats (*Capra hircus*). Behav Ecol 18:103–107

Shrader AM, Kerley GI, Brown JS, Kotler BP (2012) Patch use in free-ranging goats: Does a large mammalian herbivore forage like other central place foragers? Ethology 118:967–974

Siebert K, Langbein J, Schön PC, Tuchscherer A, Puppe B (2011) Degree of social isolation affects behavioural and vocal response patterns in dwarf goats (*Capra hircus*). Appl Anim Behav Sci 131:53–62

Silanikove N (2000) The physiological basis of adaptation in goats to harsh environments. Small Rumin Res 35:181–193

Smith BR, Blumstein DT (2008) Fitness consequences of personality: A meta-analysis. Behav Ecol 19:448–455

Sorinas J, Ferrández JM, Fernandez E (2020) Brain and body emotional responses: Multimodal approximation for valence classification. Sensors 20:313

Spigarelli C, Zuliani A, Battini M, Mattiello S, Bovolenta S (2020) Welfare assessment on pasture: A review on animal-based measures for ruminants. Animals 10:609

Špinka M (2012) Social dimension of emotions and its implication for animal welfare. Appl Anim Behav Sci 138:170–181

Stanley CR, Dunbar RIM (2013) Consistent social structure and optimal clique size revealed by social network analysis of feral goats, *Capra hircus*. Anim Behav 85:771–779

Stewart M, Webster JR, Verkerk GA, Schaefer AL, Colyn JJ, Stafford KJ (2007) Non-invasive measurement of stress in dairy cows using infrared thermography. Physiol Behav 92:520–525

Stewart M, Stafford KJ, Dowling SK, Schaefer AL, Webster JR (2008a) Eye temperature and heart rate variability of calves disbudded with or without local anaesthetic. Physiol Biol 93:789–797

Stewart M, Schaefer AL, Haley DB, Colyn J, Cook NJ, Stafford KJ, Webster JR (2008b) Infrared thermography as a non-invasive method for detecting fear-related responses of cattle to handling procedures. Anim Welf 17:387–393

Stubsjøen SM, Flø AS, Moe RO, Janczak AM, Skjerve E, Valle PS, Zanella AJ (2009) Exploring non-invasive methods to assess pain in sheep. Physiol Behav 98:640–648

Sutliff DJ (2019) Pack goats in the Neolithic Middle East. Anthropozoologica 54:45–53

Tallet C, Špinka M, Maruščáková I, Šimeček P (2010) Human perception of vocalisations of domestic piglets and modulation by experience with domestic pigs (*Sus scrofa*). J Comp Psychol 124:81–91

Tamioso PR, Rucinque DS, Taconeli CA, da Silva GP, Molento CFM (2017) Behaviour and body surface temperature as welfare indicators in selected sheep regularly brushed by a familiar observer. J Vet Behav 19:27–34

Tattersall GJ (2016) Infrared thermography: A non-invasive window into thermal physiology. Comp Biochem Physiol Part A 202:78–98

Taylor AM, Reby D (2010) The contribution of source–filter theory to mammal vocal communication research. J Zool 280:221–236

Thomas BL, Claassen N, Becker P, Viljoen M (2019) Validity of commonly used heart rate variability markers of autonomic nervous system function. Neuropsychobiology 78:14–26

Thompkins AM, Lazarowski L, Ramaiahgari B, Gotoor SSR, Waggoner P, Denney TS, Deshpande G, Katz JS (2021) Dog-human social relationship: Representation of human face familiarity and emotions in the dog brain. Anim Cogn 24:251–266

Toinon C, Waiblinger S, Rault JL (2021) Maternal deprivation affects goat kids' stress coping behaviour. Physiol Behav 239:113494

Toinon C, Waiblinger S, Rault JL (2022) Maternal deprivation affects goat kids' social behaviour before and after weaning. Dev Psychobiol 64:e22269
Travain T, Valsecchi P (2021) Infrared thermography in the study of animals' emotional responses: A critical review. Animals 11:2510
Travain T, Colombo ES, Grandi LC, Heinzl E, Pelosi A, Previde EP, Valsecchi P (2016) How good is this food? A study on dogs' emotional responses to a potentially pleasant event using infrared thermography. Physiol Behav 159:80–87
Tuyttens FAM, de Graaf S, Heerkens JL, Jacobs L, Nalon E, Ott S, Stadig L, Van Laer E, Ampe B (2014) Observer bias in animal behaviour research: Can we believe what we score, if we score what we believe? Anim Behav 90:273–280
Uddin J, McNeill DM, Lisle AT, Phillips CJ (2020) A sampling strategy for the determination of infrared temperature of relevant external body surfaces of dairy cows. Int J Biometeorol 64:1583–1592
Udell MA, Dorey NR, Wynne CD (2010) What did domestication do to dogs? A new account of dogs' sensitivity to human actions. Biol Rev 85:327–345
Valone TJ (1989) Group foraging, public information, and patch estimation. Oikos 56:357–363
Valone TJ, Templeton JJ (2002) Public information for the assessment of quality: A widespread social phenomenon. Philos Trans R Soc B 357:1549–1557
van Bergen Y, Coolen I, Laland KN (2004) Nine-spined sticklebacks exploit the most reliable source when public and private information conflict. Proc R Soc B 271:957–962
van Horik JO, Langley EJ, Whiteside MA, Laker PR, Beardsworth CE, Madden JR (2018) Do detour tasks provide accurate assays of inhibitory control? Proc R Soc B 285:20180150
Vaz J, McElligott AG, Narayan E (2022) Linking the roles of personality and stress physiology for managing the welfare of captive big cats. Anim Welf 31:99–111
Völter CJ, Call J (2017) Causal and inferential reasoning in animals. In: Call J, Burghardt GM, Pepperberg IM, Snowdon CT, Zentall T (eds) APA handbook of comparative psychology: Perception, learning, and cognition. American Psychological Association, Washington, DC, pp 643–671
von Borell E, Langbein J, Després G, Hansen S, Leterrier C, Marchant-Forde J, Marchant-Forde R, Minero M, Mohr E, Prunier A, Valance D, Veissier I (2007) Heart rate variability as a measure of autonomic regulation of cardiac activity for assessing stress and welfare in farm animals—A review. Physiol Behav 92:293–316
Wascher CA (2021) Heart rate as a measure of emotional arousal in evolutionary biology. Philos Trans R Soc B 376:20200479
Wathan J, Burrows AM, Waller BM, McComb K (2015) EquiFACS: The equine facial action coding system. PLoS One 10:e0131738
Weaver S (2021) The goat: A natural history. Ivy, Brighton
Wechsler B (1995) Coping and coping strategies: A behavioural view. Appl Anim Behav Sci 43:123–134
White RW (1959) Motivation reconsidered: The concept of competence. Psychol Rev 66:297–333
Wilson DS (1998) Adaptive individual differences within single populations. Philos Trans R Soc B 353:199–205
Wilson V, Guenther A, Øverli Ø, Seltmann MW, Altschul D (2019) Future directions for personality research: Contributing new insights to the understanding of animal behaviour. Animals 9:240
Wolf M, McNamara JM (2012) On the evolution of personalities via frequency-dependent selection. Am Nat 179:679–692
Yorzinski JL (2017) The cognitive basis of individual recognition. Curr Opin Behav Sci 16:53–57
Zebaria HM, Hidayet HM, Al-Naqshabendy AA, Hussein NJ, Kakarash NA (2021) Pain caused by ear tagging in kids of native black goats. Sci J Univ Zakho 9:15–19
Zeder MA, Hesse B (2000) The initial domestication of goats (*Capra hircus*) in the Zagros mountains 10,000 years ago. Science 287:2254–2257
Zobel G, Nawroth C (2020) Current state of knowledge on the cognitive capacities of goats and its potential to inform species-specific enrichment. Small Rumin Res 192:106208

Zobel G, Neave HW, Webster J (2019) Understanding natural behaviour to improve dairy goat (*Capra hircus*) management systems. Transl Anim Sci 3:212–224

Zulkifli I (2013) Review of human-animal interactions and their impact on animal productivity and welfare. J Anim Sci Biotechnol 4:25

Zych AD, Gogolla N (2021) Expressions of emotions across species. Curr Opin Neurobiol 68:57–66

Welfare Issues in Goat Farming: Housing and Nutrition

Inger Lise Andersen, Judit Vas, Knut Egil Bøe,
Silvana Mattiello, Luca Rapetti,
and Massimo Trabalza-Marinucci

Abstract

Goats are kept in both intensive and extensive production systems. The most intensive farms are highly industrialized with larger herd sizes. In more extensive systems with large herds, goats have access to an outdoor area in periods or the whole year round. This may decrease animal density and thus increase the welfare of goats. Goats and goat kids can be kept in an uninsulated, cold building and, to a certain extent, may also tolerate high temperatures if they have access to shade and the barns are adequate in terms of space and design. An increase in group size while keeping the floor space per individual constant, reduces aggression, increases social tolerance and locomotion, and opens new possibilities for creating a more stimulating indoor environment. Environmental enrichment improves welfare by reducing social conflicts, providing stimulating activities, and allowing for individual differences. Overall, a heterogeneous environment offering resources and activities placed on different locations in the environment makes it possible for the animals to make individual choices. Environments allowing more individual choices may enhance welfare in itself. The ideal natural pasture should be a vegetational mosaic, more suited for satisfying the behavioural and dietary needs. When evaluating a diet at pasture, the quality of a specific feed goes beyond the mere nutritional value as it also includes the pres-

I. L. Andersen (✉) · J. Vas · K. E. Bøe
Department of Animal and Aquacultural Sciences, Norwegian University of Life Sciences, Ås, Norway
e-mail: inger-lise.andersen@nmbu.no

S. Mattiello · L. Rapetti
Department of Agricultural and Environmental Sciences - Production, Landscape, Agroenergy, University of Milan, Milan, Italy

M. Trabalza-Marinucci
Department of Veterinary Medicine, University of Perugia, Perugia, Italy

S. Mattiello, M. Battini (eds.), *The Welfare of Goats*, Animal Welfare 25,
https://doi.org/10.1007/978-3-031-62182-6_4

ence of any positive nutraceutical compounds and the absence of elements that can be toxic to the animal.

Keywords

Goat housing · Group size · Space allowance · Environmental enrichment · Goat feeding · Pasture

1 Production Systems and Housing Conditions

The total goat population in 2020 was about 1.1 billion heads (FAOSTAT 2020), with goats kept in high numbers in Asia (51.3% of the world population) and Africa (43.4%; Table 1). Goats are mainly used for milk and milk products and meat (Devendra and Solaiman 2010). They contribute to producing food at socially and geographically marginal areas and often increase income for families. In several countries, goat farming has a long tradition and is an important part of culture (Morales et al. 2019). Goats are also used to prevent wildfires by browsing on shrubs, and the indirect social and cultural relevance are also considerable. Norwegian goat producers report about keeping goats for hobby, landscape management, and breed conservation besides the primary dairy/meat/fibre production, emphasizing their multiple functions (Vas et al. in prep). According to Morales et al. (2019), goat farms in both Europe and North America are more typically focused on efficient production of milk and dairy products such as cheese compared to Asia and Africa, where meat production is more important and production efficiency does not necessarily play the main role.

Goats can be kept in various production systems (Table 2), and these can be categorized according to time spent indoors. In intensive systems, goats are closed inside the whole year. These farms may be highly industrialized with larger herd sizes and with focus on high productivity and disease prevention. For instance, intensive dairy production is typically categorized with high animal density and barren indoor environments without access to a stimulating outdoors environment or any sources of environmental enrichment (Fig. 1). The goat kids in such systems are often separated from their mother at the time of birth and male kids are often culled immediately. Some farmers may keep the kids that are not recruited into the

Table 1 World goat populations and breeding purposes (Faostat 2020)

Area	Goats (million)	Percent[a]	Meat (tonnes)	Percent[a]	Milk (tonnes)	Percent[a]
Africa	489	43.4	1,407,655	22.9	4,487,005	21.8
Asia	579	51.3	4,486,883	73.1	12,219,732	59.2
Americas	39	3.5	130,103	2.1	801,285	3.9
Europe	16	1.4	88,129	1.4	3,121,548	15.1
Oceania	4	0.4	29,370	0.5	40	0.0
World	1128	100	6,142,140	100	20,629,610	100

[a]Percent of the total world production

Table 2 Reference overview for description of production systems in different countries around the world

Region	Reference
General	Escareño et al. (2013); Monteiro et al. (2018); Toussaint (1997)
Europe	Morales et al. (2019)
Mediterranean	Caroprese et al. (2016); Dubeuf et al. (2018); de Rancourt et al. (2006)
Northern Europe	Dýrmundsson (2006)
Algeria	Laouadi et al. (2018)
Bangladesh	Hasan et al. (2014)
Brazil	De Figueiredo Guilherme et al. (2017)
Ethiopia	Kefyalew (2015)
Greece	Gelasakis et al. (2017)
India	Nandi et al. (2011)
Italy	Grosso et al. (2016); Sandrucci et al. (2019)
Jordan	Khalaf Al-Khaza'leh (2018)
Lebanon	Serhan and Mattar (2018)
Norway	Muri et al. (2012); Simensen et al. (2010); Vas et al. (in prep); Ådnøy (2014)
Senegal	Sow et al. (2021)
Spain	Castel et al. (2003, 2010, 2011)
Turkey	Daskiran et al. (2018)
UK	Anzuino et al. (2010)
US	Devendra and Solaiman (2010); Hempstead et al. (2021)

breeding herd and use them for meat although these kids have a much lower weight gain than, for instance, those bred for meat production such as the Boer goat. However, feeding dairy goat kids of dairy breeds up to slaughter weight is costly and they do not produce as much meat as, for instance, the Boer goat that has been bred for meat production.

In extensive systems (Figs. 2 and 3), goats have access to an outdoor area in periods or the whole year round. This may decrease animal density and increase welfare of goats by providing a more variable environment in terms of increased physical, structural complexity, or foraging opportunities (see chapter on access to outdoor environment). These systems may also utilize natural resources in a sustainable way and contribute to landscape management. Goats can have access to an outdoor yard or pasture directly at the barn with or without daily access to it. Sometimes access to the outdoor area is limited to daytime or some seasons. In other systems, goats may have a distinctive outdoor season, such as being on pasture during summertime.

In some countries, such as in Norway, this access to pasture is a requirement with a minimum time period (e.g. minimum 16 weeks on rangeland pastures are required in Norway) for goats otherwise kept indoors. These pastures are typically located far distances from the barn, such as mountain pastures. Under extensive conditions, goats are typically kept outdoors the whole year around with or without free access to indoor facilities at night or in harsh weather conditions. In nomadic systems, the

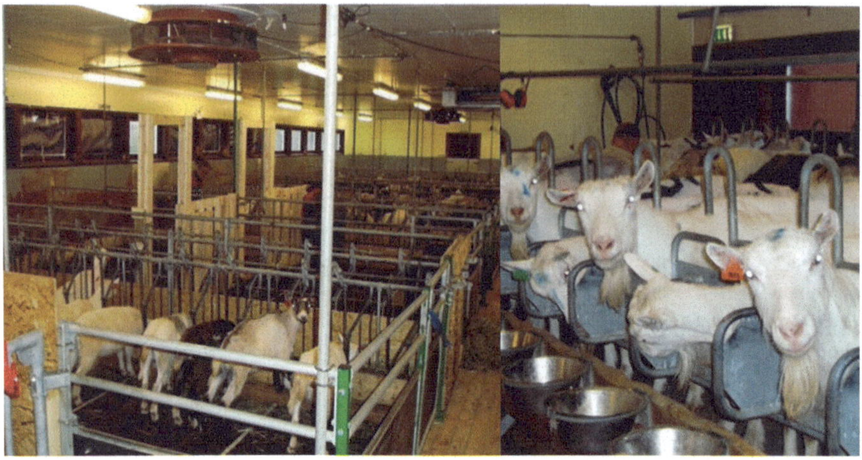

Fig. 1 Example of a traditional indoor housing system in a Norwegian farm with small groups of dairy goats on plastic slatted floors, fed with silage or hay. They are typically fed with concentrates while being milked in the milking parlour (to the right). These goats are kept in an indoor environment like this for 7–8 months but are kept on mountain pasture for around 4–5 months (Photo: Inger Lise Andersen)

Fig. 2 Example of extensive system for dairy goats with straw bedding inside and larger groups allowing more total space and access to enriched outdoor area (Photo: Knut Egil Bøe)

whole community and herds may move together from place to place, like in North Africa or in some Asian regions.

In Europe, both continuous grazing or seasonal grazing are being used. Housing of the goats is more commonly used when there is a lack of available pastures or there are other environmental challenges like snow during winter. EU has no specific regulation for indoor or outdoor housing of goats, but the Council Directive 98/58/EC of 20 July 1998 includes some general requirements concerning the protection of animals kept for farming purposes.

Fig. 3 Example of extensive system with Boer goats having access to large outdoor areas (Photo: Judit Vas)

2 Housing and Thermoregulatory Needs

When considering type of housing for dairy goats (insulated or uninsulated), it is important to have information on thermal demands in goats. There seem to be few studies providing data on lower critical temperature (LCT) in goats. Eik (1991) compared goat kids (11–100 days old) during winter raised in an insulated room (control), where the ambient air temperature was kept at 10–14 °C, with goat kids raised in an uninsulated room, where the ambient air temperature mainly was below 0 °C, and at certain days was even below –20 °C. The daily gain did not differ between treatments. At ambient air temperatures below –20 °C, the goat kids became less active and huddled together to reduce heat loss. In another experiment, when adult dairy goats in a small laboratory test were first exposed to an ambient temperature 10–12 °C for some days (control, period 1) and then to an ambient temperature of –8 to –12 °C for 14 days (period 2; Bøe et al. 2007), both free fatty acids and thyroxine (T4) increased significantly the first couple of days, indicating an increased heat production. Immediately after that, these values went back to basic level. Observations of the goats' activity for the last 48 h in both period 1 and 2 showed that lying time slightly decreased and time spent standing and eating increased when the goats were exposed to the low ambient temperatures. It thus appears from this experiment that dairy goats are well adapted to low temperatures, although feed consumption and activity may increase at lower temperatures. One option when considering type of housing is a system allowing free access to outdoor yards (Fig. 4). Bøe et al. (2012) found that the outdoor yard was less used at decreasing temperatures from 0 to –23 °C, but that the goats spent more time in the yards when roofs were offered. When feeders were positioned in the outside yards, the

Fig. 4 Farm allowing Norwegian dairy goats kept in a traditional indoor system with fully slatted floors access to an outdoor yard during winter in Norway, in an area where the temperature quite often reaches –20 °C (Photo: Inger Lise Andersen)

time spent in the yards increased significantly. Even if lying time was not influenced by weather conditions, lying time in the yards decreased at lower temperature, especially in the yards without roofs. The housing system evaluated here, including access to an outside yard, seems to provide adequate protection for the goats even at low temperatures. Also, studies in a commercial dairy goat herd (Bøe et al. 2012) confirmed that goats appreciated access to an outdoor yard. Even if the outdoor temperature was between 0 and 8 °C, the goats spent nearly 50% of the observation time outside. However, the relationship between environmental temperature and relative humidity should also be taken into account when assessing thermal comfort. This can be expressed by the Temperature Humidity Index (THI), and signs of cold stress (bristled hair and shivering) were actually observed in goats of cosmopolitan breeds when THI was below 55, highlighting that, under certain environmental conditions, goats may suffer from cold and that, in those cases, adequate protection should be provided to prevent cold stress (Battini et al. 2016).

Another and maybe bigger challenge for housing of goats might be high ambient air temperatures. Goats seem to better tolerate heat stress compared to dairy cows, thanks to a higher sweating rate and to a lower body weight:surface ratio (Salama et al. 2014). Furthermore, despite the dramatic reduction in feed intake and increase in respiratory rate and rectal temperature, heat-stressed dairy goats can keep their blood glucose and pH levels like goats under thermo-neutral conditions (see chapter "The Welfare of Goats in Adverse Environments"). However, when THI is above 70, and especially if it goes above 75, heat stress signs (elevated respiration and panting) may appear in cosmopolitan goat breeds (Battini et al. 2016). In the wild, goats often rest in caves to avoid the warm sun. Salama et al. (2021) found that lactating goats exposed to heat stress increased time spent lying, decreased time spent eating, increased drinking time and number of drinking bouts. Heat stress may reduce growth, milk and meat production and compromise immune response. Therefore, to improve welfare levels, strategies to alleviate heat stress should be encouraged, and these include the implementation of appropriate feeding (e.g. provision of water, decreasing the forage:concentrate ratio), management (e.g. avoiding keeping animals on pasture during the hottest hours choosing appropriate grazing time (e.g. avoiding to keep animals on pasture during the hottest hours), minimizing animal handling), and housing strategies. As to housing, shelters' design is of outmost importance to provide an optimal microenvironment and to protect goats from deleterious environmental impacts (Sejian et al. 2021). When goats are kept outdoors, shade should be provided, especially during summer, to minimize solar radiation-induced heat stress (Al-Tamimi 2007). If natural shelters, such as trees or shrubs, are available, they represent the most economic and effective means to protect goats from direct solar radiation; if natural shelters are not available, huts or simple roofs (e.g. straw panels or metal sheets painted with white on top) can be used to effectively protect goats (Sarangi 2018). Several additional strategies for alleviation of heat stress have been reviewed and suggested by Sarangi (2018). For example, sufficient space availability is important to allow goats to stretch out on the ground in order to maximize body contact with the floor, thus favouring the dissipation of body heat. For this reason, the choice of flooring material can play a role to maintain goats' thermal comfort, keeping in mind that goats prefer solid, dry, and clean surfaces (see Sect. 5). Additionally, in a hot environment, a good ventilation system is crucial, especially in indoor housing systems. This can be ensured by avoiding the use of fully enclosed barns and providing open walls, windows, or leaving doors open, possibly taking into account the direction of prevailing winds, to facilitate airflow. Fans and evaporative cooling systems can additionally be useful to alleviate heat stress. Site selection is also important and should contribute to maximize heat dissipation, for example, by minimization of the surface exposed to direct solar radiation and maximization of airflow (Sarangi 2018).

We can conclude that healthy dairy goats and goat kids can adjust to a wide range of temperatures, provided that relative humidity is not too high. They can be kept in an uninsulated, cold building and appreciate access to an outdoor yard even at low temperatures. In a hot climate, access to shade and the provision of barns with adequate space and design, as well as proper ventilation, are crucial. However, a potential difference in thermoregulation between different goat breeds needs to be studied with special attention to sick, recently shorn animals, and newborn kids.

3 Space Allowance and Spatial Organization

Except for extensive systems, for instance, with simple and uninsulated shelters and use of outdoor areas year-round, the indoor environment for intensive production systems is usually more limited in space and often homogenous and barren

(Simensen et al. 2010). Space requirements in the housing regulations of farm animals is to a large extent determined based on the need for physical space and the space needed to rest unrestrictedly. There are no EU-regulations that state specific floor space requirements for goats in commercial production systems, but for organic production, there is a minimum requirement of 1.5 m^2 per goat. Comparatively, the Swiss guidelines recommend 1.5–2.2 m^2 depending on weight and number of animals in the group. The New Zealand Code of Welfare: Goats (2018) recommends a minimum standard of 2.0 m^2 per individual mature housed goat, and 3.0 m^2 per goat as recommended best practice. Comparatively, The Recommended Code of Practice for the Care and Handling of Farm Animals—Goats (https://www.goats.ca/wp-content/uploads/2020/12/Goat-Code-for-Public-Comment-Period-Dec1-2020.pdf) in Canada only recommends 1.4–2.3 m^2 for does. "Personal space" can be defined as the area around an individual which it tries to keep free from other conspecifics (Keeling 1995), and this area is usually larger in front of the head (McBride 1971). Furthermore, "individual distance" can be defined as "the minimum distance that an animal keeps between itself and other members of the same species" (Drickamer et al. 2002). "Nearest neighbour" distance is another synonymous term quite often being used. While it may be advantageous to stay at some distance to avoid resource competition, it may also be beneficial to stay close when searching for food and to decrease the chances of predator attack. Animals are often associated with other individuals with similar age and rank (Clutton-Brock et al. 1976), and the individual distance is reported to increase with increasing age (Francois et al. 1999). Young animals are observed to be in closer proximity to other group members than older individuals (Pereira 1988), and one reason for this may be that younger animals are more vulnerable to predation compared to older individuals. In an experiment with heterogeneous vs. homogenous age groups of dairy goats where they had 12 m of available length in the pen (Bøe et al. 2013), the individual distance was shorter during resting than feeding in the heterogeneous age groups, but this difference was not present in the adult (>3 years) or young (<2 years) homogenous age groups (Fig. 5). Furthermore, individual distance between pairs of young goats (1.5 ± 0.2 m) was lower than between pairs of adult goats (2.1 ± 0.2 m) and pairs of young–adult goats (2.7 ± 0.1 m), irrespective of activity. Depending on group structure and how groups are housed, minimum individual distance should be taken into account when calculating space requirements during the planning process of future building design for goats.

Young goats appear to be more tolerant to having other individuals of similar age in proximity and keep the longest distance to elder goats. Aschwanden et al. (2008a) found freely chosen distances of 1.2 m for juvenile groups and 2.5 m for adult groups at feeding, and this distance was affected by grouping age and the quality of social bonds. Goats grouped as juveniles freely chose smaller distances than those grouped as adults. When goats are kept in social groups, they need space to interact and communicate, and to be able to withdraw from aggressive conflicts and avoid others. They also need enough space to pass each other on their way to feeders and drinkers. Taken into consideration the preferred distances of 2.5 m for adult goats, 1.0–1.5 m^2 of floor space per goat may thus not be sufficient to ensure

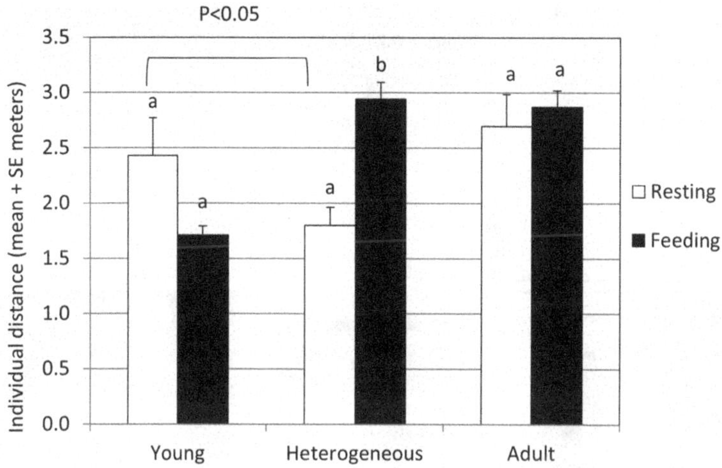

Fig. 5 Preferred individual distance (mean + SE) between individual goats when resting and feeding in homogenous and heterogeneous groups of young (<2 years) and adult (>3 years) goats (from Bøe et al. 2013). This information is important for planning housing design

well-functioning social groups. In another experiment, when 1.0, 2.0, and 3.0 m^2 per goat was compared (Vas et al. 2013), the amount of offensive and defensive behaviours increased with increasing density and the goats kept a greater individual distance when more space was available (Vas and Andersen 2014). The increase in agonistic behaviours was not reflected in blood cortisol level, weight gain, or production data, suggesting that goats easily adapt to less favourable conditions (Vas et al. 2013). Goat kids whose mothers were exposed to the highest prenatal density (and especially female offspring) were more fearful than the kids from the lower prenatal animal densities, and prenatal animal density influenced cognitive capacity as well (Vas et al. 2019). These results show that high density situations increase social stress, and that this may have consequences for the next generation of females that are recruited to the breeding herd, with potential for epigenetic effects (genotype–environment interactions) over several generations and indirect selection for fearful females. Therefore, animal density could have major and long-term effects on the goat population on a farm. When discussing effects of animal density, we should not forget that organization of space could be of equal importance for welfare. For instance, when goats are resting, they prefer resting against a wall rather than in the open space (Fig. 6), and resting time increases when extra walls are provided in the pen (Ehrlenbruch et al. 2010a).

According to Andersen and Bøe (2006), goats spent less time resting and rested less synchronously if resting space with solid floor was 0.5 m^2 compared to 0.75 m^2 or 1.0 m^2, showing that it is important to differentiate between resting space and space for other activities such as exploration, feeding, and drinking (see also Sect. 5 about goat preferences). Goats prefer to rest on solid floor that is dry and clean and may rather prefer to stand on a slatted floor area than to rest on a floor where there is urine, faeces or where the floor is wet for some other reason. Although resting

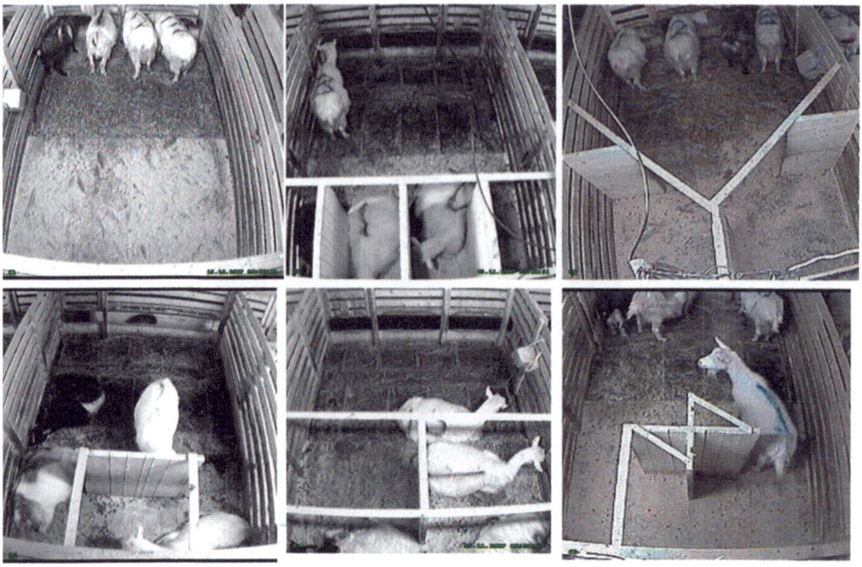

Fig. 6 Goats prefer resting towards a wall and may compete for resting space adjacent to a wall. This could be because they feel safer and/or comfortable. This picture shows an experimental set-up with the purpose of providing goats extra walls to rest against (Photos: Knut Egil Bøe)

space is most important for time spent resting, a multi-level resting area reduces the number of social conflicts compared to having the resting area on one level when total lying space is the same (Andersen et al. 2007). In addition, in small groups, the lowest ranked individuals spend less time resting, less time resting against a wall, and spend more of their resting time in a low-comfort activity area than in the more attractive resting area when one or a few goats can monopolize this space. This finding pinpoints the importance of understanding how much space each individual occupies and to what extent movement of one individual is limited by the presence and movement of other pen mates. Freedom of movement within a limited space could thus be viewed as an important resource. The distribution of specific resources such as feeding places, water, and resting places will thus affect individual differences in animal movement. The closer these resources are localized, the more aggregation and aggressive conflicts we can predict to find.

Regarding space, we need to combine floor space with quality and function of the space and relevant activities or stimuli in the environment to be able to design optimal environments for goats in the future. The design of the environment is just as important as space, but a space allowance of 1.0 m^2 or less cannot be considered sufficient to meet the needs of the goats. Information about use of space and individual distances both during feeding and resting is crucial, and social dynamics in different environments must be studied in more detail to ensure a better planning of future housing systems.

4 Group Size in Goats

Food distribution and predator risk have historically been pointed out as the two key factors affecting group size in wild mammals (reviewed by Krause and Ruxton 2002). Goats (*Capra hircus*) are no exception. Feral and wild goats often live in stable groups of 2–10 individuals, and the largest groups of 100–150 individuals are only seen in open terrain (e.g. Shank 1972). Social network analysis revealing social bonds between female feral goats has suggested optimal clique sizes of 12–13 individuals (Stanley and Dunbar 2013). Considering social bonds between females, this group size appeared to be socially most stable even when ecological and climatic conditions differed in a natural habitat. Their work showed that individuals had an active maintenance of relationships through seeking proximity with preferred partners over time. The importance of social bonds when discussing costs and benefits of group living has until recent years been quite neglected in farm animals, including goats kept in production environments. However, when discussing future housing systems, this should be taken into consideration.

Several studies involving animals across different taxonomic groups in a production environment show that, if space per animal is kept constant, implying an increased total space with increasing group size, individuals within larger groups may display less aggression and competition for limited resources (e.g. reviewed by Estevez et al. 2007; pigs: Andersen et al. 2004; poultry: Estevez et al. 1997, 2002; calves: Færevik et al. 2007), less behavioural synchrony, enhanced locomotion and movement over larger distances (calves: Færevik et al. 2007), and an increased attraction between familiar animals (calves: Færevik et al. 2007). It is not yet clear whether the increased locomotion and distance moved with increasing group size simply is a direct consequence of increased total space, or whether the goats are avoiding others to a greater extent. Behavioural synchrony can be viewed as a risk reducing anti-predator strategy, and in larger groups, vigilant behaviour can be shared among more individuals spending time watching for danger. Overall flock synchrony may thus decline with increasing group size, but not necessarily within the smaller sub-groups of individuals that stick together. Larger groups would allow a greater range of behavioural strategies than in smaller groups (Estevez et al. 2007; Andersen et al. 2004), and behavioural plasticity should allow animals to change strategies and adapt more easily to varying social and physical conditions. The effects of group size have often been studied shortly after mixing, and there is a lack of documentation in stable groups. In social species, some animals can intervene on negative interactions and thus serve as "social mediators" in conflict resolution. This behaviour is also documented in goats (Keil and Sambraus 1998) and is observed both during negative and positive social interactions such as social grooming or play in other species. Social interventions and positive affiliative behaviours are important for the stability of the social group. When social behaviours were compared in groups of 6, 12, and 24 goats (Andersen et al. 2011), a lower frequency of both positive and negative social activity in the largest group size was found, but this difference was most pronounced for butting and clashing, which can be considered the most intensive aggressive interactions. However, there was a substantial decline in agonistic interactions from day 1 of regrouping until day 5, suggesting that group size may have a stronger effect when unacquainted individuals are mixed and not so much after the group is established. This is in accordance with the results reported by Andersen (2015), who found that there were no significant differences in agonistic interactions between stable groups of 6 vs. 24 goats kept in a similar environment. As documented by Andersen et al. (2007), repeated regrouping enhances the level of social stress through an increased number of agonistic interactions, with prenatal consequences for coping style and fear in the offspring. Group stability is thus one out of many key factors for successful and welfare friendly housing of goats.

The overall findings suggest that they are negatively affected by high densities, but that they are well adapted to different group sizes. It should be mentioned that there is a lack of studies on larger groups of 50–100 or even more goats. When groups become sufficiently large, the chance of interacting within one specific individual within a certain time frame will decline. We would thus predict that the impact of dominance with respect to access to resources would decline and that the social structure will change towards smaller sub-groups of individuals developing social bonds, taking different roles in the group and that social strategies become more dynamic. This would allow individuals to seek proximity to some preferred pen mates while seeking distance to others. In small groups of, for instance, 4–6 goats, everyone must interact with each other. As discussed earlier in this chapter, larger flocks open new possibilities for creating a more varying and stimulating indoor environment, as total space increases and may be socially more stimulating. Larger groups are beneficial both because of increased social tolerance in the group, increased extent of sub-grouping and not the least because this increases the amount of total space, allowing the goats a greater freedom of movement.

5 Flooring Preferences in Goats and Quality of Rest Areas

Several different types of flooring are used in goat housing: solid wood, rubber mats, mattresses possibly covered with some kind of bedding (wood shavings or straw), and types of slatted floors like expanded metal, plastic slats, concrete slats, or wooden slats. When choosing type of flooring in goat housing it is important to know something about the goats' own preferences. While there are several experiments on the preferences for different floor types in cattle (e.g. Tucker et al. 2003, 2009) and sheep (Færevik et al. 2005; Jørgensen et al. 2017), data on flooring preferences in goats are scarce. In a laboratory test, dairy goats could choose between two different floor types in a test pen (Bøe et al. 2007). The different floor types included expanded metal, solid wood, straw, and rubber mattress. At moderate ambient temperatures (+10 to 12 °C), the goats showed a significant preference for lying on expanded metal over solid wood and straw, and a clear preference for lying on solid wood over straw. However, at low temperatures (−10 to −12 °C), the goats showed a preference for wood over expanded metal and there was no longer a clear preference for solid wood over straw. When interpreting these results, we need to take into account individual preferences and the effect of previous experience. These goats were normally kept in pens with expanded metal and might have selected the familiar floor instead of an unfamiliar type of floor. It has also been found that goats enjoy resting on sand and rubber mats (Ehrlenbruch et al. 2009), the latter being developed for cows. Goats may enjoy hard surfaces for resting if they are dry and clean. They do not rest where there are leftovers of concentrates and roughage are present, and generally avoid resting on urine and defecation, and would rather rest standing than lying in an area that is wet or dirty.

Sutherland et al. (2017) did an experiment where 32 non-lactating Saanen goats were exposed to four different floor surface treatments: Wood shavings, metal mesh, rubber mats, and plastic slats. Initially, all goats went through an acclimatization period where they had free access to all the four floor types for 7 days. Then they were exposed to just one of the floor types at a time for 10 days (restriction phase) and lastly, they had free access to all four floor types for 2 days (free access phase). During the restriction phase, there were no difference in lying time between the floor types. In the free access period, the goats spent most time lying and standing on the rubber mats, a little less on plastic slats, and least on metal mesh and wood shavings. Interestingly, the goats both urinated and defaecated more on wood shavings compared to metal mesh, plastic slats, and rubber mats. If they urinate on wood shavings, the goats more easily avoid getting urine on their legs compared to on harder surfaces. Both cattle (e.g. Tucker et al. 2009) and sheep (e.g. Jørgensen et al. 2017) show a preference for soft bedding materials. Hence, it seems reasonable to predict that goats also prefer similar materials (Sutherland et al. 2017). However, also today goats apparently thrive in steep cliff hills and perhaps may not have developed natural preferences for soft surfaces. The results in the two studies on floor preference in goats make it difficult to provide clear recommendations for the "best" type of flooring, as the goats' preference may depend on experience and ambient temperature, but the findings suggest that goats may enjoy harder surfaces if they are clean and dry. Alpine goats did not reduce their resting comfort even with minor amounts of straw and show preferences for resting with wall contact and without body contact with another goat (Celozzi et al. 2021).

6 Feed Trough Dimension and Design

Feed troughs may have different size, shapes, and design, which may obviously affect the access to food. As a rule, all animals should have free access to food, respecting their natural feeding behaviour and avoiding excessive competition with other animals. From the available literature, it seems clear that, by limiting access to food, agonistic behaviour increases (Loretz et al. 2004; Jørgensen et al. 2007). Although there is no clear consensus on feed trough size, most authors agree that each goat should have access to at least one feeding place (Jørgensen et al. 2007; Muri et al. 2013); to limit agonistic behaviour, suggested values for space at the feed trough range from 0.33 to 0.40 cm/head (Cull 1988; Toussaint 1997). These values can go up to 60 cm/head for bucks (Cull 1988); however, this space can probably be reduced if feed is available ad libitum. Furthermore, access to feed can double in case of horned goats (Loretz et al. 2004) because of their usually larger body size (Tölü and Savaş 2007). This also applies to those undesirable cases when horned and dehorned goats are kept in the same pen, because usually dominant horned goats are likely to have priority access to food. Feed barrier design can differently affect the social behaviour of goats with regard to ease of leaving the manger, backward view, and presence of physical separation and stress-related behaviours can be due to the presence or absence of horns (Nordmann et al. 2011). Palisades (wooden or metal) appear more appropriate for goats, whereas the use of neck rail and diagonal fence is highly discouraged (Nordmann et al. 2011). These types of feed barrier increase the agonistic behaviour especially in hornless goats. Both hornless and horned goats need longer time to leave the diagonal fence, but horned goats may also find difficult to leave the neck rail. The metal palisade was found to reduce chronic stress and the level of cortisol during feeding time (Nordmann et al. 2011).

Partitions between feeding spaces may contribute to improve animal welfare by decreasing agonistic behaviours, thus allowing goats to spend more time feeding (Aschwanden et al. 2009a). Neave and Zobel (2020) also noted that the level of agonistic behaviour during feeding may be affected by individual personality traits, but further research is required to understand these different individual coping strategies. Another important aspect when designing a feed trough for goats is the feeding behaviour of this species. As goats like to browse on trees and shrubs, an elevated position of the feed bunk may favour the expression of goats' natural feeding behaviour, as observed by several authors (Aschwanden et al. 2009a; Keil et al. 2017; Neave et al. 2018).These last two aspects (the presence of partitions and of elevated feed bunks) may also contribute to provide a more varied environment and can therefore be considered as environmental enrichments (see Sect. 8). According to Cellier et al. (2022), these kinds of environmental modifications are feasible on farm and could be beneficial for feeding management, allowing goats a more varied choice of feeding behaviour.

7 Feeding of Goats in Extensive and Intensive Systems: Welfare Implications

7.1 Feeding Behaviour of the Goat

The goat is an extraordinary ruminant. Goats are classified as intermediate feeders (Hofmann and Stewart 1972) but, among the domestic ruminant species, they have the biggest capability of adaptation to different environmental conditions under a nutritional ecology point of view (Van Soest 1994). Thanks to a mouth similar to that of wild intermediate feeders, with an extremely mobile upper lip, goats' feeding selection activity is more marked if compared not only to cattle, but also to sheep. With respect to eating and ruminating, goats have intermediate characteristics between those of wild concentrate feeders (e.g. roe deer), which have a relatively simple rumen and eat and ruminate frequently, and, on the opposite side, those of grazers, who have a more developed rumen and a less frequent ingestion activity (Van Soest 1994). Goats also have a very effective chewing activity: compared to sheep and cows, the size of the particles of the ingested bolus is smaller. Comparative studies showed that, with the same chewing acts, the goat is able to perform a better shredding of the fibrous particles during the first chewing, while it tends to perform less mericic chews (i.e. chewing during rumination). Moreover, the goat shows a better efficiency in the nitrogen recycling from the blood to the rumen. The amount of recycled urea is between 0.4 and 9.7 g/day, depending on the amount of ingested nitrogen and the physiological state of the animals (Brun-Bellut et al. 1991). The same authors have estimated that urea recycling is able to provide about 20% of the rapidly degraded protein required by the animal. Also known is the goat's high ability to use the fibrous component of the ration. If fed diets rich in low-quality forages, poor in nitrogenous substances, and high in fibre, goats ingest more dry matter and digest fibre better than sheep (Domingue et al. 1991). These differences in the digestive capacity of the goat species compared to other ruminants tend to be reduced when the ingestion consists of good-quality feedstuffs and the diets are in any case balanced. With respect to the digestive capacity of very fibrous diets, it must also be emphasized that within the species there are differences between breeds as a result of adaptation to the environment. For example, a comparative study highlighted the more pronounced digestive capacity of the small desert Bedouin breed compared to the cosmopolitan Saanen breed (Silanikove 2000).

7.2 Feeding and Welfare in Extensive Farming Systems

As reported above, goats are characterized by a selective and opportunistic feeding behaviour, which means that they are able to use a large variety of plants including shrubs, trees, and herbaceous species (Bartolomé et al. 1998; Manousidis et al. 2016). If goats are left free to choose, however, they usually prefer the tree/shrub component, especially when the herbaceous layer has reached maturity and senescence (Egea et al. 2014). Thanks to their physiological and anatomical features, they can feed on spiny and thorny plants and stand on the hind limbs to pick up leaves and more nutritious plant parts (Fig. 7). Accordingly, the ideal natural pasture to be grazed by goats should be a vegetational mosaic (such as the *garrigue* in the Mediterranean ecosystems), more suited to satisfying the behavioural and dietary needs of the species.

Although the nutritional value of the tree-shrub components tends to be medium to low [a low content of crude protein and digestible carbohydrates, and a high content of lignin and tannins (Ben Salem et al. 1994, 1999)], these allow the maintenance of a minimum supply of nutrients during the most critical periods of the year, when herbaceous species are in the senescence stage (Decandia et al. 1999, 2005). Because of these periodic fluctuations in nutrient content, which vary with latitude and other microclimatic factors, it is advisable that breeders perform systematic chemical analyses of the herbaceous component of the pasture. This allows a nutritional evaluation of the diet, which in turn directly affects animal welfare. Generally,

Fig. 7 Goats browsing a wild rose shrub in a mountain area (Photo: Luca Rapetti)

mixed pastures ensure a more consistent supply of nutrients than monophyte pastures, since different herbaceous species have vegetative cycles more evenly distributed over the seasons. Additionally, they show different degrees of resistance to adverse climatic factors, such as summer drought in the Northern hemisphere (Hofer et al. 2016). When evaluating a diet at pasture, the quality of a specific feed goes beyond the mere nutritional value, which is based on the level of classical nutrients (such as proteins, lipids, and carbohydrates) and its digestibility; it also includes the presence of any nutraceutical compounds and the absence of elements that can be toxic to the animal (e.g. allergens, poisonous plants, parasites, or extraneous elements) and compromise welfare (Poutaraud et al. 2017). Nutrition and psychological welfare in grazing ruminants are tightly related. The hedonic welfare can affect voluntary feed intake by producing a reward response mediated by a series of neuronal and hormonal pathways. This means that factors beyond homeostatic/nutritional needs influence ruminant interest in a particular feed. In a recent review, Beck and Gregorini (2020; Fig. 8) suggested an additional factor for consideration in the process of achieving good welfare, i.e. the possibility for the animal to select when feeding in a varied environment. When giving animals a choice, we let them "take control" of their environment in a way that depends on their individuality, improving the so-called eudaimonic welfare. Eudaimonia is a concept firstly proposed by Aristotle and more recently interpreted by the two models of psychological well-being and self-determination theory (Duignan 2020; Arasteh Gatchpazian 2022). According to this holistic theory, eudaimonic welfare

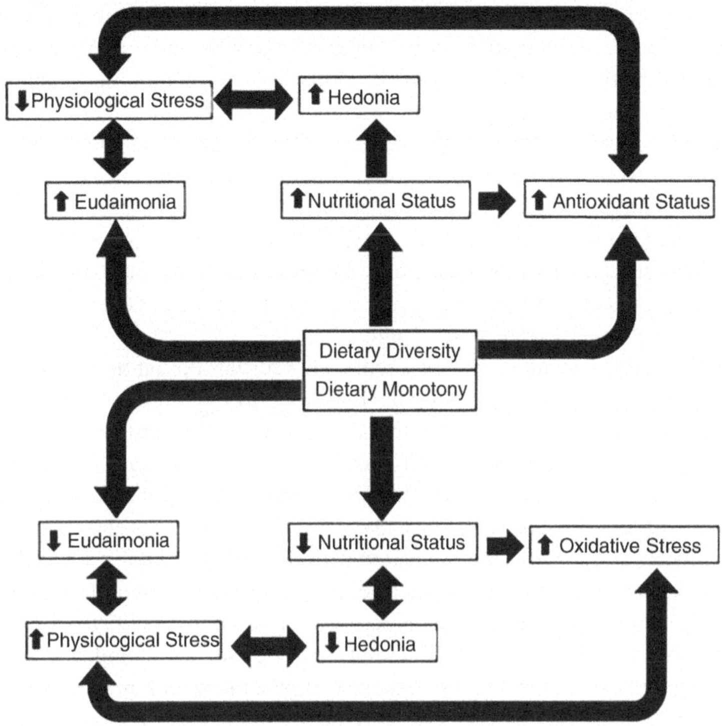

Fig. 8 Interrelationships among dietary diversity, nutritional status, and hedonic/eudaimonic welfare. Allowing feeding on one or few feeds only might affect the nutritional value of the diet and animal welfare (Beck and Gregorini 2020)

contributes to general welfare and interacts with the hedonic welfare, which is in turn more associated to the coverage of nutritional needs.

7.2.1 Supplementary Feeding at Pasture

Supplementing a diet based on pasture may not be necessary if the pasture biomass offers a balanced supply of nutrients and energy that meets the nutritional needs of the animal. However, as mentioned above, in some periods of the year, the composition of the vegetation proves insufficient to achieve this goal, especially in late pregnancy and early lactation. In the last phase of gestation and in the case of twin pregnancies, the risk of onset of pregnancy toxaemia is very high in the goat (Rook 2000). This is probably the most important metabolic disorder in the goat and needs to be adequately prevented, as it is often underestimated in the farm and diagnosed only in the last stage, putting both the survival of the mother and that of the foetuses at risk (Weaver et al. 2021). On the contrary, when the administration of supplementary feeds is excessive, it might cause the onset of other metabolic diseases such as ruminal acidosis (for concentrates rich in starch) or alkalosis (if the concentrate contains high quantities of easily fermentable protein). The risk of alkalosis is much higher when protein supplements are fed to goats grazing pastures during early vegetative growth, which contain large amounts of protein and non-protein nitrogen (Walsh and Birrell 1987; Kilcher 1981). Additional information on metabolic diseases can be found in chapter "Enhancing Animal Welfare Through Disease Prevention in Goats". Feeding high quantities of concentrate, in addition to exposing the animal to metabolic disorders, may reduce the biomass ingested at pasture, depending on the forage/concentrate replacement rate. This, in turn, relates to concentrate dosage, forage quality, and physiological stage of the animal (Goetsch 2019a). This constraint to goat's natural feeding behaviour imposed by humans, besides being not compatible with animal welfare, might be not economically valid if the nutritional quality of pasture is high.

If provision of concentrate is being reduced, a further option may be to increase the grazing area available to the animal. However, this operation may prove useless if the dispersion of trophic resources in the area is excessive and such as to prevent an adequate level of daily dry matter intake, due to the increase in grazing time (Mellado et al. 2003). In such circumstances, the shepherd's experience, his/her knowledge of the place, and the ability to use the grazing circuits appear essential to ensure welfare and coverage of nutritional needs (Manousidis et al. 2016). The shepherd's experience plays a leading role in pasture management. If the pasture is subjected to excessive grazing, plant regrowth is compromised; in contrast, the abandonment of marginal areas leads to a decrease of species abundance and diversity (Plieninger et al. 2014). In both cases, the impact on the colonization–extinction dynamics of the animal/plant species may be significant and its long-term consequences can deeply affect the environment. The decrease in nutritional quality of the pasture affects welfare and productivity (Kharrat and Bocquier 2010), with most of the risks being associated to transition goats and growing kids. Grazing management must follow local environmental gradients of temperature. It is known that ruminants tend to reduce the time spent grazing during the hottest parts of the day (when resting and ruminating activities increase) and prefer grazing at sunrise and sunset (Pokorná et al. 2013; Castro et al. 2015). Accordingly, in many Mediterranean regions, night grazing has become a common practice during summer. It is also true that nocturnal grazing exposes the flock to a greater risk of predator aggression with obvious negative consequences for the animals (Temple and Manteca 2020), from acute stress during flight to abortion or even death (either caused by the predator or self-procured in the attempt to escape). Grazing areas should therefore be adequately protected with anti-predator fencing systems in areas where this risk appears to be high.

7.2.2 Pasture Biomass as a Source of Secondary Compounds

When compared to conserved forages, such as hay or silage, pasture plant species contain large amounts of secondary metabolites. These phytochemicals are produced by plants for a number of different purposes and may help to enhance animal health and welfare. The three major classes of phytochemicals are phenolics, terpenes, and nitrogen compounds. However, depending on animal's physiological state, plant species, growth phase, and other environmental factors, these compounds can have toxic effects (Poutaraud et al. 2017). Due to their chemical characterization, and the metabolic pathway, they may vary largely for concentration in plant tissues. The same concept applies to primary compounds such as vitamins. For instance, the differences in tocopherol levels and phenolic content across plant species are more significant than those between vegetative phases within species. In the case of carotenoids, despite differences in concentrations between plant species, most species show high levels of these compounds during the vegetative stage (Maxin et al. 2020). Differences in the concentration of both primary and secondary compounds in plant tissues clearly have an impact on welfare, health, and productivity of animals. Uncertainty exists on how much the fluctuations in metabolite concentration affect goats' foraging behaviour (Mkhize et al. 2015). Several studies seem to support the existence of a "nutritional wisdom" in various animal species but, undoubtedly, more research is needed on the subject (Goetsch 2019a, b).

As for many other herbivores, plant selection allows goats to balance their diet and ingest plant secondary metabolites at adequate levels, which means obtaining positive benefits (immune stimulating, antioxidant, anti-parasitic, and antimicrobial action) and avoiding potential negative consequences (Egea et al. 2014; Delgadillo-Puga et al. 2019; Beck and Gregorini 2020). Since different compounds are metabolized through distinct detoxification and elimination pathways, it is unlikely that these become saturated and cause health disorders (Freeland and Janzen 1974; Dunson et al. 2007; Distel et al. 2020). Because of their feeding behaviour and degree of selectivity, goats are often used in successful strategies of weed control in marginal areas (Lamming 2001; Sedivec et al. 1995). Nevertheless, the ability of an animal to adapt to new diets, or select among different available feeds, is greatly influenced by previous experiences. Indeed, the consumption of a potentially harmful food is often associated to negative post-ingestive effects (Bryant et al. 1991). When only the most favoured feeds are available for consumption, herbivores are less likely to learn to combine different plants rich in potentially toxic metabolites.

Tannins

Many studies have focused on the tannin content of tree-shrub species. Tannins are high molecular weight, water-soluble polyphenols that can be classified into condensed tannins and hydrolysable tannins. They are known to depress the digestibility of certain nutrients, especially that of proteins, which they bind to form soluble or insoluble complexes. However, the presence of these phenolic compounds—used by the plant as a defence against grazers and parasites (Furlan et al. 2011)—does not represent a real threat to goats. Like many species of wild herbivores, goats are more tolerant to tannins than are other domestic ruminants (Decandia et al. 1999; Silanikove et al. 1997; Egea et al. 2014). This phenomenon appears related to the adaptive behaviour of rumen microorganisms (Makkar 2003) and the presence of proteins of salivary origin, which can bind to tannins with greater affinity than diet-derived proteins (Vaithiyanathan et al. 2001). The effect of tannins on the digestibility of the diet can be mitigated by using tannin-binding agents such as charcoal or polyethylene glycol, able to bind stably to these compounds (Silanikove et al. 2001). The increased availability of nutrients has predictable effects on animals' welfare, especially in the most critical physiological periods such as that of transition. Being a group of highly reactive chemical compounds, tannins are potentially able to interact with other toxic secondary metabolites (such as alkaloids and saponins) and to counteract their detrimental effects. In an experiment with lambs, Lisonbee et al. (2009) found that the animals infused with tannins, when given the option of choosing among different forage varieties, showed higher consumption of the high-alkaloid tall fescue than the control group. Other potential areas of interest related to the beneficial effects of tannins on small ruminants' welfare are the following (Waghorn 2008): (a) anthelmintic action; (b) prevention of rumen bloat; (c) possible beneficial effect for gastro-intestinal integrity; (d) reduced risk of laminitis and pulmonary emphysema.

As for the anti-parasitic effects exerted by tannins, many studies have been conducted on both wild and domestic animal species. However, the opinion of researchers is not univocal. French (2018) suggests that these methods based on plant bioactive molecules, known for centuries and largely adopted by rural communities, represent a valid alternative to the classical pharmacological approach. Indeed, interesting results on ruminant species were obtained using a variety of tannin-rich forage species such as sainfoin (*Onobrychis viciifolia*) (Fig. 9), *Sulla* spp., and *Lotus major*, or other browse species (Waghorn 2008). In this latter case, however, a series of detrimental effects on animals' health were observed. Athanasiadou and Kyriazakis (2004) suggested taking into consideration pros and cons of adopting diets rich in tannins. In fact, in several circumstances, the depression in digestibility and the adverse reactions determined by tannins might be not counteracted by the positive effects on animal welfare (i.e. reduction of parasite burden, antioxidant action, microbial selection within the gastro-intestinal ecosystem). It must be pointed out that most of these studies have been conducted on grazing sheep, and

Fig. 9 *Onobrychis viciifolia* at early flowering stage (Photo: Luca Rapetti)

data are not directly transposable to the goat (Paolini et al. 2004; Min and Solaiman 2018). Indeed, goats and other wild ruminants, such as deer, are more adapted to shrubs consumption and high levels of tannins. Several studies showed that feeding tanniferous plants (such as *Sericea lespedeza* and *Viscum verrucosum*) or tannins extracted from quebracho to goats infected with gastro-intestinal parasites (primarily *Haemonchus contortus, Teladorsagia circumcincta,* and *T. colubriformis*) managed to reduce egg counts and larval development (Madibela and Jansen 2003; Paolini et al. 2003). It is concluded that condensed tannins from a variety of herbaceous and browse forage species can play an important role in mitigating the effects of goats' internal parasites. It must be stressed that pasture management by experienced shepherds and appropriate training in grazing circuits is essential for goats to be able to balance the ingestion of bioactive compounds in the right amounts and proportions.

Antioxidant Compounds

Antioxidants are now recognized as having a positive impact on health and welfare because they lessen oxidative stress, which is associated to numerous diseases and metabolic disorders (Durand et al. 2013; Rahal et al. 2014). The general mechanism of an antioxidant compound consists essentially in preventing oxidative stress by reducing the formation and/or scavenging of oxidants. Three major groups of substances are integrated into the antioxidant system of an animal organism: (a) antioxidant enzymes (superoxide dismutase, glutathione peroxidase, and catalase); (b) low molecular mass compounds such as tocopherols, ascorbic acid, carotenoids, uric acid, glutathione, and phenolics; (c) proteins that bind to transition metals. The association between oxidative stress and the most common infectious pathologies and metabolic disorders (metritis, mastitis, pregnancy toxaemia, and hypocalcaemia), especially in the transition period, has been largely documented and reviewed in small ruminants by Celi et al. (2010). The generation of free radicals, which cause oxidative stress, is well documented under many breeding conditions, especially when intensification is applied (Ihejirika et al. 2017; Goetsch 2019a, b). When compared to dairy cows and sheep, the relationship between diet composition and oxidative stress has received little research interest in goats. In a study where transition Red Syrian goats were fed a diet providing either 80 or 140% of their daily energy requirements, no changes or very few modifications on redox homeostasis were observed (Celi et al. 2010). Samimi et al. (2021), in a study conducted on Saanen goats in late pregnancy and early lactation, reported symptoms of oxidative stress related to hormonal/metabolic disorders and mineral deficiencies, that were exacerbated under thermal stress. Jimoh et al. (2018) compared the oxidative status in West African dwarf sheep and goats and found that rams had superior antioxidant defences than bucks.

Among antioxidants compounds, polyphenols are present in many plant tissues and can be found in large amounts in the biomass selected by goats at pasture. Besides having antioxidant properties, phenolic compounds are known for their antimicrobial, antifungal, anti-parasitic, and anti-inflammatory actions (Di Meo et al. 2013). Again, it must be pointed out that there is a lack of studies in this field conducted with goats, and most of them focused on the productive performance and oxidative stability of meat/milk and derived products (Zhong et al. 2009; Boutoial et al. 2013; Delgadillo-Puga et al. 2019; Cimmino et al. 2018). This makes it difficult to extrapolate observations relating to animal welfare. Nudda et al. (2019) observed a decrease in milk production in Sarda ewes after dietary supplementation with 75 g of exhausted myrtle berries/day, but no negative effects were recorded on animal health. Similar results were obtained by Abbeddou et al. (2011) when olive leaves were fed to lactating ewes. The depression in milk production and dry matter intake and/or digestibility can be possibly ascribed to the negative effect exerted by polyphenols on palatability and rumen functions. For this reason, it is suggested limiting the polyphenol content in the diet to avoid negative effects on production and performance (Correddu et al. 2020; Serra et al. 2021). When evaluating these studies, however, it cannot be ignored that most of them were performed under controlled experimental conditions and the amount of polyphenols was included in the diet at fixed dosage. It is likely that goats left free to graze are able to select the right proportion of botanical species such as not to compromise performance and positively influence their oxidative state.

Toxic Plants

Besides being a serious threat to animal welfare, toxic plants represent a significant financial loss (3–5% heads affected) for the livestock sector as a whole (Panter et al. 1990). Even though goats are selective herbivores, they are still susceptible to plant poisoning. In intensive and semi-intensive systems, goats can be unintentionally intoxicated if poisonous plants are mixed with the rest of the diet. Alternatively, at pasture, goats may be forced to consume plants they generally avoid when other types of forage are available. This is likely to happen when stocking rate is higher than it should be. In addition, unexperienced animals are more likely to expose themselves to a risk of intoxication when they encounter for the first time a poisonous plant (Distel et al. 2020; Bates 2022). As already mentioned for secondary metabolites, this risk depends on plant parts and amounts eaten, and the vegetative stage. Simmonds et al. (2000) have published a comprehensive review on the degree of palatability and toxicity to goats of a large variety (more than 350) of Australian weeds. Among others, a high risk of toxicity to goats was associated to the following plant species: *Aconitum napellus, Atropa belladonna, Conium maculatum, Datura stramonium, Digitalis purpurea, Helleborus niger, Nerium oleander, Leucaena* spp., *Laburnum* spp., *Rhododendron* spp., *Swainsona* spp., and *Taxus baccata.* These species are also common in many other areas of the world.

In many circumstances, acute symptoms and sudden death can be observed in goats following ingestion of toxic plants. After consuming leaves from a felled sugar gum tree (*Eucalyptus cladocalyx*), 24 Angora goats died within a day from cyanide poisoning (Webber et al. 1985). Nwude et al. (1977), in an experimental trial with goats, demonstrated an acute intoxication followed by death after ingestion (2.2 g/kg body weight) of *Dichapetalum barteri* leaves, a plant widespread in central tropical Africa. In this case, monofluoroacetate was considered the responsible toxic principle. Several *Rhododendron* and *Pieris* species, which contain grayanotoxins, and *Prunus* species, which contain cyanogenic glycosides, are known to cause acute poisoning in goats (Bates 2022). Kikuyu (*Pennisetum clandestinum*) is a tropical grass which has been imported into many countries of the world and used for its resistance to drought, salinity, and foot traffic. However, for its high content of calcium oxalate, it can be responsible of acute intoxication and death in goats (Peet et al. 1990). It can be concluded that acute poisoning is mainly observed when goats are not correctly managed, or under experimental conditions. In these cases, treatment of intoxicated animals consists of removing them from pasture, rehydration therapy, administration of effective antidotes (if known and available), and rumenotomy (Bates 2022).

Chronic disorders and negative effects on performance and metabolism are reported as well. Most common symptoms are those associated to abortion, foetal malformations, and reproduction failure. For example, goats are known to be susceptible to the toxic and teratogenic quinolizidine and piperidine alkaloids of lupines, poisonous hemlock (*Conium maculatum*), and *Nicotiana* spp. (Panter et al. 1990). When ingested at sub-lethal doses, *Amorimia septentrionalis* may cause embryonic deaths and abortions due to the presence of sodium monofluoroacetate in its tissues. Fetotoxic effects were reported in pregnant goats after consumption of locoweed (*Astragalus lentiginosus*) (Furlan et al. 2007). Pimentel et al. (2007) described several foetal malformations in goats at all stages of pregnancy associated to the presence in the diet of *Mimosa tenuiflora*. In other cases, the ingestion of poisonous plants may cause neurological symptoms. Barbosa et al. (2007) induced an experimental intoxication in goats by feeding dried *Ipomoea sericophylla* and *I. riedelii* (Convolvulaceae) containing 0.05% and 0.01% swainsonine, respectively. The identified pathogenic mechanism was a glycoprotein storage pathology. In another study, goats were fed (50 g/kg body weight per day) fresh leaves, flowers, and stems of *Ipomoea carnea* var. fistulosa, a South American perennial plant, during a period of 43–60 days (Ríos et al. 2008). Neurological symptoms included incoordination, tremors, and paresis associated to more generic symptoms. The histopathologic evaluation showed a central nervous system damage, with the presence of cytoplasmatic vacuolation.

Additionally, goats may develop intoxication after ingestion of a variety of other common plants such as *Portulaca oleracea* (Simões et al. 2018), *Jacobaea vulgaris* (Anholt and Britton 2017), and *Buxus sempervirens* (Bates 2022). Some of them (for instance, *Ipomoea* spp.) are reported to be highly palatable (Gorniak et al. 2008).

7.3 Feeding and Welfare in Intensive Farming Systems

The goat, as reported above, is a skilled selector, with good chewing and fermentation capacity, able to use well both the fibrous components and the concentrates of the diet. In fact, its high selective capacity is well known (Morand-Fehr et al. 1991), both in the pasture and in the manger, which allows it to ingest the richest parts of nutrients, especially proteins. At the manger, before ingesting large quantities of feed, the goat explores small samples of the ration and chooses the most palatable parts first (Fig. 10). In particular, the goat tends to choose concentrated feeds at the expense of forages, demonstrating a good ability to tolerate, within certain limits, diets rich in starch without problems of acidosis. It is therefore a species capable of making good use of poor rations as well as of rich and balanced diets administered in the manger in intensive farming systems. The high adaptability and productivity that can be obtained from specialized breeds justify the breeding of this small ruminant not only in extensive but also in intensive conditions.

Similar to the cow, the goat is also characterized by a lactation with an initial phase of the ascending curve that leads to the achievement of a peak of maximum milk production (Sauvant and Giger-Reverdin 2018), generally between the fifth and eighth week of lactation for the multiparous, and between the seventh and the tenth week, for the primiparous. The concentration of the main constituents of milk (fat and protein) in turn varies according to the days of lactation, but with a trend

Fig. 10 Goats at the manger exploring and choosing the most palatable parts of the hay (Photo: Luca Rapetti)

opposite to that of the quantity of milk. The increase in energy requirements due to maintenance and, above all, to the milk yield is faster than the increase in the ingestion of dry matter, because the first weeks of lactation are characterized by a poor appetite which negatively affects feed intake. In this initial phase of lactation, the goat makes up for this energy deficit by catabolizing its body reserves, mainly fat, which causes a decrease in the animal's live weight (Sauvant et al. 2010). Subsequently, as lactation proceeds, a trend reversal occurs, passing from the catabolic to the anabolic phase with the replenishment of body fat reserves. It is a completely physiological situation, if kept within certain limits, which highlights the strong homeorhetic ability of the goat (Sauvant and Giger-Reverdin 2018) which allows it to maintain itself in a situation of "oscillating" energy balance that the farmer must assist thanks to proper nutrition in the different phases of the cycle. In intensive farming, at least from a theoretical point of view, it should be easier to feed a flock covering the energy and nutrient requirements, avoiding situations of deficiency or excess that may lead to a reduction in the welfare of the animal. In practice, although the goats physiologically have a good homeorhetic capacity regarding energy, it is very important to evaluate their body condition score (BCS) during the entire reproductive cycle for monitoring the evolution of the body's reserves of adipose tissue. The presence of subjects scored as too thin or too fat compared to the ideal ranges in the various phases of the cycle suggests that the energy allowances of the ingested diet have not always been adequate and may indicate that goats' welfare is not ideal. For example in dairy goats, according to the indication of the French Institut de l'Élevage (Idele 2011), using a 5-point scale, the optimal range of BCS of lumbar area should be 2.75–3.00 at drying off, 2–2.25 at peak of lactation, and 2.25–2.50 before the breeding period. Feeding errors related to dietary energy can have negative consequences not only for production but also for welfare status, even leading to acute or sub-acute nutritional metabolic diseases. In general, for lactating ruminants, it should be noted that the greater their productive potential, the greater the energy density of the diet must be. This derives from the fact that the ingestion capacity of dry matter resulting from the increase in milk production is proportionally lower than the corresponding increase in energy requirements. For example, according to the indicative data tabulated by Sauvant et al. (2010), passing from 2 to 4 kg of milk produced daily, it is possible to calculate that the energy concentration of the ingested ration should increase by about 17%. This explains the consequent need to increase the amount of concentrate feeds in the ration, since they are characterized by a higher energy density than forages, leading to a reduction of the forage to concentrate ratio. From a hedonistic point of view, this satisfies the goats because in confined environment they prefer a diet rich in concentrate (Abijaoudé et al. 2000). However, this leads to a decrease in chewing time, consequently reducing the amount of saliva produced and the relative intake of buffer substances, increasing the risk of ruminal acidosis overall (Sauvant and Giger-Reverdin 2018), with a detrimental effect on the welfare of the animal. This risk is also influenced by the feed composition of the concentrate, being positively related to the starch fermentation rate. Particular attention must be paid with starch from wheat and barley grain because it undergoes to ruminal fermentation faster than that from corn or sorghum grain.

The particle size of the cereals is also relevant. As already demonstrated by Orskov (1979), the goat, compared to the cow, has a more efficient digestive utilization of the whole grain of cereals, also thanks to the above-mentioned greater chewing activity during the first ingestion. In a trial with lactating Saanen goats, the inclusion in the diet of whole maize grain (33% on DM; 33% of starch on DM) compared with the corresponding maize meal, determined a slower starch ruminal fermentation, causing a lower fall down in rumen pH measured four hours after the morning meal (Rapetti and Bava 2004). The lower variation in the rumen pH allowed a higher diet Neutral Detergent Fibre (NDF) digestibility, compensating for the lower starch digestibility caused by the passage in the faeces of some small undigested pieces of grain. In dairy cattle, the sub-acute ruminal acidosis (SARA) is the most important nutritional disease (Enemark 2008); it is characterized by episodes of low ruminal pH with values below 5.6 for at least 3 h/day (Gozho et al. 2005) or below 5.8 for more than 5–6 h/day (Zebeli et al. 2008). Rumen pH variation is related to the variation of rumen volatile fatty acids (VFA). According to Sauvant and Giger-Reverdin (2018) this relationship shows the same trend for cattle and goats. In particular, for pH values lower than 5.8, the acetate to propionate ratio derived from the regression equations proposed by these authors should be lower than 2.3, due to an increase of molar proportion of propionate at the expense of that of acetate when pH is reducing. However, comparing cows and goats fed diets with similar NDF concentration or concentrate inclusion, the pH in the rumen of the goats is about 0.4 points higher than in cows, showing that goats, as intermediate feeders, seem to better tolerate nutritional factors linked to ruminal acidosis.

From a practical point of view, it is advisable that concentrates are inserted in the diet gradually and administered by dividing them into several daily meals. Regarding their composition, for what has been said previously, attention must be given to the fermentability of the cereals used and to the overall quantity of starch, bearing in mind that starch can be at least partially replaced with other non-fibrous carbohydrates such as pectins and beta glucans, or highly digestible fibre such as those contained in the husks of the seeds of some feedstuffs (e.g. soybean hulls). Finally, also the use of mineral substances with a buffering effect can be useful for reducing rumen pH variations and therefore favouring the maintenance of an adequate state of welfare of the goat. Comparatively, in dairy cows, a meta-analysis of 40 publications on various buffer substances (sodium bicarbonate, disodium carbonate, potassium bicarbonate, and magnesium oxide) showed their efficacy in increasing rumen pH, dry matter intake, milk production, and milk fat percentage (Meschy et al. 2004).

7.4 The Use of Dietary Supplements in Goat Feeding: Welfare Implications

Researchers have been trying to improve ruminant nutrition for decades and several approaches are now available to address animal needs and improve animal welfare

(Goetsch 2019a, b). One of them is the use of feed supplements, which may vary according to their origin, chemical structure, or mechanism of action. Being the goat a ruminant, these compounds are primarily used to modulate the ruminal ecosystem. Prebiotics, probiotics, enzymes, organic acids, buffers, vitamins, minerals, and phytogenic compounds are commonly used supplements (Mamuad et al. 2019). Dietary lipids are also frequently introduced in ruminant rations both for their energy density and for their polyunsaturated fatty acid (PUFA) content (Savoini et al. 2010). This broad group of bioactive compounds not only affects digestive efficiency and productivity, but also influences several aspects of the goat's metabolism (e.g. immune response, reproductive activity, and antioxidant status) and, indirectly, their welfare (Savoini et al. 2019). For this reason, many dietary supplements are classified as "nutraceuticals" ("nutritious" + "pharmaceutical"). Recent studies, and especially those which rely on omics technologies, have largely contributed to understand the metabolic aspects in which these bioactive compounds are involved. Unfortunately, the scientific literature on dietary supplements in goats is only marginally addressed to their nutraceutical activity.

Probiotics and Prebiotics

The definition of "probiotic" applies to living microorganisms (bacteria, yeast, and fungi) which have beneficial effects when ingested in appropriate amounts by the animal. Their action is mainly directed to the gastro-intestinal system, where they improve absorption, reduce the incidence of pathogen infection, and restore the intestinal ecosystem balance following diarrhoeal episodes. Additionally, ruminal functions, nutrient synthesis, productive performances, and immune response may be positively affected by probiotic administration (Ekwemalor et al. 2017). Probiotics have also been shown to have a significant impact on fibre degradation and methane production in the rumen, as well as on the prevalence of metabolic diseases including acidosis (Bidarkar et al. 2014). Several studies have demonstrated that not only kids' growth rate, but also clinical status and biochemical profiles were improved following probiotic (*Lactobacillus* spp.) supplementation (Chiofalo et al. 2004). Cai et al. (2022) showed that, after including *Saccharomyces cerevisiae* and *Clostridium butyricum* in the diet of heat-stressed goats, the ruminal concentration of vitamin B1, B2, and niacin was improved. This result appears particularly interesting when considering the impact exerted by heat stress on ruminant metabolism and ruminal biochemical synthesis, along with negative effects on vitamin production. In another study, Ma et al. (2020) reported that dietary supplementation of probiotics (*Saccharomyces cerevisiae, Bacillus subtilis, and Enterococcus faecalis*, singly or in combination) improved milk yield and composition and positively affected the intestinal microbiota in Saanen dairy goats. It must be stressed that the effectiveness of probiotics varies depending on multiple factors, such as the viability and stability of the strains, dose, length of administration, interactions with bioactive and pharmacological compounds, and the characteristics of the animal (genetics, age, stress level, health, and nutritional status) (Ekwemalor et al. 2017). In addition, not in all cases the optimal dose and mode of action have been precisely assessed (Yoon and Stern 1995). For instance, no effects whatsoever were observed when the diet of Black Bengal goat kids was supplemented with *Saccharomyces cerevisiae boulardii* (Saha et al. 2018). The term "Synbiotic" defines the association between a probiotic and a prebiotic (prebiotic = mostly non-digestible carbohydrates that selectively stimulate the growth and/or activity of some genera of probiotic microorganisms) (Pandey et al. 2015). In numerous experiments, this combination has proved more effective than the individual components administered alone. Raksasiri et al. (2019) demonstrated that the association between inulin (prebiotic) from Jerusalem artichoke and a mix of various probiotics (*Lactobacillus acidophilus, Lactobacillus plantarum, Pediococcus pentosaceus, Streptococcus faecium, Saccharomyces cerevisiae, Bacillus subtilis,* and *Bacillus licheniformis*) was able to ameliorate both faecal score and composition of the faecal bacterial population in growing kids, with an increase of the lactic acid bacteria fraction. In addition, the phagocyte activity was improved by the dietary treatment. In another study, Apas et al. (2015) evaluated the effect of the association between fermented sugarcane (prebiotic) and various bacterial species isolated from goats' faeces (*Lactobacillus reuteri, Lactobacillus alimentarius, Enterococcus faecium, Bifidobacterium bifidum* and a probiotic mixture containing equal amounts of the four probiotics) on 45-day-old Creole kids. The most encouraging results were obtained with the probiotic mix, both in terms of intestinal bacterial sub-populations (decreased concentration of harmful microorganisms and increased level of the beneficial bacteria) and in relation to the incidence of diarrhoeal episodes, which is an important cause of kids' morbidity and mortality. In Spanish x Boer goats, Ekwemalor et al. (2017) observed that dietary probiotic supplementation could affect the expression of several genes associated to immune response (both innate and adaptive) and homeostasis, with no effects on productive performances. On the contrary, when using kefir or a commercial mix of probiotics (*Lactobacillus plantarum, L. delbrueckii, L. acidophilus, L. rhamnosus, Bifidobacterium bifidum, Streptococcus salivarius, S. faecium, Aspergillus oryzae*, and *Candida pintolopesii*), Ataşoğlu et al. were not able to observe any effect whatsoever on growth and metabolism of Saanen kids. Taken together, these results suggest that probiotics have a complex mechanism of action, and their effects are not necessarily related to zootechnical performance; however, they may be a valid tool for achieving both gastrointestinal balance and a condition of general welfare.

Minerals

Minerals must be necessarily considered when formulating a ration. Excess or deficiency of most minerals may cause serious disorders that compromise animal health and welfare. Mineral requirements for goats have been extensively discussed elsewhere (NRC 2007) and will not be treated in this section. Usually, concentrate feeds contain a mineral-vitamin premix that ensures coverage of the basal nutritional needs. When goats are at pasture, it is crucial that chemical analyses include the mineral component and are done on a regular basis. Again, it must be stated that research aimed at investigating the relationship between mineral supplementation and welfare in goats is scarce. Here we will report on some specific conditions in which, beyond the coverage of nutritional needs, mineral supplementation can positively influence some metabolic aspects and, indirectly, animal welfare. Selenium (Se) deficiency in soil is known to cause muscle dystrophy ("white muscle disease") in several animal species and, for this reason, most breeders add selenium to the diet of goats, especially in the transition period. Supranutritional Se administered to young goats fed high concentrate diets was shown to be beneficial to hepatic oxidative injury and inflammation. This condition is known to affect health and welfare in highly producing animals (Shah et al. 2022). Similarly, both long- and short-term dietary Se (in the form of Se-enriched yeast) was able to ameliorate fertility and antioxidant status in male goats (Shi et al. 2010). Copper (Cu) nutritional requirement for goats is 25 mg/kg dry matter intake according to the NRC (2007); nevertheless, when higher Cu dietary levels (100–150 mg/kg) were used in goats, growth performance was improved and no toxicity symptoms were recorded (Solaiman 2010). Even if the mechanism of action was not perfectly elucidated, these supranutritional levels of copper were beneficial to the immune system (humoral response improved by day 72 and cell-mediated response improved by days 72–98 of the experiment). In a recent study, Cu dietary supplementation affected lipid metabolism of Lezhi black goat kids (muscular fat increased and serum LDL cholesterol decreased). These results were associated to a lower mRNA abundance of genes associated to long chain fatty acid metabolism (Zhang et al. 2023). Contrary to what observed in sheep, that are extremely sensitive to Cu, goats appear very tolerant to high concentrations of this element in the diet. For instance, Solaiman (2010) reported that goats fed a mineral supplement containing 3050 ppm Cu did not show signs of toxicity for at least 9 months since the beginning of the treatment. When sheep and goats are reared together, Cu supplementation should be closely monitored due to the opposite behaviour observed in the two species.

Dietary cation-anion difference (DCAD) is a nutritional index that indicates the difference between major cations (Na and K) and major anions (Cl and S) in the diet (mmol/kg dry matter). More in detail, it has been shown that feeding diets with a negative DCAD can be effective in preventing the onset of hypocalcaemia (see chapter "Enhancing Animal Welfare Through Disease Prevention in Goats"), which is one of the main metabolic disorders found in dairy ruminants (Block 1984). For this reason, the DCAD level is now considered an important parameter in formulating rations for high-producing dairy goats. Yang et al. (2021) have demonstrated that diets characterized by a negative DCAD can be fed to goats without compromising the ruminal ecosystem and can improve homeostasis while increasing blood calcium level. Finally, it is worth mentioning that dietary supplementation with zeolite (a clay composited of crystalline aluminosilicates and characterized by a high sorbent property) can have various beneficial actions in the digestive system of goats. Zeolite is commonly used in ruminant feeding because of its ability to block ammonium ions, decreasing the possibility for them to be absorbed by the rumen wall. In addition, it can be effective in absorbing toxic compounds such as mycotoxins (Jaynes et al. 2007). In a study by Valpotić et al. (2017), clinoptilolite of zeolite was shown to behave as pH-buffering agent, therefore improving ruminal metabolism. Recently, it has been demonstrated that both natural and nano-zeolite may affect ruminal fermentation and nutrients digestibility (El-Nile et al. 2021). Taken together, these observations become of interest when considering the risk of acidosis experienced by goats fed diets rich in concentrates.

Vitamins

Apart from their activity as essential nutrients in a variety of metabolic pathways, vitamins can be regarded as valid contributors to animal welfare. In goats, for instance, recent studies have shown that the inclusion of thiamine (vitamin B1) in high-energy density diets was able to mitigate the symptoms of sub-acute acidosis by increasing ruminal pH and slowing down lactate production. In addition, the ruminal microbial community appeared stabilized and the amount of toxic compounds produced was reduced (Ma et al. 2021). Similar results were reported by Ahmed and Wang (2022), in an in vitro model using goat ruminal inoculum, when thiamine associated to active dry yeast was included in the medium.

Fats

The effects of dietary lipids on productive performance and on the quality of milk and meat have been deeply investigated in all species of domestic ruminants, goats included (Savoini et al. 2010). In contrast, little research has addressed the influence of dietary fats on animal metabolism and welfare. It is generally known that n-3 PUFA [α-linolenic acid (ALA), eicosapentaenoic acid (EPA), and docosahexaenoic acid (DHA)] exert a series of beneficial actions, among which the regulation of plasma triglycerides and cholesterol, the immune system stimulation, the antioxidant activity, and the regulation of the reproductive activity (Teama and El-Tarabany 2016). Besides, dietary n-3 PUFA can alleviate the effects of the inflammatory processes, by stimulating the production of resolvins and protectins, which act as mediators (Savoini et al. 2019). In goats, Mahla et al. (2017) studied the effect of a diet supplemented with an EPA and DHA-rich fish oil administered for 72 days during the breeding season. The number of preovulatory follicles and ovulation rate were increased, and the kidding rate was improved. The dietary treatment with n-3 PUFA was also reported to improve the immune response, especially the cell-mediated type (Savoini et al. 2019; Agazzi et al. 2004; Caroprese et al. 2016). Pisani et al. (2009) showed that neutrophils isolated from dairy goats, when treated with DHA, produce less reactive oxygen species and their phagocytic activity is increased. Liu et al. (2021) conducted an experiment with cashmere goats aimed at evaluating the effects of the dietary inclusion of linseed (rich in ALA) on lipid metabolism. They observed that linseed grain had a stronger positive effect on blood lipid profile, when compared to linseed oil, and effectively decreased lipid deposition. The same authors speculated that the influence on blood lipid profile might be associated to other effects exerted on the intestinal microbial population. It is also known that omega-3 PUFA can affect the expression in the liver of several genes involved in the lipid metabolism (Savoini et al. 2010). Four-month dietary supplementation with a commercial mix of essential long chain fatty acids, under hot summer conditions, was effective in improving growth rate and blood profile in growing male Baladi kids (Teama and El-Tarabany 2016). Total cholesterol and triglycerides were decreased, and no signs of heat stress were observed. Finally, Gantner and Kompan

(2012) reported beneficial effects on the somatic cell count in milk produced by Alpine goats following dietary supplementation with ALA.

Taking into consideration the above observations, it is not surprising that the inclusion of fats in the goats' diet is relatively common during the transition period and the onset of lactation, when the depression of the immune response is coupled with a reduced dry matter intake and a negative energy balance. It must be emphasized, however, that when feeding PUFA to ruminants (for example, in the form of fish oil), they must be rumen-protected (or rumen inert) to avoid saturation by the microorganisms present in the rumen. In addition, feeding large amounts of dietary oils in a non-protected form negatively affects the activity of ruminal bacteria, namely the cellulolytic sub-population. Shi et al. (2020) reported reduced numbers of rumen protozoa and cellulolytic bacteria (associated to decreased fibre degradation and growth rate) following coconut oil dietary supplementation at a rate of 8 g/goat per day, while lower doses improved growth performance and decreased methane production.

Phytogenic Compounds

Phytogenic compounds account for a large set of chemicals with a wide range of properties, bioactivity and structure, and are commonly found in many plant species that contribute to the goat's diet at pasture (see Sect. 7.2.2). However, they can be also included in the daily ration of animals reared under intensive or semi-intensive systems (Karásková et al. 2015). They are regarded as "natural substances" by the breeder and are relatively safe for both humans and animals. Again, most studies on phytochemicals have focused on the productive performances (with special emphasis on the quality of products of animal origin), often disregarding the advantages in terms of animal welfare and health. Among their biological properties, the immune modulating, the antimicrobial, and the antioxidant activities are the most interesting to be considered in animal nutrition. Only a limited number of studies have been conducted so far focusing on the effects of phytogenic compounds on the welfare of goats. Tannins and polyphenols, already discussed in Sect. 7.3, will not be treated here. It is worth remembering that several extracts rich in these bioactive compounds are available on the market and currently used in small ruminant feeding, mainly with the objective of improving the quality of milk or meat. Some of these studies, however, confirm the antioxidant and immune modulating activity of phenolic compounds in both sheep and goats (Correddu et al. 2020; Branciari et al. 2020; Carta et al. 2022; Yaseen et al. 2022).

The following studies have focused on a variety of different plants or plant parts used in the feeding of goats. The *Boswellia* genus has a wide range of pharmacological activities and is one of the most promising sources of phytochemicals (Hashem et al. 2021). It contains resins, gums, and essential oils (mixture of monoterpenes, diterpenes, and sesquiterpenes) that are characterized by several biological properties (among which: antioxidant, anti-inflammatory, antiarthritic, antirheumatic, antidiarrhoeal, antihyperlipidaemic, antimicrobial, antifungal). The inclusion of Boswellia resin in the diet of lactating goats (2 g/day) induced beneficial effects in terms of lipid metabolism, health conditions of the reproductive apparatus, and colostrum immunoglobulin content. Wang and Wang (2016) fed adult goats with a mix of officinal Chinese plants (Chinese Mosla, Atractylodes rhizome, Officinal Magnolia bark, white hyacinth bean, and Liriope root tuber) and reported a series of beneficial effects on the nervous system, the antioxidant balance, and the immune response. In a recent study, Amsathkumar et al. (2019) observed that a phytogenic feed additive composed of *Woodfordia fruticosa* (flowers, leaves, shoots), *Solanum nigrum* (whole plant), *Trigonella foenum-graecum* (seeds), and *Withania somnifera* (roots), at a rate of inclusion of 1.5%, was able to prevent kids, treated intravenously with the *Escherichia coli* endotoxin, from body weight loss and contributed to restore animals' health status. The effects of feeding an aqueous extract of *Aloe vera* (20 or 40 g/kg dry matter) to crossbred lactating goats were studied by Banakar et al. (2021) in an experiment that lasted 100 days. Apart from increasing milk yield and quality (the higher dose only), the aloe extract improved rumen functions and the health status of the goats, with reference to the antioxidant capacity. Dubious results were obtained by Abrahamsen et al. (2021) using hempseed meal in a feeding trial conducted with 5-month-old, castrated Boer cross goats, since some positive influence on the cell-mediated, but not on the antibody-mediated immunity, was observed.

Algal bioactive compounds have recently attracted the attention of researchers, and several studies on a range of animal species have been conducted. In a 30-day trial, Yang et al. (2022) evaluated the effect of fucoidan (a polysaccharide rich in sulphates extracted from brown algae) dietary supplementation on Chuanzhong weaned male goat kids. Thanks to its antimicrobial, immune stimulating, and antioxidant properties, the fucoidan extract improved the intestinal and immune functions, caecum microbial ecosystem and antioxidant capacity. Similarly, Angulo et al. (2020) observed that, after feeding Saanen × Anglo-Nubian kids the macroalga *Sargassum* spp., the animals' antioxidant status and the immunoglobulin response (both IgA and IgG) were improved.

The large group of bioactive compounds included in the essential oil category has been deeply investigated in monogastric species. In goats, only a few studies focused on the beneficial effects on health status, such as that performed by Shokrollahi et al. (2015) on newborn kids fed rosemary aqueous extract mixed with milk, in which a positive effect on immunity and skin thickness was reported.

In conclusion, the use of phytogenic compounds in goats' feeding is going to be the subject of further research, considering the promising data obtained so far in other animal species. However, to prove efficacy and safety of phytochemicals, long-term experiments are required.

7.5 Water Restriction

In the past decades, the study of energy and nutrient needs of livestock species has received the greatest attention by researchers, with hardly any focus on water requirements and quality. Goats' water requirements vary with dry matter intake and feed water content, physiological state, and environmental factors such as temperature and humidity (Giger-Reverdin and Gihad 1991; McGregor 2004). It is also known that goats can more easily survive water and feed shortages, especially under heat stress, than other domestic ruminants. Among the adaptive mechanisms that the species has developed, the most important are the decrease in voluntary dry matter intake, the reduction in water loss through faeces and urine, and the capacity to decrease the amount of evaporative water from the skin (sweat) and respiratory tract (Reshma Nair et al. 2021). The decrease in animal metabolic rate, observed under water stress, is also considered a conservative energy-saving mechanism (Choshniak et al. 1995). When water stress is associated to heat stress, further depressive effects on feed intake are observed, with a clear negative impact on animal metabolism and welfare (Wilson 1989). For instance, when three local goat breeds from Saudi Arabia were subjected to a 3-day water deprivation experiment at an elevated environmental temperature, their live weight decreased by 21%, while a lower reduction in body weight (18%) was observed in Sudanese male goats which underwent a similar trial but at a lower temperature (Alamer 2006; El-Hadi 1986). Additionally, the type of forage available during water restriction, and its moisture content, exerts a significant effect on dry matter intake reduction by goats (Muna and Ammar 2001). Water shortage or deprivation can affect digestive metabolism, immune response, endocrine system, and metabolic homeostasis; however, there is not perfect agreement among studies mainly because of the differences in experimental conditions and breed, as reviewed by Akinmoladun et al. (2019). In a more recent study, Akinmoladun et al. (2020) evaluated the effects of water restriction on a series of hormones (tri-iodothyronine and thyroxine), digestive enzymes (amylase and lipase), and plasma cortisol (used as stress indicator) in Xhosa ear-lope goats. Concentrations of enzymes and hormones were decreased by the experimental treatment, while cortisol level was increased. Supplementation with vitamin C proved effective in mitigating the cortisol increase and minimized the negative effects on growth rate induced by the period of water scarcity. It must be emphasized that observations reported for a single goat breed cannot be extrapolated to the species as a whole. Documentation exists that certain indigenous populations of goats might have developed the capacity to maintain their level of feed intake under heat stress, as suggested by McGregor (2004) and Akinmoladun et al. (2019). A lower impact of water restriction on metabolism and lactation physiology was reported by Maltz and Shkolnik (1984): black Bedouin goat under water deprivation were able to sustain their milk production for 2 days and were able to attain 70% of their milk yield (total milk yield and milk solids) after 4 days of deprivation followed by 2 days of rehydration. In addition, the normal growth rate of the kids did not appear to be affected. Altogether, these results suggest that goats may adapt to shortage of water for short periods of time, but this is not recommended when water restriction is associated to heat stress and poor feed quality.

7.6 Water Quality

Water quality is of great importance for the health and welfare of animals. Its evaluation must include chemical, physical, and microbiological aspects. Indeed, there might be primary and secondary contaminations, the latter being most common when water is collected in tanks or drinking troughs. Water contamination is not only a direct health risk but can also decrease water intake in small ruminants and cattle (Willms et al. 2002; Al-Khaza'leh et al. 2020). The most common chemical parameter measured in water is the amount of mineral salts dissolved, indicated as "TDS" (Total Dissolved Solids) or "salinity". Major salts dissolved in water are carbonates, bicarbonates, sulphates, nitrates, chlorides, phosphates, and fluorides (Araújo et al. 2010). Goats can adapt to water salinity changes, and this relates to a number of physiological and environmental factors. In his report "Water quality and provision for goats", McGregor (2004) indicates that goats can drink saline water up to 12,500 mg TDS/L. However, a gradual adaptation to salinity changes seems necessary to avoid negative consequences on animal metabolism. These negative effects are mainly mediated by the decrease in water and feed intake. With levels of TDS up to 9500 mg/L, apparently no measurable effects on feed intake were reported. Cardoso et al. demonstrated that increasing water salinity may cause changes in the levels of urea and blood triglyceride. According to Mdletshe et al. (2017), when Nguni goats were offered drinking saline water (5.5 or 11 g total dissolved salts per litre) a decreased water and feed intake, associated to lower growth rate, was observed. In addition, their cardiac frequency increased as a response to water salinity. McGregor (2004) reported the death of Angora goats following ingestion of saline water with TDS and magnesium above the safety levels. This water derived from a dam subjected to evaporation after a period of drought. Finally, even if no conclusive evidence was found, it is likely that differences among breeds exist in both tolerance to salinity and water deprivation (Greenwald 1967; McGregor 2004).

As for calcium and magnesium, they are considered the major responsible factors of the water hardness, while a minor role is played by zinc, iron, and manganese (Beede 2012). No evidence exists that this parameter is significantly correlated to goats' health status at least for values up to 1450 mg/L, as reported by McGregor (2004). However, it is worth mentioning that water hardness can be associated to a higher susceptibility of adult males and wethers to urolithiasis. The aetiology of this pathology is relatively complex and is the result of the interaction of several organic and inorganic compounds (among these, the most important are phosphates, calcium, and magnesium). In addition, the quality of diet and the urinary pH are crucial for its pathogenesis (further information can be found in chapter "Enhancing Animal Welfare Through Disease Prevention in Goats"). Goats are generally able to cope with variations of water acidity. When offered water at a pH equal to 5.2, water intake was only slightly lower compared to control (pH 6.9), but dry matter intake was significantly decreased. Ruminal pH was decreased from 7.0 to 6.6 as the water pH was lowered from 6.9 to 3.8, but no significant effects on nutrient utilization were observed (Ali et al. 2022). After being exposed to heat stress, the animals were able to thermoregulate and increase water intake, but only when water at the highest pH (6.9) was provided. In conclusion, these data demonstrate that goats can adapt to mild acidic water. Nevertheless, if thermal stressors are present, negative effects on physiology and welfare can be observed.

Other drinking water chemical (and possibly toxic) contaminants are represented by nitrates, sulphates, heavy metals, polychlorinated biphenyls, and several compounds derived from agricultural or industrial activities (NRC 2007). However, there is a lack of knowledge related to this subject for the caprine species. The same applies to the microbiological contaminants. Environmental pollution, faecal contamination, and improper cleaning of drinking water containers are the main sources of microorganisms (Beede 2012). Most investigated pathogenic bacteria in livestock species, as water contaminants, are *Campylobacter jejuni*, *Escherichia coli*, and *Salmonella* spp. Occasionally, other pathogens found in drinking water are virus, protozoa, and helminths. When elevated concentrations of phosphorus and nitrogen are present in the water, a rapid growth of various algal species is frequently observed. Among them, cyanobacteria (known also as "blue-green algae") are often implicated in livestock intoxications (Kamphues and Ratert 2014). In conclusion, while harmful microorganisms are occasionally identified in drinking water, the level of risk for ruminant species is mostly unknown. Reliable research data are needed to establish useful recommendations for goats.

7.7 Daily Water Intake and Type of Drinkers

The daily intake of water will depend on several factors. Ehrlenbruch et al. (2010b) found that for dairy goats the intake of water was higher when the goats were fed hay (6.2 l/day) than when fed silage (4.4 l/day). Other factors such as temperature, feeding, reproductive status, and production may all affect how much the goats drink (see Sect. 7). When goats are kept inside or in confinement, drinking water can be supplied either from nipple drinkers, water bowls, or water troughs. Comparatively, studies on cows (Machado et al. 2004) and horses (Nyman and Dahlborn 2001) show that they seem to prefer to drink from a water source with a large and open surface. Both in pigs (e.g. Phillips and Phillips 1990) and sheep (Bøe 1984), however, the water intake from water nipples was higher than from water bowls, probably because of better water quality. In studies of dairy goats (Bøe et al. 2011) the water usage was higher when the goats had access to water nipples than when they had access to water bowls, most likely for the same reasons as for the pigs. However, there was no difference in actual water intake. The results therefore suggest a minor preference for nipples, but that water bowls may work just as well if they are regularly and accurately cleaned. Information/data on drinking patterns of goats is scarce. According to Rossi and Scharrer (1992), Pygmy goats drank mainly during feeding. Time spent drinking was reduced when reducing the number of water nipples available (Ehrlenbruch et al. 2010c). The number of displacements increased significantly when the number of goats per water nipple increased from 15 to 30, and with 30 goats per water nipple, the goats were queueing up for water. Regarding water intake, we can conclude that goats have a weak preference for drinking from water nipples, probably because of better water quality. The number of goats per drinking spot should not exceed 15–20 individuals.

8 Environmental Enrichment

Goats have evolved in physically, temporally, and socially complex environments (Zobel and Nawroth 2020). Wild and feral goat populations and domestic flocks on pasture are facing variable environments with respect to climate, type of vegetation, topography, etc. The present goat breeds are dispersed across different geographical regions adapted to climates from hot (subtropical, Mediterranean) to colder temperatures, from dense forests to arid desert regions, demonstrating the huge potential to adapt to different environments. Their flexibility is represented also in their foraging strategies and diet composition (Pérez-Solano et al. 2020). When ample space is provided, goats live in fission–fusion social systems (Leidy et al. 2015; Shank 1972; Shi et al. 2005; Stanley and Dunbar 2013), where the population is structured in smaller and larger groups of individuals holding closer distances, and where the exchange of group members is depending on resource distribution. It is postulated that with the neural basis adapted to this level of complexity there may be a mismatch between natural physical and social environment and the present practices of intensive, indoor-based production systems (Zobel and Nawroth 2020). Although comparisons are scarce across different housing systems in terms of welfare of goats, goats in herds kept on pasture scored higher when assessed with qualitative behaviour assessment methods than goats in herds without access to pasture (Grosso et al. 2016).

Provision of pastures may contribute to increased space per animal (see Sect. 3) allowing opportunity to withdraw from unwanted social interactions, and decisions about individual distances and formation of sub-groups. Pastures are normally more complex environments than indoor pens, thereby satisfying the basic behavioural needs of the goats to a larger extent. Physical (and social) heterogeneity in general may contribute to a broader behavioural repertoire, development of higher behavioural flexibility, increased cognitive performance, and better stress coping. Nonetheless, a complex environment may allow expression of individual differences in preferences. Pastures often include physical structures, such as slopes, rocks which challenge motoric skills, allow more locomotory play, exercise, possible hiding from others and that lead to brain-stimulating decision-making. Goats with access to an outdoor yard benefit from this possibility, with some goats being outside more than others. In general, provision of roughage outside in addition to allowing the goats access to trees and natural vegetation in the outdoor enclosure (Bøe and Ehrlenbruch 2013) would lead to an increased use of the outdoor enclosure (Stachowicz et al. 2018a). At temperate climate, dairy goats spent more time outside at higher temperatures (Bøe and Ehrlenbruch 2013), when it did not rain or snow during the winter season (Bøe and Ehrlenbruch 2013; Stachowicz et al. 2018b), at higher "warmth index" measured as the combination of temperature, solar radiation, and humidity (Stachowicz et al. 2018b), and without wind (Stachowicz et al. 2018b). Provision of roof in the outside area as protection from

Fig. 11 Example of enrichment: goats like to eat and chew on branches, leaves, and bushes. The picture shows Boer goats manipulating branches that were provided in the area (Photo: Judit Vas)

precipitation increased use of outdoor enclosure during winter (Bøe and Ehrlenbruch 2013). In indoor facilities, 82.3% of 121 Norwegian goat producers (mixture of dairy and meat goat producers) reported about giving some source of environmental enrichment to their goats.

The following, various kinds of enrichments have been applied experimentally (see examples in Figs. 11, 12, and 13):

- Structural modifications at the feeding places: among others, partitions or blinds between feeding slots (Aschwanden et al. 2009a, b; Hillmann et al. 2014; Miranda-de la Lama et al. 2013; Nordmann et al. 2015)
- Different kinds of objects, such as plastic bottles with feed, suspended tires, or wood logs (Araújo do Nascimento et al. 2022)
- Feeding provided at different heights to enable different feeding positions (Kakarash et al. 2021; Miranda-de la Lama et al. 2013; Neave et al. 2018; Neave and Zobel 2020; Rosas-Trigueros et al. 2017)
- Extra climbing opportunities such as platforms, racks, tyres, logs, straw bales (Flint and Murray 2001; Gomes et al. 2018; Kakarash et al. 2021; Miranda-de la Lama et al. 2013; Oesterwind et al. 2016; Rosas-Trigueros et al. 2017)
- Resting shelves (Andersen and Bøe 2006; Aschwanden et al. 2009b) and extra walls in the lying area (Ehrlenbruch et al. 2010a)
- Classical music (Araújo do Nascimento et al. 2022)
- Positive contact with humans, which may serve as environmental enrichment for goats being used to that and having positive associations with humans, as it is, for instance, documented in zoos (Farrand et al. 2014). Goats are curious and contact seeking, yet sensitive to the type of human contact
- Brush and extra chewing material (Vas et al. in prep)
- Twigs from trees or bushes (Vas et al. in prep)

Fig. 12 Example of resting shelves in an otherwise relatively barren environment but with large groups and with fully slatted plastic flooring (Photo: Inger Lise Andersen)

Fig. 13 Positive human contact is stimulating for the goats as they are highly motivated to seek contact with humans that handle them in a positive and predictable way (Photo: Tonje H. Walde)

Provision of these environmental additions in most cases decreased agonistic interactions (Andersen and Bøe 2006; Aschwanden et al. 2008a, b, 2009a; Holt 2018; Miranda-de la Lama et al. 2013; Nordmann et al. 2011, 2015) and displacements (Andersen and Bøe 2006; Aschwanden et al. 2008a, b, 2009a; Nordmann et al. 2011, 2015) but, when resources could be monopolized, it increased agonistic interactions (Kakarash et al. 2021) and displacements (Neave et al. 2018). If the enrichment is limited or distributed in a clumped way, the dominance order in a group will decide which of the goats have access to the enrichment and which do not (Andersen and Bøe 2006; Aschwanden et al. 2009b; Ehrlenbruch et al. 2010a Nordmann et al. 2015; Neave and Zobel 2020). Time-budget of long-lasting activities such as resting (Andersen and Bøe 2006; Ehrlenbruch et al. 2010a; Nordmann et al. 2015) and feeding (Nordmann et al. 2015) is not so affected by environmental provisions.

Dairy goats exposed to various environmental enrichments in their pens showed less interest in humans, and higher reactivity in handling situations, such as under capturing, restraint, and blood sampling (Miranda-de la Lama et al. 2013). They showed more exploration in an open field or when presented with a novel object (Langbein et al. 2004), but showed higher emotional reactivity when isolated from group mates in another study (Tölü et al. 2017). Provision of environmental enrichments appears to have no or minor effect on production. No difference was found in weight gain in some studies (Hillmann et al. 2014; Nordmann et al. 2015; Rosas-Trigueros et al. 2017), while others observed a higher weight gain in meat type castrated bucks kept in feeding lots with enrichments compared to goats kept in standard conditions (Flint and Murray 2001). No effect of enrichments was found on blood plasma (Rosas-Trigueros et al. 2017; Tölü et al. 2017) or serum cortisol levels (Oesterwind et al. 2016) of goat kids kept under conventional or enriched conditions, and in another study, in fact, a higher blood cortisol was found in adult dairy goats kept under enriched conditions (Miranda-de la Lama et al. 2013). The latter might be due to increased competition for the sources of enrichment, but it could also well be that the sympatho-adrenomedullary system is more affected than the HPA-axis in goats. Adrenal activity could be more reliable to measure than cortisol (discussed in Andersen et al. 2007).

In conclusion, environmental enrichment will in most cases improve welfare by reducing social conflicts and engaging goats to focus on relevant and stimulating activities. However, the positive effects are strongly dependent on the amount and distribution of the attractive resources. A more complex and variable environment will indeed be more stimulating for the goats and allow for more individual differences in preferences and choices.

References

Abbeddou S, Rischkowsky B, Richter EK, Hess HD, Kreuzer M (2011) Modification of milk fatty acid composition by feeding forages and agro-industrial byproducts from dry areas to Awassi sheep. J Dairy Sci 94:4657–4668

Abijaoudé JA, Morand-Fehr P, Tessier J, Schmidely P, Sauvant D (2000) Influence of forage: concentrate ratio and type of starch in the diet on feeding behaviour, dietary preferences, digestion, metabolism and performance of dairy goats in mid lactation. Anim Sci 71:359–368

Abrahamsen F, Reddy G, Abebe W, Gurung N (2021) Effect of varying levels of hempseed meal supplementation on humoral and cell-mediated immune responses of goats. Animals 11(10):2764

Ådnøy T (2014) The dairy goat industry in Norway: Challenges in a historical perspective. Small Rumin Res 122:4–9

Agazzi A, Cattaneo D, Dell'Orto V, Moroni P, Bonizzi L, Pasotto D, Bronzo V, Savoini G (2004) Effect of administration of fish oil on aspects of cell-mediated immune response in periparturient dairy goats. Small Ruminant Res 55:77–83

Ahmed G, Wang H (2022) Active dry yeast and thiamine in synergistic mode can mitigate adverse effects of in vitro ruminal acidosis model of goats. Animals 12:2333

Akinmoladun OF, Muchenje V, Fon FN, Mpendulo CT (2019) Small ruminants: farmers' hope in a world threatened by water scarcity. Animals 9(7):456

Akinmoladun OF, Fon FN, Mpendulo CT (2020) Stress indicators, carcass characteristics and meat quality of Xhosa goats subjected to different watering regimen and vitamin C supplementation. Livest Sci 238:104083

Alamer M (2006) Physiological responses of Saudi Arabia indigenous goats to water deprivation. Small Rumin Res 63:100–109

Ali AIM, Sandi S, Sahara E, Rofiq MN (2022) Effects of acid drinking water on nutrient utilization, water balance, and growth of goats under hot-humid tropical environment. Small Rumin Res 210:106689

Al-Khaza'leh J, Abdelqader A, Abuajamieh M, Hayajneh FMF (2020) Assessment of water source availability and quality for small ruminant consumption in the Northern Badia region of Jordan. Vet World 13:1073

Al-Tamimi HJ (2007) Thermoregulatory response of goat kids subjected to heat stress. Small Rumin Res 71:280–285

Amsathkumar L, Jadhav SE, Pattanaik AK, Dutta N (2019) Nutrient utilization and performance of endotoxin exposed kids supplemented with phytogenic feed additive. Anim Nutr Feed Technol 19:371–383

Andersen IL representing the rest of WP3 AWIN Team (2015) Practical implications of pre-natal and neonatal environments for welfare, health, and productivity. In: Abstract in the proceedings of the last AWIN conference, April 21st, Brussels, Belgium

Andersen IL, Bøe KE (2006) Resting pattern and social interactions in goats—The impact of size and organization of lying space. Appl Anim Behav Sci 108:89–103

Andersen IL, Nævdal E, Bakken M, og Bøe KE (2004) Aggression and group size in domesticated pigs, Sus scrofa: 'when the winner takes it all and the looser is standing small'. Anim Behav 68:965–975

Andersen IL, Roussel S, Ropstad E, Braastad BO, Steinheim G, Janczak AM, Jørgensen GM, Bøe KE (2007) Social instability increases aggression in groups of dairy goats, but with minor consequences for the goats' growth, kid production and development. Appl Anim Behav Sci 114:132–148

Andersen IL, Tønnesen H, Estevez I, Cronin GM, Bøe KE (2011) The relevance of group size on goats' social dynamics in a production environment. Appl Anim Behav Sci 134:136–143

Angulo C, Chavez-Infante L, Reyes-Becerril M, Angulo M, Romero-Geraldo R, Llinas-Cervantes X, Cepeda-Palacios R (2020) Immunostimulatory and antioxidant effects of supplemental feeding with macroalga Sargassum spp. on goat kids. Tropl Anim Health Prod 52:2023–2033

Anholt H, Britton A (2017) Presumptive chronic pyrrolizidine alkaloid poisoning in 2 pygmy goats due to ingestion of tansy ragwort (*Jacobaea vulgaris*) in southwestern British Columbia. Can Vet J 58:1171

Anzuino K, Bell NJ, Bazeley KJ, Nicol CJ (2010) Assessment of welfare on 24 commercial UK dairy goat farms based on direct observations. Vet Rec 167:774–780

Apas AL, Arena ME, Elias AG, González F, Gonzalez SN (2015) Beneficial effects of fermented sugarcane residue with goat probiotic on gut health. Int J Agric Innov and Res 3:2319–1473

Arasteh Gatchpazian MA (2022) Eudaimonia: Definition, meaning, & examples. The Berkeley Well-Being Institute. https://www.berkeleywellbeing.com/eudaimonia.html. Accessed 6 Oct 2023

Araújo do Nascimento AP, Castro MSM, de Sousa Oliveira D, Silveira RMF, Vega WHO, Nobre MEM, Andrioli A, de Vasconcelos AM (2022) Environmental enrichment in dairy goats in a semi-arid region: Thermoregulatory and behavioral responses. J Therm Biol 106:103248

Araújo GGLD, Voltolini TV, Chizzotti ML, Turco SHN, Carvalho FFRD (2010) Water and small ruminant production. Revista Brasileira de Zootecnia 39:326–336

Aschwanden J, Gygax L, Wechsler B, Keil NM (2008a) Social distances of goats at the feeding rack: Influence of the quality of social bonds, rank differences, grouping age and presence of horns. Appl Anim Behav Sci 114:116–131

Aschwanden J, Gygax L, Wechsler B, Keil NM (2008b) Cardiac activity in dairy goats whilst feeding side-by-side at two different distances and during social separation. Physiol Behav 95:641–648

Aschwanden J, Gygax L, Wechsler B, Keil NM (2009a) Structural modifications at the feeding place: Effects of partitions and platforms on feeding and social behaviour of goats. Appl Anim Behav Sci 119:180–192

Aschwanden J, Gygax L, Wechsler B, Keil NM (2009b) Loose housing of small goat groups: Influence of visual cover and elevated levels on feeding, resting and agonistic behaviour. Appl Anim Behav Sci 119:171–179

Athanasiadou S, Kyriazakis I (2004) Plant secondary metabolites: Antiparasitic effects and their role in ruminant production systems. Proc Nutr Soc 63:631–639

Banakar PS, Kumar S, Vinay VV, Dixit S, Tyagi N, Tyagi AK (2021) Supplementation of Aloe vera extract in lactating goats' diet: effects on rumen fermentation efficiency, nutrient utilization, lactation performance, and antioxidant status. Tropl Anim Health Prod 53:1–10

Bartolomé J, Franch J, Plaixats J, Seligman NG (1998) Diet selection by sheep and goats on Mediterranean heath-woodland range. Rangeland Ecol Manag/J Range Manag Arch 51:383–391

Bates N (2022) Plant poisoning in goats. Livestock 27:178–185

Battini M, Barbieri S, Fioni L, Mattiello S (2016) Feasibility and validity of animal-based indicators for on-farm welfare assessment of thermal stress in dairy goats. Int J Biometeorol 60:289–296

Beck MR, Gregorini P (2020) How dietary diversity enhances hedonic and eudaimonic well-being in grazing ruminants. Front Vet Sci 7:191

Beede DK (2012) What will our ruminants drink? Anim Front 2:36–43

Ben Salem H, Nefzaoui A, Abdouli H (1994) Palatability of shrubs and fodder trees measured on sheep and dromedaries: 1. Methodological approach. Anim Feed Sci Technol 46:143–153

Ben Salem H, Nefzaoui A, Ben Salem L (1999) Sheep and goat preferences for fodder shrubs: Relationships with the nutritive characteristics. 8th FAO-CIHEAM cooperative network on sheep and goat nutrition, Grignon (France), 3—5 Sept. 1998. Options Méditerranéennes 52:155–159

Bidarkar VK, Swain PS, Ray S, Dominic G (2014) Probiotics: Potential alternative to antibiotics in ruminant feeding. Trends Vet Anim Sci 1:1–4

Block E (1984) Manipulating dietary anions and cations for prepartum dairy cows to reduce incidence of milk fever. J Dairy Sci 67:2939–2948

Bøe KE (1984) Drikkevannsutstyr til sau (Water supply equipment for sheep). IBT-report no. 201. Inst. for bygningsteknikk, NLH, Ås. In Norwegian

Bøe KE, Ehrlenbruch R (2013) Thermoregulatory behavior of dairy goats at low temperatures and the use of outdoor yards. Can J Anim Sci 93:35–41

Bøe KE, Andersen IL, Buisson L, Simensen E, Jeksrud WK (2007) Flooring preference in dairy goats at moderate and low ambient temperature. Appl Anim Behav Sci 108:45–57

Bøe KE, Ehrlenbruch R, Andersen IL (2011) The preference for water nipples vs. water bowls in dairy goats. Acta Vet Scand 53:68

Bøe KE, Ehrlenbruch R, Andersen IL (2012) Outside enclosure and additional enrichment for dairy goats—a preliminary study. Acta Vet Scand 54:68

Bøe KE, Ehrlenbruch R, Jørgensen GHM, Andersen IL (2013) Individual distance during resting and feeding in age homogeneous vs. age heterogeneous groups of goats. Appl Anim Behav Sci 147:112–116

Boutoial K, García V, Rovira S, Ferrandini E, Abdelkhalek O, López MB (2013) Effect of feeding goats with distilled and non-distilled thyme leaves (*Thymus zygis* subp. gracilis) on milk and cheese properties. J Dairy Res 80:448–456

Branciari R, Galarini R, Miraglia D, Ranucci D, Valiani A, Giusepponi D, Trabalza-Marinucci M (2020) Dietary supplementation with olive mill wastewater in dairy sheep: Evaluation of cheese characteristics and presence of bioactive molecules. Animals 10:1941

Brun-Bellut J, Lindberg JE, Hajipanayotou M (1991) Protein nutrition and requirements of adult dairy goats. In: Morand-Fehr P (ed) Goat nutrition. Pudoc, Wageningen, pp 82–93

Bryant JP, Provenza FD, Pastor J, Reichardt PB, Clausen TP, du Toit JT (1991) Interactions between woody plants and browsing mammals mediated by secondary metabolites. Annu Rev Ecol Syst 22:431–446

Cai L, Hartanto R, Xu Q, Zhang J, Qi D (2022) *Saccharomyces cerevisiae* and *Clostridium butyricum* could improve b-vitamin production in the rumen and growth performance of heat-stressed goats. Metabolites 12(8):766

Caroprese M, Ciliberti MG, Santillo A, Marino R, Sevi A, Albenzio M (2016) Immune response, productivity and quality of milk from grazing goats as affected by dietary polyunsaturated fatty acid supplementation. Res Vet Sci 105:229–235

Carta S, Tsiplakou E, Nicolussi P, Pulina G, Nudda A (2022) Effects of spent coffee grounds on production traits, haematological parameters, and antioxidant activity of blood and milk in dairy goats. Animal 16:100501

Castel JM, Mena Y, Delgado-Pertíñez M, Camúñez J, Basulto J, Caravaca F, Guzmán Guerrero JL, Alcalde MJ (2003) Characterization of semi-extensive goat production systems in southern Spain. Small Rumin Res 47:133–143

Castel JM, Ruiz FA, Mena Y, Sánchez-Rodríguez M (2010) Present situation and future perspectives for goat production systems in Spain. Small Rumin Res 89:207–210

Castel JM, Mena Y, Ruiz FA, Camúñez-Ruiz J, Sánchez-Rodríguez M (2011) Changes occurring in dairy goat production systems in less favoured areas of Spain. Small Rumin Res 96:83–92

Castro M, Fernández-Núñez E, Castro J (2015) Management of grazing circuits in order to promote extensive livestock welfare. In: World congress on integrated crop-livestock-forestry systems and 3rd international symposium on integrated crop-livestock systems (ICLS3)

Celi P, Di Trana A, Claps S (2010) Effects of plane of nutrition on oxidative stress in goats during the peripartum period. Vet J 184:95–99

Cellier M, Nielsen BL, Duvaux-Ponter C, Freeman HBR, Hannaford R, Murphy B, O'Connor E, Cote KRL, Neave HW, Zobel G (2022) Browse or browsing: Investigating goat preferences for feeding posture, feeding height and feed type. Front Vet Sci 9:1032631

Celozzi S, Mattiello S, Battini M, Zucali M, Pirovano L, Albano C, Brasca M, Bava L (2021) Effect of a reduced amount of straw bedding on goats' comfort and hygienic characteristics of milk and straw. Italian J Anim Sci 20:995–1001

Chiofalo V, Liotta L, Chiofalo B (2004) Effects of the administration of Lactobacilli on body growth and on the metabolic profile in growing Maltese goat kids. Reprod Nutr Dev 44(5):449–457

Choshniak I, Ben-Kohav N, Taylor CR, Robertshaw D, Barnes RJ, Dobson A, Belkin V, Shkolnik A (1995) Metabolic adaptations for desert survival in the Bedouin goat. Am J Physiol 268:1101–1110

Cimmino R, Barone CMA, Claps S, Varricchio E, Rufrano D, Caroprese M, Albenzio M, De Palo P, Campanile G, Neglia G (2018) Effects of dietary supplementation with polyphenols on meat quality in Saanen goat kids. BMC Vet Res 14:181

Clutton-Brock TH, Greenwood PJ, Powell RP (1976) Ranks and relationships in Highland ponies and Highland cows. Zeitschrift Fur Tierpsychologie 41:202–216

Correddu F, Lunesu MF, Buffa G, Atzori AS, Nudda A, Battacone G, Pulina G (2020) Can agro-industrial by-products rich in polyphenols be advantageously used in the feeding and nutrition of dairy small ruminants? Animals 10:131

Council Directive 98/58/EC of 20 July 1998 concerning the protection of animals kept for farming purposes

Cull P (1988) Goat housing. Goat Vet Soc J 9(1 & 2):40–44

Daskiran I, Savas T, Koyuncu M, Koluman N, Keskin M, Esenbuga N, Konyali A, Cemal İ, Gül S, Elmaz O, Kosum N, Dellal G, Bingöl M (2018) Goat production systems of Turkey: Nomadic to industrial. Small Rumin Res 163:15–20

Decandia M, Molle G, Ligios S, Scanu G (1999) Effect of different levels of polyethylene glicol on the nutritional value of *Pistacia lentiscus* L. fed to goats. In: Proceedings of the XIII ASPA congress, Piacenza, June 21–24, pp 384–386

Decandia M, Molle G, Pinna G, Cabiddu A, Yiakoulaki M (2005) Comportamento alimentare ed ingestione di capre al pascolo su macchia mediterranea. In: Pulina G (ed) L'alimentazione della capra da latte. Ed. Avenue Media, Bologna, pp 173–207

De Figueiredo Guilherme R, De Farias DA, Alves JRA, Da Costa DF, Pinheiro RR, Alves FSF, De Azevedo SS, Alves CJ (2017) Characterization and typology of sheep and goat production systems in the State of Paraíba, a semi-arid region of northeastern Brazil. Semin Agrar 38:2163–2178

Delgadillo-Puga C, Cuchillo-Hilario M, León-Ortiz L, Ramírez-Rodríguez A, Cabiddu A, Navarro-Ocaña A, Morales Romero AM, Medina-Campos ON, Pedraza-Chaverri J (2019) Goats' feeding supplementation with Acacia farnesiana Pods and their relationship with milk composition: Fatty acids, polyphenols, and antioxidant activity. Animals 9:515

de Rancourt M, Fois N, Lavín MP, Tchakérian E, Vallerand F (2006) Mediterranean sheep and goats production: An uncertain future. Small Rumin Res 62:167–179

Devendra C, Solaiman SG (2010) Perspectives on goats and global production. In: Solaiman SG (ed) Goat science and production. Blackwell, Hoboken, pp 3–19

Di Meo F, Lemaur V, Cornil J, Lazzaroni R, Duroux JL, Olivier Y, Trouillas P (2013) Free radical scavenging by natural polyphenols: atom versus electron transfer. J Phys Chem A 117:2082–2092

Distel RA, Arroquy JI, Lagrange S, Villalba JJ (2020) Designing diverse agricultural pastures for improving ruminant production systems. Front Sustain Food Syst 4:596869

Domingue BMF, Dellow DW, Barry TN (1991) Voluntary intake and rumen digestion of a low-quality roughage by goats and sheep. J Agric Sci 117:111–120

Drickamer LC, Vessey SH, Jakob EM (2002) Animal behavior: mechanisms, ecology, evolution. McGraw-Hill, New York

Dubeuf JP, Ruiz Morales FDA, Guerrero YM (2018) Evolution of goat production systems in the Mediterranean basin: Between ecological intensification and ecologically intensive production systems. Small Rumin Res 163:2–9

Duignan B (2020) Eudaimonia. Encyclopedia Britannica, July 3, 2020. https://www.britannica.com/topic/eudaimonia. Accessed 6 Oct 2023

Dunson WT, Scott CB, Campbell ES, Taylor CA Jr, Carr MA, Callaway TR (2007) Rumen function and the ability of goats to consume redberry juniper (*Juniperus pinchottii*). In: Panter KE, Wierenga TL, Pfister JA (eds) Poisonous plants: global research and solutions. CABI, Wallingford, pp 377–385

Durand D, Damon M, Gobert M (2013) Oxidative stress in farm animals: general aspects. Cah Nutr Diet 48:218–224

Dýrmundsson ÓR (2006) Sustainability of sheep and goat production in North European countries From the Arctic to the Alps. Small Rumin Res 62:151–157

Egea AV, Allegretti L, Paez Lama S, Grilli D, Sartor C, Fucili M, Guevara JC, Passera C (2014) Selective behavior of Creole goats in response to the functional heterogeneity of native forage species in the central Monte desert, Argentina. Small Rumin Res 120(1):90–99. ISSN: 0921-4488

Ehrlenbruch R, Andersen IL, Bøe KE, Simensen E (2009) Liggeunderlag for geit: Hvilket underlag foretrekker geita? Sau og Geit 1(2009):36–37

Ehrlenbruch R, Jørgensen GHM, Andersen IL, Bøe KE (2010a) Provision of additional walls in the resting area—The effects on resting behaviour and social interactions in goats. Appl Anim Behav Sci 122:35–40

Ehrlenbruch R, Eknæs M, Pollen T, Andersen IL, Bøe KE (2010b) Water intake in dairy goats—the effect of different types of roughages. Ital J Anim Sci 9:400–403

Ehrlenbruch R, Pollen T, Andersen IL, Bøe KE (2010c) Competition for water at feeding time-The effect of increasing number of individuals per water dispenser. App Anim Behav Sci 126:105–108

Eik LO (1991) Performance of goat kids raised in a non-insulated barn at low temperatures. Tech note. Small Rumin Res 4:95–100

Ekwemalor K, Asiamah E, Osei B, Ismail H, Worku M (2017) Evaluation of the effect of probiotic administration on gene expression in goat blood. J Mol Biol Res 7:88–98

El-Hadi HM (1986) The effect of dehydration on Sudanese desert sheep and goats. J Agric Sci 106:17–20

El-Nile A, Elazab M, El-Zaiat H, El-Azrak KE-D, Elkomy A, Sallam S, Soltan Y (2021) In vitro and in vivo assessment of dietary supplementation of both natural or nano-zeolite in goat diets: effects on ruminal fermentation and nutrients digestibility. Animals (Basel) 11:2215

Enemark JMD (2008) The monitoring, prevention and treatment of sub-acute ruminal acidosis (SARA): A review. Vet J 176(1):32–43

Escareño L, Salinas-Gonzalez H, Wurzinger M, Iñiguez L, Sölkner J, Meza-Herrera C (2013) Dairy goat production systems: Status quo, perspectives and challenges. Tropl Anim Health Prod 45:17–34

Estevez I, Newberry RC, de Reyna LA (1997) Broiler chickens: a tolerant social system? Etologia 5:19–29

Estevez I, Newberry RC, Keeling LJ (2002) Dynamics of aggression in the domestic fowl. Appl Anim Behav Sci 76:307–325

Estevez I, Andersen IL, Nævdal E (2007) Group size, density and social dynamics in farm animals. Invited review-paper. Appl Anim Behav Sci 103:185–204. Top 10 ranked among the hottest, cited papers in Applied Animal Behaviour Science

Færevik G, Andersen IL, Bøe KE (2005) Preferences of sheep for different types of pen flooring. Appl Anim Behav Sci 90:265–276

Færevik G, Andersen IL, Jensen MB, Bøe KE (2007) Increased group size reduces conflicts and strengthens the preference for familiar group mates in regrouped calves (Bos taurus). Appl Anim Behav Sci 108:215–228

FAOSTAT (2020) Food and Agriculture Organization (FAO) of the United Nations. https://www.fao.org/faostat/en/#data/QCL. Accessed 4 Nov 2022

Farrand A, Hosey G, Buchanan-Smith HM (2014) The visitor effect in petting zoo-housed animals: Aversive or enriching? Appl Anim Behav Sci 151:117–127

Flint M, Murray P (2001) Lot-fed goats—the advantages of using an enriched environment. Aust J Exp Agric 41:473–476

Francois N, Mills AD, Faure JM (1999) Inter-individual distances during open-field tests in Japanese quail (*Coturnix japonica*) selected for high or low levels of social reinstatement behaviour. Behav Processes 47:73–80

Freeland WJ, Janzen DH (1974) Strategies in herbivory by mammals: the role of plant secondary compounds. Am Nat 108:269–289

French KE (2018) Plant-based solutions to global livestock anthelmintic resistance. Ethnobiol Lett 9(2):110–123

Furlan S, Panter KE, Pfister JA, Stegelmeier BL (2007) Fetotoxic effects of locoweed (*Astragalus lentiginosus*) in pregnant goats. In: International symposium on poisonous plants, pp 130–135

Furlan CM, Motta LB, Santos DYAC (2011) Tannins: What do they represent in plant life. In: Petridis GK (ed) Tannins: types, foods containing, and nutrition. Nova Science, Hauppauge, pp 1–13

Gantner V, Kompan D (2012) Milk yield, milk composition and somatic cell count of dairy goats given n-3 unsaturated fatty acids diet supplement. Acta Vet Brno 62:281–287

Gelasakis AI, Rose G, Giannakou R, Valergakis GE, Theodoridis A, Fortomaris P, Arsenos G (2017) Typology and characteristics of dairy goat production systems in Greece. Livest Sci 197:22–29

Giger-Reverdin S, Gihad EA (1991) Water metabolism and intake in goats. In: Morand-Fehr P (ed) Goat nutrition. Pudoc, Wageningen, The Netherlands, pp 37–45

Goetsch AL (2019a) Recent advances in the feeding and nutrition of dairy goats. Asian Australas J Anim Sci 32:1296

Goetsch AL (2019b) Recent research of feeding practices and the nutrition of lactating dairy goats. J Appl Anim Res 47:103–114

Gomes KAR, Valentim JK, Lemke SSR, Dallago GM, Vargas RC, Paiva ALDC (2018) Behavior of Saanen dairy goats in an enriched environment. Acta Sci Anim Sci 40:e42454

Gozho GN, Plaizier JC, Krause DO, Kennedy AD, Wittenberg KM (2005) Subacute ruminal acidosis induces ruminal lipopolysaccharide endotoxin release and triggers an inflammatory response. J Dairy Sci 88:1399–1403

Greenwald LI (1967) Water economy of the desert dwelling springbok (*Antidorcas marsupialis*). M.Sc. (Zool.) dissertation, Syracuse University, Syracuse, USA

Grosso L, Battini M, Wemelsfelder F, Barbieri S, Minero M, Dalla Costa E, Mattiello S (2016) On-farm qualitative behaviour assessment of dairy goats in different housing conditions. Appl Anim Behav Sci 180:51–57

Hasan MJ, Ahmed JU, Alam MM (2014) Reproductive performances of Black Bengal goat under semi-intensive and extensive conditions at rural areas in Bangladesh. J Adv Veter Anim Res 1:196–200

Hashem NM, Morsy AS, Soltan YA, Sallam SM (2021) Potential benefits of *Boswellia sacra* resin on immunity, metabolic status, udder and uterus health, and milk production in transitioning goats. Agriculture 11:900

Hempstead MN, Lindquist TM, Shearer JK, Shearer LC, Plummer PJ (2021) Health and welfare survey of 30 dairy goat farms in the Midwestern United States. Animals 11

Hillmann E, Hilfiker S, Keil NM (2014) Effects of restraint with or without blinds at the feed barrier on feeding and agonistic behaviour in horned and hornless goats. Appl Anim Behav Sci 157:72–80

Hofer D, Suter M, Haughey E, Finn JA, Hoekstra NJ, Buchmann N, Lüscher A (2016) Yield of temperate forage grassland species is either largely resistant or resilient to experimental summer drought. J Appl Ecol 53:1023–1034

Hofmann RR, Stewart DRM (1972) Grazer or Browser: a classification based on the stomach-structure and feeding habits of east African ruminants. Mammalia 36(2):226–240

Holt RV (2018) The effects of environmental enrichments, group size and confinement length on play behaviour in juvenile Norwegian dairy goats (Capra hircus). Norwegian Univ Life Sci Res 1:196–200

Ihejirika UGD, Adenekan OO, Jimoh OA, Jack AA, Uwaeziozi UC (2017) Serum biochemistry and oxidative stress indicators in West African dwarf goats under semi-intensive management system. Niger J Anim Prod 44:110–117

Institut de l'élevage (Idele) (2011) Etat corporel: des reperes pour noter les chèvres. In: Legarto J, Leclerc MC (eds) L'Alimentation pratique des chèvres laitières. Collection Les Incontournables. Technipel/Institut de l'élevage, Paris, France, 216 pp

Jaynes WF, Zartman RE, Hudnall WH (2007) Aflatoxin B1 adsorption by clays from water and corn meal. Appl Clay Sci 36:197–205

Jimoh OA, Adenekan OO, Jack AA, Uwaeziozi UC, Ihejirika UDG (2018) Physiological variations and oxidative status in West African Dwarf male sheep and goat in semi-intensive management system. Anim Res Int 15:2898–2905

Jørgensen GHM, Andersen IL, Bøe KE (2007) Feed intake and social interactions in dairy goats—The effects of feeding space and type of roughage. Appl Anim Behav Sci 107:239–251

Jørgensen GHM, Johanssen JRE, Bøe KE (2017) Preference in shorn sheep for different types of slatted flooring at low ambient temperatures. Small Rumin Res 153:17–22

Kakarash NA, Ramzi DO, Ismaeel DO, Marif HF (2021) Effects of environmental enrichment on behaviours and welfare of Meriz goat. Assiut Vet Med J 67:11–18

Kamphues J, Ratert C (2014) Drinking water quality in food producing animals. In: Proceedings of the international workshop "new updates in animal nutrition, natural feeding sources and environmental sustainability", Arzachena, Italy, 5–6 May 2014

Karásková K, Suchý P, Straková E (2015) Current use of phytogenic feed additives in animal nutrition: a review.Czech. J Anim Sci 60:521–530

Keeling L (1995) Spacing behaviour and an ethological approach to assessing optimum space allocations for groups of laying hens. Appl Anim Behav Sci 44:171–186

Kefyalew SF (2015) Sheep and goat production systems in Degehabur zone, Eastern Ethiopia: Challenge and opportunities. J Adv Dairy Res 3:134

Keil NM, Sambraus HH (1998) "Intervenors" in agonistic interactions amongst domesticated goats. Z Sägetierkunde 63:266–272

Keil NM, Pommereau M, Patt A, Wechsler B, Gygax L (2017) Determining suitable dimensions for dairy goat feeding places by evaluating body posture and feeding reach. J Dairy Sci 100(2):1353–1362

Khalaf Al-Khaza'leh JM (2018) Goat farming and breeding in Jordan. In: Kukovics S (ed) Goat science. IntechOpen, London, pp 367–380

Kharrat M, Bocquier F (2010) Adaptive responses at the whole lactation scale of Baladi dairy goats according to feed supply and level of body reserves in agro-pastoral feeding system. Small Rumin Res 90:120–126

Kilcher MR (1981) Plant development, stage of maturity and nutrient composition. J Range Manage 34:363–364

Krause J, Ruxton GD (2002) Living in groups. Oxford series in ecology and evolution. Oxford University Press, Oxford. 209 pp

Lamming L (2001) Successfully controlling noxious weeds with goats. Altern Weed Strateg 21:19–23

Langbein J, Nürnberg G, Manteuffel G (2004) Visual discrimination learning in dwarf goats and associated changes in heart rate and heart rate variability. Physiol Behav 82:601–609

Laouadi M, Tennah S, Kafidi N, Antoine-Moussiaux N, Moula N (2018) A basic characterization of small-holders' goat production systems in Laghouat area, Algeria. Pastoralism 8

Leidy RS, Jorge CR, Elena BR, Jordi BF (2015) Comparative study of trophic behaviour and herd structure in wild and feral goats living in a Mediterranean Island: Management implications. Appl Anim Behav Sci 165:81–87

Lisonbee LD, Villalba JJ, Provenza FD (2009) Effects of tannin on selection by sheep of forages containing alkaloids, tannins and saponins. J Sci Food Agric 89:2668–2677

Liu S, Wang X, Li Y, Shi B, Guo X, Zhao Y, Yan S (2021) Flaxseed oil and heated flaxseed supplements have different effects on lipid deposition and ileal microbiota in Albas cashmere goats. Animals 11:790

Loretz C, Wechsler B, Hauser R, Rusch P (2004) A comparison of space requirements of horned and hornless goats at the feed barrier and in the lying area. Appl Anim Behav Sci 87(3):275–283

Ma ZZ, Cheng YY, Wang SQ, Ge JZ, Shi HP, Kou JC (2020) Positive effects of dietary supplementation of three probiotics on milk yield, milk composition and intestinal flora in Sannan dairy goats varied in kind of probiotics. J Anim Physiol Anim Nutr 104:44–55

Ma Y, Wang C, Zhang H, Yu L, Dong L, Gong D, Yao J, Wang H (2021) Illumina sequencing and metabolomics analysis reveal thiamine modulation of ruminal microbiota and metabolome characteristics in goats fed a high-concentrate diet. Front Microbiol 12:653283

Machado LCP, Teixeira DL, Weary DM, von Keyserlingk MAG, Hotzel MJ (2004) Designing better water troughs: dairy cows prefer and drink more from larger troughs. Appl Anim Behav Sci 89:185–193

Madibela OR, Jansen K (2003) The use of indigenous parasitic plant (*Viscum verrocosum*) in reducing faecal egg counts in female Tswana goats. Livest Res Rural Dev 15:9

Mahla AS, Chaudhari RK, Verma AK, Singh AK, Singh SK, Singh G, Sarkar M, Dutta N, Kumar H, Krishnaswamy N (2017) Effect of dietary supplementation of omega-3 polyunsaturated fatty acid (PUFA) rich fish oil on reproductive performance of the goat (*Capra hircus*). Theriogenology 99:79–89

Makkar HPS (2003) Effects and fate of tannins in ruminant animals, adaptation to tannins, and strategies to overcome detrimental effects of feeding tannin-rich feeds. Small Rumin Res 49:241–256

Maltz E, Shkolnik A (1984) Milk composition and yield of black Bedouin goat during dehydration and rehydration. J Dairy Res 1984(51):23–27
Mamuad LL, Lee SS, Lee SS (2019) Recent insight and future techniques to enhance rumen fermentation in dairy goats. Asian Australas J Anim Sci 32:1321
Manousidis T, Kyriazopoulos AP, Parissi ZM, Abraham EM, Korakis G, Abas Z (2016) Grazing behavior, forage selection and diet composition of goats in a Mediterranean woody rangeland. Small Rumin Res 145:142–153
Maxin G, Cornu A, Andueza D, Laverroux S, Graulet B (2020) Carotenoid, tocopherol, and phenolic compound content and composition in cover crops used as forage. J Agric Food Chem 68:6286–6296
McBride G (1971) Theories of animal spacing: the role of flight, fight and social distance. In: Esser AH (ed) Behavior and environment: the use of space by animals and men. Plenum, New York, pp 53–68
McGregor BA (2004) Water quality and provision for goats. RIRDC
Mdletshe ZM, Chimonyo M, Marufu MC, Nsahlai IV (2017) Effects of saline water consumption on physiological responses in Nguni goats. Small Rumin Res 153:209–211
Mellado M, Valdez R, Lara LM, Lopez R (2003) Stocking rate effects on goats: A research observation. Rangeland Ecol Manag/J Range Manag Arch 56:167–173
Meschy F, Bravo D, Sauvant D (2004) Analyse quantitative des réponses des vaches laitières á l'apport de substances tampon. Prod Anim 17:11–18
Min BR, Solaiman S (2018) Comparative aspects of plant tannins on digestive physiology, nutrition and microbial community changes in sheep and goats: A review. J Anim Physiol Anim Nutr 102:1181–1193
Miranda-de la Lama GC, Pinal R, Fuchs K, Montaldo HH, Ducoing A, Galindo F (2013) Environmental enrichment and social rank affects the fear and stress response to regular handling of dairy goats. J Vet Behav 8:342–348
Mkhize NR, Heitkönig IM, Scogings PF, Dziba LE, Prins HH, de Boer WF (2015) Condensed tannins reduce browsing and increase grazing time of free-ranging goats in semi-arid savannas. Appl Anim Behav Sci 169:33–37
Monteiro A, Costa JM, Lima MJ (2018) Goat system productions: Advantages and disadvantages to the animal, environment and farmer. In: Kukovics S (ed) Goat science. IntechOpen, London, pp 351–366
Morales FDAR, Genis JMC, Guerrero YM (2019) Current status, challenges and the way forward for dairy goat production in Europe. Asian Australas J Anim Sci 32:1256–1265
Morand-Fehr P, Owen E, Giger-Reverdin S (1991) Feeding behaviour of goats at the trough. In: Morand-Fehr P (ed) Goat nutrition. Pudoc, Wageningen, The Netherlands, pp 3–12
Muna MMA, Ammar IES (2001) Effects of water and feed restriction on body weight change and nitrogen balance in desert goats fed high and low quality forages. Small Rumin Res 41:19–27
Muri K, Tufte PA, Skjerve E, Valle PS (2012) Human-animal relationships in the Norwegian dairy goat industry: attitudes and empathy towards goats (Part I). Anim Welf 21:535–545
Muri K, Stubsjoen SM, Valle PS (2013) Development and testing of an on-farm welfare assessment protocol for dairy goats. Anim Welf 22(3):385–400
Nandi D, Roy S, Bera S, Kesh SS, Samanta AK (2011) The rearing system of Black Bengal Goat and their farmers in West Bengal, India. Vet World 4:254–257
National Research Council (US) (NRC) (2007) Committee on Nutrient Requirements of Small Ruminants, Board on Agriculture, Division on Earth, and Life Studies. Nutrient requirements of small ruminants: sheep, goats, cervids, and new world camelids
Neave HW, Zobel G (2020) Personality of dairy goats affects competitive feeding behaviour at different feeder heights. Small Rumin Res 192:106222
Neave HW, von Keyserlingk MAG, Weary DM, Zobel G (2018) Feed intake and behavior of dairy goats when offered an elevated feed bunk. J Dairy Sci 101:3303–3310
Nordmann E, Keil MN, Schmied-Wagner C, Graml C, Langbein J, Aschwanden J, von Hof J, Maschat K, Palme R, Waiblinger S (2011) Feed barrier design affects behaviour and physiology in goats. Appl Anim Behav Sci 133:40–53

Nordmann E, Barth K, Futschik A, Palme R, Waiblinger S (2015) Head partitions at the feed barrier affect behaviour of goats. Appl Anim Behav Sci 167:9–19
Nudda A, Bua G, Atzori AS, Cappai MG, Caboni P, Fais G, Pulina G (2019) Small amounts of agro-industrial by products in dairy ewes diets affects milk production traits and hematological parameters. Anim Feed Sci Technol 251:76–85
Nwude N, Parsons LE, Adaudi AO (1977) Acute toxicity of the leaves and extracts of *Dichapetalum barteri* (Engl.) in mice, rabbits and goats. Toxicology 7:23–29
Nyman S, Dahlborn K (2001) Effect of water supply method and flow rate on drinking behaviour and fluid balance in horses. Physiol Behav 73:1–8
Oesterwind S, Nürnberg G, Puppe B, Langbein J (2016) Impact of structural and cognitive enrichment on the learning performance, behavior and physiology of dwarf goats (Capra aegagrus hircus). Appl Anim Behav Sci 177:34–41
Orskov ER (1979) Recent information on processing of grain for ruminants. Livest Prod Sci 6:335–347
Pandey KR, Naik SR, Vakil BV (2015) Probiotics, prebiotics and synbiotics—a review. J Food Sci Technol 52:7577–7587
Panter KE, Bunch TD, Keeler RF, Sisson DV, Callan RJ (1990) Multiple congenital contractures (mcc) and cleft palate induced in goats by ingestion of piperidine alkaloid-containing plants: Reduction in fetal movement as the probable cause. J Toxicol Clin Toxicol 28(1):69–83
Paolini V, Bergeaud JP, Grisez C, Prévot F, Dorchies P, Hoste H (2003) Effects of condensed tannins on goats experimentally infected with *Haemonchus contortus*. Vet Parasitol 113:253–261
Paolini V, Fouraste I, Hoste H (2004) In vitro effects of three woody plant and sainfoin extracts on 3rd-stage larvae and adult worms of three gastrointestinal nematodes. Parasitology 129:69–77
Peet RL, Dickson J, Hare M (1990) Kikuyu poisoning in goats and sheep. Aust Vet J 67:229–230
Pereira ME (1988) Effects of age and sex on intra-group spacing behaviour in juvenile savannah baboons Papiocynocephalus. Anim Behav 36:184–204
Pérez-Solano LA, Mandujano S, Bautista Z (2020) Spatio-temporal movements of goat herds in a protected area of Mexico. Eur J Wildl Res 66
Phillips PA, Phillips MH (1990) Effects of dispenser on water intake of pigs at weaning. Trans ASAE 42:1471–1473
Pimentel LEA, Correa FR, Gardner D, Panter KE, Dantas AFM, Medeiros RMT, Mota RA, Araújo JAS (2007) *Mimosa tenuiflora* as a cause of malformations in ruminants in the northeastern Brazilian semiarid rangelands. Vet Pathol 44:928–931
Pisani LF, Lecchi C, Invernizzi G, Sartorelli P, Savoini G, Ceciliani F (2009) In vitro modulatory effect of ω-3 polyunsaturated fatty acid (EPA and DHA) on phagocytosis and ROS production of goat neutrophils. Vet Immunol Immunopathol 131:79–85
Plieninger T, Hui C, Gaertner M, Huntsinger L (2014) The impact of land abandonment on species richness and abundance in the Mediterranean basin: a meta-analysis. PLoS One 9:e98355
Pokorná P, Hejcmanová P, Hejcman M, Pavlu V (2013) Activity time budget patterns of sheep and goats co-grazing on semi-natural species-rich dry grassland. Czeh J Anim Sci 58:208–216
Poutaraud A, Michelot-Antalik A, Plantureux S (2017) Grasslands: a source of secondary metabolites for livestock health. J Agric Food Chem 65(31):6535–6553
Rahal A, Kumar A, Singh V, Yadav B, Tiwari R, Chakraborty S, Dhama K (2014) Oxidative stress, prooxidants, and antioxidants: the interplay. Biomed Res Int 2014:1–19
Raksasiri BV, Paengkoum S, Poonsuk K, Thitisak P, Paengkoum P (2019) Effect of synbiotics in creep feed on productive performance and selected fecal characteristics of goat kids. Int J Agric Biol 21:1019–1023
Rapetti L, Bava L (2004) Effect of grinding of maize and level of starch on digestibility and lactation performance of Saanen goats. S Afr J Anim Sci 34(Suppl. 1):85–88
Reshma Nair MR, Sejian V, Silpa MV, Fonsêca VFC, de Melo Costa CC, Devaraj C, Krishnan G, Bagath M, Nameer PO, Bhatta R (2021) Goat as the ideal climate-resilient animal model in tropical environment: revisiting advantages over other livestock species. Int J Biometeorol 65:2229–2240

Ríos E, Cholich L, Silva J, Acosta de Pérez O (2008) Histopathological lesions in central nervous system of goats poisoned by *Ipomoea carnea*. Rev vet 19:130–134

Rook JS (2000) Pregnancy toxemia of ewes, does, and beef cows. Vet Clin North Am Food Anim Pract 16:293–317

Rosas-Trigueros AP, Candanosa-Aranda IE, Ducoing-Watty AE, Gutiérrez-Molotla J, Galindo F, Sisto-Burt AM (2017) Histological differences in the adrenal glands and cortisol levels of suckling dairy goat kids in enriched and non-enriched environments. Res Vet Sci 115:221–225

Rossi R, Scharrer E (1992) Circadian patterns of drinking and eating in pygmy goats. Physiol Behav 51:895–897

Saha S, Singha S, Ahmed SS, Toledo-Alvarado H, Khan MM (2018) Effects of yeast (*Saccharomyces cerevisiae* type *boulardii* CNCM I-1079) supplementation on growth performance and blood metabolites in Black Bengal goat kids. Veterinarski arhiv 88:661–672

Salama AAK, Caja G, Hamzaoui S, Badaoui B, Castro-Costa A, Facanha DAE, Guilhermino MM, Bozzi R (2014) Different levels of response to heat stress in dairy goats. Small Rumin Res 121:73–79

Salama AAK, Hamzaoui S, Albanell E, Such X, Caja G (2021) Metabolic and behavior responses of lactating goats under heat stress. Small Rumin Res 203:1–9

Samimi AS, Aghamiri SM, Nazifi S, Asadi Z (2021) Changes in ghrelin, microminerals, antioxidants and vitamins A, E and C levels during different physiological status in high yielding Saanen goats subjected to heat stress. J Therm Biol 100:103014. ISSN 0306-4565

Sandrucci A, Bava L, Tamburini A, Giulia G, Zucali M (2019) Management practises and milk quality in dairy goat farms in Northern Italy. Ital J Anim Sci 18:1–12

Sarangi S (2018) Adaptability of goats to heat stress: A review. Pharma Innov J 7(4):1114–1126

Sauvant D, Giger-Reverdin S (2018) Dairy and growing goats. In: INRA feeding system for ruminants. Wageningen Academic, Wageningen, The Netherlands, pp 339–374

Sauvant D, Giger-Reverdin S, Meschy F (2010) Alimentation des caprins. In: Alimentation des bovins, ovins et caprins—Besoins des animaux—Valeurs des aliments—Tables INRA 2007—mise à jour 2010. Editions Quae, Paris, France, pp 139–152

Savoini G, Agazzi A, Invernizzi G, Cattaneo D, Pinotti L, Baldi A (2010) Polyunsaturated fatty acids and choline in dairy goats nutrition: Production and health benefits. Small Rumin Res 88:135–144

Savoini G, Omodei Zorini F, Farina G, Agazzi A, Cattaneo D, Invernizzi G (2019) Effects of fat supplementation in dairy goats on lipid metabolism and health status. Animals 9:917

Sedivec K, Hanson T, Heiser C (1995) Controlling leafy spurge using goats and sheep. North Dakota State University Extension Service, Fargo

Sejian V, Silpa MV, Nair MRRR, Devaraj C, Krishnan G, Bagath M, Chauhan SS, Suganthi RU, Fonseca VFC, König S, Gaughan JB, Dunshea FR, Bhatta R (2021) Heat stress and goat welfare: adaptation and production considerations. Animals 11:1021

Serhan M, Mattar J (2018) The goat dairy sector in Lebanon. In: Kukovics S (ed) Goat science. IntechOpen, London, pp 381–395

Serra V, Salvatori G, Pastorelli G (2021) Dietary polyphenol supplementation in food producing animals: Effects on the quality of derived products. Animals 11:401

Shah T, Malhi M, Kachiwal AB, Bhutto B, Shah QA, Lei Y, Soomro SA, Soomro J, Kalhor NH, Gui H (2022) Ameliorative effects of supranutritional selenium on TLR-4-NF-kB-TNF-α-mediated hepatic oxidative injury and inflammation in goats fed high concentrate diet. Food Sci Nutr 1–13

Shank CC (1972) Some aspects of social behaviour in a population of feral goats (Capra hircus L.). Z Tierpsychol 30:488–528

Shi J, Dunbar RIM, Buckland D, Miller D (2005) Dynamics of grouping patterns and social segregation in feral goats (Capra hircus) on the Isle of Rum, NW Scotland. Mammalia 69:185–199

Shi L, Zhang C, Yue W, Shi L, Zhu X, Lei F (2010) Short-term effect of dietary selenium-enriched yeast on semen parameters, antioxidant status and Se concentration in goat seminal plasma. Anim Feed Sci Technol 157:104–108

Shi L, Zhang Y, Wu L, Xun W, Liu Q, Cao T, Hou G, Zhou H (2020) Moderate coconut oil supplement ameliorates growth performance and ruminal fermentation in Hainan black goat kids. Front Vet Sci 7:622259

Shokrollahi B, Amini F, Fakour S, Andi MA (2015) Effect of rosemary (*Rosmarinus officinalis*) extract on weight, hematology and cell-mediated immune response of newborn goat kids. J Agric Rural Dev Tropics Subtropics (JARTS) 116:91–97

Silanikove N (2000) The physiological basis of adaptation in goats to harsh environments. Small Rumin Res 35:181–193

Silanikove N, Perevolotsky A, Provenza FD (2001) Use of tannin-binding chemicals to assay for tannins and their negative postingestive effects in ruminants. Anim Feed Sci Technol 91:69–81

Simensen E, Hardeng F, Lunder T (2010) Housing of Norwegian goat herds and associations with milk yield and milk quality. Acta Agric Scand Sect A 60:187–193

Simmonds H, Bourke C, Holst P (2000) The palatability and potential toxicity of Australian weeds to goats. Rural Industries Research and Development Corporation, Kingston, Australia, p 166

Simões JG, Medeiros RMT, Medeiros MA, Olinda RG, Dantas AFM, Riet-Correa F (2018) Nitrate and nitrite poisoning in sheep and goats caused by ingestion of *Portulaca oleracea*. Pesquisa Veterinária Brasileira 38:1549–1553

Solaiman SG (2010) Goat science and production. Blackwell, Ames

Sow F, Camara Y, Traore EH, Cabaraux JF, Missohou A, Antoine-Moussiaux N, Hornick JL, Moula N (2021) Characterisation of smallholders' goat production systems in the Fatick area, Senegal. Pastoralism 11

Stachowicz J, Gygax L, Hillmann E, Wechsler B, Keil NM (2018a) Dairy goats use outdoor runs of high quality more regardless of the quality of indoor housing. Appl Anim Behav Sci 208:22–30

Stachowicz J, Lanter A, Gygax L, Hillmann E, Wechsler B, Keil NM (2018b) Under temperate weather conditions, dairy goats use an outdoor run more with increasing warmth and avoid light wind or rain. J Dairy Sci 102:1508–1521

Stanley CR, Dunbar RIM (2013) Consistent social structure and optimal clique size revealed by social network analysis of feral goats, Capra hircus. Anim Behav 85:771–779

Sutherland MA, Lowe GL, Watson TJ, Ross CM, Rapp D, Zobel GA (2017) Dairy goats prefer to use different flooring types to perform different behaviours. Appl Anim Behav Sci 197:24–31

Teama FEI, El-Tarabany AA (2016) Physiological and biochemical response to Omega-3 plus as a dietary supplement to growing goats under hot summer conditions. Revista Brasileira de Zootecnia 45:174–180

Temple D, Manteca X (2020) Animal welfare in extensive production systems is still an area of concern. Front Sustain Food Syst 4:545902

Tölü C, Savaş T (2007) A brief report on intra-species aggressive biting in a goat herd. Appl Anim Behav Sci 102:124–129

Toussaint G (1997) The housing of milk goats. Livest Prod Sci 49:151–164

Tucker CB, Weary DM, Fraser D (2003) Effects of three types of free-stall surfaces on preferences and stall usage by dairy cows. J Dairy Sci 86:521–529

Tucker CB, Weary DM, von Keyserlingk MAG, Beauchemin KA (2009) Cow comfort in tie-stalls: Increased depth of shavings or straw bedding increases lying time. J Dairy Sci 92:2684–2690

Vaithiyanathan S, Mishra JP, Sheikh Q, Kumar R (2001) Salivary gland tannins binding proteins of sheep and goat. Indian J Anim Sci 71:1131–1134

Valpotić H, Gračner D, Turk R, Đuričić D, Vince S, Folnožić I, Lojkić M, Žura Žaja I, Bedrica L, Maćešić N, Getz I, Dobranić T, Samardžija M (2017) Zeolite clinoptilolite nanoporous feed additive for animals of veterinary importance: Potentials and limitations. Period Biol 119:159–172

Van Soest PJ (1994) Nutritional ecology of the ruminant, 2nd edn. Cornell University Press, Ithaca

Vas J, Andersen IL (2014) The effects of prenatal stocking densities on the fear responses and sociality of goat (*Capra hircus*) kids. PLoS One 9:e94253

Vas J, Chojnacki RM, Kjøren MF, Lyngwa C, Andersen IL (2013) Social interactions, cortisol and reproductive success of domestic goats (Capra hircus) subjected to different animal densities during pregnancy. Appl Anim Behav Sci 117–126

Vas J, Chojnacki RM, Andersen IL (2019) Search behavior in goat (Capra hircus) kids from mothers kept at different animal densities throughout pregnancy. Front Vet Sci 6:1–14
Vas J, Aabøe Inglingstad R, Bøe KE (in prep) Indoor environment of dairy and meat goat herds in Norway and use of environmental enrichments in practice
Waghorn G (2008) Beneficial and detrimental effects of dietary condensed tannins for sustainable sheep and goat production—Progress and challenges. Anim Feed Sci Technol 147:116–139., ISSN 0377-8401
Walsh GL, Birrell HA (1987) Seasonal variations in the chemical composition and nutritive value of five pasture species in south-western Victoria. Aust J Exp Agric 27:807–816
Wang SP, Wang WJ (2016) Effects of dietary supplementation of Chinese herb medicine mixture on rumen fermentation, nutrient digestion and blood profile in goats. S Afr J Anim Sci 46:247–260
Weaver LF, Boileau MJ, Gilliam LL, Taylor JD (2021) Characterization of short-and long-term morbidity and mortality of goat kids born to does with pregnancy toxemia. J Vet Intern Med 35:1155–1163
Webber JJ, Roycroft CR, Callinan JD (1985) Cyanide poisoning of goats from sugar gums (*Eucalyptus cladocalyx*). Aust Vet J 62:28
Willms WD, Kenzie OR, McAllister TA, Colwell D, Veira D, Wilmshurst JF, Entz T, Olson ME (2002) Effects of water quality on cattle performance. Rangeland Ecol Manag/J Range Manag Arch 55:452–460
Wilson RT (1989) Ecophysiology of the Camelidae and desert ruminants. Springer, Berlin. 120 pp
Yang K, Tian X, Ma Z, Wu W (2021) Feeding a negative dietary cation-anion difference to female goats is feasible, as indicated by the non-deleterious effect on rumen fermentation and rumen microbial population and increased plasma calcium level. Animals 11:664
Yang W, Guo G, Chen J, Wang S, Gao Z, Zhao Z, Yin F (2022) Effects of dietary fucoidan supplementation on serum biochemical parameters, small intestinal barrier function, and cecal microbiota of weaned goat kids. Animals 12:1591
Yaseen A, Hussain T, Hameed A, Shahzad M, Mazhar MU, Chughtai MI (2022) Flavonoid enriched supplementation abrogates prenatal stress and enhances goat kids' performance reared in a sub-tropical environment. Res Vet Sci 146:70–79
Yoon IK, Stern MD (1995) Influence of direct fed microbials on ruminal microbial fermentation and performance of ruminants: A review. Asian Australas J Anim Sci 8:533–555
Zebeli Q, Dijkstra J, Tafaj M, Steingass H, Ametaj BN, Drochner W (2008) Modeling the adequacy of dietary fiber in dairy cows based on the responses of ruminal pH and milk fat production to composition of the diet. J Dairy Sci 91:2046–2066
Zhang YM, De AO, Lei KW, Lin XI, Spears JW, Shi HT, Huang YL, Yang FL (2023) Dietary copper supplementation modulates performance and lipid metabolism in meat goat kids. J Integr Agric 22(1):214–221. Available online 10 August 2022, ISSN 2095-3119
Zhong RZ, Tan CY, Han XF, Tang SX, Tan ZL, Zeng B (2009) Effect of dietary tea catechins supplementation in goats on the quality of meat kept under refrigeration. Small Rumin Res 87:122–125
Zobel G, Nawroth C (2020) Current state of knowledge on the cognitive capacities of goats and its potential to inform species-specific enrichment. Small Rumin Res 192:106208

Welfare Issues in Goat Farming and Management

Marianne Berthelot, Anne Aupiais, Xavier Boivin, Renée de Cremoux, Alice Fatet, Virginie Michel, Pierre Mormede, Raymond Nowak, Isabelle Palhière, Carine Paraud, Jean-Louis Poulet, Marie-Madeleine Richard, Gosia Zobel, and Alain Boissy

M. Berthelot · C. Paraud
Anses, Laboratoire de Ploufragan-Plouzané-Niort, Unité Pathologie et Bien-Être des Ruminants, Niort Cedex, France

A. Aupiais
Institut de l'Elevage, Service Bien-être et Santé des ruminants, Le Rheu, France

X. Boivin · M.-M. Richard · A. Boissy (✉)
INRAE, Université Clermont Auvergne, VetAgro Sup, UMR Herbivores, Saint-Genès Champanelle, France
e-mail: alain.boissy@inrae.fr

R. de Cremoux
Institut de l'Elevage, Service Bien-être et Santé des ruminants, Paris Cedex 12, France

A. Fatet
INRAE, UE FERLUS, Lusignan, France

V. Michel
Anses, Direction de la Stratégie et des Programmes et Pôle Recherche et Référence et du travail, Maisons-Alfort Cedex, France

P. Mormede · I. Palhière
INRAE, Département de génétique animale, Centre de recherches de Toulouse, Castanet-Tolosan, France

R. Nowak
CNRS, INRAE, Université de Tours, Physiologie de la Reproduction et des Comportements, Nouzilly, France

J.-L. Poulet
Institut de l'Elevage, Service Productions Laitières, Le Rheu, France

G. Zobel
Animal Behaviour and Welfare Team, AgResearch Ltd., Ruakura Research Centre, Hamilton, New Zealand

S. Mattiello, M. Battini (eds.), *The Welfare of Goats*, Animal Welfare 25,
https://doi.org/10.1007/978-3-031-62182-6_5

Abstract

Goats are raised in a wide range of production systems all around the world. Whatever the system, different aspects of the farming management can affect the welfare of goats, namely management of the relationships with social peers and humans, kids' management, outdoor access, and milking, reproduction and genetics management. In this chapter, each of these management aspects is discussed with regard to the risks of impairing the welfare of goats and the strategies to mitigate their stress by better fitting the farming management to the behavioural and physiological needs of goats. In addition to stress mitigation, this chapter also highlights the possibility throughout management practices to elicit positive emotional experiences to induce positive welfare in goats. In tandem, the positive effects of facilitating positive welfare of goats on the quality of life at work for the farmer are reported, illustrating the interrelationships between human wellbeing and animal welfare. Recent scientific knowledge in behavioural and social needs of goats and their cognitive abilities should be better taken into account to develop good practice guidelines and specific regulations to ensure more welfare-friendly goat farming systems.

Keywords

Goat farming management · Behavioural needs · Stress mitigation · Enrichment · Farming practices · Human–animal relationship

1 Introduction

Goats are raised in a wide range of production systems around the world. Most of them are bred in smallholder, mixed farming systems. Worldwide, goats are an important component of pastoralist herds. High-value animals are raised for specialised dairy production in temperate regions (Battini et al. 2021). For this chapter, the following farming systems were identified (EFSA 2014; Battini et al. 2021; Silva Salas et al. 2021):

- Intensive (no outdoor access or access to a paddock in some cases)
- Semi-intensive (pasture access during the day, housing similar to intensive system during the night and during non-outdoor access periods)
- Semi-extensive (systems that rely mainly on pasture, with limited use of feed supplements as needed and some provision of housing for shelter)
- Extensive (systems that rely almost exclusively on natural resources for feeding, with no or very limited access to housing)
- Pastoralism (management system of marginal areas, where the stockperson is constantly with the animals while they are grazing)
- Mixed system (combination of two or more of the previous systems over the year)

This chapter will deal with animal welfare in relation to farming management in the context of goat husbandry, excluding housing and feeding (see chapter "Welfare Issues in Goat Farming: Housing and Nutrition"). The following topics will be presented: (1) management of social relationships; (2) relationships with humans; (3) management of kids; (4) milking management; (5) reproductive management; (6) management of outdoor access; and (7) genetics management. The main aim of the chapter is to discuss the welfare risks associated with these complementary topics within and between different production systems, and to identify how to promote positive welfare of goats. Indeed, the life of an animal is more than just avoiding suffering; it is important to balance good health and physiological functions with positive experiences and the ability to perform natural behaviours, which overall contribute to a good quality of life (Fraser et al. 1997). Whatever the system, the processes used to manage goats may affect their welfare. As the behavioural and physiological needs and the cognitive abilities of goats become well known, there is a greater opportunity to better fit the farming systems to the goats in order to improve the quality of their life. Throughout the chapter, the positive and negative aspects of breeding practices are discussed from the animals' point of view (i.e. the perception of goats). This animal-centric position focuses on the animals' experience in the conditions they are kept and how that affects the quality of their life. As most of the literature available on goats refers to dairy goats, the majority of the literature cited in this chapter focuses on dairy goats.

2 Management of Social Relationships

As most domesticated farmed species, goats live in groups and have a high level of sociality and inter-individual recognition (see chapter "Goats' Natural Behaviour"). In natural conditions, they form small groups in which they build stable relationships. Stanley and Dunbar (2013) suggest that a group size of 12–13 individuals may be optimal for goats. However, depending on the farming system, group size is highly variable. While smallholders maintain small groups (e.g. Singh et al. 2013), in many farming systems group sizes are large (Fig. 1). For instance, during welfare assessments in European countries, authors have reported various group sizes in dairy goat herds (e.g. mean ± SD: 113 ± 84 goats in 30 Portuguese farms—Can et al. 2016; up to 192 goats in a single pen among 30 Italian farms—Battini et al. 2016a; median 85 (IQR: 61–124) in 24 UK farms—Anzuino et al. 2010). Large groups may lead to social stress by reducing inter-individual recognition.

As widely discussed in chapter "Goats' Natural Behaviour", the social organisation of goats is the result of different levels of relationships: between the dam and her kids, between kids and between all the individuals that make up the herd. Briefly, social organisation is mainly based on relationships of dominance, affinity, attachment, and cooperation that ensure stability and cohesion in the group. Knowledge of this network and of the group hierarchy makes it possible to ensure that social affinities and incompatibilities are respected in the event of changes in herd composition. Social stability is essential for the welfare of goats since the formalisation of

Fig. 1 Illustrations of large groups of goats (Photos: Johanna Droz-Vincent (**a**); Renée de Cremoux (**b** and **c**))

dominance–subordination relationships limits conflicts and aggressive interactions in farming conditions, as the subordinate individual recognises the dominant individual's status and responds by fleeing or avoiding even before any aggressive demonstration by the dominant. Affinity relationships are based on affiliative interactions and allow mutual tolerance between individuals. Moreover, the presence of conspecifics allows the individual to cope with anxiety-provoking situations by reducing fear reactions. It also provides the individual with a social model for acquiring skills, such as appropriate food or sexual choice, through a process of learning by observation and imitation.

Current management systems that may expose goats to different social stressors and thus compromise welfare are discussed, mainly the impacts of social isolation and regrouping (e.g. separation of individuals from conspecifics for treatment management, or regrouping for reproductive or other management reasons). We also highlight a few examples that do not receive as much attention as they should, namely feral goats captured for intensive feedlot and bucks' management.

2.1 Separation and Reintroduction of One or a Few Goats

Health management may require separating a goat from the rest of the herd (Patt et al. 2012). This can be for treatment purposes or to prevent injury to a weak animal from the other goats. This practice causes stress for the isolated goats, both at the separation and at the reintegration, and thus care is needed to reduce this stress. The welfare of individual goats of various Swiss milking breeds introduced in an unknown flock was seriously affected for at least 5 days and the presence of goats with horns amplified these consequences (Patt et al. 2012). The introduced goat displayed a longer lying period, shorter feeding times, and elevated concentrations of faecal cortisol metabolites during the introduction period. In a follow-up study, Patt et al. (2013a) replicated their results, showing that separated goats feed less and have higher concentrations of faecal cortisol metabolites. These authors also found that on the first day of reintroduction, the reintroduced goats were more likely to display agonistic and sniffing behaviours towards resident goats and persistently higher concentrations of faecal cortisol metabolites. Visual and tactile contact with the group (as opposed to only auditory) reduced the stress response. Resident goats were not affected by the separation and reintroduction of a pen mate. Patt et al. (2013b) also suggested that some of the negative effects can be mitigated by handling animals with familiar conspecifics, rather than individually, when introducing to a new unfamiliar group.

2.2 Regrouping and Mixing

Goats can be moved from one group to another for different reasons: (1) regrouping just before or after parturition with integration of primiparous animals; (2) regrouping kids or goats depending on different criteria (e.g. weight, production, mammary health, other disease status); and (3) regrouping at mating (e.g. artificial insemination, natural reproduction with bucks) (Szabò et al. 2013). Providing enough space to allow goats to hide, as well as outdoor access and environmental enrichment, may help to mitigate the stress of mixing unfamiliar individuals (Keil and Patt 2013). In French Alpine goats, Fernández et al. (2007) showed that aggressive behaviour increased after all regroupings, whereas milk production decreased only after the first regrouping; while this suggests some capacity to adapt to novel and stressful managements, this work was confounded by small group size resulting in some level of familiarity between individuals following regrouping events. Szabò et al. (2013) found that young dairy goats experience less social stress when being introduced into a herd of adult dairy goats shortly after parturition and with kids still present rather than during the dry period. Social instability increased aggression in groups of dairy goats, but consequences for kids were only minor (Andersen et al. 2008). In intensive dairy systems, as kids are generally removed from their dams just after birth, kids cannot be present when young goats are introduced in the adult herd. Therefore, while more research would be useful to streamline the best methods of transitioning goats between groups, the existing literature suggests that moving goats together (as opposed to individually), at key moments (e.g. after parturition while kids are present) are promising ways to reduce social stress.

2.3 Feral Goats

Introduced feral goats are considered a problem in many parts of the world and controlling them presents numerous challenges. In some countries (e.g. New Zealand, parts of Australia), non-harvesting culling efforts are often used, sometimes in conjunction with "judas goats" (e.g. a few individuals are equipped with tracking collars; being social, these goats are motivated to return to their herd mates, and thus contribute to efficient hunting; Lethbridge 2016). Interestingly, where the terrain lends itself to live harvesting, markets for feral goats have emerged (Thompson et al. 2002). Capturing methods include mustering (where densities are high) and live trapping (where water sources are scarce and can be used to lure goats; Sharp 2005). Trapping is considered to be less stressful; however, it results in feral animals being held in closely confined areas. In its Standard Operating Procedure for capturing feral goats (Sharp 2005), the Australian government provides numerous recommendations to avoid compromising welfare, including ensuring large enough trap areas so that all captured animals have access to water and shelter and minimising the mixing of unfamiliar animals. Furthermore, to avoid dam and kid separation, it is recommended that trapping is not done during the kidding season. Regardless of how feral goats are captured, the transition from rangeland to intensive rearing is very stressful since goats are held under semi-intensive (e.g. controlled rangeland) or intensive (e.g. feedlot) conditions and are exposed to humans and working dogs, and their social groups change dramatically. Miller et al. (2018) found that captured feral goats that received higher levels of exposure to humans were involved in fewer agonistic interactions within a few weeks of capture. These findings suggest that habituation to humans may help to manage social behaviour of feral goats. This may be linked to their domestic background, as suggested by Leidy et al. (2015) when comparing the gregarious nature of feral goats to truly wild individuals. While Flint and Murray (2001) used feral cross goats in their work, they nonetheless found that habituation to feedlot settings was key to successful rearing. Importantly, their work highlighted that increasing the space allowed per goat (11 m^2 vs. 6 m^2/goat) and providing climbing and interaction opportunities reduced feeding-related aggression and improved weight gain.

As described, it appears that careful management options may reduce negative behavioural responses of feral goats when captured and reared intensively; nonetheless, it is still important to consider that there is a dearth of information about the welfare of these animals. Lethbridge (2016) highlighted the potentially very wide home ranges (e.g. between 2 and 224 km^2) of feral goats, Stronge et al. (1997) described the large feeding variability of feral goats, and O'Brien (1984) documented the important individual differences in behaviour of feral goats and their response to kids. Therefore, it can be postulated that, while captured, feral goats may habituate to human interaction, the important reduction in space allowance, exercise, and feed variability, as well as dramatic change in social dynamics, may negatively and irreparably affect the welfare of these animals.

2.4 Buck Management

Goats are usually maintained in single-sex groups of similar age or size, the main exceptions being the dam-kid dyads, and mixed-sex groups at mating (Fig. 2). In order to prevent unwanted mating with too young female kids and to control the mating period, bucks are generally housed away from the females (Fig. 3) after weaning and are kept separate from other batches of animals outside of the breeding period (Ridler et al. 2012).

Sexual segregation seems to alter the physiological and behavioural development of young bucks. The sexual performance of mature males may be negatively affected by the lack of contact with females in the prepubertal period; the complete absence of females during male rearing disrupts their behaviour and sexual maturation (Lacuesta et al. 2018a). Young bucks reared with adult female goats in the prepubertal period prefer social contact with females compared to interactions with males, while those reared only with other males did not express social preferences (Lacuesta et al. 2018b). This underlines the importance of letting male kids interact with female kids at a young age (Fatet and Johnson 2023).

Presence of adult females in groups of male kids had short-term positive effects on scrotal circumference and testosterone concentration until females became pregnant, which occurred when bucks attained approximately 28 weeks of age (Lacuesta et al. 2015). Allowing only visual and tactile contact through a barrier with female goats during the prepubertal period may be an option for improving buck welfare (while preventing pregnancy).

Fig. 2 Male–female group at mating (Photo: Renée de Cremoux)

Fig. 3 Group of bucks (Photo: Renée de Cremoux)

In order to not alter young bucks' development, they should have contact with females, especially at a young age. In addition to that, a particular attention should be paid to the housing of the bucks when not in the mating season, to assure them to live in an environment, which meets their needs.

3 Relationships with Humans

The relationship resulting from interactions between human and animals is a dynamic process (Waiblinger et al. 2006). It is based on past interactions between individuals that know each other or are able to generalise their living experience from particular humans to unknown humans. This allows the partners of the relationship network to anticipate future interactions, emotionally connoted positively or negatively. Good relationships with humans are particularly important for managing goat welfare (Nawroth et al. 2019; Jardat and Lansade 2021). The development of the human–animal relationship is influenced by communication between humans and animals (Waiblinger et al. 2006). The quality and quantity of contact between humans and goats can be very different depending on the farm management systems. Human proximity with goats is typically more frequent in intensive farming systems compared to extensive pasture-based systems. However, beyond this broad generalisation, strong variations exist among all types of goat farms in terms of the level and nature of human–goat interactions.

As described in chapter "Where Do Goats Come from?", goats were among the earliest farm species to be domesticated. They have a long process of domestication reinforcing their coadaptation to the human environment (Celozzi et al. 2022). Goats and humans are two species that are social and used to living in groups, communicating, and building a social organisation based on affinities and dominance relationships. Goats demonstrate important cognitive capacities related to memory and ability to gather information from humans; this can promote positive human–animal communication and relationships (Nawroth and McElligott 2017, Celozzi et al. 2022). For example, experimental studies in cognition demonstrated that goats, similarly to horses and dogs, are able to use human gesturing or human gaze to solve problems (Nawroth et al. 2016, 2020). Despite selective breeding efforts, goats are known to have diverse individual temperament traits, which can affect the development of individual relationships with humans. Farmers should be aware of goats' emotional, cognitive, communicative, and memory abilities in order to treat goats as partners within these working relationships. The aim of this section is to review the various factors that could help farmers, and other people involved in goat management, to create positive human–animal relationships.

3.1 Early Handling Experience

Human contact in early life of animals has long-lasting effects on the human–animal relationship in the future. As for many other species, both the absence of contact and limited contact lead to later fearfulness towards humans in goats. It is suggested that receiving some human contact during feeding and housing management is not sufficient to prevent fearfulness in artificially reared kids (e.g. those removed from their dams soon after birth) (Boivin and Braastad 1996). The presence of the dam, and her own behaviour towards humans, can affect the development of the human–animal relationship in her kids (Lyons 1989; Lyons et al. 1988; Nowak and Boivin 2015). Toinon et al. (2021) showed that weaned kids reared with their dams had a greater avoidance distance from humans compared to artificially reared kids during human–animal relationship tests. However, in this study, artificially reared kids were in daily contact with a human until weaning, including positive physical interactions during feeding such as gentle stroking, whereas dam-reared kids were only handled for marking or veterinary interventions. While Lyons (1989) found no difference in overall milk production between goats that had been dam reared compared to those that were hand fed, they did find that the former had higher residual milk for the first 5 days after being moved into the dairy barn. The initial treatment (rearing environment as kids) was pseudo replicated in this study, and residual milk was measured using a potentially confounding approach (i.e. goats were milked by hand to extract residual milk); nonetheless, these findings suggest that early human contact with goat kids may improve their transition into a new milking herd. Importantly, the number and quality of these interactions are also essential for the establishment of the human–animal relationship (Mastellone et al. 2020; Miller et al. 2018; Yoshida and Koda 2020; Jackson and Hackett 2007).

3.2 Human Behaviour

As with the relationship with cattle, the farmer's attitudes, his empathy and more directly his behaviour towards his goats can have a significant impact on the relationship he builds with them (Mersmann et al. 2016; Muri et al. 2012). For

Fig. 4 Positive handling (**a** and **b**) and neutral handling (**c**) (Photos: Renée de Cremoux)

instance, the proportion of negative interactions during milking (e.g. shouting, pushing, or pulling strongly) results in goats having higher avoidance and lower approach behaviours in standardised reactivity tests towards an unfamiliar human (Mersmann et al. 2016). Handling can have an impact on both goats and kids. Examples of positive and neutral handling are presented in Fig. 4. Baxter et al. (2016) showed that dairy goats are sensitive to the type of handling they receive during pregnancy (e.g. aversive, minimal, or positive handling for 10-min periods twice daily, 5 days a week between days 80 and 115 of pregnancy). Foetal loss was observed only for goats in the aversive handling treatment. Placental morphology, goats' behaviour towards their young and kids' behaviour were also affected by treatment. For instance, kids received less grooming and nosing behaviour from their dams in the aversive treatment and performed play behaviour later than kids in other treatments.

Miller et al. (2018) demonstrated that more frequent interactions with humans, consisting of entering the goat enclosure daily and walking calmly, promoted habituation of feral rangeland goats to intensive farming systems. In fact, these frequent interactions led to less aggressive reactions and reduced flight responses compared to those recorded in feral rangeland goats under similar farming conditions, which had experienced a lower degree of interactions with humans. As shown in sheep (Fell and Shutt 1989), painful husbandry procedures may cause animals to associate humans with negative handling experiences, however to our knowledge no similar studies exist in goats.

3.3 Physical Environment

The physical environment in which animals live can affect the quality of the human–animal relationship. For instance, farm size potentially affects fear of humans (Battini et al. 2016a). A study conducted on 30 Italian farms showed that animals can be touched less easily on the muzzle on farms with more than 100 goats than on farms that owned less than 50 goats. This can be explained by the closer proximity and more frequent contact between the farmer and his goats on smaller farms. However, this was not confirmed in another study using another test (reaction to a stationary person) which found that latency to the first contact was lower in large farms than in small and medium farms (Can et al. 2016). This highlights the difficulties in comparing between studies at the farm level, as there are many confounding factors including the tests used, breed differences or housing type. The two tests used by Battini et al. (2016a) and Can et al. (2016) to assess human–goat relationship in farms of different size present a convergent validity on goats (Battini et al. 2016b). Nevertheless, the two test procedures reflected different components of the human–animal relationship, and studies warn about switching between tests at the herd level (Battini et al. 2016b). For instance, goats not approaching a motionless unfamiliar human or avoiding an approaching unfamiliar human could both reflect fear responses towards humans (Lyons et al. 1988); however, lack of approach to a motionless human could also reflect a lack of interest from the animal to the human in the context. On the other hand, goats' curiosity may be more important than fear, and thus may result in them approaching a person despite their potential fear (Celozzi et al. 2022). The avoidance test can be also influenced by different factors such as breed (some are more reactive) or production system (intensive vs extensive; Can et al. 2016). For instance, Saanen and Murciano-Granadina are common breeds in large farms in Portugal and are reported to be easy to manipulate, whereas in small farms a high variety of breeds was reported (e.g.: Serrana, Alpine, Malagana, Murcino-Granadina, Saanen). Furthermore, goats on small farms frequently have access more often to outdoor grazing, which might reduce the level of confidence in humans as observed in dairy cattle (Battini et al. 2011).

Some other factors, such as the presence of automatic systems (automation of food or litter delivery, of milking, camera monitoring, health monitoring through sensors), may represent risk factors for developing poor human–animal relationships by reducing direct positive human contact (Cornou 2009). Farmers will receive information directly at all times on the state of an animal on a computer or phone and may spend less time observing their animals directly. Moreover, if all routine tasks are automated, this may lead the farmer to interact with animals mainly for aversive tasks such as disbudding, vaccination, or trimming (Cornou 2009). Conversely, the free time given back to the farmer by automation may allow farmers to spend more time with their animals, reduce the time for repetitive and laborious operations, and identify earlier health issues (Cornou 2009). Impact will depend on the way automatic systems are used on farm. To the authors' knowledge, no studies have yet explored these aspects in goats.

Providing goats with an enriching environment is mentioned as a way of promoting natural behaviour (e.g. Zobel et al. 2019). However, Miranda-de la Lama et al. (2013) suggest that the introduction of enrichment objects may create distance between goats and humans by focusing the animals' attention on these objects and/or by diluting the presence of humans in a more complex environment. For instance, goats in enriched environments (e.g. varied food presentation, use of physical barriers to limit social aggression, use of elevated areas) remained further away from a stationary experimenter, compared to control goats. This may not mean they are more fearful, but maybe they are less interested in humans as a source of stimulation. The enriched environment may be sufficient by itself to reduce exploration of another stimulus such as humans.

3.4 Individual Traits

Behavioural differences have been shown among breeds (Rosenberger et al. 2022) and also depending on individual personality traits (some individuals are particularly bold or shy; Lyons et al. 1988; Ndou et al. 2010; Finkemeier et al. 2019; Neave and Zobel 2020). Such differences could be considered as selection traits, as already observed in sheep (Tamioso et al. 2018). Genetic traits may affect human–animal relationships; for example, individual differences in temperament will influence goats' attitude towards humans, and goats' behavioural reactions (Finkemeier et al. 2019; Neave and Zobel 2020). Genetic differences also exist, for example, among goat breeds; in fact, it has been observed that some breeds respond more negatively to human manipulation than others (Molale et al. 2016; Ndou et al. 2010; Can et al. 2016).

Social factors such as dominance rank might affect goats' relationship with humans. For instance, lower-ranking goats have a smaller distance from an experimenter positioned in the centre of a bow for two minutes compared to higher-ranking goats (Miranda de la Lama et al. 2013). Nevertheless, environmental enrichment tends to limit the effect of social dominance on relationships with humans. Furthermore, social facilitation may also promote different responses of goat kids towards humans; the presence of a bold dam or bold non-affiliated adult resulted in kids being more willing to approach humans (Lyons et al. 1988).

3.5 Interconnection Between Goat and Human Welfare

As demonstrated in a wide variety of species, a poor human–animal relationship can strongly affect animal welfare, including affective state and production (Hemsworth et al. 2000; Zulkifli and Siti Nor Azah 2004; Ivemeyer et al. 2011; Mellor et al. 2020; Pol et al. 2021). Improving relational practices with goats should really be a strategy to both improve animal welfare and facility of handling (Beaujouan et al. 2021). Likewise, improving goat welfare can be beneficial for humans as it increases job satisfaction and reduces tediousness for humans (Celozzi et al. 2022). For farmers, building good human–goat relationships should therefore be considered as important as any other husbandry tasks. Pinillos et al. (2016) defined the concept of One Welfare as "the interconnections between animal welfare, human wellbeing and the environment". This concept has become more and more popular in recent years, even if it has not been deeply investigated so far: many systems that aim to assess sustainability of livestock systems do not incorporate measures of animal welfare (Pręgowski 2019). It is particularly important to explore not only where synergies exist, but also where antagonisms occur by defining trade-offs for developing robust perspectives for future husbandry systems.

4 Management of Kids

This part of the chapter focuses on management practices from the colostrum phase to weaning, across different farming systems. Several factors have to be considered to assure good colostrum and milk management: method of distribution, quality, quantity, and appropriate timing of feeding. Concerning kids rearing systems, in intensive dairy systems, kids are generally separated from their mothers (Fig. 5a) at birth and fed with milk replacer (Delgado-Pertíñez et al. 2009) for economic and health reasons. In less intensive dairy systems, in regions where kids from dairy breeds are fattened up on farm, or in meat production systems, kids are raised with their dams (Fig. 5b; Berthelot et al. 2022; Panzuti 2018; Torres et al. 2021). Weaning, which is defined here as the process of getting to zero milk, is a stressful period, as the kids have to adapt to a change, that can be progressive or abrupt, of feeding from a mixed (liquid and solid) to a solid diet only. All these aspects will be detailed in this section through their impacts on kids' welfare.

4.1 Colostrum Management

Colostrum available at birth is crucial for kids' survival and robustness since, in addition to nutrients, it allows the transmission of immunoglobulins (IgG). IgG are necessary for the establishment of passive immunity, since maternal immunoglobulins do not cross the blood–placenta barrier in goats (Nowak and Poindron 2006). Colostrum-feeding methodology differs widely among systems according to farm practices; if not given correctly, failure of passive transfer of immunity could occur. The most important factors for good passive transfer of immunity are the quality, the appropriate timing of feeding, as well as the amount of colostrum (Turini et al. 2020; Zamuner et al. 2023).

The manner colostrum is delivered to the newborn can be very different across farming systems. The dam can be directly the source of colostrum for the kids. However, kids can be separated from their dams (Bélanger-Naud and Vasseur 2021;

Fig. 5 Kids reared without their dams (Photos: Renée de Cremoux (**a** and **b**)) or without their dams (Photos: Ludivine Derboule (**c**); Renée de Cremoux (**d**))

Zamuner et al. 2023). In case of separation, colostrum can be hand fed, generally using a nipple bottle (Zamuner et al. 2023). Another method of hand colostrum delivery is the use of an oesophageal tube (Godden et al. 2019), especially in kids with weak sucking activity. Kids receive milk directly in the stomach (Kerr n.d.). This method has been reported as timesaving and efficient to accomplish successful IgG passive transfer. However, this method is invasive and requires technical skill, as there are risks with improper placement and damage to the oesophagus or liquid aspiration in the lungs (Lopez and Heinrichs 2022). This must be done by a trained person and with adapted equipment in order to not harm the kids. In calves, the oesophageal groove reflex is not stimulated using this method (Lateur-Rowet and Breukink 1983). Research is needed to examine the prevalence of this method and its impacts on goat kids' welfare.

The quality of colostrum can vary between goats, with important inter-individual differences in IgG concentrations. When kids are hand fed, the source of colostrum may be diverse. It can come from the dam or another goat, from a pool of colostrum from different goats, from cows or be commercial colostrum replacer (Bélanger-Naud and Vasseur 2021). The IgG concentration in kids' serum does not seem to be affected by the type of colostrum whether it comes from goats, cows or is lyophilized as long as IgG concentration is equivalent in the different types of colostrum (Bélanger-Naud and Vasseur 2021). It is the quality (IgG/g colostrum) of the colostrum, which seems to be the most important, regarding IgG serum concentration and passive immunity. Poor concentration in IgG in the colostrum results in lower concentration of IgG in kids' serum (Castro et al. 2005). Colostrum quality should always be tested prior to feeding; this can be done using various on-farm equipment such as hydrometers or refractometers (Kessler et al. 2021; Zobel et al. 2020a). However, as cautioned by these authors, a threshold for "good" colostrum has yet to be established for goat colostrum.

Regardless of the method of distribution, the origin and the quality of the colostrum, the key point is also to give colostrum as soon after birth as possible. Recommendations for the timing of the first intake of colostrum varies from immediately after birth to 6 h later (Bélanger-Naud and Vasseur 2021), in order to allow the absorption of the IgG through the small intestine. In calves, the intestinal permeability is reduced at 12 h and stopped at 24 h (Lopez and Heinrichs 2022). If dams are permitted to suckle their young, kids should still be monitored to ensure they are thriving and fed within the first few hours of life. If colostrum is obtained from the mother and hand fed indirectly to the young, it should be collected as soon as possible after parturition since Yang et al. (2009) showed a fourfold decrease in colostrum quality after 24 h, which highlights the importance of collecting colostrum immediately after birth. In cattle, the highest IgG concentration in calf blood was achieved when calves received their main colostrum meal 2 h after birth; if they received it later, calves required significantly more colostrum (Chigerwe et al. 2008). Such work has not been conducted in goat kids. Recommendations concerning the duration for the colostrum-feeding period after the first intake varied between 24 h and 3 days, but studies are needed to see the impact of this duration on transmission of nutrients, fat or growth hormones, which are also important for kids' health (Bélanger-Naud and Vasseur 2021).

The quantity of colostrum given seems also important, but to the authors' knowledge, there are no robust evaluations linking colostrum amount and kid serum IgG concentrations. A commonly cited value (10–12% of body weight) divided into 2–4 meals exists; however, this threshold appears to refer to calves (e.g. Bélanger-Naud and Vasseur 2021; Chigerwe et al. 2008; Godden et al. 2019). Nonetheless, this information shows that the amount of colostrum, combined with appropriate timing and quality, contributes significantly to successful passive transfer of immunity; thus, similar research should be conducted in goat kids. In addition, as colostrum gives energy to avoid hypothermia for the kids, the quantity required may also depend on temperatures in the barn.

When the herd is infected with caprine arthritis-encephalitis virus (CAEV), goat colostrum should be heated in order to prevent transmission (Zamuner et al. 2023). Colostrum heating up to 56 °C for 30 min negatively affected some immunological parameters in kids' plasma (lower total protein, γ-globulins, and IgG levels), but these changes did not affect the clinical status or performance of the kids during the experiment (Fernández et al. 2006). Colostrum treated at temperatures exceeding 59 °C undergoes denaturation of immunoglobulins and develops clumps according to Rowe and East (1997), but these effects on kids have not been studied yet. In cattle, a meta-analysis conducted by Malik et al. (2022) showed a reduction in IgG concentration in colostrum and in calves' serum when colostrum was heated above 60 °C, but not at 60 °C or below. Heat treatment of colostrum at 56 °C seems to have little impact on kids' immunity and heat treatment of 56 °C during 60 min allows the elimination of viruses like CAEV (Rowe and East 1997). However, further research is needed to better evaluate the effect of colostrum heat treatment (duration and temperature) on IgG and other immunological parameters, such as cytokines and immune cells and kids' welfare.

Several factors must be taken into account to assure proper transmission of immunity: feeding method, quality and origin of colostrum, time and duration of first feed, quantity and heating treatment. However, a key point is that most research focused on immunoglobulins, but colostrum is much more than that: it also contains growth hormones, fat, and other nutrients that are important for kids' health (Zamuner et al. 2023). The impacts of these different factors on health and kids' development should now be investigated.

4.2 Separation Between Kids and Dams

One important feature of the behaviour of goats is their marked sociality and the bond formation between mother and young (Poindron et al. 2007). This attachment is based on early mother–young individual recognition. During development, kids extend their social world to their siblings, and their activities are then synchronised.

However, kids from dairy breeds are generally separated from their dams soon after birth (Bélanger-Naud and Vasseur 2021).

Early separation between dam and kids can be stressful for both partners (Miranda-de la Lama and Mattiello 2010; Winblad von Walter 2010). However, in their review on cattle, Meagher et al. (2019) emphasise that early separation (less than 24 h after birth) compared to a later separation (4, 7, or 14 days) minimises the acute stress response on cows and calves. Nevertheless, Meagher et al. (2019) underline that long-time contacts may have positive impacts on welfare by promoting social behaviour and reducing abnormal behaviour in calves.

Saanen goat kids' reactivity to behavioural tests differs depending on their rearing conditions. Artificially reared kids showed a more active coping strategy than dam-reared kids in different behavioural tests (novel arena, novel goat, and novel object tests). They vocalised, ran and jumped more, whereas dam-reared kids self-groomed and urinated more than artificially reared kids. Artificially reared kids also initiated contact more quickly with the novel object and the novel goats than dam-reared kids (Toinon et al. 2021). This suggests that artificially fed kids are bolder in anthropised contexts compared to dam-reared kids. However, according to Winblad von Walter et al. (2021), the effects of dam's presence during rearing on kids' response to isolation and to an arena test are not so clear. In their experiment, kids were either kept permanently or partly with dams, or completely separated. Differences in reactions were observed between treatments, but the authors are not able to say which treatment triggered the highest response in terms of fear and stress, as all kids show fear responses and their responses vary depending on the tests. This highlights the difficulty of assessing stress responses and underlines the need to combine physiological and behavioural measurements with the contextual environment.

Contact with the dam can modify social behaviour development by comparison with kids reared artificially and with same-age peers. Before weaning, artificially reared kids spend less time allogrooming and express more play-fighting than kids reared with their dams (Toinon et al. 2022). According to the authors, higher occurrences of play-fighting would be a strategy to compensate for the lack of maternal care, such as suckling and allogrooming. Immediately after weaning and mixing, dam-reared kids initiate more and receive less agonistic interactions from artificially reared kids. However, the number of agonistic interactions displayed by a dam-reared kid towards an artificially reared kid decreases whereas the number of agonistic interactions given by artificially reared kids to dam-reared kids increases over time. Toinon et al. (2022) state that artificially reared kids develop their agonistic behavioural repertoire in contact with dam-reared kids and orientated it towards dam-reared kids maybe because they were less familiar.

In pasturing or more extensive systems, where goats had an outdoor access, the presence of the dam may be even more relevant than in indoor conditions, especially in the development of learning abilities concerning pasturing and foraging. Indeed, early mother–young separation may not allow kids to learn appropriate grazing behaviour from their dams in extensive systems. Kids will not have any experience the first time they will access outdoors and have to learn by themselves. The presence of dams would favour earlier diversification in solid food intake by imitation and the recognition of plants in pastures and facilitate learning of pasturing behaviour (Arviv et al. 2016).

Another behavioral consequence of leaving kids with their dams is a potentially less developed relationship with humans, resulting in shyer and less easy-to-handle kids (Toinon et al. 2021). Attention should be given to accustom the kids to being in contact with their caregiver and create a bond based on frequent and gentle stroking rather than only on feeding (Boivin and Braastad 1996).

Torres et al. (2021) found that Canarian goat kids that are reared indoors by their dam and weaned at 9 weeks of age have a better growth performance at 15 weeks than artificially reared kids weaned at the same age. This may be explained by two factors. First, the composition of goat milk may be more appropriate for proper growth than the milk replacer. Second, dams (and other adults) are strong social partners and demonstrators to explore new solid food. The beneficial effect of maternal presence on kids' growth is confirmed by some authors for Criollo and Damascus goat kids (Paez Lama et al. 2013; Tsiplakou et al. 2016), but not for other breeds (Florida goat kids; Delgado-Pertíñez et al. 2009; Saanen goat kids: De Palo et al. 2015). However, the results of these studies cannot be compared, because differences may be due to the several factors, such as different breeds and production (meat vs dairy), goat kid ages, or the quantitative and qualitative characteristics of maternal milk and additional solid feed, as well as different methods of administration of milk replacers (Tsiplakou et al. 2016).

One of the main reasons to separate kids at birth from dams is to prevent pathogens transmission, such as the ones responsible for CAEV or paratuberculosis. These pathogens can be transmitted at birth, through colostrum and milk from CAEV-infected dams (Blacklaws et al. 2004), and also by horizontal contamination (Adams et al. 1983), and through faecal-oral contamination for paratuberculosis (Idris et al. 2022; Windsor 2015). These two diseases can have a severe negative impact on goat welfare (see chapter "Painful Procedures, Transport and End-of-Life Management"), their clinical expression being observed much later in the animal's life (Ghanem et al. 2009). As these diseases are common, a particular attention has to be given, especially for replacement females. In meat breeds, as the lifespan of kids is much shorter than in dairy breeds, this issue is less important. It is important to note that, if disease prevention is one of the main reasons for removing kids from their dams and feeding heated colostrum manually, precautions should be taken, as the success of removal of goat kids prior to suckling from the dam is variable (Zobel et al. 2016).

Another health issue to take into account when kids are dam-reared and if they have outdoor access with their dams is parasitism, as goats are very sensitive to gastrointestinal nematodes (Hoste et al. 2012). Kids have not had the opportunity to develop any immunity against gastrointestinal parasites before the first access to pasture. It is therefore recommended to let them go outside on a healthy pasture (newly sown pasture, pasture not used for over a year) without adults. Quiédeville et al. (2020) even indicate that one of the alternative gastrointestinal nematode control practices is the non-access to pastures for kids and young goats (up to one-year old) to reduce exposure to parasites. However, this practice was the least adopted by the eight farmers interviewed in this study, due to factors such as farmer beliefs, social values, and conflict with the organic principle of access to pasture. No studies assess the impact of dam-rearing systems on parasitism prevalence (gastrointestinal nematodes) of kids. This aspect should be monitored to identify possible parasitism contamination from adults to kids.

One question raised by early separation from the mother is: are there feasible alternative solutions to promote social enrichment for kids? In some dairy cattle systems, lactating cows called "nurse" cows are chosen in order to foster and suckle calves (Neave et al. 2022). This represents a social enrichment while at the same time such cows provide natural milk to the calves rather than artificial milk. In sheep, an alternative solution to replace the missing mother was tested (Nowak et al. 2018). Artificially fed lambs were reared in the presence of non-lactating adult ewes and compared to lambs reared by their mothers or reared in a standard artificial feeding system. At 21 days, the anal region of lambs reared with adult ewes or mothers was less soiled than for lambs reared in an artificial feeding system, and only the latter displayed episodes of diarrhoea. At two months of age, lambs expressed a preference for the familiar adult ewe compared to an unknown ewe. Transposing such situations to goats would be worth trying to see if they can have positive impacts on kids' welfare. The choice of lactating or non-lactating goats should be carefully done in order to avoid transmission of diseases; a feasibility assessment of such hybrid systems should be performed in goats as it has been tested in dairy cows (e.g. Neave et al. 2022) and sheep.

The presence of an older weaned companion is another interesting possibility. De Paula Vieira et al. (2012) demonstrated that dairy calves reared with an older social companion increased pre-weaning solid feed intake and body weight gains before and after weaning compared to calves housed in groups of similar age. This can be due to a different feeding behaviour: during the pre-weaning period, the number and duration of visits to the hay feeder, as well as the amount of hay consumed, were higher for calves housed with older companions, compared to those housed in groups of similar age. This higher feed intake may reflect social learning from the older calves and may have allowed calves to develop a better functioning rumen (by increasing rumen capacity and rumination). However, births are generally synchronised in goats; therefore, the use of older kids may be difficult. Potential health hazards due to parasitism contamination also have to be monitored.

4.3 Milk Feeding Strategies

When kids are separated from their dams, whole milk (raw, pasteurised, or acidified) or milk replacer is provided. Depending on the practices and the number of kids being fed, delivery methods may range from bottles with nipples, troughs, buckets with or without nipples, multiple nipple units, tanks without nipples, or automatic feeding units (Fig. 6; Vickery et al. 2022). The use of nipples allows kids to perform suckling behaviour, reducing occurrence of abnormal oral behaviour

Fig. 6 Kids fed with a bucket with multiple nipples (Photo: Marianne Berthelot (**a**)), or with an automatic feeding unit (Photos: Marianne Berthelot (**b**); Renée de Cremoux (**c**))

(Mattiello 2016). Similarly, in calves, teat feeding reduces cross-sucking (Jensen and Budde 2006). Moreover, the use of nipple feeding systems has been suggested to promote a more natural feeding position (e.g. kids lifts its head up to suckle); this promotes milk bypassing the rumen, thus potentially avoiding gastric issues (Bélanger-Naud and Vasseur 2021).

Automatic feeding units are generally used in intensive systems, where large numbers of kids are typically reared at one time. Automatic feeding unit is the most common feeding system in commercial dairy systems in the UK (Anzuino et al. 2019), Canada (Bélanger-Naud et al. 2021) and USA (Hempstead et al. 2021). Automatic feeding units allow kids to suckle at any time, promoting more natural feeding bouts (e.g. rather than a few large feeds, kids can have many small feeds during the day, as they would with their dam). However, with this system, the number of nipples can vary from five kids per teat (37.7% over 72 farmers who reported using an ad libitum feeding machine) to more than 21 kids per teat (2.8%; Vickery et al. 2022). As kids tend to display synchronous behaviours, a large kid-to-teat ratio could lead to frustration and aggression at the teats. When seven or nine kids have access to one milk teat with an automatic feeding unit, Vickery et al. (2023) indicate that they observed few attempted and successful displacements from the milk station, as well as minimal amount of time spent queuing for milk access. However, there are no other studies, which assess the effect of the number of nipples available on kids' behaviour and welfare.

The use of buckets or tanks with an adapted number of nipples or adapted length allows kids to drink at the same time. However, in this situation, milk administration is generally restricted to two or three meals per day. Mattiello et al. (2008) observed that restricted milk administration led to an increased number of agonistic and abnormal oral behaviour in goat kids compared to ad libitum system. Very few studies investigate the effect of milk feeding strategies on kids' welfare. Importantly, to the authors' knowledge, no science-based recommendations exist for the quantity of milk that kids should be provided. In their review, Bélanger-Naud and Vasseur (2021) indicate that recommendations suggest giving kids between 0.5 and 1.5 L/day in the two first weeks, and then between 1.8 and 2 L/day until weaning. However, these recommendations do not appear to be science-based. The authors suggest that this is a much needed area of future research in goats.

During the milking-feeding period, in order to prepare kids for weaning (cessation of milk consumption), it is indispensable to give them access to solid food (forage, concentrates) as soon as possible after birth (Zamuner et al. 2023). This would promote the development of the digestive system and accustom kids to consume solid feed, which will help the transition from liquid to solid diet. Another important aspect before weaning is to provide kids with water and teach them how to use the troughs. In their review, Bélanger-Naud and Vasseur (2021) suggest that this should be done as early as possible (e.g. 1 week of age), particularly if the dispenser requires learning of an input (e.g., pushing a lever to gain access to water).

4.4 Weaning

Among wild goats, weaning appears to take place between 3 and 6 months of age (Collias 1956, quoted by Miranda-de la Lama and Mattiello 2010), while kids can remain with their mothers until 11–12 months of age (Fabre-Nys 2000, quoted by Miranda-de la Lama and Mattiello 2010). Weaning age and weight depend on the farming system. In the study conducted by Wang et al. (2014), Cashmere goat kids were weaned at the age of 4 months (20 kg). However, in intensive dairy systems, kids are often weaned as young as 6 weeks after birth; however, a later weaning age (e.g. 10–12 weeks; Todd et al. 2019) can result in better growth. Indeed, in addition to age, weight is also an important criterion to decide when to wean kids. Sirohi, Marwari, and Kutchi kids can be weaned at 2 months (about 13 kg) compared to 3 months of age (about 16 kg) without detrimental effects on growth measured when 180 days old (Nagpal et al. 1995). Similar results were found with early weaning in Alpine goat kids: weaning at 12.2 kg (40 days) compared to 17.7 kg (60 days) did not lead to significant consequences on growth at 200 days of age (Panzuti et al. 2018). Early weaning did not affect the age at first lactation either (Panzuti et al. 2018). However, impacts on kids' behaviour, cognitive capacities, and physiological parameters were not studied; these aspects matter in terms of welfare and should be investigated.

Weaning can be abrupt or progressive. The effect of three weaning strategies (abrupt vs gradual milk reduction vs dilution of milk with water) was assessed on feed intake and behaviour of artificially reared Saanen kids at three months of age (Zobel et al. 2020b). With the exception of higher pellet intake prior to weaning in kids receiving diluted milk, and initially lower pellet intake immediately following weaning in kids receiving a gradual reduction in milk, the reported pellet intake and average daily gain (ADG) were similar among the weaning treatments. The treatments had no effect on lying behaviour and vocalisations (all kids vocalised more after weaning). It was concluded that these three weaning strategies seem suitable for three-month-old kids, an age at which kids are commonly weaned in New Zealand systems (Todd et al. 2019). However, as kids are generally weaned around the age of two months in European and North American intensive systems, a similar experiment should be done at this age.

Ugur et al. (2007) compared three different weaning methods linked to goat milking management in Saanen goats raised with their kids:

- Kids stayed with their dam at night and dams were only milked after separation at 08.00 a.m.
- Kids stayed with the dams at night, and dams were milked at 17.00 p.m., just before kids' reintegration, and at 08.00 a.m.
- Kids in the last treatment were subjected to the first method for the first 4 weeks, and then switched to the second method described above for the following 4 weeks.

All kids were weaned at eight weeks of age. ADG in the week 1–8 period was greater for kids in the first than in the second method described, and no significant differences were found between the last method and the other two methods. ADG did not differ among treatments after weaning. The different weaning methods based on goat milking management resulted in similar outcomes of ADG for Saanen kids; unfortunately, as already mentioned, welfare measures (e.g. behaviour and physiology) were not considered in this study.

Based on the actual knowledge, it cannot be said which weaning method is the least stressful, since most studies did not include behavioural and physiological indicators of welfare. When weaning occurs at three months of age in artificial system, it seems that the method adopted does not affect kids' welfare (Zobel et al. 2020b), but this should be confirmed for younger kids. Moreover, when kids are reared with their does, one should not minimise the mother–young bond: weaning is usually associated with dam separation, representing another source of stress, in addition to the feeding stress, whose effect on kids' (and the mother) welfare should be assessed.

4.5 Nonreproductive Kids' Management

A specific issue concerns young bucks, which are not kept in the herd but often transported to specialised fattening facilities. Concerning transportation, the European Union legislation states that it is forbidden to transport newborn mammals whose umbilical cord is not completely healed (Council Regulation (EC) No 1/2005 of 22 December 2004). It is also the case in Canada where an animal is inapt to transportation if he has an unhealed and infected cord (Gouvernement du Canada 2020).

In general, numerous young kids from different farms are transported and then mixed, either as soon as they leave the farms or later on in the fattening centres. This practice has many implications, particularly in terms of stress response, and transmission of pathogens and therefore infectivity, especially as kids of different ages, vigour and weight may be grouped together. As risks of transmission of infectious diseases are increased due to the mixing of kids from different farms and age, kids must receive good quality colostrum, umbilical disinfection immediately after birth to provide them a healthy start (Meijer et al. 2021). Some kids may not be used to artificial feeding if they were only fed by their dams on the farm where they were born. A specific attention should be given to provide optimal sanitary management, to reduce social stress, and to make sure that all the kids can suckle correctly from the milk feeder. The fate of male goat kids is still a huge welfare issue as they do not have any commercial value in most western countries, even if improvements are being made (establishment of charters and good practice guidance in particular). Sexing semen before artificial insemination and extended lactation may be potential ways to reduce the number of male goats being born on farms. On-farm fattening may also be a solution to reduce stress for kids, but this is not possible in all farms due to practical issues (house design), economic reasons, supplementary working time needed (Meijer et al. 2021), absence of local outlets and suitable and local slaughter infrastructure.

5 Milking Management

In dairy systems, milking is one of the main components of farm management practice. Standard length lactation in goats is 9 months, but this can vary considerably depending on breed, genetic, inter-individual capacities, and management system. Milking can be performed by hand or using a mechanical milking machine. Hand-milking is sometimes used in more extensive or pastoral systems. However, the use of a mechanical milking machine is generally the norm, with an important diversity in the milking systems (number of animals that can be milked at one time, automatic systems, movable machine, or presence of milking parlour). Several parameters of mechanised milking management have to be considered to promote goat welfare. Thus, in this section, we will focus on mechanical milking, which is widely used on commercial goat farms (Fig. 7). Different practices linked to milking management will also be discussed: dry period, once-daily milking, and extended lactation.

5.1 General Milking Management

In addition to the impact on production, proper milking management influences the udder health and welfare of females as well as the working comfort of the farmer (Poulet 2021).

A well setup milking parlour, well-maintained milking equipment, and properly trained staff (Fig. 8) may combine to result in cooperative animal behaviour and optimised milk emission kinetics as discussed in dairy cows (Ujita et al. 2021) as well as better working conditions for the farmer. Positive tactile stimulation and pre-partum milking routine training have been shown to have a positive impact on cows during milking (Ujita et al. 2021). Similarly, good milking conditions (clusters are well-positioned and not annoying for the goat) may contribute to calm goats, thus probably offering improved working conditions for the milker. Thus, it could

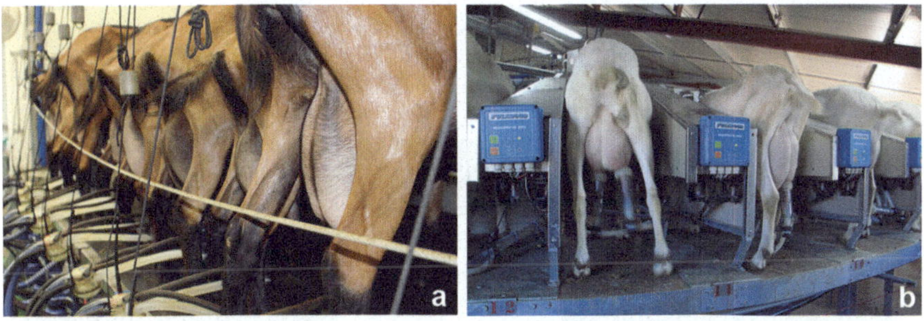

Fig. 7 Mechanical milking (Photos: Jean-Louis Poulet (**a**); Renée de Cremoux (**b**))

Fig. 8 Milking surveillance (Photo: Jean-Louis Poulet)

be considered that there is a One Welfare loop in the milking parlour (Poulet 2021; Fig. 9).

For a complete, fast, and not painful milking procedure, a fine balance between emptying of the udder, but at the same time avoiding over-milking, is necessary to maintain the integrity and health of udders and teats (FAO 2019). Suitable conditions for non-painful, complete, and fast milking are reported in Table 1. Milking pace is important as it would prevent impatience and keeps animals calm (Ujita et al. 2021). Finally, milking does not have to be painful and has to guarantee the integrity of the teats for following milking. Milking should therefore be at least neutral with respect to goat welfare. It can even impact goats' welfare positively by

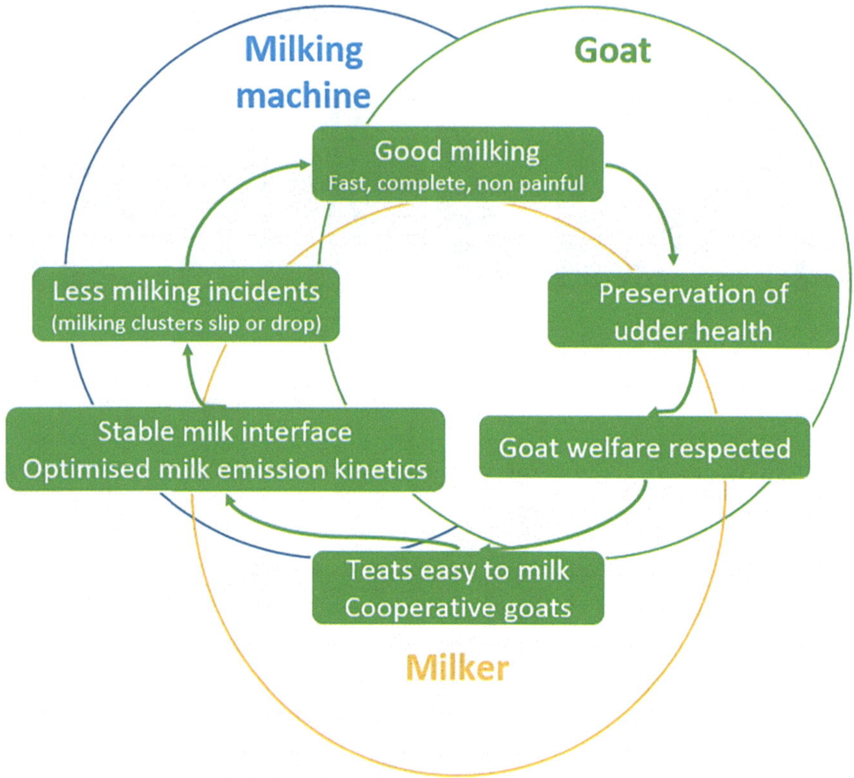

Fig. 9 One welfare loop in milking parlour (Poulet 2021)

limiting the intramammary pressure generated by the quantity of milk stored in the udder and teat cisterns and reinforcing positive human–animal relationships, provided milker behaviour (including milking practices) and milking management are adapted to the goat needs and particularities (Mersmann et al. 2016).

Milking conditions have a major impact on intramammary infections. In goats, the dominant aetiology is staphylococcal (Koop et al. 2012; Menzies 2021) whose main reservoirs are mammary (intramammary and mucocutaneous sources). The teat end is both the primary defence (physical barrier of the teat canal) and the gateway to bacterial invasion of the udder and needs to be protected from both contamination and mechanical aggression (i.e. preservation of its integrity). Thus, there is a need to reduce the risks of contamination (passive or active transfer of bacteria and associated mechanisms) or teat end trauma by a better adjustment of the animal-machine-milker triptych. For this purpose, it is first necessary to evaluate milking design and parameters (vacuum, pulsation, milking liner, cleaning; Table 1) and their adaptation to the animal (including udder and teat conformation, and milk ejection characteristics). A second set of adjustments concerns the fluctuations of the vacuum and their potential impacts on bacterial transfer.

Table 1 Milking conditions for non-painful, complete, and fast milking in dairy goats

Topic		Objectives	Recommendations or standards[a]
Design of the milking system			
Milk line	Position	Stable milk flow	Increased risks if high line (more vacuum loss) (Hubert et al. 2018; Manzur et al. 2012; Romero et al. 2022)
	Slope		Slope between 0.5 and 2%[a], depending on the milk line diameter
	Looped	Less air leakage impact	Looped milk line preferable
Milking liners		Massage on the entire teat tissues (reduction of the risk of oedema or congestion)	Fit liners-teats: diameters, length, mouthpiece, and barrel
Number of clusters		Over-milking limitation	10–20 clusters per milker simultaneously depending on milking systems, level of automation (automatic cluster remover) and practices (da Borso et al. 2022; Hubert et al. 2017)
Milking parameters and maintenance of the milking system			
Vacuum	Level	Sphincters opening, maximum milk flow, hold of liners on teats Not too high vacuum level required (otherwise: increased milking rate, but teat integrity not granted) (Lu et al. 1991)	Adaptation to milking configuration, to liners and possibly to udder characteristics (Hubert et al. 2018; Sinapis et al. 2000; Sevi et al. 2009; Caria et al. 2021) Low line: 36–44 kPa, High line: 38–48 kPa (Caria et al. 2021; Romero et al. 2021; Sinapis et al. 2006; Zucali et al. 2019)
	Fluctuation	Vacuum regulation to avoid too large vacuum fluctuations (depending on milking machine configuration)	In accordance with animal milk flows
Pulsation	Frequency	Alternation of milking phases with open liners, vacuum under the teats and milk flow, and massage phases with closed liners, with vacuum off and teat compression to favour blood and lymph circulation and avoid oedema and congestion	90 (Lu et al. 1991) to 120 cycles/min for productive herds (Romero et al. 2021)
	Ratio		Mainly 60:40 (Lu et al. 1991; Romero et al. 2021) or 70:30/65:35 for rustic breeds (Sinapis et al. 2000)

(continued)

Table 1 (continued)

Topic		Objectives	Recommendations or standards[a]
Milking liners	Renewal	Limitation of material contamination and good material working conditions	Change as often as expected, i.e. at least once a year and more frequently in case of deformation or premature wear (manufacturer guidelines have to be followed) Use within ranges of vacuum and pulsation levels, for optimal performance (Sinapis et al. 2000; Hubert et al. 2018; Romero et al. 2021; Bueso-Ródenas et al. 2023)
	Condition		Not twisted, not cracked, not ovalized, clean
Cleaning		Good device functioning	Adapted cleaning (at minimum and in most cases: sufficient temperature above 35–40 °C at the end of the cycle, adapted product concentration, turbulence) for normal devices work and no contamination during milking
		No udder contamination	
Milking practices			
Clusters manipulation		Avoiding irregular vacuum fluctuation under teat	Avoid air leakage during milking cups positioning
Milking supervision		Avoiding aggressive milking that may lead to teat lesions	Ensure soft removal of the clusters, avoid over-milking and machine stripping; avoid double cluster attachment; limit the contact between the machine and the teat to the effective milking time

[a]Item subject to an ISO standard

The importance of managing the end of milking is often neglected. At this stage, the milk flow decreases to zero and, almost proportionally, the vacuum increases again, under a teat with less resistivity because it is empty. The negative effects of milking (including milking incidents as milking clusters slip or drop) are therefore often increased even if the final phase is appropriately managed. Automatic cluster removal systems based on milk flow seem to be more adapted to the particularities of each animal than those based on time. However, thresholds have to be set in accordance with milk flow kinetics (Poulet et al. 2018; Dzidic et al. 2019).

The definition of indicators of proper milking at the animal level is indispensable to evaluate milking conditions as a whole. Observing the physical state of the udder and teats provides a set of indicators to assess both the presence of chronic infections (damage to the mammary parenchyma) and the quality of milking conditions (injuries of the teat and sphincter; de Cremoux et al. 2018). Figure 10 illustrates lesions that can be observed on goats' teats at the end of inadequate milking. Frequency, intensity, and persistence of lesions have to be monitored. Rarely described in other dairy ruminant species (or with different locations), the presence

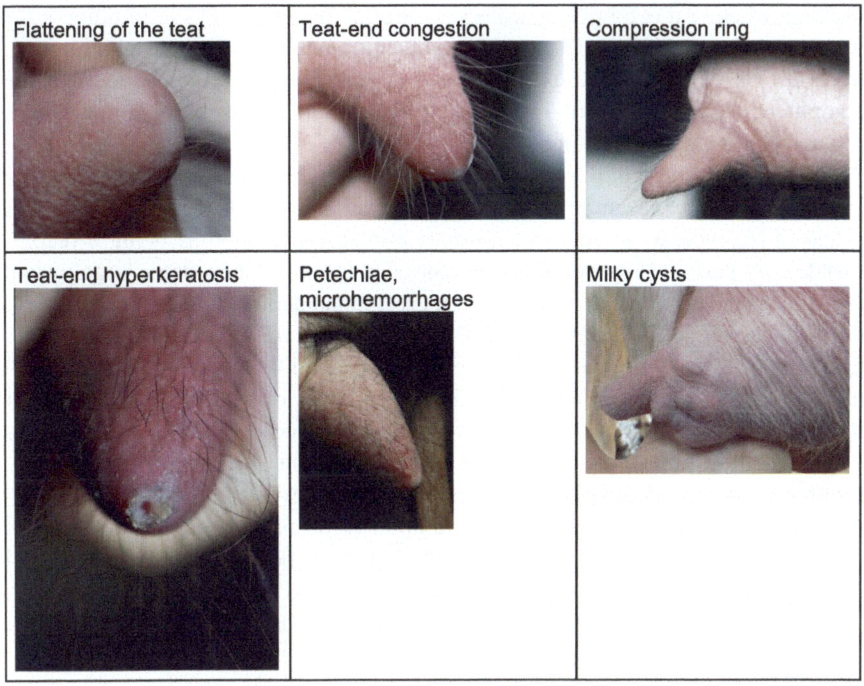

Fig. 10 Outcomes of incorrect milking technique on the tip or body of the teat (Photos: Renée de Cremoux and Justine Simonet)

of milk cysts at the tip of the teat, isolated or in clusters, is worth noting even though its origin remains poorly understood. It is sometimes associated with defects known as "weeping teats" (Gazzola et al. 2021), and a genetic origin has been suggested (Yeruham et al. 2005). It reveals a particular response of the teat to the interface with the liner (probably repeated trauma) and, in the most severe cases, it degrades milking conditions by preventing the correct positioning of the milking liner (de Cremoux et al. 2024). Wet milking (residual milk on the teat body at the end of milking, indicating that it has been covered by milk during milking) is also symptomatic of engorgement or reverse flow phenomena, considered as risk factors for intramammary infections. The duration of exposure to poor milking management (repetition of deleterious conditions) must be taken into account to better assess the risks to the animals.

In addition, goat behaviour may also provide valuable information on the quality of milking (Simitzis et al. 2021). The following indicators can be measured: agitation or prostration, abrupt withdrawal of the clusters by kicking or absence of rumination. The merycic-type reflex driven by the discharge of oxytocin observed in the event of good milking conditions is indeed inhibited when the animals are disturbed or stressed, oxytocin itself being considered as a hormone that regulates the antistress system (Andersson et al. 1958; Marnet 1997; Svennersten-Sjaunja 2004; Uvnäs-Moberg et al. 2001). Importantly, there is a learning and adaptation phase to milking, whether it is for primiparous goats during the first weeks of milking or for the adaptation of all goats in case of modification of the milking installation (e.g. management of entries and exits, restraint system, change or adjustments of the equipment itself or of the milking routine). Such practices are observed in the field; references on goats are lacking.

Finally, a dry period, the time when goats are forced to cease lactation before parturition, seems relevant to maintain positive welfare of goats. Technical resources (Idele 2016) indicate a length of 60 days prior to parturition as being a recommended dry period. However, this time-period recommendation stems largely from cow literature, which itself is varied in its recommendations (e.g. minimum of 30 days: Bachman 2002; Gulay et al. 2003; minimum of 45 days: Kuhn et al. 2005; El-Tarabany 2015). For goats, references are lacking for the dry period duration and methodology (abrupt vs gradual cessation of milking), as well as on the impact of the dry period on goats' welfare and affective state (Zobel et al. 2015a). Regardless of the length of time, dry-off has been shown to present different benefits in dairy ruminants. In cows, a dry period has been shown to prevent a decrease in milk production for next lactation by promoting a turnover of milk epithelial cells, but this has not been shown in goats (Zobel et al. 2015a). A dry period is also recommended as it is important to assure good colostrum quality (IgG concentration) (Caja et al. 2006; Klusmeyer et al. 2009) and to address udder health (ewes: Petridis et al. 2013; cows: Vilar and Rajala-Schultz 2020). In goats, subclinical intramammary infections have been reported across the dry period (group of 49 goats: 44%—Contreras et al. 2003; 400 does on 10 farms: 49 ± 3%—Zobel et al. 2014) and the dry period represents a period where infections could be treated or heal spontaneously (Fox et al. 1992). A dry period appears to have a positive impact on the foetuses' growth (Caja et al. (2006). The authors mentioned in this regard the competition between lactation and pregnancy, especially during the last third of gestation when the dry period is omitted), but these results have to be consolidated as a very small number of goats were used in this study. Forcing a cessation of lactation may be stressful, as in some situations it may be paired with changes in management and reductions in the quantity, and sometimes also in the quality, of feed provided (Zobel et al. 2015a), but technical recommendations insist on the fact that feeding, in particular, has to be adapted to goats' needs (Idele 2016). The objectives are twofold. They are both to allow the recovery of the body condition, essential for the start of the next lactation, and to accompany the decrease in production with a reduction in feed intake. This goes through a gradual reduction in the supply of concentrates, a reduction of high protein feeds inputs, an adaptation of the energy level of the ration so as to avoid the mobilisation of body reserves, cover the needs of minerals, trace elements and vitamins, take into account the evolution of ingestion capacity and always ensure food transitions and permanent access to water at all time. Reducing milk production prior to dry-off is highly recommended, especially in high-yielding animals; in cows, Odensten et al. (2007) showed that high milk production resulted in increased risk of intramammary infections and metabolic challenges. Zobel et al. (2015b) reported that high producing goats that were allowed to have a shorter (or even omitted) dry period prior to parturition had a reduced risk of metabolic issues in the next lactation; however, we do caution that this was a field observation on one farm. In cows, Jolicoeur et al. (2014) reported that shorter dry period lengths promoted increased feed intake in the next lactation, reducing metabolic challenges, without negatively impacting milk production. A dry period is considered an important practice when goats are managed on an annual lactation, but how the dry period is achieved (e.g., method, length, milk production at dry-off) is very important to ensure welfare is not compromised.

5.2 Once-Daily Milking

Once-daily milking is a traditional practice in Europe, as described by Marnet and Komara (2008), and can have an impact on goats' health and welfare. This practice has been described for Canaries breed (Capote et al. 1999), Murciano-Granadina breed (Salama et al. 2003), and Damascus breed (Papachristoforou et al. 1982). In some countries, especially in the context of on-farm cheese processing, a focus on workload simplification means that the demands of twice-daily milking are decreased by adopting a once-daily milking strategy. Once-daily milking seems to be able to be applied from the beginning of lactation (at the end of the colostral period, at least for multiparous goats), or over a sequence of several weeks, to meet a temporary need to reduce the farmers' work.

In the study conducted by Marnet et al. (2005), the milkers did not report increases in vocalisation nor increased agitation when milking frequency was changed from twice to once-daily when goats were at 2–6 weeks in milk and producing 4 kg/day. However, these observations need to be taken with caution, as they were not registered using a specific grid of measures. These authors specified that no leakage of milk on the ground was observed and no clinical mastitis was detected over three years of trials. They highlighted the maintenance of the integrity of the secretory epithelium of the mammary gland despite a temporary increase in filtration phenomena (leakage of blood protein into the alveolar lumen over the first 8 days after the change of milking regimen). Komara et al. (2009) also reported goats not having any behavioural changes in response to the transition from twice- to once-daily milking. However, these observations were not quantified and were only based on experimenters' experience, and thus we caution that more work examining the behavioural responses of goats in response to this transition is needed, with a consideration for the amount of milk being produced at the time of transition.

The conformational adaptation of goats to once-daily milking has been confirmed by many authors and appears as another particularity of this species (high proportion of cisternal milk and udder compliance), accentuated in certain breeds (Marnet and Komara 2008): goats are found to adapt to once-daily milking and once-daily milking is suitable due to their udder. Torres et al. (2016) have revealed in particular that the resistance of mammary epithelial tissue to paracellular leakage of monovalent ions was an important component in the adaptation of goats to the milk stasis and its disruptive effect. However, the change from two to one milking per day inevitably led to milk loss, even if this was much more limited than in cattle:

loss of milk production estimated at 18% for Murciano Granadina, around 15–20% for Alpine, 7% for Dasmacus, and 6% for Canaries (Komara et al. 2009). As discussed earlier, reducing milk production prior to the cessation of milking is advantageous for a number of reasons. The decrease in milk production associated with a reduction in feed needs and a reduced milking frequency can be used to assist in dry-off. In high producing cows, abrupt cessation of milking has been reported to cause pain (Silanikove et al. 2013), as well as frustration and milk leakage (Zobel et al. 2013). The latter has been suggested to increase the risk of developing clinical mastitis during the dry period (Schukken et al. 1993). In ewes, Petridis et al. (2013) did not find an increased risk of subsequent mammary infection and development of mastitis; however, this work did not report the milk yield of the animals, and the sample size was likely too small to determine treatment differences in infection rate. In goats, scientific references examining milk cessation methods are still lacking.

Once-daily milking impact on udder health should be questioned. Indeed, significant increases in somatic cell counts (SCC) have been observed when milking once daily in multiparous (Komara et al. 2009) or primiparous animals (Lefrileux et al. 2008), suggesting for the latter that twice-daily milking might favour udder development at the beginning of lactation (greater stimulation of the udder, impact on the endocrine determinism of terminal mammogenesis). An analysis of the French national milk-recording database confirmed the significantly higher SCC in farms using once-daily milking over the whole year compared to those using twice-daily milking (de Cremoux et al. 2018). However, direct effects of once-daily milking could not be differentiated from indirect effects related to, for example, an inadequacy of the milking system to the changes in milk quantities, milking times, and milk ejections rates induced by the switch to once-daily milking (Poulet et al. 2016). Indeed, controlled studies are needed, and should focus on identifying intramammary infections, in addition to SCC.

5.3 Extended Lactation

Extended lactation is a management strategy used both at individual and flock levels that can affect goats' welfare. A proposed definition of an extended lactation is a continuous lactation lasting more than 16 months without kidding. According to recent data established in French dairy goat production contexts, the duration of an extended lactation reaches an average of 717 days (Astruc et al. 2021). Originally, extended lactation was applied in out-of-season breeding systems to buffer lower reproduction performances and avoid goat culling. Nowadays, extended lactation strategy is on the rise because of economic and social developments and expectations, including a better regularity in milk production, an increase in the share of winter milk production, and a reduced number of kids, especially males that have a low economic value (Meijer et al. 2021). Farmers often decide to extend the lactation of primiparous goats for different objectives: being able to group them with adults, maintaining lactation without disrupting their growth, and not exhausting high-yielding producers. Extended lactation has also become a strategy to manage goats at the end of their career (i.e. they are still productive, but considered at risk of parturition-related diseases). In addition to reducing mortality or morbidity at kidding (Douhard et al. 2012, 2013), extending lactation allows farmers to reduce their workload and consequently improve animal care. Extended lactation management therefore has potential direct and positive repercussions on longevity and indirect effects on the quality of care provided to the replacement animals. However, it seems that not all goats have the potential to be kept in extended lactation. As reported by Chastin et al. (2001) and de Cremoux et al. (2016a, b), extended lactation goats have a high production level and an above-average lactation persistence. They seem to differ from others from a physiological point of view: according to Douhard et al. (2013), the ability of goats to achieve extended lactation depends on their potential to partition energy from the diet towards milk production instead of body tissue gain. It is assumed that during a 24-month period, extended lactation goats produce as much milk as in successive normal lactations (Chastin et al. 2001; Douhard et al. 2013; Salama et al. 2005; Astruc et al. 2021). This supports the hypothesis that the milk produced by extended lactation during the dry period of normal lactation can compensate for the smaller amplitude of the second milk peak phase. At the career level, milk performance (daily and cumulative production) and longevity (as measured by number of days in milk production) of the goats are generally in favour of extended lactation with some variability associated with the positioning of extended lactation in the goat career (Astruc et al. 2021) (Table 2).

Data and observations concerning milk performance or quality for extended lactation need to be considered with caution insofar as farmers always use a set of criteria to define which goats will undergo extended lactation. Indeed, future goats

Table 2 Number of milking days and milk production of goats with at least one extended lactation (EL) during their career compared to goats of the same first parturition cohort

	Number and positioning of EL in goat careers						
Type of career	One EL only[a]	EL at beginning of the career[b]	EL in the middle of the career[b]	EL at the end of the career[b]	2 ELs or more	At least one "normal" lactation[c]	Others[d]
Mean number of days in milk in the career	841	1130	1520	1345	1845	810	274
Cumulated production at the career level: adjusted mean (l)	2350	3252	4539	4048	5704	2242	642
Daily production: adjusted mean (l)	2.73	2.81	2.91	2.90	3.01	2.68	2.36

From Astruc et al. (2021)

[a]One EL only: EL is the unique lactation in the goat career

[b]EL happens at either the beginning, the middle, or the end of goat careers

[c]At least one "normal" lactation of more than 250 days in the goat career and zero extended lactation

[d]All lactations of duration less than 250 days (zero "normal" lactation)

who will be selected for an extended lactation can be chosen either before reproduction (absence of reproduction: notion of chosen extended lactation), or because of reproductive failure. In all cases, production performance, the milk persistence of milk production, and/or animal health are also considered. Due to the lack of recording of farmers' choices (identification of selected animals and selection criteria), the reported results in terms of milk performance, milk quality, or longevity, take little or no account of induced biases and of the possibility that some goats selected for extended lactation fail to produce milk beyond 16 months.

Attention must also be paid to the udder health. De Cremoux et al. (2016b) demonstrated the impact of an extension of the lactation on SCC, suggesting either a physiological effect (effect of concentration and/or of the stage of lactation, particularity of cell renewal) or the role of the lack of rest of the udder on its recovery. Goats with high SCC have less frequent extended lactation, as they may produce less milk and have a less persistent lactation (Arnal et al. 2018). Conversely, analyses currently being carried out on French farms show that extended lactations often have favourable genetic indexes for SCC as well as higher milk production indexes than others. The duration of their lactation seems to be increased in farms with good mammary health management and good technical skills. Under these conditions, SCC does not appear to be significantly increased compared to those of other goats (de Cremoux et al., unpublished data). However, at the flock level, goats in extended lactation with intramammary infections can represent a source of bacteria and might increase the infectious risk for other lactating females (Bergonier et al. 2010). Thus, the choice of goats undergoing extended lactation always needs to take into account the health status of the udder.

The use of extended lactation appears as an effective answer to farmers' expectations for a simplification of husbandry practices. This management decision can also influence both the genetic selection process (Douhard et al. 2013) and the health status of the flock with probable middle- and long-term consequences. A comprehensive approach to health and welfare has yet to be realised. Nevertheless, even if risks exist in terms of udder health, benefits are expected at the animal and herd level in terms of longevity of the goats and proper management and care of replacement females.

6 Reproductive Management

Reproduction management, from mating to parturition, is an important component of commercial goat farms. Success at mating and initiation of lactation are keys to production performance in both dairy and meat systems. If either fails, goats may be prematurely culled.

As goats have a seasonal reproductive cycle, whose onset is triggered by a decreasing photoperiod, production is naturally clustered over a limited period of the year. While many systems work around the seasonality, to face market demand for milk throughout the year, numerous techniques, such as photoperiodic treatments, male effect or hormonal treatments, have been developed to allow year-round reproduction in dairy goat systems. Although natural and unassisted mating is predominant in goat production worldwide, techniques allowing grouped kidding and out-of-season reproduction are becoming increasingly common, particularly in dairy systems. These manipulations are used in conjunction with both natural mating and Artificial Insemination (AI). While these reproduction techniques have improved reproduction and ensured more consistent production (Fatet et al. 2011), many parts of reproductive management may negatively affect goats. However, their potential impacts on goat welfare have received little attention in the literature. In this section, the welfare issues of reproduction management are discussed in relation to oestrus induction and synchronisation (including seasonality control), assisted reproduction techniques, and pregnancy management. Based on the social and cognitive skills of goats, some behavioural strategies such as animal training and good human–animal relationship can be recommended for mitigating negative experiences of goats during reproduction management.

6.1 Oestrus Induction and Synchronisation

The onset and the duration of breeding season depend on many factors such as latitude and climate, breed, physiological status, presence of males, farming system, and photoperiod; the latter is highlighted when considering breed differences. For instance, breeds from lower latitudes (Creole, Murciana-Granadina, etc.) are less seasonal. This is an important point if a manager is interested in maintaining natural mating; such breeds could be employed to expand the breeding period. While these breeds tend to become more seasonal in higher latitudes, the duration of their breeding season tends to remain longer than in breeds from temperate climates (Chemineau et al. 2004). However, if reproductive management is deemed necessary, various strategies can be used to induce or maintain cyclic ovarian activity and sufficient spermatogenesis in out-of-season breeding schemes: hormonal synchronisation, buck effect, and photoperiodic treatments. Roger (2012) reports that these controlled breeding techniques may involve potentially aversive techniques, and thus compromise animal welfare.

6.1.1 Hormonal Synchronisation

Hormonal synchronisation of oestrus and ovulations is an efficient tool for reproduction management, but involves individual manipulation of the animals. Roger (2012) argues that manipulation of the hormonal cycle and associated animal handling is an adverse experience. Specifically, injections of PMSG[1] (Pregnant Mare Serum Gonadotropin) or insertion of cloprostenol and intra-vaginal sponges, and the related need for human handling, may cause stress and/or local pain. Local infection and/or the formation of tissue adhesion of the device to the vaginal mucosa

[1] Hormonal synchronisation treatment including PMSG is still one of the most used protocols before insemination at least in Europe and America, despite the welfare issues due to the production conditions of PMSG in pregnant mares.

may also arise from the use of vaginal sponges, possibly causing vaginitis (Gonzalez-Bulnes et al. 2020). Interestingly, shifting away from these techniques should help meet growing societal demands for more naturalness in animal production and address welfare concerns related to the mares involved in the production of PMSG. For instance, EU countries have adopted restrictions on the use of synthetic hormones in livestock (Delgadillo et al. 2014). In response, protocols for oestrus synchronisation using only the buck effect or vaginal sponges combined with the buck effect are available to farmers willing to inseminate their goats without using hormone injections. It is of course possible that because of the welfare concerns of vaginal sponges, these options will also become more heavily regulated.

6.1.2 Male Effect

The male effect, or buck effect, is the deliberate introduction of males among a group of females that have been previously isolated from the males, in order to induce and synchronise the onset of oestrus. As this approach more closely resembles sexual behaviour, it is considered to be more "natural" for managing small ruminant reproduction compared to the use of hormones or artificial photoperiods (Scaramuzzi and Martin 2008). Males must be in good physical condition before reproduction, as loss of body condition is common due to increased physical activity and reduced resting and feeding time. Supplementary feeding and exercise may be provided to bucks before using them for breeding (Fatet and Johnson 2023). Bucks should be suitably selected for the age and size of the females being serviced. For instance, bucks used for buck effect to stimulate oestrus onset in nulliparous goats should be of the same age to avoid physical harm. First-mating does weigh about 50% as much as adult goats, while adult males weigh about 1.5 times more than an adult goat; the size differential makes adult males unsuitable to mount nulliparous does.

Farmers may use marking aprons to prevent bucks from mating when they are used to stimulate the onset of oestrus (buck effect) or to detect heats before insemination. Aprons consist of a tarp covering the abdomen to prevent mating and a crayon holder over the chest for the buck to leave marks on the back of oestrous females being mounted. This technique involves handling the bucks to attach the apron; the potential consequences on bucks' welfare of this procedure are not well known. Vasectomy is an alternative technique, but this requires an invasive surgery; potential negative consequences may result from the general anaesthesia, as well as surgery related issues (e.g. inflammatory decompensation, tissue adhesions, and cicatrisation issues). However, it does allow for display of natural behaviours; vasectomised males can fully interact with females and express the whole range of sexual behaviours (Batista et al. 2002).

6.1.3 Photoperiodic Treatments

Seasonality can be controlled by using photoperiodic treatments. Photoperiodic treatments may be used in combination with natural mating, buck effect, or hormonal synchronisation to obtain reproduction outside the natural breeding season. The photoperiodic treatment consists in submitting the goats alternatively to long days (LD = 16 h of continuous lighting) and short days (SD = 8–12 h of continuous lighting) at precise periods of the year (Pellicer-Rubio et al. 2019). While many dairy operations may achieve this lighting by containing the animals continuously indoors, providing additional light at dawn and dusk is also sufficient. This makes it possible to breed goats outside their natural breeding season by making the females receptive to the buck effect and stimulating the buck's sexual activity.

Anecdotally, long lighting periods upwards of 20 h may be observed on-farm, but this technique is less efficient in inhibiting sexual activity and may cause tiredness by disrupting natural sleep cycles. Long days may reduce hair growth due to the effect of photoperiod on moulting (Sumner and Bigham 1993).

When days are naturally short (8–12 light hours/day), they can be used as the "short day" following a long-day treatment. Otherwise, it is necessary to use melatonin implants that induce hormonal changes similar to short days (Pellicer-Rubio et al. 2019). Implant insertion may cause discomfort during handling and mild local pain of short duration during the insertion itself. Issues may arise at the site of implant depending also on the skills of the operator. "Darkening" the barns was recommended in the past by using tarps on openings to ensure low light exposure during short days. However, this practice is no longer recommended, as it may cause degradation of the environment/atmosphere inside the barn, and therefore lead to respiratory issues. When out-of-season breeding is used, it is important to ensure that feed and water availability match goat nutritional requirements; this is particularly important for goats being managed on range and pastures that may be dormant or low growth (and thus lower energy) during gestational and lactational peaks.

6.2 Assisted Reproduction Techniques

6.2.1 Artificial Insemination

One of the most used assisted reproduction techniques is artificial insemination (AI). Handling the animals and lifting their rear end to perform AI is stressful. Depending on the available equipment (e.g. headlocks, portative restrainer) and the number of technicians, the duration of restraint and consequent discomfort may vary. To limit the stress caused to the goat, prior habituation and positive reinforcement (gentle touching, hand feeding; Celozzi et al. 2022) should be implemented as well as the presence of familiar and trained persons to handle the animals.

In the standard technique (Baril et al. 1993a), the goat is upside down in anterior bi-pedal position, and a Duckbill speculum is inserted to locate the cervix. The Brazilian technique allows the goat to stay in a standing position, a Collins speculum is used, and the cervix is immobilised with Allis forceps. In both cases, the semen is deposited using an AI gun emptying the content of a semen dose at the entrance of the os cervix (Fonseca et al. 2017). Local injuries may happen due to the handling of the AI gun and defensive movements of the animal during handling. The potential stress and/or pain caused by these procedures has not been scientifically evaluated. Positively, new insemination tools with embedded cameras initially developed for cattle are now under development for small ruminants and should allow inseminating animals in a standing position, with fewer manipulations of the genital tract.

6.2.2 Semen Production

Semen production from bucks is necessary for AI. Buck semen can be collected by using an artificial vagina (AV) and a "teaser" female, electro-ejaculation (EE), or trans rectal ultrasound-guided massage of the accessory sex glands (TUMASG). Training of the males for semen collection requires good human–animal relationship and time. Positive reinforcement during the first steps of training (clicker training, accustom the buck to be isolated from the group, exploration of the path to and the mounting stable itself…) is essential. When using the AV, teaser females are restrained for the duration of semen collections. When a male mounts the teaser female, its penis is derived into the artificial vagina by a familiar and trained operator to collect its sperm. The implementation of the clicker-training method to train males to mount a dummy is a promising technique to avoid using restrained teaser females.

Electro-ejaculation and TUMASG techniques result in various negative impacts on animal welfare and in reduced semen quality. The electro-ejaculation technique may be used to collect sperm from animals of seasonal breeding species out of the normal breeding season, from animals not trained to ejaculate into an artificial vagina, from prepubertal males or males that have been isolated from females. Electro-ejaculation is stressful and painful; it has been shown to induce increases in plasma cortisol concentration, rectal temperature, respiratory and heart rates, as well as modifications in haematological and biochemical variables indicative of the stress response (Abril-Sánchez et al. 2019). Some countries have strict regulations about this procedure. For instance, in New Zealand, minimum standards require that the procedure be performed by a veterinarian or by a trained and competent technician being supervised by a veterinarian. The buck must also be provided with pain relief (Animal Welfare (Care and Procedures) Amendment Regulations 2020). The TUMASG procedure requires fewer or even no electrical pulses to achieve ejaculation in comparison to EE. Guerrero-Gutiérrez et al. (2021) suggest that TUMASG is less stressful and painful than electro-ejaculation, since bucks vocalise fewer times during TUMASG than during EE. These results are in coherence with those of Abril-Sánchez et al. (2017), who found that EE provokes more vocalisations and cortisol concentrations than TUMASG.

6.2.3 Embryo Transfer

Embryo transfer techniques have been developed in small ruminants, but they are hardly used in commercial systems. These techniques may be used as an alternative to import/export live animals or as part of breeding programmes. In vivo embryo production requires a hormonal treatment to stimulate superovulation, followed by AI (potential welfare impacts of both these procedures have been previously discussed) and embryo recovery, either surgically under general anaesthesia via laparoscopy or laparotomy, or trans-cervically (Baril et al. 1993b; Fonseca et al. 2013).

Both procedures raise welfare concerns. Although the trans-cervical procedure is less invasive and faster, it may not be possible on all females due to the anatomy of the cervix. Moreover, Santos et al. (2020) reported that the manipulation of the cervix of a conscious female might be stressful and even painful, as determined by the increase of heart rate shortly after the end of the collection and the maintenance of elevated cortisol concentrations three hours after embryo collection. Surgical collection involves fasting the animal, affecting energy levels, potentially negatively affecting the quality of the embryo. In addition, the surgery itself, anaesthesia, and the postoperative period are stressful and painful for the animals. Appropriate pain mitigation strategies must be implemented.

Reproduction management involves a number of potentially stressful and/or painful invasive procedures. The negative effects of these manipulations can be reduced or even suppressed by the participation of a human familiar with the animals concerned and well trained for the procedure to be carried out, and by getting the animals used to the manipulation beforehand, or even by training with positive reinforcement (Celozzi et al. 2022).

6.3 Pregnancy Management

Particular attention should be given to goats during pregnancy. Stressful events during this period could affect the development of the foetus and the health and welfare of future kids; it is therefore important to continuously monitor pregnant females. For example, Baxter et al. (2016) report that aversive handling in pregnant females results in poorer placental quality, which represents a potential risk for foetal loss, and affects goats' and kids' behaviour. Kids from goats that received aversive handling took 25 min longer to perform play behaviour than kids from goats that did not receive specific interactions or were submitted to gentle interactions. According to Chojnacki et al. (2014), high stocking density (1 m^2/goat) during pregnancy also affects kids' behaviour, as kids from such environments were more fearful during an isolation test than kids whose dams were raised at lower density (3 or 2 m^2/goat).

Stress linked to feeding management also has an impact on goats during pregnancy. Both the metabolism and behaviour are modified by food restriction during the last third of gestation (Laporte-Broux et al. 2011). In this study, goats subjected to nutrition stress gained less weight, their body condition scores were lower after three weeks of restriction compared to control goats, and four out of 30 of them developed pregnancy toxaemia. Maternal feed restriction also affected both feeding behaviour and the stress physiology of female offspring for up to 2 years (Laporte-Broux et al. 2012). Live weight and body condition were lower for kids from goats subjected to restricted feeding than for kids from the control group. Considering physiological parameters, no modification in the glucose metabolism was found, but the HPA axis response to an ACTH injection was greater in kids born from feed restricted goats than from control goats. Evidence from a multi-farm study suggests that monitoring lying behaviour in late gestation could be used as an early indicator of pregnancy toxaemia (Zobel et al. 2015b); this same work also described the potential link between dry period length and pregnancy toxaemia risk, and while it was an observation from a single herd, it nonetheless provides an interesting hypothesis.

7 Watchful Management of Outdoor Access to Promote Positive Animal Welfare

The behavioural benefits for goats of having access to an outdoor area have been widely discussed in the previous chapter. Outdoor access promotes wide range of explorative, climbing and feeding behaviours, including browsing (Figs. 11 and 12), increases locomotor and positive social behaviours (Bøe et al. 2012; Stachowicz et al. 2018; Negretti et al. 2020), as well as goats' cognitives abilities (Zobel et al.

Fig. 11 Goats having access to a rocky environment (Photo: Renée de Cremoux)

Fig. 12 Goats expressing browsing behaviour (Photos: Renée de Cremoux)

2019). It also limits agonistic interactions (Aschwanden et al. 2009) according to the type of outdoor access. Outdoor access can be occasional access to pasture (Fig. 13) or to outdoor pens in semi-intensive systems, regular or permanent access to pasture or grazing lands in extensive and pastoralist systems (Fig. 14). With an indoor and outdoor space, goats can choose where to perform different activities, such as resting activities. Providing animals with the possibility to make choices can be a positive experience (Lay et al. 2011). Additionally, the outdoor environment may also provide the opportunity to express "sun-bathing behaviour", as observed in other species (domestic chicken: Lay et al. 2011; cows: Arnott et al. 2017). Bezoar goats are reported to prefer shaded slopes in the summer, but sunny slopes in the winter (Esfandabad et al. 2010; Naderi et al. 2013). This suggests that wild goats not only enjoy sunlight, but they might also look for warm areas.

Health benefits deriving from outdoor access in goats have been scarcely investigated so far. In their review on the welfare of dairy cows in continuously housed and pasture-based production systems, Arnott et al. (2017) reported that outdoor access has health benefits on cows. Cows on pasture-based systems have lower levels of lameness, hoof pathologies, hock lesions, mastitis, uterine disease, and

Fig. 13 Goats with occasional access to pasture (Photos: Marianne Berthelot (**a** and **b**) and Renée de Crémoux (**c**))

Fig. 14 Goats in extensive system (Photo: Renée de Cremoux)

mortality compared with cows housed indoors only. As to goats, according to Zobel et al. (2019), giving an outdoor access to goats might also have a positive impact on their hoof health, especially when they can climb or walk on hard surfaces, as this may reduce the prevalence of overgrown hooves. Other positive benefits of having outdoor access could be found and need to be studied in goats, as in cows. For instance, Crump et al. (2019) reported that cows with an overnight access to pasture had longer overnight lying duration, fewer up or down transitions, and a more synchronous lying behaviour at pasture compared to cows housed indoors only. However, whatever the type of outdoor access, potential challenges have to be considered to prevent negative effects, mainly on health.

Points of attention include direct impacts of the environment on the animals (feed intake, watering, exposure to sunlight and varying environmental conditions), exposure to direct or vector-borne pathogens, and interactions with potential predators. Some conditions have to be fulfilled so that the outside pen is effectively used when goats can get access both to inside and outside pens. Stachowicz et al. (2018) reported that the outdoor pens were used even if the inside conditions were good. However, the weather conditions influence the way the outside pen is used. Bøe and Ehrlenbruch (2013) showed that when temperatures decreased and/or when precipitations occurred, goats spent less time outdoors. Stachowicz et al. (2019) reported that goats go preferentially outside when temperature increases; conversely, during windy and rainy days they prefer to stay inside. To promote outdoor access and the welfare of goats, it is therefore essential to provide shelter and options for them to seek refuge from sun exposure and adverse weather conditions whatever the type of outdoor access. Alvarez et al. (2013) studied the effects of putting shades over feeders to protect goats against solar radiations and heat stress on goat kids feeding behaviour. They reported that goat kids spent more time feeding when the feeders were shaded than when they were unshaded. Exposure of the skin to sunlight, particularly in animals with depigmented skin, can have a deleterious impact on goat health. In particular, Saanen, Saanen crossbreeds, and goat breeds that lack pigmented skin and live in sunbelt areas, are at high risk for papillomatosis according to Theilen et al. (1985). Freeman (2018) reports erythema on the udder of Saanen goats with an outdoor access. Proliferative lesions on the udder or teat skin due to sun exposure, even if benign, can cause pain (bleeding, torn scabs) during milking and deteriorate milking conditions, also accompanied by an increase in milk somatic cell counts. In this case, the choice of the breeds raised outdoors can be questioned from a welfare point of view, according to the conditions of sun exposure. More generally, the risks of photosensitisation, especially induced by plants (some of which toxic), must be considered when managing access to the outdoors (Chen et al. 2019; Jacob and Peet 1987; Mazni et al. 1985; Opasina 1985; Santos et al. 2017; see chapter "Welfare Issues in Goat Farming: Housing and Nutrition" for more precisions about toxic plants).

Water is of the utmost importance for goats as well as for other farm animals. Nevertheless, goats' ability to tolerate water shortage is often reported (Al-Ramamneh et al. 2012; Kaliber et al. 2016; Silanikove 1989, 2000a). As an example, Kaliber et al. (2016) recorded the ability of crossbred goats to cope with a substantial reduction in water intake of up to 90% for more than one month. No significant adverse physiological effects were then noted on energy balance and thermoregulation capacity. Given that metabolic adaptation, it is often suggested that the access to watering spots once or twice a day would be sufficient for goats at pasture (Silanikove 2000b; Ganter et al. 2012). However, these physiological capacities to tolerate conditions of low water availability do not mean that animals do not feel thirsty, altering their positive emotional state. Numerous welfare codes or best practices guides around the world make continuous access to water mandatory (e.g. Canada, NFACC 2022; New Zealand, Code of Welfare: Goats 2018; RSPCA n.d.).

One major health issue with grazing goats is internal parasitism by cestodes (e.g. tapeworms), trematodes (e.g. flukes), and gastrointestinal nematodes. Among these parasites, gastrointestinal nematodes are particularly of concern, especially in grazing systems. The natural ability of goats to browse, which prevents contact between goats and gastrointestinal nematode larvae, explains the low development of their immunity against these parasites (Hoste et al. 2001). Therefore, kids are particularly exposed to early infestation if they go outdoors with their dams. Hoste et al. (2001) report that Angora goats co-grazing the same area with Saanen goats showed significantly higher parasitic infections compared to Saanen. There is a variability between breeds concerning parasitism sensitivity, which may be linked to the ability to perform a flexible foraging behaviour. Goats can be infected all along their life by gastrointestinal nematodes and these parasites have negative impacts on their welfare. They induce anorexia and a loss of nutriments. Consequently, parasitized goats lose weight and milk production is reduced. In case of high parasite burden or in more susceptible animals (e.g. kids during their first grazing season or high-yielding goats), parasites can cause diarrhoea and/or mortality. The detrimental effects of parasitism on sheep behaviour have been demonstrated: around weaning, lambs in the high parasite exposure group displayed shorter daily lying time and lower average daily motion index (total activity) compared to the control group (Högberg et al. 2021), but these aspects have been scarcely investigated in goats. The control of parasite infections is usually done by means of anthelmintic treatment. However, due to the over-use of these products, parasites have been developing resistance, which can impair a correct control of these parasites and threaten the possibility of raising goats outdoors in good health conditions (Paraud and Chartier 2017). This situation calls for a new approach to managing gastrointestinal parasitism. It involves the implementation of integrated strongyle management, which combines pasture management, improved goat resilience or resistance (including genetic selection), and selective targeted treatment (Maqbool et al. 2017). Goats raised outside can also be infected by external parasites, like ticks which can transmit several pathogens, such as the haemoparasites *Babesia*, *Anaplasma*, *Trypanosoma*, *Theileria*, and *Ehrlichia* (Stuen 2017).

Another sanitary issue of grazing is the impact on udder health. Smistad et al. (2021) mentioned an effect of the pasture season on the increase of milk SCC. Different hypotheses are given to explain this effect, such as walking activity, heat stress, increased infection pressure as suggested for cows (Coulon et al. 1998; Lamarche et al. 2000), or endocrine function related to phytoestrogens intake (Poutrel et al. 2012). The observations made by de Cremoux et al. (2016c) in an experimental station suggest a non-infectious and transient phenomenon (a little more than three weeks) depending on the initial udder status (less increase of SCC for the goats with the lowest SCC, i.e. <750,000 cells/ml in this study). This increase of SCC was more pronounced in the period immediately following the start of the grazing period during feed transition compared to a period with full access to pasture (>10 h/day). Further studies are required to better elucidate the effect of outdoor access on goat welfare and udder health.

Outdoor access inevitably highlights the issue of sharing territories, whether by different herds of the same or different species, or through interactions with wildlife. Herd mixing or contacts with closely related species increase health hazards. Non-exhaustively, shared grazing can favour the transmission of zoonotic diseases such as brucellosis (even in officially free countries, when wildlife reservoirs exist) or Q-fever, particularly if kidding or abortion take place on the grazing or transhumance spaces. Thus, collective health control programmes must take into account the management of these areas and the risks of cross-species transmission. In terms of animal health, mycoplasmosis or caprine arthritis-encephalitis virus (CAEV) infection (Nord et al. 1998; Peterhans et al. 2004) can be mentioned. In the latter case, the mixing of herds (Al-Qudah et al. 2006) favours the transmission of the virus between infected animals, sometimes asymptomatic carriers, and healthy animals (direct contact, droplet from respiratory system, sexual contacts). Furthermore, walking long distances or jumping, by straining the joints, could trigger the clinical expression of the disease (enlarged knees) and cause pain to the goats. Transit conditions (Teles et al. 2023) to the grazing areas also need to be taken into account from a sanitary point of view. Peterhans et al. (2004) underlined the importance of contact prevention during transhumance (movement of animals between different climatic regions) which, according to them, plays an important role in the transmission of the disease in certain European countries. When goats have outdoor access, the risks associated with predation are important and may affect the survival of the goats, as well as their welfare, which might be reduced as a consequence of the stress generated by the presence of predators (Fardell et al. 2020). Those risks vary according to geographical areas (mountain vs flat areas; communal vs private pastures; specific risk of drinking areas, where predators could more easy attack groups of goats who gather to drink), predator density, and farming systems (Durawo et al. 2017; Pimenta et al. 2018).

To reduce the impacts of predation, the protection of young animals and the surveillance of the herds are essential (Pimenta et al. 2017; Fig. 15). This means promoting the presence of shepherds and using a sufficient number of guard dogs with regard to the size of the goat herds and, if necessary, with regard to the size of predators' packs (number of wolves per wolfpack, for instance). This also implies

Fig. 15 Farmers looking at their goats (Photos: Renée de Cremoux)

upstream reflection on the management of shared spaces, on the formation and working conditions of shepherds and also on the selection and education of guard dogs associated with an awareness of both farmers and of all persons using these territories. In the future, flock management in pastoral areas, particularly without guarding, could undoubtedly be based on on-board technologies combining accelerometers and GPS tags. According to Grisot et al. (2022), the definition of alerts in relation to predator attacks still seems insufficiently specific and data are still lacking to define which animals should be equipped within the flock. Nevertheless, such equipment seems promising both for better locating flocks and for intervening if necessary, and for optimising the use of pastoral areas.

Finally, when kids are separated from their dams at birth in systems that are more intensive and reared indoors when they are young, they do not learn appropriate grazing strategies from their dams. They generally lack foraging experience and have to learn by themselves (or with peers of the same age) how to graze when they go outside for the first time in pasture-based systems. A particular attention should be given to ensure that goats raised in these conditions are able to feed themselves enough.

Outdoor areas have the potential to provide goats with diversified, complex, and stimulating environments, if designed and managed correctly. However, attention should be given to critical aspects (including shelter, water access, parasitism, sun exposure, and predation) to ensure welfare is not compromised. Biosecurity measures have to be implemented and the characteristics of the environment and contact risks (predators, wildlife, neighbouring herds, people, visitors…) should be taken into account. Less investigated than risks and hazards, positive health impacts still need to be objectified and better described. Finally, particularly in a context of an increased anthropisation of territories and of constant evolution of spaces as well as environmental or sanitary constraints, the use of outdoor spaces is a subject that requires reinvestment and joint reflection with all territorial stakeholders, taking into account the benefits for livestock as well as, more broadly, the ecosystem services rendered by the territories and herds.

8 Genetics Management

8.1 Impact of Current Breeding Programmes on Goat Welfare

During the last decades, production levels of livestock species have considerably increased in many countries. For example, the amount of milk produced per goat and per lactation increased from 644 to 989 kg between 1990 and 2020 in French goats (Piacere et al. 2000; Idele 2020). It has been reported recently that the milk production per goat on 24 English farms varied from 300 to 1600 kg/goat (Anzuino et al. 2010), and that individual records for 2194 goats in the United States varied from 268 to 2394 kg/goat (Zobel et al. 2019). Besides the improvement of environmental factors such as housing, feeding strategies, and reproduction techniques, genetic selection has strongly contributed to this progress with the implementation of successful breeding programmes. In the world, few well-organised breeding programmes have been developed for dairy goats (Montaldo and Manfredi 2002). Selection has been mainly focused on milk production traits, for which significant genetic gains have been achieved in selected populations (Montaldo and Manfredi 2002; Larroque et al. 2011).

In this context, we may wonder whether the combination of genetic selection for production traits and changes in production systems has contributed to a deterioration of goat welfare. Many studies in pigs, cattle, or poultry (e.g. Rauw et al. 1998; Rauw 2007; Canario et al. 2013) show that animals selected for high production efficiency are more at risk for behavioural, physiological, and immunological problems. The main explanation deals with the genetic antagonism between production and "functional" traits, i.e. health and metabolic, reproduction, behavioural, and other adaptation traits. Selection for improving production efficiency may result in correlated unfavourable responses in other traits. To assess this antagonism, one way is to estimate genetic correlations between traits. In dairy goats, unfavourable genetic correlations were found between milk production and some functional traits, such as subclinical mastitis resistance (Rupp et al. 2011), internal parasitism resistance (Heckendorn et al. 2017), udder type (Manfredi et al. 2001), reproduction traits (Montaldo et al. 2010), and heat stress sensitivity (Menéndez-Buxadera et al. 2012). However, in most cases, the genetic correlation values are weak to moderate (from –0.10 to –0.30), indicating that selection for milk yield weakly deteriorates these traits.

In order to counterbalance such undesirable degradation of functional traits, these are usually introduced into the breeding goals, together with production traits. The genetic response on functional traits depends on their heritability and on their relative weight in the global selection index as compared to milk production traits. The objective may be to stop the degradation of functional traits, or even to improve them. Indeed, the breeding goals for French dairy goats have greatly evolved in recent years to give an increasing importance to functional traits. First, eleven traits, mainly related to udder morphology, have been considered in the official genetic evaluation and introduced as selection criteria since 2006. Then, milk somatic cell counts as a predictor of resistance to mammary infections (Rupp et al. 2019) have been selected since 2013. Finally, fertility after artificial insemination has been introduced as a breeding goal since 2022. Currently, all these functional traits together account for 50% of the breeding goals (Larroque et al. 2011). Other initiatives are emerging in other countries and/or breeds (Ziadi et al. 2021; Desire et al. 2018), showing that selection for milk production traits will gradually give way to more selection for functional traits; however, the link of these traits to goat welfare is not well studied, beyond health outcomes.

8.2 Using New Genetic Management Strategies to Improve Goats' Robustness

Instead of considering individual functional traits, we can consider the robustness of the animal as a whole. Robustness may be seen as a global measure of the sensitivity of the animal to the climatic, physical, feeding, infectious, and social environment, and to the metabolic load of its genetic potential for production traits (Mormède et al. 2011). Such traits are important for animal health and welfare (Rauw 2009). Functional longevity is a synthetic trait reflecting the ability of the animal to avoid premature culling. In dairy cattle, many countries and breeds have been selected for functional longevity since the early 2000s (Sasaki 2013). In goats, some recent studies in the USA, Germany, and France show that functional longevity is a trait of interest and responsive to genetic selection (Castañeda-Bustos et al. 2014, 2017; Palhiere et al. 2018; Wolber et al. 2021; Arnal et al. 2022; Ithurbide et al. 2022).

With the development of molecular technologies and the availability of a SNP chip[2] in goats since 2011 (Tosser-Klopp et al. 2014), it is now possible to detect genes or genomic regions associated to specific traits: health traits (Martin et al. 2018; Mucha et al. 2018; Silva et al. 2018), milking speed (Palhiere et al. 2014), and genetic defects (Martin et al. 2016). Selection on these gene polymorphisms can therefore be introduced into the breeding schemes and is frequently very effective to improve these traits.

A first possible application to improve the health and welfare of goats is resistance to scrapie. Scrapie is a fatal neurodegenerative disease that affects goats worldwide, belonging to the group of transmissible spongiform encephalopathies. Recently, a regulation from the European Commission (Regulation (EC) No 2020/772 of June 11, 2020, amending Regulation (EC) No 999/2001) has introduced the possibility to use genetic strategy to better control the disease (Migliore et al. 2020), as an alternative to a complete culling of the infected herds. Indeed, a major gene, called PRNP, allows for the discrimination of susceptible and resistant animals to classical scrapie (Barillet et al. 2009; Vaccari et al. 2009). The European Food Safety Authority

[2] Dense Single Nucleotide Polymorphism (SNP) genotyping arrays provide extensive information on variation across the genome of species of interest. Such information can be used in studies of the genetic architecture of quantitative traits and to improve the accuracy of selection in breeding programmes.

(EFSA) provided a ranking of the best alleles to select in order to improve genetic resistance (Ricci et al. 2017). Simulation studies on Italian (Sacchi et al. 2018) and French breeds illustrate different strategies to select for scrapie resistance. In France, all artifical insemination bucks are genotyped for the PRNP gene and a pool of semen doses from resistant bucks has been created to help infected flocks. In Italy, different scenarios were modelled and compared the expected evolution of the "resistant" allele for the Saanen and the Chamois coloured breeds.

Another possible application would be the gene selection for the absence of horns. It could represent an alternative to painful disbudding practices in kids. In goats, the absence of horns is controlled by the PIS (Polled Intersex Syndrome) mutation (Asdell 1944; Vaiman et al. 1996; Pailhoux et al. 2001). Animals carrying one copy of the mutated allele do not have horns. But the mutation also triggers intersexuality in females carrying two copies of the mutated allele. The use of this mutation is therefore difficult to imagine on a large scale today, but could be used in the future if the context changes (e.g. if disbudding becomes no longer socially acceptable, or large-scale genotyping of females for this mutation becomes available) to discover other genes influencing horn development, as in cattle (Grobler et al. 2021).

The genetic lever could also be used to select directly on welfare indicators. This genetic approach requires a set of proxies that can be measured easily on a large scale. The AWIN project (Battini et al. 2015) developed and tested animal-based indicators for welfare assessment in several species, including goats. Phenotyping a large number of goats, reared in various production systems, is the aim of the H2020 EU SMARTER project (https://www.smarterproject.eu/). A set of 11 welfare indicators, derived from the AWIN and GOATwell projects (Berthelot et al. 2021), was used to quantify the absence of disease, injuries, and pain. The objective is to evaluate the welfare status in commercial goat farms on a large scale and to assess if some of these welfare indicators (e.g. bag-shaped udder, arthritis…) are under genetic control and amenable to selection.

Genomic selection based on phenotypic, genotypic, and pedigree data opens new perspectives for breeding programmes. In goats, genomic evaluations have been applied in some countries and breeds (Carillier et al. 2013, 2014; Mucha et al. 2015; Molina et al. 2018; de Sousa et al. 2021; Massender et al. 2022). Even if the rate of genetic gain has not increased as much as in dairy cattle, the incorporation of genomic information has improved the efficiency of goat breeding programmes. The selection of new traits, and more particularly functional traits, which were difficult to select in the pedigree-based breeding programmes, is now easier. It represents a good opportunity to improve welfare-related traits in goats.

During the last decades, genetic selection has increased production levels of goats. However, breeding goals have evolved, as shown by the French example, and have focused more on functional traits, such as udder morphology, milk somatic cell counts, fertility, and behavioural skills. Genetic management can be a lever to improve goat welfare by selection based on robustness to factors such as climate change or disease.

9 Conclusion

The farming conditions under which goats are kept vary widely (e.g. permanently confined, entirely outdoors, or a combination), depending on the type of production (e.g. goats reared for dairy, meat, or fibre production), the age of the animals, and the management practices. Goat management deals with various dimensions, from relationships with social partners, dams and kids, and humans, to genetics, milking and reproduction management. Goat management strives to satisfy the needs and expectations of the animals as much as possible, but some conditions may compromise goat welfare. This chapter shows that goat welfare issues are different between farming systems: some offer more possibilities for the animals to express their physiological and behavioural needs, while others are more focused on health aspects. Further studies evaluating both goat health and other welfare metrics are needed. This chapter also highlights the need to consider the long-term effects of management on reducing stress and inducing positive emotional experiences. Management is strongly linked with farmers' choices, which affect their own working conditions and quality of life, confirming the importance of the One Welfare concept, which states that human wellbeing and animal welfare are closely linked, and that improving one will often improve the other. As we have seen throughout this chapter, scientific knowledge about behavioural and social needs of goats and their cognitive abilities is not complete, but good practice guidelines and regulations can already benefit from recent developments in technical-scientific knowledge. The current revision of the European legislation on animal welfare, including the development of a specific regulation for small ruminants, and by extension the impact on the international legislation, should make it possible to promote managements which not only reduce the goats' stress but also promote their positive welfare.

References

Abril-Sánchez S, Freitas-de-Melo A, Beracochea F, Damián JP, Giriboni J, Santiago-Moreno J, Ungerfeld R (2017) Sperm collection by transrectal ultrasound-guided massage of the accessory sex glands is less stressful than electroejaculation without altering sperm characteristics in conscious goat bucks. Theriogenology 98:82–87

Abril-Sánchez S, Freitas-de-Melo A, Giriboni J, Santiago-Moreno J, Ungerfeld R (2019) Sperm collection by electroejaculation in small ruminants: A review on welfare problems and alternative techniques. Anim Reprod Sci 205:1–9

Adams DS, Klevjer-Anderson P, Carlson JL, McGuire TC, Gorham JR (1983) Transmission and control of caprine arthritis-encephalitis virus. Am J Vet Res 44(9):1670–1675

Al-Qudah K, Al-Majali AM, Ismail ZB (2006) Epidemiological studies on caprine arthritis-encephalitis virus infection in Jordan. Small Rumin Res 66(1):181–186

Al-Ramamneh D, Riek A, Gerken M (2012) Effect of water restriction on drinking behaviour and water intake in German black-head mutton sheep and Boer goats. Animal 6:173–178

Alvarez L, Guevara N, Reyes M, Sánchez A, Galindo F (2013) Shade effects on feeding behavior, feed intake, and daily gain of weight in female goat kids. J Vet Behav 8(6):466–470

Andersen IL, Roussel S, Ropstad E, Braastad BO, Steinheim G, Janczak AM, Jørgensen GM, Bøe KE (2008) Social instability increases aggression in groups of dairy goats, but with minor consequences for the goats' growth, kid production and development. Appl Anim Behav Sci 114(1):132–148

Andersson B, Kitchell R, Persson N (1958) A study of rumination induced by milking in the goat. Acta Physiol Scand 44(1):92–102

Animal Welfare (Care and Procedures) Amendment Regulations (2020) (LI 2020/172) schedule consequential amendments to animal codes of welfare—New Zealand Legislation. https://www.legislation.govt.nz/regulation/public/2020/0172/26.0/LMS331176.html. Accessed 21 Sept 2023

Anzuino K, Bell NJ, Bazeley KJ, Nicol CJ (2010) Assessment of welfare on 24 commercial UK dairy goat farms based on direct observations. Vet Rec 167(20):774–780

Anzuino K, Knowles TG, Lee MRF, Grogono-Thomas R (2019) Survey of husbandry and health on UK commercial dairy goat farms. Vet Rec 185(9):267–267

Arnal M, Robert-Granié C, Larroque H (2018) Diversity of dairy goat lactation curves in France. J Dairy Sci 101(12):11040–11051

Arnal M, Palhiere I, Clément V (2022) Maturity, a new indicator to improve longevity of Saanen dairy goats in France. In: Proceedings of 12th world congress on genetics applied to livestock production (WCGALP). https://www.wageningenacademic.com/doi/abs/10.3920/978-90-8686-940-4_738. Accessed 21 Sept 2023

Arnott G, Ferris CP, O'Connell NE (2017) Review: Welfare of dairy cows in continuously housed and pasture-based production systems. Animal 11(2):261–273

Arviv A, Muklada H, Kigel J, Voet H, Glasser T, Dvash L, Ungar ED, Landau SY (2016) Targeted grazing of milk thistle (*Silybum marianum*) and Syrian thistle (*Notobasis syriaca*) by goats: Preference following preconditioning, generational transfer, and toxicity. Appl Anim Behav Sci 179:53–59

Aschwanden J, Gygax L, Wechsler B, Keil NM (2009) Loose housing of small goat groups : Influence of visual cover and elevated levels on feeding, resting and agonistic behaviour. Appl Anim Behav Sci 119(3):171–179

Asdell SA (1944) The Genetic Sex of Intersexual Goats and a Probable Linkage with the Gene for Hornlessness. Science 99(2563):124–124

Astruc JM, de Boissieu C, Buisson D, Clément V, de Cremoux R, Doucet M, Larroque H, Palhière I, Rupp R, Arnal M, Bossis N, Coppin S, Gautier JM, Gousseau V, Jousseins C, Lagriffoul G, Legris M, Loywyck V, Martin P, Morin E, Robert-Granié C, Rostellato R, Tortereau F (2021) RUSTIC—Vers une approche intégrée de la robustesse des petits ruminants. Innov Agronomiques 82:369–386

Bachman KC (2002) Milk production of dairy cows treated with estrogen at the onset of a short dry period. J Dairy Sci 85(4):797–803

Baril G, Chemineau P, Guérin Y, Leboeuf B, Orgeur P, Vallet JC (1993a) Manuel de formation pour l'insemination artificielle chez les ovins et les caprins (Numéro 83, p 231). FAO. https://hal.inrae.fr/hal-02852710

Baril G, Chemineau P, Cognie Y, Guerin Y, Leboeuf B, Orgeur P, Vallet J-C (1993b) [Training manual on artificial insemination in sheep and goats]. Etude FAO: Production et Sante Animales (FAO). https://scholar.google.com/scholar_lookup?title=%5BTraining+manual+on+artificial+insemination+in+sheep+and+goats%5D&author=Baril%2C+G.&publication_year=1993

Barillet F, Mariat D, Amigues Y, Faugeras R, Caillat H, Moazami-Goudarzi K, Rupp R, Babilliot JM, Lacroux C, Lugan S, Schelcher F, Chartier C, Corbière F, Andréoletti O, Perrin-Chauvineau C (2009) Identification of seven haplotypes of the caprine PrP gene at codons 127, 142, 154, 211, 222 and 240 in French Alpine and Saanen breeds and their association with classical scrapie. J Gen Virol 90(Pt 3):769–776

Batista M, Prats N, Calero P, González F, Cabrera F, Medrano A, Gracia A (2002) Semen characteristics and plasma levels of testosterone after bilateral vasectomy in bucks. Reprod Domest Anim 37(6):375–378

Battini M, Andreoli E, Barbieri S, Mattiello S (2011) Long-term stability of Avoidance Distance tests for on-farm assessment of dairy cow relationship to humans in alpine traditional husbandry systems. Appl Anim Behav Sci 135(4):267–270

Battini M, Stilwell G, Vieira A, Barbieri S, Canali E, Mattiello S (2015) On-farm welfare assessment protocol for adult dairy goats in intensive production systems. Animals 5(4):Article 4

Battini M, Barbieri S, Vieira A, Stilwell G, Mattiello S (2016a) Results of testing the prototype of the AWIN welfare assessment protocol for dairy goats in 30 intensive farms in Northern Italy. Ital J Anim Sci 15(2):283–293

Battini M, Barbieri S, Waiblinger S, Mattiello S (2016b) Validity and feasibility of Human-Animal Relationship tests for on-farm welfare assessment in dairy goats. Appl Anim Behav Sci 178:32–39

Battini M, Renna M, Giammarino M, Battaglini L, Mattiello S (2021) Feasibility and reliability of the AWIN Welfare Assessment Protocol for dairy goats in semi-extensive farming conditions. Front Vet Sci 8. https://www.frontiersin.org/article/10.3389/fvets.2021.731927

Baxter EM, Mulligan J, Hall SA, Donbavand JE, Palme R, Aldujaili E, Zanella AJ, Dwyer CM (2016) Positive and negative gestational handling influences placental traits and mother-offspring behavior in dairy goats. Physiol Behav 157:129–138

Beaujouan J, Cromer D, Boivin X (2021) Review : From human–animal relation practice research to the development of the livestock farmer's activity: an ergonomics–applied ethology interaction. Animal 15(12):100395

Bélanger-Naud S, Vasseur E (2021) Graduate student literature review: Current recommendations and scientific knowledge on dairy goat kid rearing practices in intensive production systems in Canada, the United States, and France. J Dairy Sci 104(6):7323–7333

Bélanger-Naud S, Cinq-Mars D, Julien C, Arsenault J, Buczinski S, Lévesque J, Vasseur E (2021) A survey of dairy goat kid-rearing practices on Canadian farms and their associations with self-reported farm performance. J Dairy Sci 104(9):9999–10009

Bergonier D, de Cremoux R, Berthelot X (2010) Spécificités du traitement au tarissement chez les Petits ruminants, ovins, caprins, janvier 1. SNGTV. https://www2.sngtv.org/article-bulletin/specificites-du-traitement-au-tarissement-chez-les-petits-ruminants-ovins-caprins/. Accessed 21 Sept 2023

Berthelot M, Caramelle-Holtz E, Bezançon C, Mialon M-M, Aubert C, Mourier K, de Cremoux R, Lollivier V, Merlot E, Aupiais A, Tabouret M, Boissy A, Baudry C, Michel V (2021) Rapport final du projet GOATWELL, p 140. Report, ANSES. https://hal.inrae.fr/hal-03558599

Berthelot M, Aubert C, Ehrhardt N, Baudry C, Paraud C (2022) L'élevage des chevrettes de renouvellement sous les mères en élevage caprin laitier en France : description des pratiques. Rencontres Recherches Ruminants 26:271. https://www.journees3r.fr/IMG/pdf/texte_15_affiche_elevage_et_valorisation_des_jeunes_animaux_m-berthelot.pdf. Accessed 21 Sept 2023

Blacklaws BA, Berriatua E, Torsteinsdottir S, Watt NJ, De Andres D, Klein D, Harkiss GD (2004) Transmission of small ruminant lentiviruses. Vet Microbiol 101(3):199–208

Bøe KE, Ehrlenbruch R (2013) Thermoregulatory behavior of dairy goats at low temperatures and the use of outdoor yards. Can J Anim Sci 93(1):35–41

Bøe KE, Ehrlenbruch R, Andersen IL (2012) Outside enclosure and additional enrichment for dairy goats—a preliminary study. Acta Vet Scand 54(1):68

Boivin X, Braastad B (1996) Effects of handling during temporary isolation after early weaning on goat kids' later response to humans. Appl Anim Behav Sci 48:61–71

Bueso-Ródenas J, Romero G, Roca A, Moya F, Alejandro M, Díaz JR (2023) Optimal parameters to milk Murciano-Granadina goats in mid and low-line milking parlours. Animals 13(7):Article 7

Caja G, Salama AAK, Such X (2006) Omitting the dry-off period negatively affects colostrum and milk yield in dairy goats. J Dairy Sci 89(11):4220–4228

Can E, Vieira A, Battini M, Mattiello S, Stilwell G (2016) On-farm welfare assessment of dairy goat farms using animal-based indicators : The example of 30 commercial farms in Portugal. Acta Agric Scand A Anim Sci 66(1):43–55

Canario L, Mignon-Grasteau S, Dupont-Nivet M, Phocas F (2013) Genetics of behavioural adaptation of livestock to farming conditions. Animal 7(3):357–377
Capote J, Lopez J, Caja G, Peris S, Arguello A, Darmanin N (1999) The effects of milking once or twice daily throughout lactation on milk production of Canarian dairy goats. In: Barillet F, Zervas NP (eds) Milking and milk production of dairy sheep and goats. EAAP Publ. Wageningen Pers, Wageningen, The Netherlands, pp 267–273
Caria M, Boselli C, Calvo A, Grande S, Todde G (2021) Influence of milking units and working vacuum level on the mechanical milking of goats. Anim Sci J 92(1):e13667
Carillier C, Larroque H, Palhière I, Clément V, Rupp R, Robert-Granié C (2013) A first step toward genomic selection in the multi-breed French dairy goat population. J Dairy Sci 96(11):7294–7305
Carillier C, Larroque H, Robert-Granié C (2014) Comparison of joint versus purebred genomic evaluation in the French multi-breed dairy goat population. Genet Sel Evol 46(1):67
Castañeda-Bustos VJ, Montaldo HH, Torres-Hernández G, Pérez-Elizalde S, Valencia-Posadas M, Hernández-Mendo O, Shepard L (2014) Estimation of genetic parameters for productive life, reproduction, and milk-production traits in US dairy goats. J Dairy Sci 97(4):2462–2473
Castañeda-Bustos VJ, Montaldo HH, Valencia-Posadas M, Shepard L, Pérez-Elizalde S, Hernández-Mendo O, Torres-Hernández G (2017) Linear and nonlinear genetic relationships between type traits and productive life in US dairy goats. J Dairy Sci 100(2):1232–1245
Castro N, Capote J, Álvarez S, Argüello A (2005) Effects of lyophilized colostrum and different colostrum feeding regimens on passive transfer of immunoglobulin G in Majorera goat kids. J Dairy Sci 88(10):3650–3654
Celozzi S, Battini M, Prato-Previde E, Mattiello S (2022) Humans and goats: improving knowledge for a better relationship. Animals 12(6):Article 6
Chastin P, Martin J, Monod F, Manfredi E (2001) Extended lactations at Moissac Goat Station (France). In: Proceedings of the 8th meeting "Rencontres autour des recherches sur les ruminants." INRA and Institut de l'Elevage, Paris, France, p 271
Chemineau P, Daveau A, Cognié Y, Aumont G, Chesneau D (2004) Seasonal ovulatory activity exists in tropical Creole female goats and Black Belly ewes subjected to a temperate photoperiod. BMC Physiol 4(1):12
Chen Y, Quinn JC, Weston LA, Loukopoulos P (2019) The aetiology, prevalence and morbidity of outbreaks of photosensitisation in livestock: A review. PLoS One 14(2):e0211625
Chigerwe M, Tyler JW, Schultz LG, Middleton JR, Steevens BJ, Spain JN (2008) Effect of colostrum administration by use of oroesophageal intubation on serum IgG concentrations in Holstein bull calves. Am J Vet Res 69(9):1158–1163
Chojnacki RM, Vas J, Andersen IL (2014) The effects of prenatal stocking densities on the fear responses and sociality of goat (Capra hircus) kids. PLoS One 9(4)
Code of Welfare: Goats (2018) New Zealand Government. https://www.mpi.govt.nz/dmsdocument/46033-Code-of-Welfare-Goats. Accessed 21 Sept 2023
Collias NE (1956) The analysis of socialization in sheep and goats. Ecology 37(2):228–239
Contreras A, Luengo C, Sánchez A, Corrales JC (2003) The role of intramammary pathogens in dairy goats. Livest Prod Sci 79(2):273–283
Cornou C (2009) Automation systems for farm animals : potential impacts on the human—animal relationship and on animal welfare. Anthrozoös 22(3):213–220
Coulon JB, Pradel P, Cochard T, Poutrel B (1998) Effect of extreme walking conditions for dairy cows on milk yield, chemical composition, and somatic cell count. J Dairy Sci 81(4):994–1003
Council Regulation (EC) No 1/2005 of 22 December 2004 on the protection of animals during transport and related operations and amending Directives 64/432/EEC and 93/119/EC and Regulation (EC) No 1255/97, 003 OJ L (2004). http://data.europa.eu/eli/reg/2005/1/oj/eng. Accessed 21 Sept 2023
Crump A, Jenkins K, Bethell EJ, Ferris CP, Arnott G (2019) Pasture access affects behavioral indicators of wellbeing in dairy cows. Animals 9(11)
da Borso F, Kic P, Kante J (2022) Analysis of management, labor and economics of milking systems in intensive goat farms. Agriculture 12(4):Article 4

de Cremoux R, Bossis N, Clément V, Ribaud D (2016a) Conduite des chèvres en lactation longue : diversité des pratiques et incidence sur les concentrations cellulaires des laits. In: Proceedings of the national days of veterinary technical groups, Nantes, France, pp 344–349. https://www2.sngtv.org/article-bulletin/conduite-des-chevres-en-lactation-longue-diversite-des-pratiques-et-incidence-sur-les-concentrations-cellulaires-des-laits/. Accessed 21 Sept 2023

de Cremoux R, Bossis N, Clément V, Ribaud D (2016b) Extended lactation management in dairy goats: diversity of management strategies and impacts on milk somatic cell count. In: 12th international conference on goats, September 25–30. https://www.iga-goatworld.com/uploads/6/1/6/2/6162024/12th_icg_abstracts.pdf. Accessed 21 Sept 2023

de Cremoux R, Gilbert FB, Coquantin C, Brun-Lafleur L, Fouilloux MN, Pommaret A, Lefrileux Y (2016c) Impact de la mise à l'herbe sur les concentrations en cellules somatiques du lait dans un troupeau de chèvres conduites en monotraite. Rencontres Recherches Ruminants 23:293–296. https://www.journees3r.fr/IMG/pdf/Texte_2__Sante__R-De_Cremoux.pdf. Accessed 21 Sept 2023

de Cremoux R, Lagriffoul G, Allain C, Allaoui-Sossé L, Astruc J-M, Batut E, Bergonier D, Brun-Lafleur LL, Clément V, Couzy C, Foucras GG, Fouilloux M-N, Frappat BB, Huau C, Lecomte C, Legris M, Lefrileux Y, Lopez C, Marnet P-G, Martin P, Palhière I, Poulet J-L, Ribaud D, Rupp R (2018) MAMOVICAP -Vers des outils innovants d'intervention et d'aide à la décision pour la maîtrise des mammites en élevage de petits ruminants laitiers. Innov Agronomiques 63:99

de Cremoux R, Legris M, Clément V, Bailly-Salins A, Minier M (2024) Study of the relationship between the presence of milky cysts, udder imbalances, udder morphological traits and somatic cell counts. Small Rumin Res 231:107–203

Delgadillo JA, Flores JA, Duarte G, Vielma J, Hernández H, Bedos M, Fitz-Rodríguez G, Fernández IG, López-Sebastián A, Gómez-Brunet A, Santiago-Moreno J, Zarazaga LA, Keller M, Chemineau P (2014) Out-of-season control of reproduction in subtropical goats without exogenous hormonal treatments. Small Rumin Res 121(1):7–11

Delgado-Pertíñez M, Guzmán-Guerrero JL, Mena Y, Castel JM, González-Redondo P, Caravaca FP (2009) Influence of kid rearing systems on milk yield, kid growth and cost of Florida dairy goats. Small Rumin Res 81(2):105–111

De Palo P, Maggiolino A, Centoducati N, Tateo A (2015) Effects of different milk replacers on carcass traits, meat quality, meat color and fatty acids profile of dairy goat kids. Small Rumin Res 131:6–11

De Paula Vieira A, von Keyserlingk MAG, Weary DM (2012) Presence of an older weaned companion influences feeding behavior and improves performance of dairy calves before and after weaning from milk. J Dairy Sci 95(6):3218–3224

Desire S, Mucha S, Coffey M, Mrode RA, Broadbent J, Conington J (2018) Deriving genomic breeding values for feed intake and body weight in dairy goats. https://cgspace.cgiar.org/handle/10568/98466

de Sousa DR, do Nascimento AV, Lôbo RNB (2021) Prediction of genomic breeding values of milk traits in Brazilian Saanen goats. J Anim Breed Genet 138(5):541–551

Douhard F, Tessier J-A, Friggens NNC, Martin O, Tichit M, Sauvant D (2012) Caractérisation des principales variations zootechniques observées au cours de la lactation longue chez la chèvre. Rencontres Recherches Ruminants 19:411. https://hal.science/hal-01173588/document. Accessed 21 Sept 2023

Douhard F, Friggens NC, Tessier J, Martin O, Tichit M, Sauvant D (2013) Characterization of a changing relationship between milk production and liveweight for dairy goats undergoing extended lactation. J Dairy Sci 96(9):5698–5711

Durawo C, Zindove TJ, Chimonyo M (2017) Influence of genotype and topography on the goat predation challenge under communal production systems. Small Rumin Res 149:115–120

Dzidic A, Rovai M, Poulet JL, Leclerc M, Marnet PG (2019) Review : Milking routines and cluster detachment levels in small ruminants. Animal 13:s86–s93

EFSA Panel on Animal Health and Welfare (AHAW) (2014) Scientific Opinion on the welfare risks related to the farming of sheep for wool, meat and milk production. EFSA J 12(12):3933

El-Tarabany MS (2015) Effects of non-lactating period length on the subsequent calving ease and reproductive performance of Holstein, Brown Swiss and the crosses. Anim Reprod Sci 158:60–67

Esfandabad BS, Karami M, Hemami M-R, Riazi B, Sadough M-B (2010) Habitat associations of wild goat in central Iran : Implications for conservation. Eur J Wildl Res 6(56):883–894

Fabre-Nys C (2000) Le comportement sexuel des caprins: Contrôle hormonal et facteurs sociaux. INRAE Prod Anim 13(1):Article 1

FAO (2019) IDF guide to good animal welfare in dairy production 2.0: Bulletin of the International Dairy Federation 498/2019. FAO https://www.fao.org/publications/card/en/c/CA4286EN/. Accessed 21 Sept 2023

Fardell LL, Pavey CR, Dickman CR (2020) Fear and stressing in predator–prey ecology : Considering the twin stressors of predators and people on mammals. PeerJ 8:e9104

Fatet A, Johnson L (2023) L'élevage des boucs : du choix de renouvellement à la reproduction. Fiche 2a—Groupe Reproduction Caprine, Réf : 0021 302 030, ISBN : 978-2-7148-0189-0, Mars 2023 https://idele.fr/?eID=cmis_download&oID=workspace%3A%2F%2FSpacesStore%2F3f4411ac-8393-49be-8a0d-ef526116839a&cHash=a27d063b4b109b34e9c6efc227796897. Accessed 21 Sept 2023

Fatet A, Pellicer-Rubio M-T, Leboeuf B (2011) Reproductive cycle of goats. Anim Reprod Sci 124(3):211–219

Fell LR, Shutt DA (1989) Behavioural and hormonal responses to acute surgical stress in sheep. Appl Anim Behav Sci 22(3–4):283–294

Fernández A, Ramos JJ, Loste A, Ferrer LM, Figueras L, Verde MT, Marca MC (2006) Influence of colostrum treated by heat on immunity function in goat kids. Comp Immunol Microbiol Infect Dis 29(5):353–364

Fernández MA, Alvarez L, Zarco L (2007) Regrouping in lactating goats increases aggression and decreases milk production. Small Rumin Res 70(2):228–232

Finkemeier M-A, Oesterwind S, Nürnberg G, Puppe B, Langbein J (2019) Assessment of personality types in Nigerian dwarf goats (Capra hircus) and cross-context correlations to behavioural and physiological responses. Appl Anim Behav Sci 217:28–35

Flint M, Murray PJ (2001) Lot-fed goats—The advantages of using an enriched environment. Aust J Exp Agric 41(4):473–476

Fonseca JF, Zambrini FN, Alvim GP, Peixoto MGCD, Verneque RS, Viana JHM (2013) Embryo production and recovery in goats by non-surgical transcervical technique. Small Rumin Res 111(1):96–99

Fonseca JF, Alvim GP, Souza-Fabjan JMG, Oliveira MEF, Brair VL, Brandão FZ, Facó O (2017) Reproductive features and use of an anti-inflammatory drug in estrus-induced dairy goats artificially inseminated in a standing position with cervix immobilization. Reprod Biol 17(3):268–273

Fox LK, Hancock DD, Horner SD (1992) Selective intramammary antibiotic therapy during the nonlactating period in goats. Small Rumin Res 9(3):313–318

Fraser D, Weary D, Pajor E, Milligan B (1997) A scientific conception of animal welfare that reflects ethical concerns. Anim Welf 6:187–205. https://www.wellbeingintlstudiesrepository.org/ethawel/1. Accessed 21 Sept 2023

Freeman H (2018) Providing outdoor space for indoor-housed dairy goats. Thesis, ResearchSpace@Auckland. https://researchspace.auckland.ac.nz/handle/2292/45632. Accessed 21 Sept 2023

Ganter M, Benesch C, Bürstel D, Ennen S, Kaulfuß H, Mayer K, Moog U, Moors E, Seelig B, Spengler D, Strobel H, Tegtmeyer P, Voigt K, Wagner H, Schaf S, Ziege, & Tierschutz. (2012) Empfehlung für die Haltung von Schafen und Ziegen der Deutschen Gesellschaft für die Krankheiten der kleinen Wiederkäuer, Fachgruppe der DVG Teil 1. Tierärztliche Praxis. Ausgabe G, Grosstiere/Nutztiere 40:314–325

Gazzola A, Minozzi G, Biffani S, Mattiello S, Bailo G, Piccinini R, (2021) Effect of Weeping Teats on Intramammary Infection and Somatic Cell Score in Dairy Goats. Front. Vet. Sci. 8

Ghanem YM, El-Khodery SA, Saad AA, Elragaby SA, Abdelkader AH, Heybe A (2009) Prevalence and risk factors of caprine arthritis encephalitis virus infection (CAEV) in Northern Somalia. Small Rumin Res 85(2):142–148

Godden SM, Lombard JE, Woolums AR (2019) Colostrum management for dairy calves. Vet Clin N Am Food Anim Pract 35(3):535–556

Gonzalez-Bulnes A, Menchaca A, Martin GB, Martinez-Ros P, Gonzalez-Bulnes A, Menchaca A, Martin GB, Martinez-Ros P (2020) Seventy years of progestagen treatments for management of the sheep oestrous cycle : Where we are and where we should go. Reprod Fertil Dev 32(5):441–452

Gouvernement du Canada, Agence canadienne d'inspection des aliments (2020) Transport d'animaux inaptes ou fragilisés [Matériel de référence], février 19. https://inspection.canada.ca/sante-des-animaux/animaux-terrestres/transport-sans-cruaute/transport-d-animaux-inaptes-ou-fragilises/fra/1582045810428/1582045810850. Accessed 21 Sept 2023

Grisot P-G, Llaria A, Philibert AA, Guinamard C, Menassol J-B, Flégar M, Jouven M (2022) CLOChèTE—Caractérisation du Comportement et Localisation des Ovins et des Caprins grâce aux Technologies Embarquées : Aide à la conduite des troupeaux et à la valorisation des surfaces pastorales. Innov Agronomiques 85(février):47

Grobler R, van Marle-Köster E, Visser C (2021) Challenges in selection and breeding of polled and scur phenotypes in beef cattle. Livest Sci 247:104479

Guerrero-Gutiérrez M, Ungerfeld R, Kako Rodriguez MG, Santiago-Moreno J, Giriboni J (2021) Using transrectal ultrasound-guided massage of the accessory sex glands for buck semen collection yields semen with greater cryoresistance than electroejaculation alone during the breeding season. Theriogenology 172:142–149

Gulay MS, Hayen MJ, Bachman KC, Belloso T, Liboni M, Head HH (2003) Milk production and feed intake of Holstein cows given short (30-d) or normal (60-d) dry periods. J Dairy Sci 86(6):2030–2038

Heckendorn F, Bieber A, Werne S, Saratsis A, Maurer V, Stricker C (2017) The genetic basis for the selection of dairy goats with enhanced resistance to gastrointestinal nematodes. Parasite 24:32

Hempstead MN, Lindquist TM, Shearer JK, Shearer LC, Plummer PJ (2021) Health and welfare survey of 30 dairy goat farms in the Midwestern United States. Animals 11(7):Article 7

Hemsworth PH, Coleman GJ, Barnett JL, Borg S (2000) Relationships between human-animal interactions and productivity of commercial dairy cows. J Anim Sci 78(11):2821–2831

Högberg N, Hessle A, Lidfors L, Enweji N, Höglund J (2021) Nematode parasitism affects lying time and overall activity patterns in lambs following pasture exposure around weaning. Vet Parasitol 296:109500

Hoste H, Leveque H, Dorchies P (2001) Comparison of nematode infections of the gastrointestinal tract in Angora and dairy goats in a rangeland environment : Relations with the feeding behaviour. Vet Parasitol 101(2):127–135

Hoste H, Manolaraki F, López CA (2012) Spécificités des risques parasitaires des chèvres au pâturage : Conséquences sur les modes de gestion. Fourrages 212:319–328. https://hal.inrae.fr/hal-02651169

Hubert A, de Cremoux R, Poulet J-L, Doutart E, Chandler J (2017) Les installations de traite caprines françaises,spécificités régionales et entretien. Quelles influences sur les concentrations en cellules somatiques des laits de tank ? Collection Résultats. Ed Institut de l'Elevage Réf 00 17 402 008. ISSN 1773-4738, 11 pp. https://idele.fr/?eID=cmis_download&oID=workspace%3A%2F%2FSpacesStore%2F69523033-ed16-4f06-8aa3-c82fdf303946&cHash=4d7cfc05533b3678223f83f0bd165cf4. Accessed 21 Sept 2023

Hubert A, Aubert C, de Cremoux R, Poulet J-L (2018) Variability of vacuum in goat clusters under different controlled conditions of pipeline height, vacuum and pulsation settings and flow rate. Small Rumin Res 163:65–67

Idele (2016) Gérer l'alimentation de la phase de tarissement à la reprise de la lactation. Available at: https://idele.fr/umt-gpr/?eID=cmis_download&oID=workspace%3A%2F%2FSpacesStore%2Fcd1c89a4-2ca7-4131-934c-74e4f8ae9d15&cHash=545037fb9ef32887701dccc355bebbc9. Accessed 21 Sept 2023

Idele (2020) Résultats de contrôle laitier—Espece caprine, France 2020. Available at: https://idele.fr/?eID=cmis_download&oID=workspace%3A%2F%2FSpacesStore%2Fcdfbd36a-8458-4e15-8ddb-87934cdbb1b5&cHash=7f7f971de4c326a94cc6c26e86437f15. Accessed 21 Sept 2023

Idris SM, Eltom KH, Okuni JB, Ojok L, Elmagzoub WA, El Wahed AA, Eltayeb E, Gameel AA (2022) Paratuberculosis: the hidden killer of small ruminants. Animals 12(1):Article 1

Ithurbide M, Huau C, Palhière I, Fassier T, Friggens NC, Rupp R (2022) Selection on functional longevity in a commercial population of dairy goats translates into significant differences in longevity in a common farm environment. J Dairy Sci 105(5):4289–4300

Ivemeyer S, Knierim U, Waiblinger S (2011) Effect of human-animal relationship and management on udder health in Swiss dairy herds. J Dairy Sci 94(12):5890–5902

Jackson KMA, Hackett D (2007) A note: The effects of human handling on heart girth, behaviour and milk quality in dairy goats. Appl Anim Behav Sci 108(3):332–336

Jacob RH, Peet RL (1987) Poisoning of sheep and goats by Tribulis terrestris (caltrop). Aust Vet J 64(9):288–289. Scopus

Jardat P, Lansade L (2021) Cognition and the human-animal relationship: A review of the socio-cognitive skills of domestic mammals toward humans. Anim Cogn 25(2):369–384

Jensen MB, Budde M (2006) The effects of milk feeding method and group size on feeding behavior and cross-sucking in group-housed dairy calves. J Dairy Sci 89(12):4778–4783

Jolicoeur MS, Brito AF, Santschi DE, Pellerin D, Lefebvre D, Berthiaume R, Girard CL (2014) Short dry period management improves peripartum ruminal adaptation in dairy cows. J Dairy Sci 97(12):7655–7667

Kaliber M, Koluman N, Silanikove N (2016) Physiological and behavioral basis for the successful adaptation of goats to severe water restriction under hot environmental conditions. Animal 10(1):82–88

Keil N, Patt A (2013) Intégration et séparation des chèvres. Comment réduire le stress ? Rapport ART 765 https://ira.agroscope.ch/fr-CH/publication/32341. Accessed 21 Sept 2023

Kerr S (n.d.) Tube feeding neonatal small ruminants: an essential skill for sheep and goat farmers. https://extension.wsu.edu/wam/tube-feeding-neonatal-small-ruminants-an-essential-skill-for-sheep-and-goat-farmers/. Accessed 21 Sept 2023

Kessler EC, Bruckmaier RM, Gross JJ (2021) Short communication : Comparative estimation of colostrum quality by Brix refractometry in bovine, caprine, and ovine colostrum. J Dairy Sci 104(2):2438–2444

Klusmeyer TH, Fitzgerald AC, Fabellar AC, Ballam JM, Cady RA, Vicini JL (2009) Effect of recombinant bovine somatotropin and a shortened or no dry period on the performance of lactating dairy cows. J Dairy Sci 92(11):5503–5511

Komara M, Boutinaud M, Ben Chedly H, Guinard-Flament J, Marnet PG (2009) Once-daily milking effects in high-yielding Alpine dairy goats. J Dairy Sci 92(11):5447–5455

Koop G, De Vliegher S, De Visscher A, Supré K, Haesebrouck F, Nielen M, van Werven T (2012) Differences between coagulase-negative Staphylococcus species in persistence and in effect on somatic cell count and milk yield in dairy goats. J Dairy Sci 95(9):5075–5084

Kuhn MT, Hutchison JL, Norman HD (2005) Minimum days dry to maximize milk yield in subsequent lactation. Anim Res 54(5):351–367

Lacuesta L, Orihuela A, Ungerfeld R (2015) Reproductive development of male goat kids reared with or without permanent contact with adult females until 10 months of age. Theriogenology 83(1):139–143

Lacuesta L, Giriboni J, Orihuela A, Ungerfeld R (2018a) Rearing bucks isolated from females affects negatively their sexual behavior when adults. Anim Reprod 15(2):114–117

Lacuesta L, Giriboni J, Orihuela A, Ungerfeld R (2018b) Bucks reared in close contact with adult does prefer to interact with females than with males. Small Rumin Res 162:22–24

Lamarche A, Martin B, Hauwuy A, Coulon J-B, Poutrel B (2000) Evolution of milk somatic cell count of cows grazing an alpine pasture according to the infection of udder by pathogens. Ann Zootech 49(1):45–54

Laporte-Broux B, Duvaux-Ponter C, Roussel S, Promp J, Chavatte-Palmer P, Ponter AA (2011) Restricted feeding of goats during the last third of gestation modifies both metabolic parameters and behaviour. Livest Sci 138(1):74–88

Laporte-Broux B, Roussel S, Ponter AA, Giger-Reverdin S, Camous S, Chavatte-Palmer P, Duvaux-Ponter C (2012) Long-term consequences of feed restriction during late pregnancy in goats on feeding behavior and emotional reactivity of female offspring. Physiol Behav 106(2):178–184

Larroque H, Astruc J-M, Barbat A, Barillet F, Boichard D, Bonaiti B, Clément V, David I, Lagriffoul G, Palhière I, Piacère A, Robert-Granié C, Rupp R (2011) National genetic evaluations in dairy sheep and goats in France. In: 62. Annual meeting of the European Federation of Animal Science (EAAP), 2011-08-29, Stavanger, Norway. https://scholar.google.com/scholar_lookup?title=National+genetic+evaluations+in+dairy+sheep+and+goats+in+France&author=Larroque+%2C+Helene%0A++++++++%28INRA%0A++++++++%2C+Auzeville+%28France%29.%0A+++++++++UR+0631+Station+d%27Am%C3%A9lioration+G%C3%A9n%C3%A9tique+des+Animaux%29&publication_year=2011. Accessed 21 Sept 2023

Lateur-Rowet HJ, Breukink HJ (1983) The failure of the oesophageal groove reflex, when fluids are given with an oesophageal feeder to newborn and young calves. Vet Q 5(2):68–74

Lay DC, Fulton RM, Hester PY, Karcher DM, Kjaer JB, Mench JA, Mullens BA, Newberry RC, Nicol CJ, O'Sullivan NP, Porter RE (2011) Hen welfare in different housing systems. Poult Sci 90(1):278–294

Lefrileux Y, Pommaret A, Raynaud S (2008) Impacts de la monotraite dans une exploitation caprine fromagère à haut niveau de production. Rencontres Recherches Ruminants 15:167–170. https://www.journees3r.fr/IMG/pdf/2008_05_systemes_09_Lefrileux.pdf. Accessed 21 Sept 2023

Leidy R-S, Jorge CR, Elena BR, Jordi BF (2015) Comparative study of trophic behaviour and herd structure in wild and feral goats living in a Mediterranean island : Management implications. Appl Anim Behav Sci 165:81–87

Lethbridge MR (2016) Insights into feral goat movement in Australia using dynamic Brownian Bridges for movement analysis. Rangeland J 38(4):343–359

Lopez AJ, Heinrichs AJ (2022) Invited review : The importance of colostrum in the newborn dairy calf. J Dairy Sci 105(4):2733–2749

Lu CD, Potchoiba MJ, Loetz ER (1991) Influence of vacuum level, pulsation ratio and rate on milking performance and udder health in dairy goats. Small Rumin Res 5(1):1–8

Lyons DM (1989) Individual differences in temperament of dairy goats and the inhibition of milk ejection. Appl Anim Behav Sci 22(3):269–282

Lyons DM, Price EO, Moberg GP (1988) Individual differences in temperament of domestic dairy goats : Constancy and change. Anim Behav 36(5):1323–1333

Malik MI, Rashid MA, Raboisson D (2022) Heat treatment of colostrum at 60°C decreases colostrum immunoglobulins but increases serum immunoglobulins and serum total protein : A meta-analysis. J Dairy Sci 105(4):3453–3467

Manfredi E, Piacere A, Lahaye P, Ducrocq V (2001) Genetic parameters of type appraisal in Saanen and Alpine goats. Livest Prod Sci 70(3):183–189

Manzur A, Díaz J-R, Mehdid A, Fernández N, Peris C (2012) Effect of mid-line or low-line milking systems on milking characteristics in goats. J Dairy Res 79(3):375–382

Marnet P-G (1997) Ewe management for improved milk yield and quality. In: Proceedings of the third Great Lakes dairy sheep symposium, April, 1997, pp 5–11. https://www.researchgate.net/publication/238081818_Ewe_management_for_improved_milk_yield_and_quality. Accessed 21 Sept 2023

Marnet PG, Komara M (2008) Management systems with extended milking intervals in ruminants : Regulation of production and quality of milk. J Anim Sci 86(suppl_13):47–56

Marnet P-G, Gomis B, Guinard-Flament J, Boutinaud M, Lollivier V (2005) Effet d'une seule traite par jour (monotraite) sur les performances zootechniques et les caractéristiques physico-chimiques du lait chez les Chèvres Alpine à haut potentiel. Rencontres Recherches Ruminants 12:225–228

Martin P, Palhière I, Tosser-Klopp G, Rupp R (2016) Heritability and genome-wide association mapping for supernumerary teats in French Alpine and Saanen dairy goats. J Dairy Sci 99(11):8891–8900
Martin P, Palhière I, Maroteau C, Clément V, David I, Klopp GT, Rupp R (2018) Genome-wide association mapping for type and mammary health traits in French dairy goats identifies a pleiotropic region on chromosome 19 in the Saanen breed. J Dairy Sci 101(6):5214–5226
Massender E, Brito LF, Maignel L, Oliveira HR, Jafarikia M, Baes CF, Sullivan B, Schenkel FS (2022) Single-step genomic evaluation of milk production traits in Canadian Alpine and Saanen dairy goats. J Dairy Sci 105(3):2393–2407
Mastellone V, Scandurra A, D'Aniello B, Nawroth C, Saggese F, Silvestre P, Lombardi P (2020) Long-term socialization with humans affects human-directed behavior in goats. Animals 10(4):Article 4
Mattiello S (2016) Husbandry practices and animal health. Chapter 3. In: Tsakalidou E, Papadimitriou K (eds) Non-bovine milk and milk products. Elsevier, Amsterdam, pp 39–59
Mattiello S, Villa S, Cioccarelli G (2008) Effetto delle modalità di gestione dell'alimentazione sul benessere del capretto. Large Anim Rev 4:202–208
Maqbool I, Wani ZA, Shahardar RA, Allaie IM, Shah MM (2017). Integrated parasite management with special reference to gastro-intestinal nematodes. J Parasit Dis 41;1–8
Mazni OA, Sharif H, Khusahry MYM, Vance HN (1985) Photosensitization in goats grazed on Brachiaria decumbens. Mardi Res Bull 13(2):203–206. Scopus. http://jtafs.mardi.gov.my/jtafs/13-2/Goats%20grazed.pdf. Accessed 21 Sept 2023
Meagher RK, Beaver A, Weary DM, von Keyserlingk MAG (2019) Invited review : A systematic review of the effects of prolonged cow–calf contact on behavior, welfare, and productivity. J Dairy Sci 102(7):5765–5783
Meijer E, Goerlich VC, van den Brom R, Giersberg MF, Arndt SS, Rodenburg TB (2021) Perspectives for buck kids in dairy goat farming. Front Vet Sci 8. https://www.frontiersin.org/article/10.3389/fvets.2021.662102
Mellor DJ, Beausoleil NJ, Littlewood KE, McLean AN, McGreevy PD, Jones B, Wilkins C (2020) The 2020 five domains model: including human–animal interactions in assessments of animal welfare. Animals 10(10):Article 10
Menéndez-Buxadera A, Molina A, Arrebola F, Clemente I, Serradilla J, m. (2012) Genetic variation of adaptation to heat stress in two Spanish dairy goat breeds. J Anim Breed Genet 129(4):306–315
Menzies P (2021) Udder health for dairy goats. Vet Clin N Am Food Anim Pract 37(1):149–174
Mersmann D, Schmied-Wagner C, Nordmann E, Graml C, Waiblinger S (2016) Influences on the avoidance and approach behaviour of dairy goats towards an unfamiliar human—An on-farm study. Appl Anim Behav Sci 179:60–73
Migliore S, Puleio R, Loria GR (2020) Scrapie control in EU goat population: has the last gap been overcome? Front Vet Sci 7. https://www.frontiersin.org/article/10.3389/fvets.2020.581969
Miller DW, Fleming PA, Barnes AL, Wickham SL, Collins T, Stockman CA (2018) Behavioural assessment of the habituation of feral rangeland goats to an intensive farming system. Appl Anim Behav Sci 199:1–8
Miranda-de la Lama GC, Mattiello S (2010) The importance of social behaviour for goat welfare in livestock farming. Small Rumin Res 90(1):1–10
Miranda-de la Lama GC, Pinal R, Fuchs K, Montaldo HH, Ducoing A, Galindo F, (2013. Environmental enrichment and social rank affects the fear and stress response to regular handling of dairy goats. Journal of Veterinary Behavior 8, 342–348
Molale G, Antwi MA, Lekunze JN, Luvhengo U (2016) General linear model analysis of behavioural responses of Boer and Tswana goats to successive handling. Indian J Anim Res 51(4):781–784
Molina A, Muñoz E, Díaz C, Menéndez-Buxadera A, Ramón M, Sánchez M, Carabaño MJ, Serradilla JM (2018) Goat genomic selection : Impact of the integration of genomic information in the genetic evaluations of the Spanish Florida goats. Small Rumin Res

Montaldo HH, Manfredi E (2002) Organisation of selection programmes for dairy goats. In: Proceedings of the 7th world congress on genetics applied to livestock production, Montpellier, France, August, 2002. Session 1, 1–8

Montaldo HH, Valencia-Posadas M, Wiggans GR, Shepard L, Torres-Vázquez JA (2010) Short communication : Genetic and environmental relationships between milk yield and kidding interval in dairy goats. J Dairy Sci 93(1):370–372

Mormède P, Foury A, Terenina E, Knap PW (2011) Breeding for robustness: The role of cortisol. Animal 5(5):651–657

Mucha S, Mrode R, MacLaren-Lee I, Coffey M, Conington J (2015) Estimation of genomic breeding values for milk yield in UK dairy goats. J Dairy Sci 98(11):8201–8208

Mucha S, Mrode R, Coffey M, Kizilaslan M, Desire S, Conington J (2018) Genome-wide association study of conformation and milk yield in mixed-breed dairy goats. J Dairy Sci 101(3):2213–2225

Muri K, Tufte P, Skjerve E, Valle P (2012) Human-animal relationships in the Norwegian dairy goat industry : Attitudes and empathy towards goats (Part I). Anim Welf 21(4):535–545

Naderi M, Riazi B, Aref N, Khalatbari M, Mohammadi S, Lahoot M, Kamran M (2013) Habitat preferences of Bezoar wild goats (Capra aegagrus) in Agh-Dagh protected area, Iran. North West J Zool 9:99–102. https://biozoojournals.ro/nwjz/content/v9n1/nwjz.131702.Naderi.pdf. Accessed 21 Sept 2023

Nagpal AK, Singh D, Prasad VSS, Jain PC (1995) Effect of weaning age and feeding system on growth performance and carcass traits of male kids in three breeds in India. Small Rumin Res 17(1):45–50

Nawroth C, McElligott AG (2017) Human head orientation and eye visibility as indicators of attention for goats (Capra hircus). PeerJ 5:e3073

Nawroth C, von Borell E, Langbein J (2016) 'Goats that stare at men'—revisited : Do dwarf goats alter their behaviour in response to eye visibility and head direction of a human? Anim Cogn 19(3):667–672

Nawroth C, Langbein J, Coulon M, Gabor V, Oesterwind S, Benz-Schwarzburg J, von Borell E (2019) Farm animal cognition—linking behavior, welfare and ethics. Front Vet Sci 6. https://www.frontiersin.org/article/10.3389/fvets.2019.00024

Nawroth C, Martin ZM, McElligott AG (2020) Goats follow human pointing gestures in an object choice task. Front Psychol 11. https://www.frontiersin.org/article/10.3389/fpsyg.2020.00915

Ndou SP, Muchenje V, Chimonyo M (2010) Behavioural responses of four goat genotypes to successive handling at the farm. Afr J Biotechnol 9(47):8118–8124

Neave HW, Zobel G (2020) Personality of dairy goats affects competitive feeding behaviour at different feeder heights. Small Rumin Res 192:106222

Neave HW, Sumner CL, Henwood RJT, Zobel G, Saunders K, Thoday H, Watson T, Webster JR (2022) Dairy farmers' perspectives on providing cow-calf contact in the pasture-based systems of New Zealand. J Dairy Sci 105(1):453–467

Negretti P, Bianconi G, Meo Zilio D, Noè L, Bartocci S, Di Giovanni S, Terramoccia S (2020) Effect of additional outdoor yard on behaviour of lactating domestic goats (Capra hircus) in different seasons. J Ethol 38(1):61–69

NFACC (2022) Code of practice for the care and handling of goats. ISBN 978-1-988793-42-9. Available at https://www.nfacc.ca/pdfs/codes/goat/Goat%20Code%2022_Final.pdf. Accessed 21 Sept 2023

Nord K, Holstad G, Eik LO, Grønstøl H (1998) Control of caprine arthritis-encephalitis virus and Corynebacterium pseudotuberculosis infection in a Norwegian goat herd. Acta Vet Scand 39(1):109–117

Nowak R, Boivin X (2015) Filial attachment in sheep : Similarities and differences between ewe-lamb and human-lamb relationships. Appl Anim Behav Sci 164:12–28

Nowak R, Poindron P (2006) From birth to colostrum : Early steps leading to lamb survival. Reprod Nutr Dev 46(4):431–446

Nowak R, Kraimi N, Chaillou E, Cornilleau F, Devaux M, Levy F, Parias C, Boissy A, Boivin X, Mialon MM, Bouvier F, Guilloteau LA (2018) Allaitement artificiel ovin et enrichissement social : un bénéfice potentiel de la présence de brebis adultes sur la santé et le bien-être du jeune. Rencontres Recherches Ruminants 24:209–212. https://hal.inrae.fr/hal-02738240

O'Brien PH (1984) Leavers and stayers : Maternal post-partum strategies in feral goats. Appl Anim Behav Sci 12(3):233–243

Odensten MO, Berglund B, Waller KP, Holtenius K (2007) Metabolism and udder health at dry-off in cows of different breeds and production levels. J Dairy Sci 90(3):1417–1428

Opasina BA (1985) Photosensitization jaundice syndrome in West African dwarf sheep and goats grazed on Brachiaria decumbens. Trop Grassland 19(3):120–123. Scopus. https://cgspace.cgiar.org/bitstream/handle/10568/27930/Vol19%28120_123%29%20.pdf?sequence=1&isAllowed=y. Accessed 21 Sept 2023

Paez Lama S, Egea V, Grilli D, Fucili M, Allegretti L, Guevara JC (2013) Growth and economic performance of kid production under different rearing systems and slaughter ages in arid areas of Argentina. Small Rumin Res 110(1):9–14

Pailhoux E, Vigier B, Chaffaux S, Servel N, Taourit S, Furet J-P, Fellous M, Grosclaude F, Cribiu EP, Cotinot C, Vaiman D (2001) A 11.7-kb deletion triggers intersexuality and polledness in goats. Nat Genet 29(4):Article 4

Palhiere I, Larroque H, Clément V, Tosser-Klopp G, Rupp R (2014) Genetic parameters and QTL detection for milking speed in dairy Alpine and Saanen Goats. In: 10th world congress of genetics applied to livestock production (WCGALP), Vancouver, Canada, 2014-08-17–2014-08-22, p 892. http://prodinra.inra.fr/ft/D40CE408-4C9B-4E62-BCFA-B66BF7E9DCD8. Accessed 21 Sept 2023

Palhiere I, Oget C, Rupp R (2018) Functional longevity is heritable and controlled by a major gene in French dairy goats. In: World congress on genetics applied to livestock production https://hal.inrae.fr/hal-02736063

Panzuti C (2018) Sevrage précoce et alimentation post-sevrage chez la chevrette : Impacts sur les performances zootechniques et sur le développement mammaire. PhD thesis, Agrocampus Ouest. https://theses.hal.science/tel-02315768

Panzuti C, Mandrile G, Duvaux-Ponter C, Dessauge F (2018) Early weaning and high feeding level in post-weaning period did not impact milk production in Alpine dairy goats. J Dairy Res 85(3):277–280

Papachristoforou C, Roushias A, Mavrogenis AP (1982) The effect of milking frequency on the milk production of Chios ewes and Damascus goats. Ann Zootech 31(1):37–46

Paraud C, Chartier C (2017) Facing anthelmintic resistance in goats. In: Simões J, Gutiérrez C (eds) Sustainable goat production in adverse environments: Volume I: Welfare, health and breeding. Springer, Cham, pp 267–292

Patt A, Gygax L, Wechsler B, Hillmann E, Palme R, Keil NM (2012) The introduction of individual goats into small established groups has serious negative effects on the introduced goat but not on resident goats. Appl Anim Behav Sci 138(1):47–59

Patt A, Gygax L, Wechsler B, Hillmann E, Palme R, Keil NM (2013a) Factors influencing the welfare of goats in small established groups during the separation and reintegration of individuals. Appl Anim Behav Sci 144(1):63–72

Patt A, Gygax L, Wechsler B, Hillmann E, Palme R, Keil NM (2013b) Behavioural and physiological reactions of goats confronted with an unfamiliar group either when alone or with two peers. Appl Anim Behav Sci 146(1):56–65

Pellicer-Rubio M-T, Boissard K, Grizelj J, Vince S, Fréret S, Fatet A, López-Sebastian A (2019) Vers une maîtrise de la reproduction sans hormones chez les petits ruminants. INRAE Prod Anim 32(1):Article 1

Peterhans E, Greenland T, Badiola J, Harkiss G, Bertoni G, Amorena B, Eliaszewicz M, Juste RA, Krassnig R, Lafont J-P, Lenihan P, Pétursson G, Pritchard G, Thorley J, Vitu C, Mornex J-F, Pépin M (2004) Routes of transmission and consequences of small ruminant lentiviruses (SRLVs) infection and eradication schemes. Vet Res 35(3):257–274

Petridis IG, Mavrogianni VS, Fragkou IA, Gougoulis DA, Tzora A, Fotou K, Skoufos I, Amiridis GS, Brozos C, Fthenakis GC (2013) Effects of drying-off procedure of ewes' udder in subsequent mammary infection and development of mastitis. Small Rumin Res 110(2):128–132

Piacere A, Boue P, Roguet JM (2000) Objectives and organization of the French selection program. In: 7th International conference of goats, Tours, France, p 1049

Pimenta V, Barroso I, Boitani L, Beja P (2017) Wolf predation on cattle in Portugal : Assessing the effects of husbandry systems. Biol Conserv 207:17–26

Pimenta V, Barroso I, Boitani L, Beja P (2018) Risks a la carte: Modelling the occurrence and intensity of wolf predation on multiple livestock species. Biol Conserv 228:331–342

Pinillos RG, Appleby MC, Manteca X, Scott-Park F, Smith C, Velarde A (2016) One welfare—a platform for improving human and animal welfare. Vet Rec 179(16):412–413

Poindron P, Terrazas A, de la Luz Navarro Montes de Oca M, Serafín N, Hernández H (2007) Sensory and physiological determinants of maternal behavior in the goat (Capra hircus). Horm Behav 52(1):99–105

Pol F, Kling-Eveillard F, Champigneulle F, Fresnay E, Ducrocq M, Courboulay V (2021) Human–animal relationship influences husbandry practices, animal welfare and productivity in pig farming. Animal 15(2):100103

Poulet J-L (2021) Pour la santé des mamelles, la traite 3.0. Webinaire idele, dia. 7. https://idele.fr/detail-article/pour-la-sante-des-mamelles-la-traite-30

Poulet J-L, Nouveau E, Fassion S, de Cremoux R (2016) Effects of transition from twice to once a day milking on milk flow kinetics of Alpine goats. In: 12th international conference on goats, Antalya, Turkey, 25th–30th September, 2016

Poulet J-L, Dzidic A, Ginsberg R, Rasmussen M, Manninen E, Sigurdsson S, Bruckmaier R, Tancin V, Sánchez Mainar M (2018) Teat-cup and cluster removal strategies for cattle and small ruminants review and recommendations. https://www.researchgate.net/publication/340886007_TEAT-CUP_AND_CLUSTER_REMOVAL_STRATEGIES_FOR_CATTLE_AND_SMALL_RUMINANTS_Review_and_recommendations. Accessed 21 Sept 2023

Poutrel B, Saumande J, Cochard T, Lefrileux Y (2012) Influence de l'induction de l'oestrus sur la concentration cellulaire du lait de chèvre. SNGTV 64:105–110. https://www2.sngtv.org/article-bulletin/influence-de-linduction-de-loestrus-sur-la-concentration-cellulaire-du-lait-de-chevre/

Pręgowski MP (2019) One welfare: a framework to improve animal welfare and human well-being. Anthrozoös 32:837–839. Rebeca García Pinillos. CAB, Wallingford, 2018. 112 pages. ISBN: 9781786393845 (hardback), 9781786393852 (paperback)

Quiédeville S, Hoste H, Frileux YL, Athanasiadou S, Werne S, Moakes S (2020) Cost-benefit and acceptance analysis of alternative nematode control practices on organic dairy goat farms : a case study in France. J Rural Soc Sci 35(2):Article 6. https://egrove.olemiss.edu/jrss/vol35/iss2/6. Accessed 21 Sept 2023

Rauw W (2007) Physiological consequences of selection for increased performance. In: Proceedings of the seventeenth conference of the Association for the Advancement of Animal Breeding and Genetics, Armidale, Australia, pp 240–247

Rauw W (2009) Resource allocation theory applied to farm animal production. CABI, Wallingford, UK, pp 1–21

Rauw WM, Kanis E, Noordhuizen-Stassen EN, Grommers FJ (1998) Undesirable side effects of selection for high production efficiency in farm animals; a review. Livest Prod Sci 56(1):15–33

Ricci A, Allende A, Bolton D, Chemaly M, Davies R, Fernández Escámez PS, Gironés R, Herman L, Koutsoumanis K, Lindqvist R, Nørrung B, Robertson L, Ru G, Sanaa M, Skandamis P, Speybroeck N, Simmons M, Kuile BT et al (2017) Genetic resistance to transmissible spongiform encephalopathies (TSE) in goats. EFSA J 15(8):e04962

Ridler AL, Smith SL, West DM (2012) Ram and buck management. Anim Reprod Sci 130(3):180–183

Roger PA (2012) Welfare issues in the reproductive management of small ruminants. Anim Reprod Sci 130(3):141–146

Romero G, Bueso-Ródenas J, Alejandro M, Moya F, Díaz JR (2021) Effect of vacuum level and pulsation parameters on milking efficiency and animal welfare of Murciano-Granadina goats milked in mid-line and low-line milking machines. Animals (Basel) 12(1):40

Rosenberger K, Simmler M, Langbein J, Nawroth C, Keil N (2022) Responsiveness of domesticated goats towards various stressors following long-term cognitive test exposure. PeerJ 10:e12893

Rowe JD, East NE (1997) Risk factors for transmission and methods for control of caprine arthritis-encephalitis virus infection. Vet Clin North Am Food Anim Pract 13(1):35–53

RSPCA (n.d.) Goats: introduction to welfare and ownership. Goats animal care bookletQXD. https://science.rspca.org.uk/documents/1494935/9042554/goats.pdf/dca7674d-20b6-b9a8-1f56-2bd9eb932064?t=1553171230073. Accessed 21 Sept 2023

Rupp R, Clément V, Piacere A, Robert-Granié C, Manfredi E (2011) Genetic parameters for milk somatic cell score and relationship with production and udder type traits in dairy Alpine and Saanen primiparous goats. J Dairy Sci 94(7):3629–3634

Rupp R, Huau C, Caillat H, Fassier T, Bouvier F, Pampouille E, Clément V, Palhière I, Larroque H, Tosser-Klopp G, Jacquiet P, Rainard P (2019) Divergent selection on milk somatic cell count in goats improves udder health and milk quality with no effect on nematode resistance. J Dairy Sci 102(6):5242–5253

Sacchi P, Rasero R, Ru G, Aiassa E, Colussi S, Ingravalle F, Peletto S, Perrotta MG, Sartore S, Soglia D, Acutis P (2018) Predicting the impact of selection for scrapie resistance on PRNP genotype frequencies in goats. Vet Res 49(1):26

Salama AAK, Such X, Caja G, Rovai M, Casals R, Albanell E, Marín MP, Martí A (2003) Effects of once versus twice daily milking throughout lactation on milk yield and milk composition in dairy goats. J Dairy Sci 86(5):1673–1680

Salama AAK, Caja G, Such X, Casals R, Albanell E (2005) Effect of pregnancy and extended lactation on milk production in dairy goats milked once daily. J Dairy Sci 88(11):3894–3904

Santos DS, Silva CCB, Araújo VO, de Fátima Souza M, Lacerda-Lucena PB, Simões SVD, Riet-Correa F, Lucena RB (2017) Primary photosensitization caused by ingestion of Froelichia humboldtiana by dairy goats. Toxicon 125:65–69

Santos JDR, Ungerfeld R, Balaro MFA, Souza-Fabjan JMG, Cosentino IO, Brair VL, de Souza CV, Pinto PHN, Bade ALC, da Fonseca JF, Brandão FZ (2020) Transcervical vs. laparotomy embryo collection in ewes : The effectiveness and welfare implications of each technique. Theriogenology 153:112–121

Sasaki O (2013) Estimation of genetic parameters for longevity traits in dairy cattle : A review with focus on the characteristics of analytical models. Anim Sci J 84(6):449–460

Scaramuzzi R, Martin G (2008) The importance of interactions among nutrition, seasonality and socio-sexual factors in the development of hormone-free methods for controlling fertility. Reprod Domest Anim 43(s2):129–136

Schukken YH, Vanvliet J, Vandegeer D, Grommers FJ (1993) A randomized blind trial on dry cow antibiotic infusion in a low somatic cell count herd. J Dairy Sci 76(10):2925–2930

Sevi A, Casamassima D, Pulina G, Pazzona A (2009) Factors of welfare reduction in dairy sheep and goats. Ital J Anim Sci 8(sup1):81–101

Sharp T (2005) Trapping of feral goats. Standard Operating Procedure. PestSmart website. (s. d.). Consulted March 30th 2023, at https://pestsmart.org.au/toolkit-resource/trapping-of-feral-goats/. Accessed 21 Sept 2023

Silanikove N (1989) Interrelationships between water, food and digestible energy intake in desert and temperate goats. Appetite 12(3):163–170

Silanikove N (2000a) The physiological basis of adaptation in goats to harsh environments. Small Rumin Res 35(3):181–193

Silanikove N (2000b) Effects of heat stress on the welfare of extensively managed domestic ruminants. Livest Prod Sci 67(1):1–18

Silanikove N, Merin U, Shapiro F, Leitner G (2013) Early mammary gland metabolic and immune responses during natural-like and forceful drying-off in high-yielding dairy cows. J Dairy Sci 96(10):6400–6411

Silva FF, Bambou JC, Oliveira JA, Barbier C, Fleury J, Machado T, Mandonnet N (2018) Genome wide association study reveals new candidate genes for resistance to nematodes in Creole goat. Small Rumin Res 166:109–114
Silva Salas MÁ, Mondragón-Ancelmo J, Jiménez Badillo MDR, Rodríguez Licea G, Napolitano F (2021) Assessing dairy goat welfare in intensive or semi-intensive farming conditions in Mexico. J Dairy Sci 104(5):6175–6184
Simitzis P, Seferlis I, Goliomytis M, Bizelis I, Politis I (2021) Handling and milking frequency affect milk yield and behavior in dairy ewes. Small Rumin Res 198:106351
Sinapis E, Hatziminaoglou I, Marnet PG, Abas Z, Bolou A (2000) Influence of vacuum level, pulsation rate and pulsator ratio on machine milking efficiency in local Greek goats. Livest Prod Sci 64(2):175–181
Sinapis E, Diamantopoulos K, Abas Z, Vlachos I (2006) Effect of vacuum level on milking efficiency, somatic cell counts (SCC) and teat end wall thickness in ewes of Greek mountain Boutsiko breed. Livest Sci 104(1):128–134
Singh MK, Dixit AK, Roy AK, Singh SK (2013) Goat rearing: a pathway for sustainable livelihood security in Bundelkhand region. Agric Econ Res Rev 26
Smistad M, Sølverød L, Inglingstad RA, Østerås O (2021) Distribution of somatic cell count and udder pathogens in Norwegian dairy goats. J Dairy Sci 104(11):11878–11888
Stachowicz J, Gygax L, Hillmann E, Wechsler B, Keil NM (2018) Dairy goats use outdoor runs of high quality more regardless of the quality of indoor housing. Appl Anim Behav Sci 208:22–30
Stachowicz J, Lanter A, Gygax L, Hillmann E, Wechsler B, Keil NM (2019) Under temperate weather conditions, dairy goats use an outdoor run more with increasing warmth and avoid light wind or rain. J Dairy Sci 102(2):1508–1521
Stanley CR, Dunbar RIM (2013) Consistent social structure and optimal clique size revealed by social network analysis of feral goats, Capra hircus. Anim Behav 85(4):771–779
Stronge DC, Fordham RA, Minot EO (1997) The foraging ecology of feral goats Capra hircus in the Mahoenui Giant Weta Reserve, Southern King Country, New Zealand. N Z J Ecol 21(1):81–88. https://www.jstor.org/stable/24054527. Accessed 21 Sept 2023
Stuen S (2017) Haemoparasitism of goats and sheep. In: Simões J, Gutiérrez C (eds) Sustainable goat production in adverse environments: Volume I: Welfare, health and breeding. Springer, Cham, pp 293–315
Sumner RMW, Bigham ML (1993) Biology of fibre growth and possible genetic and non-genetic means of influencing fibre growth in sheep and goats—A review. Livest Prod Sci 33(1):1–29
Svennersten-Sjaunja K (2004) The science behind milk ejection. In: NMC annual meeting proceedings, pp 215–228. https://cmapspublic3.ihmc.us/rid=1XFWSCL1B-3FY7ZS-DSQ/Milk%20Ejection%20Mechanism%20%2B%20Importance.pdf. Accessed 21 Sept 2023
Szabò S, Barth K, Graml C, Futschik A, Palme R, Waiblinger S (2013) Introducing young dairy goats into the adult herd after parturition reduces social stress. J Dairy Sci 96(9):5644–5655
Tamioso PR, Maiolino Molento CF, Boivin X, Chandèze H, Andanson S, Delval É, Hazard D, da Silva GP, Taconeli CA, Boissy A (2018) Inducing positive emotions : Behavioural and cardiac responses to human and brushing in ewes selected for high vs low social reactivity. Appl Anim Behav Sci 208:56–65
Teles JAA, Nascimento SA, Melo EX, Almeida EC, Marvulo MFV, Rizzo H, Nogueira DB, de Azevedo SS, Silva JCR, Castro RS (2023) Factors associated with small ruminant lentivirus infection in goat herds from Pernambuco state, Northeast region of Brazil. Prev Vet Med 211:105814
Theilen G, Wheeldon EB, East N, Madewell B, Lancaster WD, Munn R (1985) Goat papillomatosis. Am J Vet Res 46(12):2519–2526. https://pubmed.ncbi.nlm.nih.gov/3002215/. Accessed 21 Sept 2023
Thompson J, Riethmuller J, Kelly D, Miller E, Scanlan JC (2002) Feral goats in south-western Queensland : A permanent component of the grazing lands. Rangeland J 24(2):268–287
Todd CG, Bruce B, Deeming L, Zobel G (2019) Short communication : Survival of replacement kids from birth to mating on commercial dairy goat farms in New Zealand. J Dairy Sci 102(10):9382–9388

Toinon C, Waiblinger S, Rault J-L (2021) Maternal deprivation affects goat kids' stress coping behaviour. Physiol Behav 239:113494

Toinon C, Waiblinger S, Rault J-L (2022) Maternal deprivation affects goat kids' social behavior before and after weaning. Dev Psychobiol 64(4):e22269

Torres A, Castro N, Suárez-Trujillo A, Argüello A, Capote J (2016) Interrelationships among the length of milk stasis, tight junction permeability to lactose and monovalent cations, rate of milk secretion and composition in dairy goats traditionally milked once a day. Small Rumin Res 137:85–90

Torres A, Capote J, Fresno M, Eguiza A, Barba E, Molina JM, Ruiz A (2021) Impact of different feeding systems on cost-effectiveness and Eimeria spp. Infections in Canarian goat kids. Small Rumin Res 204:106518

Tosser-Klopp G, Bardou P, Bouchez O, Cabau C, Crooijmans R, Dong Y, Donnadieu-Tonon C, Eggen A, Heuven HCM, Jamli S, Jiken AJ, Klopp C, Lawley CT, McEwan J, Martin P, Moreno CR, Mulsant P, Nabihoudine I, Pailhoux E, International Goat Genome Consortium (2014) Design and characterization of a 52K SNP chip for goats. PLoS One 9(1):e86227

Tsiplakou E, Papadomichelakis G, Sparaggis D, Sotirakoglou K, Georgiadou M, Zervas G (2016) The effect of maternal or artificial milk, age and sex on three muscles fatty acid profile of Damascus breed goat kids. Livest Sci 188:142–152

Turini L, Conte G, Bonelli F, Sgorbini M, Madrigali A, Mele M (2020) The relationship between colostrum quality, passive transfer of immunity and birth and weaning weight in neonatal calves. Livest Sci 238:104033

Ugur F, Atasoglu C, Tolu C, Diken F, Pala A (2007) Effects of different weaning programs on growth of Saanen kids. Anim Sci J 78(3):281–285

Ujita A, El Faro L, Vicentini RR, Pereira Lima ML, de Oliveira Fernandes L, Oliveira AP, Veroneze R, Negrão JA (2021) Effect of positive tactile stimulation and prepartum milking routine training on behavior, cortisol and oxytocin in milking, milk composition, and milk yield in Gyr cows in early lactation. Appl Anim Behav Sci 234:105205

Uvnäs-Moberg K, Johansson B, Lupoli B, Svennersten-Sjaunja K (2001) Oxytocin facilitates behavioural, metabolic and physiological adaptations during lactation. Appl Anim Behav Sci 72(3):225–234

Vaccari G, Panagiotidis CH, Acin C, Peletto S, Barillet F, Acutis P, Bossers A, Langeveld J, van Keulen L, Sklaviadis T, Badiola JJ, Andréoletti O, Groschup MH, Agrimi U, Foster J, Goldmann W (2009) State-of-the-art review of goat TSE in the European Union, with special emphasis on PRNP genetics and epidemiology. Vet Res 40(5):48

Vaiman D, Koutita O, Oustry A, Elsen J-M, Manfredi E, Fellous M, Cribiu EP (1996) Genetic mapping of the autosomal region involved in XX sex-reversal and horn development in goats. Mamm Genome 7(2):133–137

Vickery HM, Neal RA, Meagher RK (2022) Rearing goat kids away from their dams 1. A survey to understand rearing methods. Animal 16(6):100547

Vickery HM, Meagher RK, Stergiadis S, Neal RA (2023) A preliminary investigation of the feeding behaviour of dairy goat kids reared away from their dams on a computerised ad libitum milk feeding system. Appl Anim Behav Sci 261:105898

Vilar MJ, Rajala-Schultz PJ (2020) Dry-off and dairy cow udder health and welfare : Effects of different milk cessation methods. Vet J 262:105503

Waiblinger S, Boivin X, Pedersen V, Tosi M-V, Janczak AM, Visser EK, Jones RB (2006) Assessing the human–animal relationship in farmed species : A critical review. Appl Anim Behav Sci 101(3):185–242

Wang Z, Wang R, Li J, Zhang W, Wuriliga, Su R, Liu Z, Zhou J, Wei Y, Meng R, Zhang Y (2014) Modeling genetic covariance structure across ages of fleece traits in an Inner Mongolia cashmere goat population using repeatability and multivariate analysis. Livest Sci 161:1–5

Winblad von Walter L (2010) Physiological and Behavioural responses to fear and discomfort in dogs and goats. Licentiate thesis ISBN 978-91-576-9009-8. https://pub.epsilon.slu.se/22421/1/winblad_von_walter_l_210218.pdf. Accessed 21 Sept 2023

Winblad von Walter L, Forkman B, Högberg M, Hydbring-Sandberg E (2021) The effect of mother goat presence during rearing on kids' response to isolation and to an arena test. Animals 11(2):Article 2
Windsor PA (2015) Paratuberculosis in sheep and goats. Vet Microbiol 181(1):161–169
Wolber M-R, Hamann H, Herold P (2021) Genetic analysis of lifetime productivity traits in goats. Arch Anim Breed 64(2):293–304
Yang X-Y, Chen J-P, Zhang F-X (2009) Research on the chemical composition of Saanen goat colostrum. Int J Dairy Technol 62(4):500–504
Yeruham I, Sharir B, Friedman S, Perl S, (2005) Cystic dilation of the teat sinuses in doe goats. Veterinary Record 156;844–844
Yoshida N, Koda N (2020) Goats' performance in unsolvable tasks is predicted by their reactivity toward humans, but not social rank. Front Psychol 11
Zamuner F, Leury BJ, DiGiacomo K (2023) Review : Feeding strategies for rearing replacement dairy goats—from birth to kidding. Animal 17(6):100853
Ziadi C, Muñoz-Mejías E, Sánchez M, López MD, González-Casquet O, Molina A (2021) Selection criteria for improving fertility in Spanish goat breeds : estimation of genetic parameters and designing selection indices for optimal genetic responses. Animals 11(2):409
Zobel G, Leslie K, Weary DM, von Keyserlingk MAG (2013) Gradual cessation of milking reduces milk leakage and motivation to be milked in dairy cows at dry-off. J Dairy Sci 96(8):5064–5071
Zobel G, Leslie K, Weary D, von Keyserlingk MAG (2014) Producer concern and prevalence of subclinical intramammary infections between lactations on 10 dairy goat farms in Ontario, Canada. In: American Association of Bovine Practitioners conference proceedings, pp 178–179
Zobel G, Weary DM, Leslie KE, von Keyserlingk MAG (2015a) Invited review : Cessation of lactation: Effects on animal welfare. J Dairy Sci 98(12):8263–8277
Zobel G, Leslie K, Weary DM, von Keyserlingk MAG (2015b) Ketonemia in dairy goats : Effect of dry period length and effect on lying behavior. J Dairy Sci 98(9):6128–6138
Zobel G, Tan BL, Deeming L (2016) Brief communication: Immediate removal of dairy goat kids from the doe as a strategy to manage colostrum intake. Proc N Z Soc Anim Prod 76:169–171. https://www.nzsap.org/system/files/proceedings/%2321%20Gosia%20Zobel.pdf. Accessed 21 Sept 2023
Zobel G, Neave HW, Webster J (2019) Understanding natural behavior to improve dairy goat (Capra hircus) management systems. Transll Animl Sci 3(1):212–224
Zobel G, Rodriguez-Sanchez R, Hea SY, Weatherall A, Sargent R (2020a) Validation of Brix refractometers and a hydrometer for measuring the quality of caprine colostrum. J Dairy Sci 103(10):9277–9289
Zobel G, Freeman H, Watson T, Cameron C, Sutherland M (2020b) Effect of different milk-removal strategies at weaning on feed intake and behavior of goat kids. J Vet Behav 35:62–68
Zucali M, Tamburini A, Sandrucci A, Gislon G, Bava L (2019) Effect of vacuum level on milk flow and vacuum stability in Alpine goat milking. Small Rumin Res 171:1–7
Zulkifli I, Siti Nor Azah A (2004) Fear and stress reactions, and the performance of commercial broiler chickens subjected to regular pleasant and unpleasant contacts with human being. Appl Anim Behav Sci 88(1):77–87

Painful Procedures, Transport, and End-of-Life Management

Genaro C. Miranda-de la Lama

Abstract

The capacity of goats to adjust to handling and husbandry in a wide variety of production systems, even under extreme climatic conditions, has made these animals a valuable resource in ancient and contemporary human societies. Every year, millions of goats worldwide are subjected to routine husbandry practices (injections, ear-tagging, hoof trimming, castration, disbudding, or dehorning), transport (land, sea, and air), and slaughter (for human consumption, euthanasia, and population control). Some of these practices are carried out because they have been shown to be effective in preserving animal health and welfare, or improving production performance and ensuring the quality or safety of goat milk and meat. However, even under good conditions, these practices can cause a stress response in the animals, and states of pain, fear, anxiety, and frustration. At the same time, there is a growing public concern for improving the conditions under which farm animals are raised, transported, and slaughtered. Therefore, the aim of this chapter is to describe the current scientific evidence on these practices and possible strategies to optimise, refine, and/or replace them to improve and preserve goat welfare.

Keywords

Castration · Dehorning · Euthanasia · Slaughter · Stunning · Lameness

G. C. Miranda-de la Lama (✉)
Department of Animal Production & Food Science, Veterinary Faculty, Agri-Food Institute of Aragon (IA2), University of Zaragoza, Zaragoza, Spain
e-mail: genaro@unizar.es

S. Mattiello, M. Battini (eds.), *The Welfare of Goats*, Animal Welfare 25,
https://doi.org/10.1007/978-3-031-62182-6_6

1 Introduction

Farm animal welfare is a growing public concern and is considered a priority for an increasing number of citizens around the world (Miranda-de La Lama et al. 2017). These concerns are the result of the increasing degree of industrialisation of animal production, socio-ethical considerations, human bonds with companion animals, and especially recent scientific evidence on the physiological and emotional states of animals (Robbins et al. 2016). Public opinion is especially concerned about practices that cause some degree of pain and fear in animals, such as husbandry procedures, transport, and slaughter (Browning and Veit 2020). These concerns have stimulated national and international debates among politicians, legislators, animal scientists, business persons, non-governmental organisations, and citizens which have led to the emergence of a large number of new regulations, public and private standards (Estévez-Moreno et al. 2022). These changes have encouraged the development of tools and protocols for monitoring and evaluating animal welfare at the farm level, during transport and slaughter (e.g. Welfare Quality and Animal Welfare Indicators, AWIN), but have also shown the need to design and validate new techniques to optimise, refine, or replace husbandry, transport, and slaughter practices to improve and preserve goat welfare. Unlike other livestock species, the scientific literature on the relationship between husbandry, transport, and slaughter practices in goats is limited and scattered. This chapter provides an overview of the available knowledge on current husbandry, transport, and slaughter practices in goats, as well as the available scientific evidence on the impact of these practices on animal welfare and possible strategies to optimise, refine, and/or replace them in order to improve and preserve goat welfare.

2 On-Farm Painful Procedures

Pain is defined as a complex, subjective experience caused by actual or potential injury that elicits progressive motor and vegetative reactions (Merskey 1994), which depends not only on the severity of tissue damage, but also on the duration of the animals' exposure to the painful stimulus (Fitzpatrick et al. 2006). Pain processing involves three essential mechanisms: sensation, perception, and cognition. Sensation refers to the lower level of neurological and biochemical components (nociception), while perception is the higher cognitive processing or interpretation of the damage (Millman 2013). Thus, experienced pain produces learned avoidance behaviour and can induce changes in behavioural, physical, and physiological patterns that are evidence of poor welfare (Molony and Kent 1997). It has been amply demonstrated that goats, like other livestock species, are capable of experiencing pain (Moussu and Dollar 1905). Pain experienced by goats may be the result of husbandry procedures such as ear-tagging, castration, dehorning, and hoof trimming (Miranda-de la Lama 2019). Many of these procedures are carried out at farm level without the use of analgesics or anaesthetics to control pain. In parallel, throughout their lives, goats are susceptible to pain from medical conditions related to accidents, productive and infectious diseases (Smith and Sherman 2009; Dawson 2010). In addition, during transport and slaughter operations, goats can suffer blows, electric shocks, trapping, falls, and mistreatment during handling that not only compromise the biological integrity of the animal, but also cause pain and suffering (Pulido et al. 2019; Valkova et al. 2021).

Pain assessment in goats is a major challenge because as a prey species they tend to instinctively mask pain as a passive anti-predatory defence strategy (Kats and Dill 1998). This would explain some difficulties for handlers or veterinarians to easily detect pain in goats, as they are perceived as potential predators by the animals (Plummer and Schleining 2013). At the experimental level, physiological and clinical parameters such as cortisol concentrations, glucose, lactate, respiration rate, heart rate, and body temperature have been studied to understand the response to pain in goats subjected to painful practices. However, most of these indicators are direct or indirect measures of pain-associated stress that stimulate the hypothalamic–pituitary–adrenal axis, but may also have their limitations, such as responding to other types of stimuli like circadian rhythms, handling, and restraint (Ajuda et al. 2020). Behavioural indicators of acute pain (high-pitched vocalisations, head and tail shaking, rubbing of the affected area, cage biting, and struggling) and chronic pain (low-pitched vocalisations, depression, loss of appetite) are commonly used and corroborated with physiological constants (Plummer and Schleining 2013).

Recently, the grimace scale has been proposed as a non-invasive and real-time assessment method for acute pain that identifies and evaluates changes in facial expression after husbandry procedures in different domestic species (Viscardi et al. 2017). Pain-induced changes in facial expression are controlled by involuntary muscle movements within major regions of the face, known as facial action units, which include orbital tension, ear position, cheek tension, and nose flaring (Häger et al. 2017). Lou (2020) developed a grimace scale for kids subjected to cautery disbudding. The results of this scale suggest that orbital clenching and downward or backward position of the ears indicate severe pain and can be considered as better facial action units to measure compared to lip clenching and nostril dilation. In addition, he found that administration of xylazine and lidocaine alone (or in combination) during the procedure caused changes in facial expression before and after disbudding. For other types of painful procedures in kids and adult goats, grimace scales have not yet been developed. The most common painful practices to which goats are subjected during their life on the farm are discussed below.

2.1 Identification Procedures and Injections

Individual identification is a routine husbandry procedure performed on goats for traceability and management purposes, which includes temporary methods (collars and bracelets) and permanent methods (tattoos and ear markers). Temporary methods do not cause pain, although they can cause accidents due to poor placement and the restless behaviour characteristic of goats, and can cause friction injuries, although collars can cause asphyxia due to over-fitting or snagging on pen fences, while bracelets can cause limb ischaemia due to over-fitting. Permanent methods may cause temporary pain during the marking process. For example, tattooing is performed on the thin part of the lower region of the ear avoiding the veins, and the needles have to go completely through the ear for the tattoo to be fixed in the tissue. For the breeds with minimal external ear, tattooing is performed on the underside of the tail (Smith and Sherman 2009). Tattoos are often used on pedigree goats or small herds, the application is laborious and the tattoos can be damaged or difficult to read, especially on dark pigmented ears, so green ink is recommended (Stanford et al. 2001). It is only recommended for animals older than 3 months of age; if done too early, the tattoo dots comprising numbers or letters expand and separate as the ear grows and the markings become difficult to interpret (Smith and Sherman 2009). During application, kids often vocalise and attempt to run away, in addition to a transient tachycardia (Orihuela and Ungerfeld 2020). It is therefore recommended to use a topical anaesthetic within an EMLA cream (Eutectic Mixture of Local Anaesthetics, composed of 2.5% prilocaine and 2.5% lidocaine) minutes before tattooing; this cream has shown to be effective in rabbits and piglets (Keating et al. 2012).

Ear-tagging consists of piercing the pinna to allow placement of a plastic earring (Lomax et al. 2017). There are two types of ear-tagging, one where a puncture has to be made with a punch in the ear with an anchor of a plastic earring then inserted; and the other, where the ear is pierced with a similar metal earring that has a sharp tip and is bent with the same applicator clamp to put it in position, hugging the ear in the manner of a staple (Orihuela and Ungerfeld 2020). This procedure is usually performed without the use of local anaesthesia or analgesia. Hussein and Hidayet (2019) studied the position of the ears as indicators of pain after ear-tagging, finding that when kids experience pain they point their ears further back and decrease the time they spend with their ears flat. Another study based on infrared thermography during ear-tagging found that a decrease in ear and eye temperature immediately after application and maintained for 30 min simultaneously decreased the frequency of suckling and standing behaviour, and increased the frequency of head shaking, vocalisation, and abnormal standing (Zebari et al. 2021). A possible alternative, to reduce pain during ear-tagging, is cryo-anaesthesia through the use of vaporising sprays. These sprays induce local anaesthesia and significantly reduce the pain response to ear-tagging in neonatal ruminants (Lomax et al. 2018). Ear notching is a long-standing practice of identification in livestock and is still widely used today in goats in many developing countries (Mbuku et al. 2010). The cuts are made with a blade, are not random and there are patterns to differentiate between owners. However, from an animal welfare point of view, this practice is considered cruel and, in some cases, can lead to necrosis and ear breakage, as well as the development of fly worms on the wounds (Caja et al. 2004).

The administration of vaccines or drugs by injection (subcutaneous, intramuscular, or intravenous) is a necessary practice in the treatment of multiple ailments and even to prevent pain (Smith and Sherman 2009). However, injections have the potential to cause momentary acute pain; furthermore, when the wrong drug is applied to the region, or the needle is contaminated or the injection technique is incorrect, they can cause bleeding, abscesses, cellulitis, necrosis, granuloma, muscle fibrosis, contractures, bruising, and injury to blood vessels, bones, and peripheral nerves (Dawson 2010). In feedlot goats, it is recommended to administer injections in the neck area in front of the shoulder and not in the hind legs because painful muscle lesions can form and decrease carcass value (Van Donkersgoed et al. 1997; Sebsibe 2008).

2.2 Hoof Trimming

Lameness is a painful condition that prevents normal gait and is one of the most serious welfare problems faced especially by goats raised indoors (Deeming et al. 2018; see also chapter "Enhancing Animal Welfare Through Disease Prevention in Goats"). There are few published studies on lameness in goats: for example, Anzuino et al. (2010) reported a 3.2% prevalence of severe lameness in 24 dairy farms in the UK, while Battini et al. (2016) reported a 3% prevalence in 30 intensive farms in Italy, and Can et al. (2016) found a 2.1% prevalence in 30 Portuguese farms. Due to the nature of lameness, treatment recommendations include a multimodal approach comprising trimming, administration of systemic or local antibiotics and analgesics (Winter 2011). The causes of lameness are foot rot, arthritis, improper hoof trimming, interdigital dermatitis, and arthritis-encephalitis (Plummer et al. 2021), although hoof overgrowth (with or without deformation) is perhaps the major cause of lameness in goats (Anzuino et al. 2010). Goat hooves have evolved in semi-arid, stony, and rough regions where uniform hoof horn wear is guaranteed (Sailer et al. 2021). In extensive systems, the prevalence of overgrowth is very low or practically absent (Salas et al. 2021). On the contrary, in intensive or indoor systems, hoof overgrowth has a prevalence of between 58 and 95%, due to lack of exercise and pens with wet and soft soils (i.e. straw bedding) that do not allow adequate and/or uniform wear of the hoof horns (Fig. 1a–d). Overgrowth can cause impaired locomotion with an irregular gait that can result in long-term irreparable damage to the foot and lower leg bones (Plummer et al. 2021). The type of hoof overgrowth (unilateral or bilateral) and the degree of wear can be different between hooves of the same animal (Ajuda et al. 2019). In addition, such hooves often trap particles of sand, twigs, manure, and other detritus (Fig. 1, 1–4) and this can lead to lameness (Vincent 2018).

Therefore, foot and leg problems in goats are caused by lack of routine foot care, poor prophylaxis, or improper trimming techniques (Prado et al. 2022). Therefore, hoof trimming is a necessary practice within a herd foot health programme (Hempstead et al. 2021), because it is useful to reduce peak pressure, expand the ground contact area of the hoof, improve balance between the medial and lateral hooves, and prevent the development of hoof, joint, and tendon disorders, improving locomotion and exercise capacity (Ibrahim et al. 2018). Moreover, trimming promotes healing in cases of hoof rot and abscesses, puncture wounds by allowing aeration and drainage of affected tissues (Smith and Sherman 2009). Using infrared thermography, Ajuda et al. (2014) emphasised that hoof overgrowth and hoof

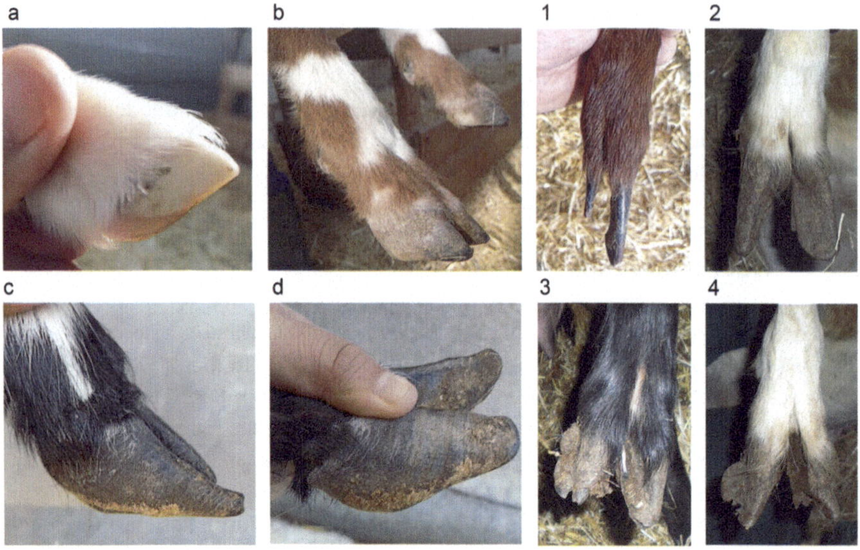

Fig. 1 In indoor systems, hoof overgrowth is gradual and becomes chronic in the absence of periodic trimming, causing severe lameness: (**a**) normal appearance in proportion and growth of the hoof, in this case of a 2-month-old kid; (**b**) kid at 6 months of age, note a slight overgrowth in the tips of the hoof; (**c**) first female with unilateral overgrowth of the left hoof and (**d**) the same animal with bilateral overgrowth of the right hoof. (**1–4**) Chronic overgrowth of hooves (Photos **a**, **b**, **c**, **d**, **1** by Genaro Miranda-de la Lama; photos **2**, **3**, **4** by Silvana Mattiello)

deformation might cause deep inflammation detected by regional hyperthermia and that it healed upon trimming and for 15 days thereafter. The basic steps of hoof trimming include removing the hoof wall that has overgrown the sole, shortening the toe, and levelling the sole and heel; ideally, a well-trimmed hoof should look like the hoof of a new-born kid (Fig. 1a) (Smith and Sherman 2009; Dawson 2010). Tools for hoof trimming include a set of sharp hoof shears, a hoof knife, work gloves, rasp, silver nitrate sticks, and copper sulphate (Dawson 2010). Recently, Prado et al. (2022) have proposed the use of an angle grinder to refine and finish trimming with scissors for severely overgrown or deformed hooves. These authors found that the additional use of the angle grinder further reduced the heel length and sole width of the hooves, and also helped to improve the pressure distribution of the hooves by shifting the point of maximum pressure from the heels to the toes. The frequency of trimming depends on many factors, such as type of terrain, age, activity level, nutrition, and genetics (Dawson 2010). However, in intensive and indoor production systems, it is advisable to trim at least twice a year (Battini et al. 2014), because regular hoof trimming is associated with a low prevalence of lameness in goats (Hill et al. 1997). Anzuino et al. (2010) found an evidence of correlation between overgrown hooves and lameness score in dairy goat farms.

2.3 Disbudding and Dehorning

Horns have evolved in goats as an instrument of protection and deterrence against predators and conspecifics, in addition to being a useful ornament for thermoregulation, body self-care, sexual selection, and mating success (Simon et al. 2022). However, disbudding or dehorning procedures are practised on goats as a requirement by breed show organisations and for safety reasons for handling and husbandry (Still-Brooks et al. 2021). Hornless goats are safer to handle, with no risk of horns breaking (Fig. 2a–c) and cause less damage to other animals, require less trough space, and have a lower negative impact on housing (Alvarez et al. 2015). Additionally, horned and hornless goats should not be housed together because there can be increased aggression, which may be directed towards those without horns, and this behaviour may lead to injuries (Plummer et al. 2021). However, keeping animals with or without horns is not merely a management detail, but has its origin in the different opinions of the farmers. Mersmann et al. (2022) found that farmers who did not disbud their kids assigned more "positive traits" to their goats and recognised more "goat needs" than farmers who disbudded their kids. Disbudding, the destruction of horn bud germinal tissue to prevent horn growth, is a common practice on goat dairy farms worldwide (Alvarez et al. 2019). The most commonly used method is by a hot-iron, although there are other alternatives such

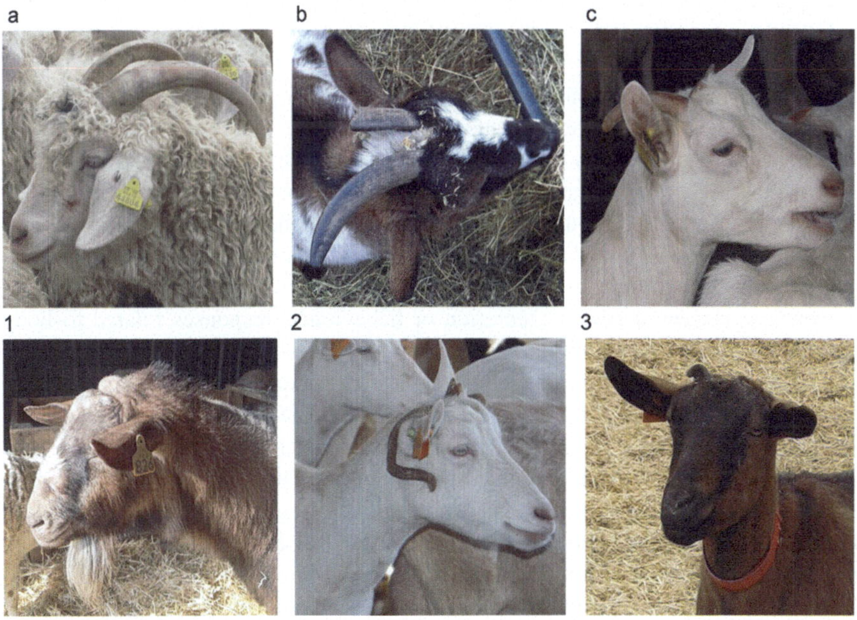

Fig. 2 Integrity of horns or quality of disbudding are indicators of good farm practice: (**a**) Goat with intact horns; (**b**) unilateral medial horn fracture; and (**c**) unilateral complete horn fracture. (**1**) Properly disbudded adult goat; (**2**) bilateral residual horns; and (**3**) unilateral scurs horn (Photos **a**, **b**, **1**, **3** by Genaro Miranda-de la Lama; photos **c**, **2** by Silvana Mattiello)

as caustic paste, cryo-surgery, and clove oil injection (Miranda-de la Lama 2019). Ideally, kids should be disbudded by 2–4 weeks of age or once the horn buds are palpable (Hartnack et al. 2018). Several studies have shown that any method of disbudding causes pain, discomfort, distress and occasionally compromises the health of kids by causing tetanus, meningoencephalitis, and brain abscesses (Smith and Sherman 2009).

Behavioural indicators of pain during hot-iron disbudding are leg shaking and intense high-pitched vocalisations (Alvarez and Gutierrez 2010). After the procedure, the animals experience head and body shaking, as well as episodes of head scratching. At the physiological level, there is an increase in plasma concentrations of cortisol and β-endorphin, which returns to basal levels approximately 30 min after disbudding (Hempstead et al. 2017). Anecdotally, some producers believe that the disbudding is not painful because goat kids seem to rapidly return to performance of normal behaviours after disbudding. However, there is a large body of evidence available highlighting the pain associated with hot-iron disbudding and supporting the use of pain mitigation strategies to improve goat kid welfare (Plummer et al. 2021). For example, hot-iron disbudding wounds in goat kids take 7 weeks to re-epithelialise, and the damaged tissue present during this period is more sensitive than newly re-epithelialised tissue, indicating that wounds are painful during the healing process (Alvarez et al. 2019). Although some countries (i.e. England, Wales, Scotland, Switzerland) have mandated the use of pain mitigation strategies for hot-iron disbudding of goat kids, many countries (i.e. European Union (EU) and United States) do not have regulations or recommendations for using pain mitigation strategies when disbudding goat kids, and this may limit their use on commercial dairy goat farms (Hempstead et al. 2020). In recent years, European veterinary associations have advocated the use of anaesthesia and analgesia in all goat disbudding procedures to minimise unnecessary pain and suffering. They also approve the use of local anaesthesia only, but under the condition that it causes an effective blockade of the four nerves to desensitise the bud (Mills 2018).

Although the use of local anaesthetics, such as lidocaine (with or without cornual nerve block), is recommended for disbudding, recent some studies show that it does not prevent increased cortisol secretion or behavioural responses related to body restriction and acute pain (Alvarez et al. 2015). In a study by Ajuda et al. (2020), two pain mitigation strategies were tested for iron-hot disbudding and compared to a control group. The first strategy used local anaesthesia (lidocaine) and the second used local anaesthesia (lidocaine) plus an analgesic (flunixin meglumine). The results showed that kids that received both pain mitigation treatments spent more time engaging in active (i.e. standing) and positive behaviours (i.e. exploring, playing, feeding). However, the incidence of pain and discomfort-related behaviours (i.e. head scratching) was not consistently reduced. Similarly, Martínez et al. (2023) found that administration of flunixin meglumine is effective in reducing pain-related behaviours when administered at the time of hot-iron disbudding. According to Hempstead et al. (2020), oral administration of meloxicam 60 min before the procedure does not reduce acute pain in the 30 min following hot-iron disbudding. These same authors also found that the application of EMLA cream does not reduce or eliminate the pain associated with disbudding if applied 60 min beforehand, and may intensify or prolong the pain. Only sedation (with or without local anaesthesia) has been shown to be effective (Nfor et al. 2016); general anaesthesia with gases such as isoflurane (with or without meloxicam) has also been shown to be effective (Hempstead et al. 2018a). A study by Hempstead et al. (2018b) indicated that caustic paste and cryosurgical disbudding appeared to cause more pain compared to hot-iron method; therefore, these methods may not provide good alternatives to hot-iron disbudding.

As an alternative to hot-iron disbudding, Schoiswohl et al. (2021) tested the regional injection of clove oil and its synthetic analogue isoeugenol. These authors found that the highest disbudding efficiency was achieved with thermal disbudding and isoeugenol, followed by clove oil. Although all three methods appeared to elicit a similar pain response, isoeugenol and clove oil caused less tissue damage than thermal disbudding.

Improper disbudding can lead to regrowth of residual horns or scurs (Fig. 2d–f), which can get stuck in housing structures and break off or grow into the head or eyes, causing pain and injury. The prevalence of animals with scurs is an indicator of problems in dehorning efficiency (Plummer et al. 2021) and is included as animal-based welfare measure in the AWIN protocol for welfare assessment of goats (AWIN 2015). When farmers are unable to disbud kids at an early age or residual horns are present, surgical dehorning is necessary (Hartnack et al. 2018). Surgical dehorning with primary wound closure in goats was first described by Mobini (1991); prior to this, incisions were bandaged and left to heal by secondary wound closure, resulting in an open sinus increasing the risk of sinusitis, lengthening healing time and increasing the amount of aftercare required (Hull 1995). Hartnack et al. (2018) monitored complications following surgical dehorning in 239 goats over a 10-year period, finding that 35% of the animals had minor complications such as redness or swelling of the incision, nasal discharge, incisional discharge, scurs, loss of appetite, or perceived decreased performance, while 4% had major complications, such as incisional dehiscence and death related to the surgical procedure. These results indicate that this procedure has a low probability of complications, as long as it is performed by a veterinarian. However, this procedure should only be performed as an exceptional action (i.e. animals with dangerous, broken, or deformed horns) and not as a routine measure. In any case, a farm's efforts to improve management and avoid horn accidents must be based on an efficient programme of disbudding of kids (Alvarez et al. 2019).

2.4 Castration

Castration is a routine practice in many countries and consists of the removal or disabling of testicular function (Adcock 2021). The aim of this procedure is to reduce aggressive and sexual behaviour, prevent accidental breeding of an inferior buck, avoid undesirable odours associated with billy goats, and improve the organoleptic quality of the meat (Dawson 2010). The most commonly used techniques include surgical and mechanical or ischaemic techniques such as the rubber ring, *burdizzo*, or emasculator. Another alternative for kids is short scrotum castration, where the testes are pushed up into the abdominal cavity by applying a rubber ring around the distal scrotum; the higher temperature within the abdomen affects testicular function and induces infertility (Rault et al. 2011). However, all these techniques cause some degree of physiological stress, pain, inflammatory reactions, immunosuppression, and behavioural changes (Karademir et al. 2016). In goats, castration is usually performed as soon as the testicles descend into the scrotum, which is between 7 and 20 days of age. It is common for producers to castrate kids without anaesthesia during the first month of life using the knife method or an ischaemic method (Smith and Sherman 2009). When the kid is several months old, surgical removal is the technique of choice, unless environmental conditions make open wounding unacceptable. At this age, the use of anaesthesia and non-steroidal anti-inflammatory drugs is mandatory due to the high likelihood of post-surgical mortality when pain control is not effective (Ajadi et al. 2012). Surgical castration in unweaned kids may result in post-procedure maternal rejection due to fresh wound odour and antiseptic solution. In these cases, the dam terminates lactation and other maternal care leading to stress and immunosuppression in the kid (Miranda-de la Lama 2019).

Castration without analgesia and/or local anaesthesia is prohibited or regulated in many countries (Zeng et al. 2022). For example, in the EU, there are recommendations adopted by the Standing Committee of the European Commission at its 25th session (6 November 1992), which discourage castration at any age; when it has to be done, it can be performed only by surgical techniques (preceded by anaesthesia) or with haemostatic tongs, and exceptionally by rubber ring (if the legislation of the member country allows it) (Peris et al. 2016). Whereas for the Australian goat standards and guidelines, the use of analgesics and/or local anaesthetics is only mandatory for kids older than 6 months of age (Animal Health Australia 2020). The effect of castration in lambs has been extensively studied, while for kids' information is limited, and the results for lambs are not directly comparable with those for kids. For example, Mellor et al. (1991) showed that, according to the stress response (cortisol and behaviour), castration with rubber rings is less painful and stressful for kids compared to lambs. The same authors propose that restlessness, lateral recumbency, and immobility are considered useful indices of pain and stress for rubber ring castration in kids. On the other hand, Ahmed and Ahmed (2011) studied the effects of castration by *burdizzo* and rubber ring methods on 7-, 21-, and 42-day-old kids, finding that the rubber ring method causes more pain and reduces suckling bouts of kids. The authors concluded that castration with the *burdizzo* produces less pain at any age.

Finally, immunocastration in goats has been proposed as an alternative to surgical or ischaemic castration, because its administration and effects are painless and do not cause behavioural or emotional changes (Zeng et al. 2022). Although there are currently no commercial vaccines for small ruminants, studies using vaccines for pigs and cattle, or experimental vaccines, have shown effectiveness in goats (Zeng et al. 2022). For example, Wassie et al. (2020) used a recombinant injectable

DNA vaccine (Kisspeptin-54) in male goats and found that immunisation did not affect performance, daily weight gain, carcass weight, carcass, and meat quality. Another study, using an orally administered vaccine (KISS1 DNA), found that it produced a strong antibody response that inhibited testosterone secretion and testicular growth, and reduced male odour (Han et al. 2020). Other methods based on temporary and reversible suppression of gonadal function by administration of synthetic progesterone (altrenogest used commercially in boars) have been tested, but results are inconsistent (Mihsler-Kirsch et al. 2021).

2.5 On-Farm Euthanasia

In goat production, on-farm assisted death is applied when an illness, injury, or accident cannot be practically treated or for an excess of male neonates in a dairy herd (Plummer et al. 2021). According to the AVMA Euthanasia Panel, acceptable methods of euthanasia for goats include a lethal overdose of a barbiturate, gunshot, and penetrating or non-penetrating captive bolt devices (Underwood and Anthony 2020). The use of barbiturates is a recommended method because an overdose induces a smooth transition to unconsciousness and death, although it can only be administered by or under the direct supervision of a veterinarian, which may increase costs associated with euthanasia and carcass incineration (Plummer et al. 2018). In the case of gunshot or captive bolt, both cause immediate loss of consciousness followed by death, but the effective use of these methods depends on several factors that are discussed at length in Sect. 3.1 of this chapter. At the field practice level, veterinarians typically euthanize adult goats and kids in two steps, using a penetrating captive bolt to stun the animal, followed by a supplemental method (intravenous potassium or magnesium salts, exsanguination, pithing, or 1 or 2 additional captive bolt stuns) to create terminal cerebral hypoxia or physical disruption of the cortex and brainstem (Still-Brooks 2019).

3 Transport

Traditionally, livestock transport has been defined as the "movement of animals from one place to another". However, from the agri-food industry logic, transport is conceived as a system that optimises the flow of animals within a geographical space and can be assessed in terms of journey time, distance journeyed, journey conditions, infrastructure and its impact on animal welfare (Miranda-de La Lama et al. 2010). Therefore, transport should be understood as a system, where its composition is given by the type of animals, vehicles, means, and facilities at the origin-destination, while its structure is the spatial and logistic organisation of the journeys, and the functioning is the interaction between composition and structure. There are logistically simple systems where journeys involve origin-destination, while others involve greater complexity in multiple scales and/or vehicles between origin-destination (Miranda-de la Lama 2019). In this context, unlike what happens with

Fig. 3 Some specialised types of trucks used to transport goats. (**a**) Double-decker gooseneck trailer (USA). Note the removable ramp attached to the side of the chassis, the roll-up access door to the trailer. Each deck is divided into two compartments with a dividing door to balance the number of animals being transported to prevent overturning of the trailer. (**b**) Four-deck chassis made of wood and iron (Mexico). This truck has configurable passive ventilation (casement windows), they also have three compartments per-deck and are totally dependent on loading/unloading docks because they are not equipped with their own portable ramps. (**c**) Triple-deck rigid chassis truck with active and passive ventilation (Spain). The design of this truck favours autonomous loading/unloading, as well as having folding compartments and decks to allow progressive loading and efficient cleaning (Photos: Genaro Miranda-de la Lama)

other vertically reared and integrated species, goat transport is highly diversified due to the great heterogeneity and territorial dispersion of goat production systems that may include transhumance (in vehicles or walking), journey between farms, or from farm to market, to feedlot, to slaughterhouse, to abattoir or even for export (Alam and Yoshino 2022). There is also a diverse range of facilities (modern or not) and vehicles used to transport goats by road, ranging from small common trucks to small trailers and specialised double or triple-decker rigid chassis trucks (Fig. 3), in addition to transportation by sea or air.

3.1 Road Transport

During the transport period (loading-journey-unloading), even under favourable conditions, goats are exposed to potential stressors such as novel visual, olfactory, and auditory stimuli, gases and dust, abrupt changes in temperature, changes in altitude, a significant decrease in individual space, difficulties in standing and/or lying down, loss of balance, social competition for space, water and feed deprivation (Miranda-de la Lama et al. 2022) (Fig. 4). The interaction between these stressors, pre-existing welfare conditions and individual differences, can limit or affect the fitness and capacity of an animal during the journey (Cockram 2019).

Fig. 4 Goat transport issues: (**a**) Poorly designed loading ramps that force operators to load animals; (**b**) low densities can cause instability and animals falling during the journey, it is also advisable that kids maintain eye contact with adult animals; (**c**) travel of adult goats with inadequate head-ceiling space that does not allow them to travel upright; (**d**) under optimal conditions goats during loading/unloading often have marked browsing and curiosity behaviour; (**e**) high densities and violent handling cause fear, frustration, discomfort, and suffering; (**f**) High densities and long-distance journeys often fatigue vulnerable individuals (sick, poor body condition, remaining) causing some to fall or seek space between the legs of their companions, and they may be trampled or gored (Photos **a**, **c**, **e**, **f**, by Genaro Miranda-de la Lama; photos **b**, **d** by Andrés Conesa Ros)

Consequently, the welfare of goats during the journey is highly dependent on the physical and mental fitness of the animals to face the journey as a challenge (Miranda-de La Lama et al. 2018). According to EFSA (2022), when assessing fitness for transport, signs of disease, such as sickly behaviour, increased and reduced respiratory function, reduced exercise tolerance, impaired ability to walk and maintain posture, and reduced ability to thermoregulate and respond to periods of food and water restriction, should be assessed. Even when they are assessed as fit for transport and transported to slaughter, signs of pathology, which may or may not have been clinically apparent prior to loading, may be detected at slaughter and may or may not have affected their ability to respond to the challenges of transport (Cockram and Velarde 2022). EFSA (2022) identified 11 welfare consequences in small ruminants during transport by severity, duration, and frequency, which are: (1) group stress, (2) handling stress, (3) heat stress, (4) injury, (5) movement stress, (6) predation stress, (7) prolonged hunger, (8) prolonged thirst, (9) movement restriction, (10) resting problems, and (11) sensory overstimulation.

While there are numerous regulations around the world regarding animal welfare during transport, most of them are generic and do not always cover goats. However, the EU has regulations for small ruminants, stating that the maximum allowable journey time is 29 hours for adult sheep and 19 h for unweaned lambs, with a tolerance in all cases of an additional 2 h to reach the final destination. After these periods, it is stipulated that animals must be unloaded to rest for at least 24 h (12 h in relation to sea transport) or more at a checkpoint (CP) before continuing the journey (European Commission 2005). These posts must be used exclusively to receive, feed, water, rest, shelter, care for, and dispatch transient animals, be located in an area free of any animal health restrictions, and be under the authority of an official veterinarian. There are 61 approved control points in the EU for receiving goats and sheep. These points are dispersed across 15 of the 27 EU members, although the majority of EU control posts (70%) are located in France (12), Italy (12), Germany (10), and Poland (10). The fact that these nations serve as required transit hubs between the Iberian Peninsula, northern and southern Europe explains the concentration of control posts in those nations (Padalino et al. 2020). Although the majority (about 60% of the total) of small ruminants are transported across large distances in the EU, only a small portion (5% of them) is transported for longer than 29 h (Padalino et al. 2018), necessitating the usage of checkpoints (EFSA 2022).

3.1.1 Adult Goats

There is a widespread belief among farmers and hauliers that goats have a high tolerance for transportation, resulting in goats being carried in less than ideal conditions compared to other domestic species (Miranda-de la Lama 2019). However, even in journeys of not more than 200 km in hot humid conditions in India, in adult goats transport significantly affected the levels of total protein, albumin, globulin, total calcium, phosphorus, and glucose, which returned to their pre-loading values in the 1-week period after unloading, except glucose, which decreased significantly in the second week after transport, but was within the normal physiological range for goats (Panchal et al. 2022). Transport stress has also been associated with live weight loss in goats; Vignesh et al. (2018) found that, on journeys of less than 20 km, animals lost 0.95% of their live weight, while animals subjected to journeys of 150 km lost 3% of their live weight. In this type of journeys, losses are related to intestinal and urinary evacuation; however, as the distance of the journey increases, losses can be up to 10% of the live weight due to fatigue and especially to the catabolic effect of stress (Kannan et al. 2000).

In a pioneering study conducted under experimental conditions with 3-h journeys, Nwe et al. (1994) found that at the beginning of the journey goats tended to display restlessness and nervous behaviours, and an acute transient tachycardia associated with the onset of the truck movement. However, this tachycardia did not induce an increase in respiratory rate, 15 and 30 min into the journey, plasma cortisol and glucose concentrations tended to rise sharply to four times the baseline values, which recovered at least 3 h after the end of the journey. They also found a decrease in the number of eosinophils as the journey progressed and a return to baseline values 15 h after the end of the journey. In another study, Nwe et al. (1996) transported adult goats (catheterised in the jugular and femoral veins) for 6 h, finding that plasma epinephrine levels increased immediately after the start of the journey and were maintained throughout the transport, and returned to basal value at the end of the journey, while the plasma level of norepinephrine did not change throughout the transport. Plasma cortisol and glucose levels increased immediately after the start of the journey and continued to increase until reaching a maximum level after 3 h of transport, although plasma cortisol and glucose levels decreased rapidly and returned to basal values 6 h after the end of transport. While the number of eosinophils decreased immediately after the start of transport, it reached a minimum level at the end of the journey and returned to the baseline value 12 h later. Both studies showed that at the onset of transport stress, the sympathetic nervous system is activated and then the medulla and adrenal cortex are simultaneously activated, followed by immediate physiological changes. It appears that the initial phase is the most important and, therefore, controlling the reaction of animals at the onset of transport may reduce their stress response.

The position that goats adopt during the journey can influence their endurance and ability to maintain stability, avoid falls, and attenuate fatigue. Available scientific evidence shows that the position of the animals during the journey is dependent on its duration: in a study with adult goats (1–3 years) that were subjected to a 50-min journey it was observed that the animals spent most of the time parallel to the direction of the vehicle, but with frequent changes of diagonal and perpendicular orientation to maintain balance (Das et al. 2001). However, Minka and Ayo (2014) reported that in long-distance journey (12 h) after the first hour of journey goats changed their position from parallel to perpendicular (relative to the direction of vehicle movement) and only laid down after 7–8 h of journey. According to the same study, goats who stood or laid perpendicular to or opposite to the direction of traffic had lower blood glucose and urea levels, as well as lower levels of creatine kinase (CK), alanine aminotransferase, and aspartate aminotransferase, indicating less muscle damage and fatigue as a result of the good journey position. In any case, the minimum vertical space suggested for sheep and goats by EFSA (2022) should be at least 15 cm above the animals in mechanically ventilated vehicles and 30 cm in naturally ventilated vehicles.

In a stress transcriptome study using RNA-seq technique, Naldurtiker et al. (2023) found that transported goats (30 min or 3 h) compared to non-transported goats had differentially expressed genes related to inflammatory pathways, caspases and apoptosis such as IL1R2, CASP14, CD14, TLR4, and MAPK14. These results show that transport induces in the species a series of events at the cellular and molecular level that cause inflammation and apoptosis. Thus, the journey contributes to the development of oxidative stress; this is because the interaction of emotional stressors and fasting stimulates the mobilisation of fatty acids and alpha-amylase for use as metabolic fuel, which increases the production of mitochondrial reactive oxygen species (ROS). In addition, fasting combined with fatigue can stimulate an inflammatory response, which results in the production of ROS by phagocytic immune cells and as a by-product of eicosanoid biosynthesis (Deters and Hansen 2020). Erythrocytes are highly susceptible to oxidative stress and their oxidation has been used as a model system for the investigation of oxidative damage of biomembranes. For example, Fazio et al. (2016) found in goats transported for

3 h that erythrocyte osmotic fragility (EOF) showed a significant increase after the journey and a decrease 24 h after the journey. This effect has also been observed in goats subjected to 7 h of transport in hot and humid tropical environments, finding haematological alterations due to oxidative stress and haematological derangements, and needing 3–16 days to recover (Polycarp et al. 2016).

Oxidative stress also can stimulate the production of many heat shock proteins (HSPs) in muscular and parenchymatous tissues. HSPs are a family of highly conserved and ubiquitously stress response proteins that are expressed at low levels under normal physiological conditions (Kalmar and Greensmith 2009). For example, Zulkifli et al. (2010) investigated the influence of two different stocking densities (0.20 m^2/animal and 0.40 m^2/animal) in three-hour rides under hot and humid tropical conditions on the induction of HSP70 in 60 female Boer goats. The results of this study indicate that, regardless of stocking density, transport significantly increased HSP70 levels in kidneys and that the HSP 70 response in kidneys was more profound compared to that in heart tissues. Another study by Hu et al. (2020) discovered that HSP27 and HSP90 are preferentially raised after short road journeys (2 h) and HSP70 is preferentially increased during prolonged journeys (>6 h). The heart may be the most resilient to transport stress, whereas the lung may be the most vulnerable. These findings were supported by Zheng et al. (2021), who discovered that journeys of more than 6 h can result in morphological damage to the lung, trachea, and bronchi, as well as a change in HSP70 expression. This lung damage caused by oxidative stress can lead to the onset of the bacterial disease pleuropneumonia commonly known as "transport fever". This multifactorial infectious disease is caused by a complex interaction between different viral and bacterial pathogens and stress during transport which suppresses host defence mechanisms, facilitates pathogen spread, or both (Lowie et al. 2021). In this sense, Taunde et al. (2019) discovered that goats showed clinical indications of pleuropneumonia caused by non-haemolytic strains of *Mannheimia haemolytica* during long-distance transportation (3700 km) in tropical humid climates, including dyspnoea, coughing, anorexia, mucopurulent ocular and nasal discharge. In another study, Karabasanavar et al. (2016) found the occurrence of transport fever (*Mannheimia haemolytica* and *Pasteurella multocida*) in a herd of 256 goats transported by road in India, especially during the first week post-arrival with a morbidity of 60% and mortality of 27%.

It is acknowledged that even in optimal conditions, road transport can be a source of stress and specially shift rumen fermentation, an effective practice to mitigate these effects is to supplement with water and feed on arrival (Xu et al. 2023). However, to prevent the effects of transport stress in goats, several studies have explored the possibility of including in the diet some component or plant extracts or a vitamin supplement with antioxidant properties for weeks or days before the journey. Antioxidants are compounds that can reduce the generation of O_2 radicals, such as superoxide, and protect the immune system by preventing or slowing cell damage caused by free radicals (Moyo et al. 2012). For example, Batchu et al. (2021) administered a perennial legume (*Lespedeza cuneata*) with antioxidant properties for 8 weeks before a journey (90 min duration) and found an anti-inflammatory effect and a reduction in the microbial count of the gastrointestinal tract and rectum. In another study, Kannan et al. (2007) supplemented adult goats with an extract of brown algae (*Ascophyllum nodosum*) for 3 weeks before subjecting them to a 6-h drive. The results of this study showed that the supplementation did not alter the physiological stress response and that stress levels increased after the journey, although it helped the animals recover from the transport, as evidenced by a simultaneous decrease in oxidative stress-induced lipid damage. Ascorbic acid is an antioxidant that has also been used to ameliorate the effect of transport stress. Minka and Ayo (2013) studied 12-h transport in goats and a post-transport recovery in grazing with or without ascorbic acid supplementation. Overall, the results showed that goats that were allowed to graze after transport needed 3 days to return to baseline values, but those that were given ascorbic acid before transport and grazed after transport needed only 2 days to recover. It is likely that giving animals ascorbic acid on its own is insufficient to help them cope with the stress of transport. In this regard, a recent study by Gupta et al. (2021) examined the comparative ameliorative effects of ascorbic acid, electrolyte, and jaggery on transportation of adult goats for 8 h at various stocking densities in hot, humid, and winter seasons. They discovered that an ascorbic acid + electrolyte combination was more effective in lowering transportation stress in both seasons compared to ascorbic acid + electrolyte supplementation alone in reducing transportation stress. Although these four studies show that the use of antioxidants improves post-transport recovery, they do not usually decrease the stress response during the journey. This makes it necessary to direct efforts to improve logistics, driving quality, route planning, and animal handling as general methods to improve the welfare of transported animals (Miranda-de la Lama et al. 2014; EFSA 2022).

3.1.2 Kids

Goat kids are usually transported for three main reasons: restocking, herd replacement, or slaughter. In the meat and dairy goat industry, female kids are used to replenish the herd or to restock other farms (domestically or for export). On the other hand, male kids can be sent to feedlots where they can be fattened during the first 3 or 4 weeks of life (dairy industry) or until puberty or adulthood (meat industry). In Europe, especially the dairy goat industry discards thousands of male goats per year that are normally sent to slaughter to obtain meat for humans or pet food (Meijer et al. 2021). According to EU legislation (European Commission 2005), it is forbidden to transport kids from the farm to the slaughterhouse whose navel is not fully healed, and kids less than 7 days old must not be transported for more than 100 km. However, a report by Looije and Moffat (2020) on the conditions of transporting male kids from commercial dairy systems to slaughter in the Netherlands found that, despite the short distances, the journeys took between 6 and 7 h, because only a small number of kids were collected from each farm, which meant that the truck had to make many stops before filling up. During this time, the animals had no access to water, electrolytes, or milk, and although on arrival at the slaughterhouse they were offered water or electrolytes the animals showed behavioural signs of hunger (vocalisations, searching, and attempts to suckle). This report emphasises the importance that water or electrolytes should not be considered as a substitute for milk, so milk should be given on arrival. For kids older than 7 days of age, EU regulations stipulate that unweaned kids receiving milk feed shall be given a sufficient rest period of at least one hour after 9 h of transport, in particular to provide them with water and, if necessary, feed. After this rest period, they may be transported for a further 9 h (European Commission 2005).

Studies on the effects of transport on behaviour and physiology in kids are very limited, although, as in adult goats, there is evidence of a strong adrenocortical response (Ali 2020). Kannan et al. (2003) investigated the physiological stress reactions of young goats (6–12 months of age) and older goats (24–30 months of age) after a 2-h journey. Although the magnitudes of blood cortisol and glucose responses to stress were greater in mature goats, these authors discovered that transportation lowered muscle glycogen concentrations, with a more prominent effect in young vs. older goats. In another study, Alcalde et al. (2017) found that, regardless of the duration of the journey (2 h vs. 6 h), transport induced an increase in adrenocortical activity and its catabolic effect and muscle damage. In addition, kids that journeyed 6 h had a greater loss of carcass fat cover compared to animals transported for 2 h, which could be attributed to lipolysis and release of fatty acids from adipose tissue due to the longer journey time. In the case of long-distance transport, a study on kids (8–10 kg) in a triple-decker trailer from the Netherlands to Spain (1400 km journey) investigated the effects of three loading densities (5, 7.5, or 10 animals/m^2) and found no significant differences in behaviour or physiological parameters of kids, although kids lost between 2.4 and 5.9% (0.2–0.5 kg) of their body weight (Hindle et al. 2013). Finally, administration of antioxidants to kids does not always have the expected effect compared to adult goats and even lambs. Saribey and Karaca (2018) supplied ascorbic acid in drinking water to lambs before a 3-h journey, finding that the adverse effects of transport stress were partly reduced in lambs, whereas in kids there was no clear effect.

3.2 Air and Sea Transport

Large volumes of goats are transported primarily by land, while intercontinental exports may necessitate air and marine transport, which is usually combined with land transport (Abbas et al. 2014). The growing demand for live goats in many Muslim countries has stimulated exports by sea. Shipping between the Horn of Africa (Djibouti, Ethiopia, Somalia, Eritrea) to Saudi Arabia and the United Arab Emirates moves at least five million goats and sheep annually (Musa et al. 2020). Romania, Spain, and Portugal are the EU countries that export the most goat-sheep (mixed journeys) by sea to Lebanon, Jordan, Israel, Libya, and Saudi Arabia (Boada-Saña et al. 2021). Australia was one of the world's largest exporters of breeding goats and feral goats (as part of feral goat depopulation programmes) by sea, although it now only exports them by air due to high mortality during the sea journey (Phillips 2019). These mortalities were especially due to the refusal of feral goats to consume pelleted feed, because they were accustomed to grazing. This inappetence and its interaction with other stressors (high temperatures, vessel movement and oscillation, crowding, and poor ventilation) during the sea journey resulted in decreased metabolic rate, weight loss, weakness, kidney damage, liver damage, and ultimately death (Phillips and Santurtun 2013). The most effective method was to keep goats in an assembly centre for 7–14 days before shipping and feed them straw and hay, gradually replacing it with pellets (Gherardi and Johnson 1994). Another welfare issue on sea journeys is ammonia saturation on cargo decks; the critical level of atmospheric ammonia in goats is considered to be 25 ppm, and strategies to reduce it are increased ventilation, bedding replacement, and feeding with lower protein diets containing acid salts (Costa et al. 2003). Goats are also subjected to long-distance truck journeys that may use roll-on/roll-off vessels for one or several short journeys. However, there are no studies that allow us to know the impact of the stress of both transports on physiology and behaviour in goats.

The volume of air transport in goats is usually marginal and has been used for the export of breeding animals and recently for exporting goats for slaughter (especially feral goats). Goats are usually transported in small numbers in plywood, metal, or plastic containers; goat crates are usually stacked in threes and are designed to retain excrement by means of a plastic sheet on the floor and walls; bedding may be provided, but some receiving countries do not allow this for safety reasons (Phillips 2019). Australian air transport standards permit the travel of kids heavier than 14 kg and weaned at least 14 days prior to the journey; adult horned goats should only be transported if their horns are not turned in, in order to prevent damage to the head or eyes, risks for other animals, and restricted access to feed or water. Tipping may be applied in some cases: for bucks, horns should be tipped within 2.5–5 cm from the tip (no lower than 2 cm diameter of the horn) and for does, less than 2 cm from the tip. Tipping should be done at least 7 days prior to transport or export (ASEL 2019). The largest exporter of goats by air is Australia: in 2021, it shipped 15,002 animals to Malaysia (53%), Philippines (24%), and China (21%), 58% of which were breeding animals and 42% were destined for slaughter (LiveCorp 2021).

Mortality rate is the most commonly used welfare indicator to assess the performance of live animal exports, where air transport tends to have lower associated mortality rates than sea transport. For example, in Australia in 2007 (the last year of ship-only goat transport) the reported mortality of goats transported by ship was 0.69% (179 animals out of a total of 25,000), whereas in 2021, when goats were transported only by air, the reported mortality was 0.01% (1 animal out of a total of 15,002). Australian standards for livestock exports require mortality investigation reports to be published for any flight if the goat mortality rate exceeds 2% (LiveCorp 2021). The main causes of mortality are ventilation problems or exposure to carbon dioxide and ammonia which contribute to respiratory diseases, asphyxiation, and heat stress (Fleming et al. 2020). Collins et al. (2020) identified other risks, such as pre-flight road journey, high density confinement in cages, food and water deprivation time, especially when flights are delayed, all imposed for varying durations from 12 to 48 h. However, the effects of journeys by air and road may have long-term effects. For example, in a long-distance mixed international shipment (road 20 h—air 8 h—road 12 h) of Boer kids between Australia and Indonesia, Shahrom and Zamri-Saad (2012) found that mortality attributed to ship fever was concentrated in the first 7 weeks after journey. The mortality rate started at 5% in the first week, increased to 35% in the third week, and gradually decreased to 3% in the seventh week.

4 Slaughter

Pre-slaughter management includes arrival—unloading from the truck, lairage and handling and moving of animals from lairage to the stunning/bleeding area. Handling, loading, transportation, and unloading at the abattoir adversely can affect the electrolyte balance of goats and result in respiratory alkalosis, dehydration, and muscular damage. At arrival, goats should be unloaded without delay and those showing signs of severe pain, signs of illness, or those unable to move independently should be inspected and a procedure for emergency slaughter should be applied without delay (EFSA 2021). Severely fatigued goats can be easily detected during unloading operations due to their inability to get out of the truck or down the ramp, in addition to showing tachypnoea with an abdominal component to breathing, muscle tremors, lameness, reluctance to move, and prostration (Chulayo and Muchenje 2013). Factors that can exacerbate the prevalence of fatigued animals are long journey duration, rough driving, position within the truck, very high or low densities, poor quality roads, and truck driver skills and individual predisposition (Miranda-de la Lama et al. 2022). To counteract the effects of fatigue and post-transport stress, a lairage period has been suggested. Lairage is a common commercial practice that can be defined as a period of time in which animals remain in the pens from unloading to the start of the transfer to the point of stunning or slaughter; in these pens, animals have access only to water (Liste et al. 2011). From a welfare point of view, animals should be slaughtered upon arrival at the slaughterhouse to avoid exposure to prolonged stress (Knowles 1998). However, lairage allows for the planning of slaughter and provides the time needed for veterinary inspection of live goats; additionally, it aids gastrointestinal emptying, thus limiting possible contamination of carcasses. It has been proposed that lairage allows animals to reduce dehydration and carcass weight loss, and to regenerate muscle glycogen levels, favouring the most affected animals (Liste et al. 2011). For example, Kannan et al. (2000) and Othman et al. (2021) agree that the positive effects of lairage are achieved at least after three hours of stay. However, lairage periods longer than 18 h lead to increased stress and decreased live weight (Kannan et al. 2002). But the positive effects of lairage could be affected by stressors such as social mixing, fasting, environmental conditions (temperature, humidity, noise), pre-slaughter incidents, and individual susceptibility to stress (Minka and Ayo 2013). To assess stress during lairage, the most important physiological indicators are plasma creatine kinase concentration, lactate dehydrogenase, glucose, cortisol, and packed cell volume (Ekiz et al. 2012).

4.1 Stunning and Bleeding

Stunning is any intentional action that causes painless loss of consciousness and sensibility, including any process that causes instantaneous death (EFSA 2021). In order for stunning and exsanguination to be administered accurately, animals must be restrained to keep them in the correct position. Restraint involves physical restraint and commonly social separation, the combination of both of which can result in behavioural responses such as escape attempts (jumping and scurrying), lying down, vocalising and in some cases aggression towards the handler (ramming and bumping) (Miranda-de la Lama et al. 2022). Therefore, in many slaughterhouses, animals are stunned in small groups in a small room, where the operator stuns animal by animal: this system, with calm herding and an experienced operator, avoids separation stress (EFSA 2021). On the contrary, if the restraint is inadequate or violent, it can cause injuries, haematomas, leg and horn fractures in goats. In industrial slaughterhouses specialising in small ruminants, restraint has been automated by installing V-shaped restraint belts that deliver goats in a continuous flow to the point of stunning, where they are dropped onto another conveyor and then shackled, hoisted, and bled (Small and Hewitt 2017). This system prevents animals from injury and prevents them from losing visual, auditory, and tactile contact with each other. Its effectiveness depends on adjusting the width of the conveyors and the inclination of the loading system to the shape and size of the goats (which have longer legs and narrower girth than sheep) to prevent the animals from falling or having the visual perception of the "cliff edge" (EFSA 2021). In goats (as in sheep), the stunning methods commonly employed at slaughter can be grouped into mechanical and electrical.

Mechanical stunning methods provoke concussion resulting in unconsciousness through the impact produced by penetrating or a non-penetrating captive bolt device (EFSA 2021). For kids, the use of non-penetrating bolt devices is recommended due to their high effectiveness (95–100%), provided that the specific firing position is applied, in the midline, between the ears with the chin tucked into the neck, along with a 0.22″ cartridge of a grain (Grist et al. 2018). In goats, two studies have shown that unconsciousness with these devices is effective at 28 J of calculated kinetic energy (Sutherland et al. 2016) and death is induced at 47 J (Grist et al. 2018). For adult goats (horned or hornless), either penetrating captive bolt devices or a non-penetrating captive bolt are effective in producing unconsciousness (Collins et al. 2017). The penetrating captive bolt uses a pointed bolt that is activated by a spring mechanism, or by pressurised air, or a blank cartridge (calibres 0.22, 0.25, and 9 mm). A recent study has confirmed that effective damage to the brainstem and key adjacent structures can be achieved by firing a penetrating bolt perpendicular to the skull over the identified anatomical site at the intersection of two lines, each drawn from the lateral canthus of one eye to the middle of the base of the opposite ear (Plummer et al. 2018). Signs of effective stunning with captive bolt devices are: immediate collapse, absence of righting reflex, absence of rhythmic breathing, absence of jaw muscle tension, absence of corneal reflex, absence of palpebral reflex, absence of eyeball rotation, presence of pupil dilation, absence of nystagmus, and presence of leg kicks (Gibson et al. 2012).

Electrical stunning, or electronarcosis, is commonly used in goats and sheep and is based on the application of a sufficient current through the brain to induce generalised epileptiform activity in the brain, so that the animal is immediately rendered unconscious (EFSA 2021). Goats are usually stunned using head-only (HO) and head-to-body (HB) electrical methods. In the case of HO stunning, the current is delivered by a scissor device with a pointed electrode placed on the side of the head or a hand piece in the form of a handle that holds the electrodes; the device often has a mechanism that wets the animal's head to improve conductivity. This is a reversible method (Lambooij 2014). In general, the recommended parameters for effective HO stunning of goats are a voltage between 70 and 250 V and a minimum current of 1.0 A for 2–4 s with 50 Hz (Sabow et al. 2018). Effective HO stunning must be followed by bleeding within 8 s (EFSA 2021, 2022). HB stunning is a combination of head-body electric current passage. In addition to causing widespread epileptiform activity in the brain, this method induces heart fibrillation or cardiac arrest, making it an irreversible technique (Farouk et al. 2014). According to the EU Regulation (European Commission 2009), a minimum current of 1.0 A (regardless of the type of device) must be applied to obtain effective electrical stunning in adult goats, whereas current intensities of 0.7 A are recommended for kids (Llonch et al. 2015). Sabow et al. (2016a) examined the influence of low and high frequency HB stunning (50 vs. 850 Hz) on bleeding efficiency in goats, finding that blood loss was higher in goats stunned at 850 Hz compared to 50 Hz. The efficiency of HO or HB electrical stunning with currents of 0.3, 0.5, and 0.7 A has also been compared, reporting immediate loss of consciousness with all three current intensities, similar to that produced with 1.0 A (Sabow et al. 2016a, b).

Bleeding is the final part of the slaughter procedure that is performed immediately after stunning and consists of severing the main blood vessels of the animal to allow blood to leave the body, resulting in death by cerebral anoxia (Miranda-de la Lama et al. 2022). Poorly performed bleeding has been associated with prolonged brain death, leading the animal to regain awareness and suffer pain, anxiety, and distress (Terlouw et al. 2016). Although there are no studies in goats that have shown the incidence of recovery of consciousness due to poor bleeding, this possibility cannot be ruled out (EFSA 2013). According to the Humane Slaughter Association (HSA 2000), specifically in goats, it is recommended to make an incision in the jugular groove at the base of the neck, directing the knife towards the entrance of the chest to cut the major blood vessels leaving the heart. As an alternative to this technique, a deep transverse cut can be made across the throat near the head to cut the four major blood vessels in the neck. Cutting into the neck is inappropriate if the stab or cut is too small to completely sever the carotid arteries and jugular veins, preventing rapid bleeding, and/or if bleeding is hindered by blood clotting in the stab wound (EFSA 2013).

Finally, monitoring animal welfare indicators at the slaughterhouse level allows valuable information on the conditions in which animals were raised, transported, and slaughtered to be obtained (Grandin 2017). In goats, specific indicators, which can be easily measured at the slaughterhouse, have been proposed to highlight problems during rearing: body condition, lameness, mammary lesions, presence of scars, horn condition (imperfect dehorning, asymmetrical antlers), abscesses, cleanliness and appearance of the coat, presence of ectoparasites, pallor of the ocular conjunctivae, oculo-nasal discharges, and diarrhoea (Mgasa and Arnbjerg 1993; Idahor 2013; Didarkhah et al. 2019). To highlight welfare problems during transport and pre-slaughter handling, the presence of non-ambulatory animals, dead animals, fractured animals (limbs or antlers), active wounds, and contusions have been proposed (Regassa et al. 2013; Pulido et al. 2019; Valkova et al. 2021).

4.2 Non-stunning Slaughter

In goats, non-stunning methods of slaughter are used in many societies around the world, where they represent important elements of rural culture and livelihoods (Miranda-de la Lama et al. 2022). Non-stunning, which may (or may not) be motivated and regulated by religious motives, consists of methods of abolishing life by bleeding conscious animals by cutting blood vessels in specific body regions. In general, two main categories of non-stunning slaughter are distinguished: religious ritual slaughter and traditional slaughter. Ritual slaughter is regulated by certifiers and authorities in denominational and non-denominational countries. The three most commonly used methods are shechitàh/kosher (Jewish), halal (Muslim), and jhatka (Sheik). The halal and kosher methods are based on elevating the body before neck cutting, or placing it manually on its side, or in an upright position, and executing a large clean cut in the upper part of the neck in its ventral projection affecting the trachea, oesophagus, carotid arteries, and jugular veins (Rosen 2004). From an animal welfare perspective, the main concerns towards these methods are related to fear and pain during physical restraint, pain caused by cutting and stimulation of nociceptors by the wound, dexterity associated with delayed loss of consciousness, and bronchial aspiration of blood (Gallo et al. 2022). If the cut is performed effectively, the animal usually loses consciousness within 10–15 s post-cutting, and the absence of signs of life should be ascertained within 30–40 s (Farouk et al. 2014), although it has been reported that between bleeding and loss of consciousness there may be an unacceptable interval of up to 1 min (Johnson et al. 2015). Additionally, electro-encephalographic changes in goats have shown that neck cutting causes noxious stimulation, regardless of whether the animals were conscious or unconscious (Sabow et al. 2016a, b).

The main difference between the two methods is that while for the halal ritual the animal must be alive, regardless of whether it is conscious or not, for the kosher ritual the animal must be conscious. Therefore, any type of stunning that makes the animal insensitive to pain before slaughter is not acceptable for the kosher rite (Bozzo et al. 2017). In the case of halal, non-stunning is preferred, although HO electrical stunning is widely accepted and practiced as it does not cause cardiac fibrillation and it is therefore reversible (Farouk et al. 2014). There are no studies on the current to be used on halal goats, although the Malaysian standard specifies a current between 0.70 and 1.0 A for 2–3 s (MS1500 2009). Conversely, HB electrical stunning is not generally accepted by halal certifiers, because it induces cardiac fibrillation and promotes epileptiform activity in the brain leading to irreversible loss of consciousness (Fuseini et al. 2022). Stunning by penetrating and non-penetrating captive bolts have been used in halal ritual, although both methods have been shown to produce fractures in the occipital and interparietal region causing cerebral haemorrhages (Collins et al. 2017). These injuries are causative of neuro-centric death, which is why most halal authorities do not approve mechanical stunning. In addition, some halal standards require the animal to remain "intact" after stunning, and carcasses are rejected if any damage to the skulls such as clefts, fractures, and holes are found after inspection (i.e. Malaysian halal standard MS1500/2009; Fuseini et al. 2022).

Jhatka is a method of killing an animal according to Hindu law to obtain meat for Hindu consumption, and is limited only to goats or sheep, because for Sikhs and Hindus cattle are sacred animals (Mortas et al. 2022). The method is a decapitation with a single axe or sabre blow below the nape of the neck, for which the goat's head is attached to a stable column and the hind legs are stretched in the opposite direction (Aghwan 2019). The blow causes sudden death because the spinal cord is severed and blood circulation to the brain is stopped, resulting in brain failure within seconds, which is why this method is considered to cause minimal pain and suffering to the animal compared to other ritual methods (Aghwan and Mac Regenstein 2019). From an animal welfare point of view, the success of this method is affected by the inexperience or incompetence of the slaughterer, poor sharpening or maintenance of the cutting instrument and violent handling; a behavioural indicator of efficiency is the absence of vocalisations during the strike (McDonough and Holoyda 2018).

Traditional slaughter contemplates those ancestral methods that are not part of a religious practice or motivation. They are not usually regulated and are related to self-consumption or for small-scale local markets (Sierra et al. 1997). The most commonly used traditional methods are throat slitting, Spanish "puntilla", and cardiac punctures with spears or daggers. Research on the effect of these methods on goat welfare is still very limited. Throat slitting is perhaps the most widespread method of traditional goat slaughter and there are variations among regions or countries in the way it is performed, such as the position of the animal for the procedure (lateral or ventral decubitus or dorsal or lifting the animal by its hind legs from a tree or post), the type of restraint of the animal (held by helpers or tied by both or all four legs), and the type of cut (clean cut of the ventral part of the neck or stabbing of the throat region and rotation of the knife) (Mdletshe et al. 2020; Qekwana et al. 2017). The puntilla method is a traditional stunning–slaughter method that consists of severing the spinal cord with a dagger or puntilla at the level of the atloid-occipital joint, between the occipital and the first cervical vertebra (Grandin 2010). For this, the goat must be standing and with the head in flexion, in order to favour vertebral joint separation. Once the cut is achieved, a general paralysis is produced, immediately followed by the bleeding by disgorging. Although this method is still used in many developing countries, the World Organization for Animal Health (OIE) considers it cruel and advises against its use, while the EU has banned it in its territory; both decisions are based on the fact that it has been demonstrated that animals are usually conscious during and after its application (Limon et al. 2012).

4.3 Feral Population Control

The widespread release of domestic goats for over 200 years has resulted in the establishment of numerous feral goat populations around the world, although their devastating effects have been widely documented in fragile ecosystems such as those of islands (O'Brien 1988). According to the New Zealand and Australian experience, population control of feral goats should be based on a combination of strategies such as: (1) monitoring and restricting migratory flows of herds, (2) group isolation and trapping, and (3) culling. The first strategy involves a series of techniques including helicopter surveys, foot transects, camera traps, scat counting transects, opportunistic community observations and especially the use of decoy or Judas goats. These goats are local or introduced animals that are usually sterilised (epididymectomy in males and tubal occlusion in females, to maintain sexual motivation), and fitted with a radiotelemetry or GPS collar to keep them located in the area (Campbell et al. 2005). This is an effective method for locating and eliminating individuals that are difficult to find or have low population densities. Currently, adult females are often used because they are more successful than males in locating and associating with other females, males and young, with which they form new groups relatively easily (Heriot et al. 2019).

The second strategy is aimed at isolating populations through fencing and trapping. Fencing is a non-lethal way to protect areas of high agricultural, livestock, or environmental value from feral goats. These fences are mesh (i.e. "hinge-hinge" or "ring-lock" fences), about 1.8 m high, and usually have a taut 30–50 cm apron extending from the base of the fence along the ground, although sometimes this apron is buried (Smith et al. 2020). Fences can be useful as a tactical technique either to create short-term manageable units during control, or to limit recolonisation during sustained control. However, such fencing is impractical on a large scale because of its cost and the need for constant maintenance. Furthermore, even in well-maintained fences, some goats will escape, so the usefulness of this method is limited in eradication campaigns (Parkes 1990). In the case of trapping, the technique is based on placing large traps near bodies of water or herding animals to concentrate them in large corrals; once the herds are captured, they are usually transported to feedlots or commercial slaughterhouses. Motorcycles, horses, helicopters, and dogs are often used for herding. When dogs are used, they must be muzzled to prevent them from injuring the animals (Cruz et al. 2009). It has been calculated that this technique usually has an effectiveness of no more than 40% of the populations (Parkes et al. 1996).

The third strategy is the elimination of goats with firearms fired from ground or aerial positions. Ground shooting is considered the most reliable humane technique for eliminating goats, although its quality is highly dependent on the skill and training of the shooter. Ideally it should be a single shot to the head (brain) or, if this fails, that to the chest (heart/lung); shots to other parts of the body are unacceptable from a welfare point of view. Wounded animals should immediately receive a second shot to the head. According to Sharp (2012a), the horns of adult goats make temporal (sideways) or back of the head shots the preferred shooting points; shots to the front of the head can be used in kids, but this method is not recommended for mature goats, as the brain is located well back in the skull. In the case of chest shooting, if the shot is in a frontal direction, it should be directed horizontally at the midpoint between the front legs and immediately below the base of the throat: this type of shot should only be used for animals in a "head high" position. In side shots, the bullet is fired so that it pierces the chest at a point behind the foreleg, slightly above and immediately behind the elbow joint (Smith 1999).

Aerial shooting consists of the use of small helicopters for locating, tracking, and shooting in large and difficult to access areas. The priority target is the chest area (heart and lungs), because it is an easier area to hit at long range and from a helicopter with unusual angles when shooting from above (Sharp 2012b). The shot in the chest does not induce immediate insensibility, causing pain and fear as part of an agonising process, so return flights and reshooting are recommended to accelerate the death process and minimise the probability of non-fatal injuries (Hampton et al. 2021). Another important point is that goat populations that have been decimated with this technique develop avoidance behaviour and reactivity to the presence, sound, and movement of the helicopters. This behavioural response affects the success rates of population control, going from 40% effectiveness in first-time herds to 21% in herds with a history of previous raids (Bayne et al. 2000).

5 Conclusions

This chapter has shown the different scenarios, consequences, and animal welfare risk factors related to husbandry, transport, and slaughter of goats around the world. It is important to note that goats play an important role in the livelihoods of farmers and smallholders around the world and that goat husbandry, handling, transport, and slaughter conditions are very heterogeneous compared to other livestock species. Therefore, the improvement of goat welfare must involve many stakeholders (scientists, farmers, middlemen, retailers, consumers) and consider the various sociocultural aspects of the region. A wide range of problems associated with stress during handling, husbandry practices, transport, and other pre-slaughter operations can be solved by controlling risk factors that affect the animals' fitness to adapt. During the application of these procedures, welfare assessment using appropriate indicators should be encouraged to monitor possible negative effects. Incorporating animal welfare at the centre of an operational quality programme has effects beyond the effective reduction of pain and suffering of animals during rearing, transport, and slaughter operations. It will also have an indirect impact on product quality and the competitiveness of the agri-food sector, such as traceability, safety, loss reduction, risk prevention, and consumer confidence.

References

Abbas B, Yousif MA, Nur HM (2014) Animal health constraints to livestock exports from the Horn of Africa. Sci Tech OIE Rev 33:711–721

Adcock SJJ (2021) Early life painful procedures: long-term consequences and implications for farm animal welfare. Front Anim Sci

Aghwan ZA (2019) Ritual and traditional slaughter practices for meat production. J Islamic Soc Econ Dev 4:224–230

Aghwan ZA, Mac Regenstein J (2019) Slaughter practices of different faiths in different countries. J Anim Sci Technol 6:111–121

Ahmed SA, Ahmed EA (2011) Behavioral responses of castrated buck kids at different ages by using different methods of castration. J Am Sci 7(5):200–209

Ajadi RA, Owanikin AO, Martins MM, Gazal OS (2012) Effect of epidural tramadol and lignocaine on physiological and behavioural changes in goats subjected to castration with a high tension band. N Z Vet J 60(6):344–348

Ajuda IG, Vieira A, Stilwell G (2014) Are there differences in dairy goats claws' temperature, before and after trimming? Int Symp Med Meas Appl 1–5

Ajuda IDGG, Battini M, Stilwell GT (2019) The role of claw deformation and claw size on goat lameness. Vet Anim Sci 8:100080

Ajuda I, Battini M, Mattiello S, Arcuri C, Stilwell G (2020) Evaluation of pain mitigation strategies in goat kids after cautery disbudding. Animals 10(2):277

Alam MK, Yoshino K (2022) The bargaining reality and safety issues of goat marketing at the marketplace in Bangladesh. Small Ruminant Res 210:106678

Alcalde MJ, Suárez MD, Rodero E, Álvarez R, Sáez MI, Martínez TF (2017) Effects of farm management practices and transport duration on stress response and meat quality traits of suckling goat kids. Animal 11(9):1626–1635

Ali KA (2020) The physical and sensory quality of goat meat under the influence of different factors. Int J Agric Sci Vet Med 8(3):45–49

Alvarez L, Gutierrez J (2010) A first description of the physiological and behavioural responses to disbudding in goat kids. Anim Welf 19:55–59

Alvarez L, De Luna JB, Gamboa D, Reyes M, Sánchez A, Terrazas A, Galindo F (2015) Cortisol and pain-related behavior in disbudded goat kids with and without cornual nerve block. Physiol Behav 138:58–61

Alvarez L, Adcock SJ, Tucker CB (2019) Sensitivity and wound healing after hot-iron disbudding in goat kids. J Dairy Sci 102(11):10152–10162

Animal Health Australia (2020) Australian industry welfare standards and guidelines for goats, Canberra, Australia

Anzuino K, Bell NJ, Bazeley KJ, Nicol CJ (2010) Assessment of welfare on 24 commercial UK dairy goat farms based on direct observations. Vet Rec 167(20):774–780

ASEL Review Technical Advisory Committee (2019) Review of the Australian standards for the export of livestock: air transport—final report. Department of Agriculture, Canberra, September CC BY 4.0

AWIN (2015) AWIN welfare assessment protocol for goats. European Animal Welfare Indicators Project (AWIN). Accessed 29 July 2022

Batchu P, Hazard T, Lee JH, Terrill TH, Kouakou B, Kannan G (2021) High-condensed tannin diet and transportation stress in goats: Effects on physiological responses, gut microbial counts and meat quality. Animals 11(10):2857

Battini M, Vieira A, Barbieri S, Ajuda I, Stilwell G, Mattiello S (2014) Invited review: Animal-based indicators for on-farm welfare assessment for dairy goats. J Dairy Sci 97(11):6625–6648

Battini M, Barbieri S, Vieira A, Stilwell G, Mattiello S (2016) Results of testing the prototype of the AWIN welfare assessment protocol for dairy goats in 30 intensive farms in Northern Italy. Ital J Anim Sci 15(2):283–293

Bayne P, Harden B, Pines K, Taylor U (2000) Controlling feral goats by shooting from a helicopter with and without the assistance of ground-based spotters. Wildl Res 27(5):517–523

Boada-Saña M, Kulikowska K et al (2021) Research for ANIT Committee—Animal welfare on sea vessels and criteria for approval of livestock authorisation. European Parliament, Policy Department for Structural and Cohesion Policies, Brussels

Bozzo G, Di Pinto A, Bonerba E, Ceci E, Mottola A, Roma R et al (2017) Kosher slaughter paradigms: Evaluation of slaughterhouse inspection procedures. Meat science 128:30–33

Browning H, Veit W (2020) Is humane slaughter possible? Animals 10(5):799

Caja G, Ghirardi JJ, Hernández-Jover M, Garín D (2004) Diversity of animal identification techniques: From 'fire age' to 'electronic age'. ICAR Tech Ser 9:21–39

Campbell KJ, Baxter GS, Murray PJ, Coblentz BE, Donlan CJ, Carrion V (2005) Increasing the efficacy of Judas goats by sterilisation and pregnancy termination. Wildl Res 32(8):737–743

Can E, Vieira A, Battini M, Mattiello S, Stilwell G (2016) On-farm welfare assessment of dairy goat farms using animal-based indicators: the example of 30 commercial farms in Portugal. Acta Agric Scand A Anim Sci 66(1):43–55

Chulayo AY, Muchenje V (2013) The effects of pre-slaughter stress and season on the activity of plasma creatine kinase and mutton quality from different sheep breeds slaughtered at a smallholder abattoir. Asian Australas J Anim Sci 26(12):1762

Cockram MS (2019) Fitness of animals for transport to slaughter. Can Vet J 60(4):423

Cockram MS, Velarde A (2022) Sheep. In: Faucitano L (ed) Preslaughter handling of livestock, 1st edn. Wageningen Academic, Wageningen, The Netherlands, pp 267–310

Collins SL, Caldwell M, Hecht S, Whitlock BK (2017) Comparison of penetrating and nonpenetrating captive bolt methods in horned goats. Am J Vet Res 78(2):151–157

Collins T, Stockman C, Hampton JO, Barnes A (2020) Identifying animal welfare impacts of livestock air transport. Aust Vet J 98(5):197–199

Costa N, Accioly J, Cake M (2003) Determining critical atmospheric ammonia levels for cattle, sheep and goats-a. Meat & Livestock Australia, Brisbane

Cruz F, Carrion V, Campbell KJ, Lavoie C, Donlan CJ (2009) Bio-economics of large-scale eradication of feral goats from Santiago Island, Galapagos. J Wildl Manag 73(2):191–200

Das KS, Srivastava BB, Das N (2001) Standing orientation and behaviour of goats during short-haul road transportation. Small Ruminant Research 41(1):91–94

Dawson LJ (2010) Preferred management practices. In: Solaiman SG (ed) Goat science and production. Blackwell, Ames, IA

Deeming LE, Beausoleil NJ, Stafford KJ, Webster JR, Zobel G (2018) The development of a reliable 5-point gait scoring system for use in dairy goats. J Dairy Sci 101(5):4491–4497

Deters EL, Hansen SL (2020) Invited Review: Linking road transportation with oxidative stress in cattle and other species. Appl Anim Sci 36(2):183–200

Didarkhah M, Vatandoost M, Dirandeh E, Dadashpour Davachi N (2019) Characterization and pattern of culling in goats. Arch Razi Inst 74(4):441–446

EFSA Panel on Animal Health and Welfare (AHAW) (2013) Scientific opinion on monitoring procedures at slaughterhouses for sheep and goats. EFSA J 11(12):3522

EFSA Panel on Animal Health and Welfare (AHAW), Nielsen SS, Alvarez J, Bicout DJ, Calistri P, Canali E et al (2021) Welfare of sheep and goats at slaughter. EFSA J 19(11):e06882

EFSA Panel on Animal Health and Welfare (AHAW), Nielsen SS, Alvarez J, Bicout DJ, Calistri P, Canali E et al (2022) Welfare of small ruminants during transport. EFSA J 20(9):e07404

Ekiz B, Ekiz EE, Kocak O, Yalcintan H, Yilmaz A (2012) Effect of pre-slaughter management regarding transportation and time in lairage on certain stress parameters, carcass and meat quality characteristics in Kivircik lambs. Meat Sci 90(4):967–976

Estévez-Moreno LX, Miranda-de la Lama GC, Miguel-Pacheco GG (2022) Consumer attitudes towards farm animal welfare in Argentina, Chile, Colombia, Ecuador, Peru and Bolivia: A segmentation-based study. Meat Sci 187:108747

European Commission (2005) Council Regulation (EC) No 1/2005 of 22 December 2004 on the protection of animals during transport and related operations and amending Directives 64/432/EEC and 93/119/EC and Regulation (EC) No 1255/97

European Union, Council Regulation (EC) (2009). No 1099/2009 of 24 September on the protection of animals at the time of killing. Official Journal of the European Union, 303, 1–30., n.d

Farouk MM, Al-Mazeedi HM, Sabow AB, Bekhit AED, Adeyemi KD, Sazili AQ, Ghani A (2014) Halal and kosher slaughter methods and meat quality: A review. Meat Science 98(3):505–519

Fazio F, Casella S, Giannetto C, Giudice E, Piccione G (2016) Erythrocyte osmotic fragility in response to a short road transport in cattle, horses, and goats. Journal of Veterinary Behavior 12:82–84

Fitzpatrick J, Scott M, Nolan A (2006) Assessment of pain and welfare in sheep. Small Ruminant Res 62(1–2):55–61

Fleming PA, Wickham SL, Dunston-Clarke EJ, Willis RS, Barnes AL, Miller DW, Collins T (2020) Review of livestock welfare indicators relevant for the Australian live export industry. Animals 10(7):1236

Fuseini A, Teye M, Lever J (2022) An update on halal slaughter: Current methods and ongoing research on halal meat production techniques and their implications for animal welfare. Anim Welf 31(2):269–276

Gallo C, Schwartzkopf-Genswein K, Gibson T (2022) Chapter 2: Cattle. In: Faucitano L (ed) Preslaughter handling and slaughter of meat animals. Wageningen Academic, Wageningen, pp 187–193

Gherardi SG, Johnson TJ (1994) Period of lot-feeding of feral goats before live export by ship. In: Proceedings of the Australian Society of Animal Production, Perth, Australia, pp 194–197

Gibson TJ, Ridler AL, Lamb CR, Williams A, Giles S, Gregory NG (2012) Preliminary evaluation of the effectiveness of captive-bolt guns as a killing method without exsanguination for horned and unhorned sheep. Anim Welf 21(S2):35–42

Grandin T (2010) Auditing animal welfare at slaughter plants. Meat Sci 86(1):56–65

Grandin T (2017) On-farm conditions that compromise animal welfare that can be monitored at the slaughter plant. Meat Sci 132:52–58

Grist A, Lines JA, Knowles TG, Mason CW, Wotton SB (2018) Use of a non-penetrating captive bolt for euthanasia of neonate goats. Animals 8(4):58

Gupta D, Kashyap G, Ashutosh M, Ashutosh P, M., Para, I., ... & Patel, B. (2021) Minerals response to transportation stress at different flocking densities in hot humid and winter seasons and the assessment of ameliorative effect of pretreatment with vitamin C, electrolyte and jaggery in goats. Biological Rhythm Research 52(2):171–185

Häger C, Biernot S, Buettner M, Glage S, Keubler LM, Held N et al (2017) The Sheep Grimace Scale as an indicator of post-operative distress and pain in laboratory sheep. PLoS One 12(4):e0175839

Hampton JO, Bengsen AJ, Pople A, Brennan M, Leeson M, Forsyth DM (2021) Animal welfare outcomes of helicopter-based shooting of deer in Australia. Wildl Res 49(3):264–273

Han YG, Wu JY, Na RS, Si WJ, Han YQ, Zeng Y et al (2020) Effects of novel oral and plasmid-injected KISS1 DNA. Vaccines on immunocastration and in inhibiting mutton odour of male goats. Pak J Zool 52(4):1603–1606

Hartnack AK, Jordan ME, Roussel AJ (2018) Complications associated with surgical dehorning in goats: A retrospective study of 239 cases. Vet Surg 47(2):188–192

Hempstead MN, Waas JR, Stewart M, Cave VM, Sutherland MA (2017) Behavioural response of dairy goat kids to cautery disbudding. Appl Anim Behav Sci 194:42–47

Hempstead MN, Waas JR, Stewart M, Dowling SK, Cave VM, Lowe GL, Sutherland MA (2018a) Effect of isoflurane alone or in combination with meloxicam on the behavior and physiology of goat kids following cautery disbudding. J Dairy Sci 101:3193–3204

Hempstead MN, Waas JR, Stewart M, Cave VM, Sutherland MA (2018b) Evaluation of alternatives to cautery disbudding of dairy goat kids using physiological measures of immediate and longer-term pain. J Dairy Sci 101(6):5374–5387

Hempstead MN, Lindquist TM, Shearer JK, Shearer LC, Sutherland MA, Plummer PJ (2020) Acute cortisol and behavior of dairy goat kids administered local anesthesia, topical anesthesia or systemic analgesia prior to cautery disbudding. Physiol Behav 222:112942

Hempstead MN, Lindquist TM, Shearer JK, Shearer LC, Plummer PJ (2021) Health and welfare survey of 30 dairy goat farms in the Midwestern United States. Animals 11(7):2007
Heriot S, Asher J, Williams MR, Moro D (2019) The eradication of ungulates (sheep and goats) from Dirk Hartog Island, Shark Bay World Heritage Area, Australia. Biol Invasions 21(5):1789–1805
Hill NP, Murphy PE, Nelson AJ, Mouttotou N, Green LE, Morgan KL (1997) Lameness and foot lesions in adult British dairy goats. Vet Rec 141(16):412–416
Hindle VA, Reimert HGM, Van der Werf JTN, Lambooij E (2013) Loading density and welfare of goat kids during long distance road transport. Animal Welfare 22(3):345–356
HSA (Humane Slaughter Association) (2000) Electrical stunning of red meat animals, guidance notes no 4, guidance on the use of electrical stunning equipment including electrical theory, restraint, effective stunning, bleeding and safety. Wheathampstead, UK, HSA
Hull BL (1995) Dehorning the adult goat. Vet Clin N Am Food Anim Pract 11(1):183–185
Hussein NJ, Hidayet HM (2019) Changes in ear postures of kid goats in response to ear tagging. Basrah J Agric Sci 32(1):25–33
Hu W, Ye T, Yang Y, Liu B, Zheng W (2020) Effects of transport stress on pathological injury and expression of main heat shock proteins in the caprine stomach. BMC veterinary research 16:1–10
Ibrahim A, Mahmoud UT, Khalil NSA, Hussein HA, Ali MM (2018) A pilot study on surgical trimming impact on severely overgrown claws in sheep: Behavioral, physiological, and ruminal function aspects. J Vet Behav 23:66–75
Idahor KO (2013) Sheep and goats slaughtered at Keffi abattoir: health status, carcass yield and foetal deaths. J Anim Sci Adv 3(6):276–283
Johnson CB, Mellor DJ, Hemsworth PH, Fisher AD (2015) A scientific comment on the welfare of domesticated ruminants slaughtered without stunning. N Z Vet J 63(1):58–65
Kalmar B, Greensmith L (2009) Induction of heat shock proteins for protection against oxidative stress. Adv Drug Deliv Rev 61(4):310–318
Kannan G, Terrill TH, Kouakou B, Gazal OS, Gelaye S, Amoah EA, Samake S (2000) Transportation of goats: effects on physiological stress responses and live weight loss. J Anim Sci 78(6):1450–1457
Kannan G, Terrill TH, Kouakou B, Gelaye S, Amoah EA (2002) Simulated preslaughter holding and isolation effects on stress responses and live weight shrinkage in meat goats. J Anim Sci 80(7):1771–1780
Kannan G, Kouakou B, Terrill TH, Gelaye S (2003) Endocrine, blood metabolite, and meat quality changes in goats as influenced by short-term, preslaughter stress. J Anim Sci 81(6):1499–1507
Kannan G, Terrill TH, Kouakou B, Galipalli S (2007) Blood metabolite changes and live weight loss following brown seaweed extract supplementation in goats subjected to stress. Small Rumin Res 73(1–3):228–234
Karabasanavar N, Madhavaprasad CB, Manjunatha SS, Bagalkote PS, Sajjan SA, Sundareshan S (2016) Shipping fever outbreak in goats: an opportunistic association of Mannheimia haemolytica and Pasteurella multocida. Indian Vet J 93(6):16–19
Karademir U, Akin I, Erdogan H, Ural K, Asici GSE (2016) Effect of Ketoprofen on acute phase protein concentrations in goats undergoing castration. BMC Veterinary Research 12:1–7
Kats LB, Dill LM (1998) The scent of death: chemosensory assessment of predation risk by prey animals. Écoscience 5:361–394
Keating SC, Thomas AA, Flecknell PA, Leach MC (2012) Evaluation of EMLA cream for preventing pain during tattooing of rabbits: changes in physiological, behavioural and facial expression responses. PLoS One 7:1–11
Knowles TG (1998) A review of the road transport of slaughter sheep. Vet Rec 143(8):212–219
Lambooij B (2014) Stunning: Electrical stunning. In: Encyclopedia of meat sciences. Elsevier, pp 407–412
Limon G, Guitian J, Gregory NG (2012) A review of the humaneness of puntilla as a slaughter method. Anim Welf 21(S2):3–8
Liste G, Miranda-de La Lama GC, Campo MM, Villarroel M, Muela E, María GA (2011) Effect of lairage on lamb welfare and meat quality. Anim Prod Sci 51(10):952–958

LiveCorp (2021) 2020–2021 Annual report. Australian Livestock Export Corporation Limited, North Sydney, Australia

Llonch P, Rodríguez P, Casal N, Carreras R, Muñoz I, Dalmau A, Velarde A (2015) Electrical stunning effectiveness with current levels lower than 1 A in lambs and kid goats. Research in veterinary science 98:154–161

Lomax S, Witenden E, Windsor P, White P (2017) Effect of topical vapocoolant spray on perioperative pain response of unweaned calves to ear tagging and ear notching. Vet Anaesth Analg 44(1):163–172

Lomax S, Hall E, Oehlers L, White P (2018) Topical vapocoolant spray reduces nociceptive response to ear notching in neonatal piglets. Vet Anaesth Analg 45(3):366–373

Looije M, Moffat LA (2020) The welfare of male goat-kids: An investigation into the welfare of male-kids on the farm, during transport and slaughter in The Netherlands. Eyes on Animals. Report. https://www.eyesonanimals.com/wp-content/uploads/2020/12/2020-EonA-summary-report-Welfare-of-male-goat-kids.pdf

Lou ME (2020) The grimace scale as a method of acute pain assessment for neonatal livestock. MSc. dissertation, University of Minnesota, USA

Lowie T, Callens J, Maris J, Ribbens S, Pardon B (2021) Decision tree analysis for pathogen identification based on circumstantial factors in outbreaks of bovine respiratory disease in calves. Prev Vet Med 196:105469

Martínez GM, Suárez VH, Hempstead M, Alfaro E, Caro LC, Alfaro J (2023) Effect of flunixin meglumine on pain-related behaviours following cautery disbudding in dairy goat kids. Vet Res Commun 47(1):159–165

Mbuku SM, Kosgey IS, Kahi AK (2010) Identification systems and selection criteria for small ruminants among pastoralist communities in northern Kenya: prospects for a breeding programme. Tropl Anim Health Prod 42:1487–1492

McDonough SP, Holoyda B (2018) Ritualistic animal killing. In: Brooks JW (ed) Veterinary forensic pathology. Springer, Cham, pp 129–138

Mdletshe ZM, Marufu MC, Chimonyo M (2020) Effect of indigenous slaughter methods on the behavioural response, bleeding efficiency and cardiac arrest of Nguni goats. Animals 10(2):247

Meijer E, Goerlich VC, Brom RVD, Giersberg MF, Arndt SS, Rodenburg TB (2021) Perspectives for buck kids in dairy goat farming. Front Vet Sci 8:662102

Mellor DJ, Molony V, Robertson IS (1991) Effects of castration on behaviour and plasma cortisol concentrations in young lambs, kids and calves. Res Vet Sci 51(2):149–154

Merskey H (1994) Part III pain terms, a current list with definitions and notes on usage. Classification of chronic pain. Descriptions of chronic pain syndromes and definitions of pain terms. IASP, Washington, DC, pp 207–214

Mersmann D, Schmied-Wagner C, Waiblinger S (2022) The relationships between attitudes, personal characteristics and behaviour of stockpeople on dairy goat farms. Anim Welf 31(4):529–544

Mgasa MN, Arnbjerg J (1993) Occurrence of lameness and digital lesions in Tanzanian goats. Small Rumin Res 10(1):55–62

Mihsler-Kirsch L, Wagner H, Failing K, Wehrend A (2021) Downregulation of testicular function in the goat by altrenogest. BMC Vet Res 17(1):1–7

Millman ST (2013) Behavioral responses of cattle to pain and implications for diagnosis, management, and animal welfare. Vet Clin Food Anim Pract 29(1):47–58

Mills G (2018) Pain relief must be given for goat disbudding. Vet Rec 183(9):283

Minka NS, Ayo JO (2013) Physiological and behavioral responses of goats to 12-hour road transportation, lairage and grazing periods, and the modulatory role of ascorbic acid. J Vet Behav 8(5):349–356

Minka NS, Ayo JO (2014) Effects of transportation positions and orientations on muscular damage of goats transported by road for 12 h during the hot-dry conditions. J Vet Med Anim Health 6(3):75–82

Miranda-de la Lama GC (2019) Goat handling and transport. In: Grandin T (ed) Livestock handling and transport. CABI, Wallingford, UK, pp 271–289

Miranda-de La Lama GC, Villarroel M, Liste G, Escós J, María GA (2010) Critical points in the pre-slaughter logistic chain of lambs in Spain that may compromise the animal's welfare. Small Rumin Res 90(1–3):174–178

Miranda-De La Lama GC, Villarroel M, María GA (2014) Livestock transport from the perspective of the pre-slaughter logistic chain: a review. Meat Science 98(1):9–20

Miranda-de La Lama GC, Estévez-Moreno LX, Sepulveda WS, Estrada-Chavero MC, Rayas-Amor AA, Villarroel M, María GA (2017) Mexican consumers' perceptions and attitudes towards farm animal welfare and willingness to pay for welfare friendly meat products. Meat Sci 125:106–113

Miranda-de La Lama GC, Rodríguez-Palomares M, Cruz-Monterrosa RG, Rayas-Amor AA, Pinheiro RSB, Galindo FM, Villarroel M (2018) Long-distance transport of hair lambs: effect of location in pot-belly trailers on thermo-physiology, welfare and meat quality. Tropl Anim Health Prod 50:327–336

Miranda-de la Lama GC, Mamani-Linares W, Estévez-Moreno LX (2022) Chapter 9: Species destined for non-traditional meat production: 2. Goats and South American domestic camelids. In: Faucitano L (ed) Preslaughter handling and slaughter of meat animals. Wageningen Academic, Wageningen, pp 349–389

Mobini S (1991) Cosmetic dehorning of adult goats. Small Rumin Res 5(1–2):187–191

Molony V, Kent JE (1997) Assessment of acute pain in farm animals using behavioral and physiological measurements. J Anim Sci 75(1):266–272

Mortas M, Awad N, Ayvaz H (2022) Adulteration detection technologies used for halal/kosher food products: an overview. Discov Food 2(1):1–23

Moussu G, Dollar JA (1905) Diseases of cattle, sheep, goats and swine. WR Jenkins, New York, p 22

Moyo B, Oyedemi S, Masika PJ, Muchenje V (2012) Polyphenolic content and antioxidant properties of Moringa oleifera leaf extracts and enzymatic activity of liver from goats supplemented with Moringa oleifera leaves/sunflower seed cake. Meat Sci 91(4):441–447

MS1500 Malaysian Standard (2009) Halal food—Production, preparation, handling and storage—General guideline. Department of Standards Malaysia, Cyberjaya, pp 1–13

Musa AM, Wasonga OV, Mtimet N (2020) Factors influencing livestock export in Somaliland's terminal markets. Pastoralism 10(1):1–13

Naldurtiker A, Batchu P, Kouakou B, Terrill TH, McCommon GW, Kannan G (2023) Differential gene expression analysis using RNA-seq in the blood of goats exposed to transportation stress. Sci Rep 13(1):1984

Nfor ON, Chan JPW, Kere M, Peh HC (2016) Disbudding pain: The benefits of disbudding goat kids with dexmedetomidine hydrochloride. Small Rumin Res 139:60–66

Nwe TM, Tsukahara Y, Manda M, Watanabe S (1994) Initiation of the simulate and real transportation stress in the goat. Anim Sci Technol 65(11):1008–1017

Nwe TM, Hori E, Manda M, Watanabe S (1996) Significance of catecholamines and cortisol levels in blood during transportation stress in goats. Small Rumin Res 20(2):129–135

O'Brien PH (1988) Feral goat social organization: a review and comparative analysis. Appl Anim Behav Sci 21(3):209–221

Orihuela A, Ungerfeld R (2020) Prácticas zootécnicas dolorosas. Evaluación y alternativas para el bienestar animal. Colegio de Postgraduados y Universidad Autónoma del Estado de Morelos, Mexico

Othman AH, Goh YM, Mohamed Mustapha N, Raghazli R, Kaka U, Imlan JC et al (2021) Physiological and electroencephalographic changes in goats subjected to transportation, lairage, and slaughter. Anim Sci J 92(1):e13610

Padalino B, Tullio D, Cannone S, Bozzo G (2018) Road transport of farm animals: Mortality, morbidity, species and country of origin at a Southern Italian control post. Animals 8(9):155

Padalino B, Barrasso R, Tullio D, Zappaterra M, Nanni Costa L, Bozzo G (2020) Protection of Animals during transport: analysis of the infringements reported from 2009 to 2013 during on-road inspections in Italy. Animals 10(2):356

Panchal P, Dalvi S, Ambade R, Dighe D, Deshmukh R (2022) Serum biochemical changes during transportation stress in adult goats. Indian J Vet Sci Biotechnol 18(1):119–121

Parkes JP (1990) Feral goat control in New Zealand. Biol Conserv 54(4):335–348

Parkes J, Henzell R, Pickles G (1996) Managing vertebrate pests: feral goats. Australian Government Publishing Service, Canberra, p 51

Peris B, de Heredia IB, Ruiz R (2016) Animal welfare legislative regulations in small ruminants farms in Spain. Small Ruminant Res 142:58–60

Phillips C (2019) Transport of cattle, sheep and other livestock by sea and air. Livestock handling and transport, p 427

Phillips CJ, Santurtun E (2013) The welfare of livestock transported by ship. Vet J 196(3):309–314

Plummer PJ, Schleining JA (2013) Assessment and management of pain in small ruminants and camelids. Vet Clin Food Anim Pract 29(1):185–208

Plummer PJ, Shearer JK, Kleinhenz KE, Shearer LC (2018) Determination of anatomic landmarks for optimal placement in captive-bolt euthanasia of goats. Am J Vet Res 79(3):276–281

Plummer PJ, Hempstead MN, Shearer JK, Lindquist TM (2021) Evaluating the welfare of small ruminants: practical management advice. Vet Clin North Am Food Anim Pract 37(1):33–54

Polycarp TN, Obukowho EB, Yusoff SM (2016) Changes in haematological parameters and oxidative stress response of goats subjected to road transport stress in a hot humid tropical environment. Comp Clin Pathol 25(2):285–293

Prado VCM, Bassoto Filho J, Yasuoka MM, Ollhoff RD, Gallo SB, Birgel Junior EH (2022) Effect of trimming of overgrown and deformed claws in goats on morphometric measurements. Vet Res Commun 46(3):947–954

Pulido MA, Estévez-Moreno LX, Villarroel M, Mariezcurrena-Berasain MA, Miranda-de la Lama GC (2019) Transporters knowledge toward preslaughter logistic chain and occupational risks in Mexico: An integrative view with implications on sheep welfare. J Vet Behav 33:114–120

Qekwana DN, McCrindle CM, Oguttu JW, Grace D, Cenci-Goga BT (2017) Assessment of welfare issues during traditional slaughter of goats in Pretoria, South Africa. J Appl Anim Welf Sci 20(1):34–41

Rault JL, Lay DC Jr, Marchant-Forde JN (2011) Castration induced pain in pigs and other livestock. Appl Anim Behav Sci 135(3):214–225

Regassa A, Moje N, Megersa B, Beyene D, Sheferaw D, Debela E et al (2013) Major causes of organs and carcass condemnation in small ruminants slaughtered at Luna Export Abattoir, Oromia Regional State, Ethiopia. Prev Vet Med 110(2):139–148

Robbins JA, Franks B, Weary DM, Von Keyserlingk MA (2016) Awareness of ag-gag laws erodes trust in farmers and increases support for animal welfare regulations. Food Policy 61:121–125

Rosen SD (2004) Physiological insights into shechita. Veterinary Record 154(24):759–765

Sabow AB, Zulkifli I, Goh YM, Ab Kadir MZA, Kaka U, Imlan JC et al (2016a) Bleeding efficiency, microbiological quality and oxidative stability of meat from goats subjected to slaughter without stunning in comparison with different methods of pre-slaughter electrical stunning. PLoS One 11(4):e0152661

Sabow AB, Goh YM, Zulkifli I, Sazili AQ, Kaka U, Ab Kadi MZA et al (2016b) Blood parameters and electroencephalographic responses of goats to slaughter without stunning. Meat Sci 121:148–155

Sabow AB, Goh YM, Zulkifli I, Ab Kadir MZ, Kaka U, Adeyemi KD et al (2018) Electroencephalographic and blood parameters changes in anaesthetised goats subjected to slaughter without stunning and slaughter following different electrical stunning methods. Animal Production Science 59(5):849–860

Sailer LM, Holinger M, Burla JB, Wechsler B, Zanolari P, Friedli K (2021) Influence of housing and management on claw health in Swiss dairy goats. Animals 11(7):1873

Salas MÁS, Mondragón-Ancelmo J, Badillo MDRJ, Licea GR, Napolitano F (2021) Assessing dairy goat welfare in intensive or semi-intensive farming conditions in Mexico. J Dairy Sci 104(5):6175–6184

Sarıbey M, Karaca S (2018) Effects of pre-slaughter ascorbic acid administration on some physiological stress response and meat quality traits of lambs and kids subjected to road transport. Anim Prod Sci 59(5):954–964

Schoiswohl J, Stanitznig A, Sigmund M, Kneissl S, Thaller D, Frahm S et al (2021) Comparison of alternative disbudding methods with hot-iron dehorning of goat kids. J Vet Behav 46:31–39
Sebsibe A (2008) Sheep and goat meat characteristics and quality. Sheep and goat production handbook for Ethiopia. Ethiopian Sheep and Goats Productivity Improvement Program (ESGPIP), Addis Ababa, Ethiopia, pp 323–328
Sharp T (2012a) Ground shooting of feral goats. Standard Operating Procedure. PestSmart website. https://pestsmart.org.au/toolkit-resource/ground-shooting-of-feral-goats. Accessed 18 May 2022
Sharp T (2012b) Aerial shooting of feral goats. Standard Operating Procedure. PestSmart website. https://pestsmart.org.au/toolkit-resource/aerial-shooting-of-feral-goats. Accessed 18 May 2022
Shahrom MS, Zamri-Saad M (2012) A retrospective study on post-arrival mortality rate of Australian Boer goats in a breeder farm in Malaysia. Pertanika Journal of Tropical Agricultural Science 35(4):737–742
Sierra A, Molina A, Delgado J, Hernández J, Rivera M (1997) Zootechnical description of the creole goat of the Oaxaca region (Mexico). Anim Genet Resour 21:61–70
Simon R, Drögemüller C, Lühken G (2022) The complex and diverse genetic architecture of the absence of horns (Polledness) in domestic ruminants, including goats and sheep. Genes 13(5):832
Small A, Hewitt L (2017) Transport and pre-slaughter management. In: Ferguson DM, Lee C, Fisher A (eds) Advances in sheep welfare. Woodhead, Kidlington
Smith G (1999) A guide to hunting and shooting in Australia. Regency, South Australia
Smith MC, Sherman DM (2009) Goat medicine. Wiley, Hoboken
Smith D, Waddell K, Allen BL (2020) Expansion of vertebrate pest exclusion fencing and its potential benefits for threatened fauna recovery in Australia. Animals 10(9):1550
Stanford K, Stitt J, Kellar JA, McAllister TA (2001) Traceability in cattle and small ruminants in Canada. Revue Scientifique et Technique-Office International des Epizooties 20(2):510–522
Still-Brooks KM (2019) Field euthanasia techniques for small ruminants. In: Proceedings of the annual conference. American Association of Bovine Practitioners, pp 187–191
Still-Brooks KM, Hempstead MN, Anderson JL, Parsons RL, Sutherland MA, Plummer PJ, Millman ST (2021) Characterization of efficacy and animal safety across four caprine disbudding methodologies. Animals 11(2):430
Sutherland MA, Watson TJ, Johnson CB, Millman ST (2016) Evaluation of the efficacy of a non-penetrating captive bolt to euthanase neonatal goats up to 48 hours of age. Anim Welf 25(4):471–479
Taunde PA, Argenta FF, Bianchi RM, Cecco BSD, Vielmo A, Lopes BC et al (2019) Mannheimia haemolytica pleuropneumonia in goats associated with shipping stress. Ciência Rural 49(01):1–6
Terlouw C, Bourguet C, Deiss V (2016) Consciousness, unconsciousness and death in the context of slaughter. Part II Evaluation methods. Meat Sci 118:147–156
Underwood W, Anthony R (2020) AVMA guidelines for the euthanasia of animals: 2020 edition. Retrieved on March, 2013(30), 2020-1
Valkova L, Vecerek V, Voslarova E, Kaluza M, Takacova D (2021) The welfare of cattle, sheep, goats and pigs from the perspective of traumatic injuries detected at slaughterhouse postmortem inspection. Animals 11(5):1406
Van Donkersgoed J, Dixon S, Brand G, VanderKop M (1997) A survey of injection site lesions in fed cattle in Canada. Can Vet J 38(12):767
Vignesh P, Babu NR, Abraham RJ, Gnanaraj T, Govind V, Rao AV (2018) Assessment of losses in sheep and goat during pre-slaughter transportation. J Anim Res 8(3):473–480
Vincent B (2018) Farming meat goats: breeding, production and marketing. CSIRO, Collingwood
Viscardi AV, Hunniford M, Lawlis P, Leach M, Turner PV (2017) Development of a piglet grimace scale to evaluate piglet pain using facial expressions following castration and tail docking: a pilot study. Front Vet Sci 4:51

Wassie T, Zeng F, Jiang X, Liu G, Kasimu H, Ling S, Girmay S (2020) Effect of Kisspeptin-54 immunization on performance, carcass characteristics, meat quality and safety of Yiling goats. Meat Sci 166:108139

Winter AC (2011) Treatment and control of hoof disorders in sheep and goats. Vet Clin Food Anim Pract 27(1):187–192

Xu K, Yang K, Yang Y, Wu W, Zhou C (2023) Supplementation of feed and water after long-duration road transportation: The effects on welfare and rumen fermentation in goats. Front Vet Sci 10:1135666

Zebari HM, Hidayet HM, Al-Naqshabendy AA, Hussein NJ, Kakarash NA (2021) Pain caused by ear tagging in kids of native black goats. Sci J Univ Zakho 9(1):15–19

Zeng FM, Yi DING, Wassie T, Jing HJ, Ahmed S, Liu GQ, Jiang XP (2022) Recent advances in immunocastration in sheep and goat and its animal welfare benefits: A review. J Integr Agric 21(2):299–309

Zheng W, Liu B, Hu W, Cui Y (2021) Effects of transport stress on pathological injury and main heat shock protein expression in the respiratory system of goats. J Anim Physiol Anim Nutr 105(1):1–13

Zulkifli I, Norbaiyah B, Cheah YW, Soleimani AF, Sazili AQ, Goh YM, Rajion MA (2010) A note on heat shock protein 70 expression in goats subjected to road transportation under hot, humid tropical conditions. Animal 4(6):973–976

The Welfare of Goats in Adverse Environments

Veerasamy Sejian, Mullakkalparambil Velayudhan Silpa, Chinnasamy Devaraj, Natesan Ramachanndran, Duraisamy Thirunavukkarasu, Chikamagalore Gopalakrishna Shashank, Aradotlu Parameshwarappa Madhusoodan, Rajendran Umaya Suganthi, Roman Mylostyvyi, Gundula Hoffmann, João Carlos Caetano Simões, and Raghavendra Bhatta

V. Sejian (✉)
Rajiv Gandhi Institute of Veterinary Education and Research, Kurumbapet, Pondicherry, India

Institute of Animal Breeding and Genetics, Justus-Liebig-Universität Gießen, Giessen, Germany

M. V. Silpa
Institute of Animal Breeding and Genetics, Justus-Liebig-Universität Gießen, Giessen, Germany

C. Devaraj · N. Ramachanndran · C. G. Shashank · R. U. Suganthi · R. Bhatta
ICAR-National Institute of Animal Nutrition and Physiology, Adugodi, Bangalore, India

D. Thirunavukkarasu
Farm Science Centre (Krishi Vigyan Kendra), Tamil Nadu Veterinary and Animal Sciences University, Chinnasalem, Tamil Nadu, India

A. P. Madhusoodan
ICAR-Indian Veterinary Research Institute, Mukteshwar Campus, Mukteshwar, Uttarakhand, India

R. Mylostyvyi
Dnipro State Agrarian and Economic University, Dnipro, Ukraine

G. Hoffmann
Department of Sensors and Modelling, Leibniz Institute for Agricultural Engineering and Bioeconomy (ATB), Potsdam, Germany

J. C. C. Simões
Universidade de Trás-os-Montes e Alto Douro Quinta de Prados, Vila Real, Portugal

S. Mattiello, M. Battini (eds.), *The Welfare of Goats*, Animal Welfare 25,
https://doi.org/10.1007/978-3-031-62182-6_7

Abstract

Goats are considered to be ideal climate resilient animals in the changing climate scenario. They are the most widely dispersed domestic animals worldwide, encompassing dry tropical, sub-tropical, and arid regions of developing and developed countries. Goat farming plays a pivotal role in livelihood and socio-economic aspects of small, marginal, and landless livestock farmers worldwide. Goats can survive in different environmental conditions, from sub-tropical arid and Mediterranean climates to continental climates, as well as in high-altitude environments. Goat farming in urban, peri-urban, and fringes of cities is reported in countries such as Nigeria, Burkina Faso, Mali, Uganda, Pakistan, India, Sri Lanka, etc. Policy makers perceive these animals as small-holder livestock and thus as a possible means to end poverty among poor and marginal farmers. In the changing climate scenario, goats possess the ability to survive in multiple locations and environments, which are very challenging in nature. However, the adaptation efforts to these challenging environments might significantly hamper goat's welfare, especially under extensive conditions. The challenges pertaining to animal welfare posed by exposure to various environmental conditions necessitate the application of several solutions concerning shelter management, dietary interventions, and disease control. A genetic approach may provide permanent solutions in developing the excellent climate resilient potential in goats.

Keywords

Climate change · Genetics · Goat · Heat stress · Shelter · Welfare

1 Introduction

Climate change poses serious threats for the survival of the ecosystems on earth. The adverse impacts of climate change are diversified, cutting across different sectors. The livestock sector is one of the important sectors in determining the gross domestic product (GDP) of several countries, particularly in developing parts of the world. Among the agricultural sector, livestock is considered extremely climate resilient; nonetheless, the adverse impacts of multiple stressors may seriously hamper it. Therefore, sustaining livestock output in the context of a changing climate is currently one of the greatest challenges for farming communities.

One approach to improve the livelihood of poor and marginal farmers is to identify and promote the adoption of appropriate climate resilient genotypes. The most recent literature on modelling approaches suggests that goats can be regarded as the ideal climate resilient farm animals (Carvajal et al. 2021). Therefore, promoting goat farming could offer suitable solutions for sustaining the livelihood of poor and marginal farmers. Although goats are considered a climate resilient species, there are still numerous differences among indigenous goat breeds in terms of climate resilience (Aleena et al. 2018). Thus, it is essential to identify the best climate resilient goat breeds and disseminate them among poor and marginal farmers for improving their economy and, at the same time, guaranteeing a high welfare to the animals.

Given the significance of goats to the economy of poor and marginal farmers, sustaining goat production in the face of climate change should therefore be a priority. Such efforts should be multi-fold involving both genetic aspects and housing and management aspects, such as provision of shelter, nutritional interventions, and disease management. While management strategies could only provide temporary relief, genetic strategies could offer a permanent solution by identifying more climate resilient goat breeds. This chapter is therefore an attempt to collate and synthesize information on welfare threats to goats when subjected to extreme environmental conditions, as well as to identify different ameliorative strategies to endure goat welfare in the changing climate scenario.

2 Goat as the Ideal Climate Resilient Species

The recent literature review suggests that goats are the ideal climate resilient animals (Rahimi et al. 2021; Carvajal et al. 2021; Nair et al. 2021). These authors expressed the scientific reasons behind their assertion that goats possess a superior ability to cope with the adverse impacts of climate change. Goats present several advantages which allow them to survive in extreme climatic conditions. They are considered to be more thermo-tolerant as well as drought tolerant than most other farmed species, which gives them a remarkable advantage to cope with heat stress in a tropical climate, especially in extensive rearing systems (Sejian et al. 2021b). Their unique feeding behaviour and potential for bipedal stance help goats to survive in areas with limited pasture availability. Furthermore, goats possess the ability to desiccate both urine and faeces in an effort to conserve body water (Nair et al. 2021); this offers them an additional advantage to survive in tropical climate. Several researchers observed these advantages in various indigenous goat breeds and have suggested that differences even exist among these breeds for their thermotolerance (Aleena et al. 2018; Pragna et al. 2018).

3 Distribution Pattern of World Goat Population

According to the latest FAO (2020) report, the goat population is distributed predominantly in African, Indian, and Australian subcontinents. These are mainly hot arid or semi-arid tropical environments. This distribution pattern of goats clearly demonstrates that this species has a unique potential to survive in extremely challenging environmental conditions. The world distribution of goats in relation to environmental temperature shows that the goat population is predominantly concentrated in tropical regions. Furthermore, goat farming is common in areas where a high proportion of pasture, uncultivable/marginal land, and/or cultivable fallow land makes up a sizable portion of the total land area, as well as in regions with limited access to irrigation (Kumar and Pant 2003; Lebbie 2004). Therefore, goat farming is preferable in challenging areas, such as deserts/arid grassland/savannah/semi-arid areas of Asiatic and African countries. These regions are often located in poor countries, where human health and welfare are threatened by poverty and where the welfare of goats might not be adequately taken into account.

4 Physiological Mechanisms Associated with Compromised Welfare of Goats in Adverse Environment

Physiological mechanisms play a vital role in animals to cope with extreme environmental conditions. For example, animal exposure to high ambient temperature and humidity induces an increase in the dissipation of excess body heat to the surrounding environment to reduce heat load in the body (Al-Haidary 2004). Further, dissipation of excess body heat is brought by various physiological mechanisms, including increase of respiratory rate, rectal temperature, pulse rate, skin temperature, and sweating rate (Sharma et al. 2013). However, exposure of goats to high ambient temperatures and humidity can have adverse effects on animal health and welfare (Bernabucci et al. 2010; Nardone et al. 2010).

4.1 Respiratory Rate

Respiratory rate is the first foremost physiological mechanism compromised when goats are exposed to extremely hot and humid climates in tropical countries. One important function of the respiratory system is to maintain a constant animal's body temperature through evaporation of moisture from the respiratory tract. This physiological mechanism is crucial for preventing hyperthermia during thermal/heat stress conditions (da Silva et al. 2017). However, respiration rate may vary depending on the magnitude of thermal stress and also on the goat breed (Sejian et al. 2021b). When the ambient air temperature exceeds the normal threshold level (i.e., it raises above the thermal comfort zone of the goats), especially during the summer season in tropical areas, a marked increase in the respiratory rate will facilitate evaporative heat loss, thereby contributing to maintaining the thermal balance (Hamzaoui et al. 2013). Goats exposed to low, medium, high, and severe heat stress will exhibit various degrees of increased respiratory rate, namely 40–60, 60–80, 80–120, and more than 200 breaths/min, respectively (Sejian et al. 2021b). This physiological evaporative cooling mechanism fails during extreme conditions, the heat dissipation by evaporation cooling mechanism is impaired and the core body temperature may increase, resulting in compromised goats' welfare. Elevated respiratory rates in goats indicate that they are subjected to heat stress. Conversely, goats suffering from cold stress were observed to lower their respiration rate (−5 breaths/min on average; Coloma-García et al. 2020). Further, in higher altitudes, goats significantly increased their respiration rate by 1% for every 100-meter increase in altitude. Cloud cover also influences the respiration rate in goats, where less clouds lead to increased breathing due to a radiant heat load effect (Sarmento et al. 2019). All these fluctuations during adverse environments can compromise the welfare of the goats.

4.2 Rectal Temperature

Rectal temperature is representative of the core body temperature. It is a good indicator of heat stress and may be used to assess the adaptive potential of goats. Goats are equipped to maintain their body temperature within a certain range. This is important for their homeostasis. However, under certain conditions, they may fail to do so. For example, the rectal temperature is high when the evaporating cooling mechanisms (sweating and respiration) fail to maintain thermal balance in goats (Kumar et al. 2018). Goats employ the increase in rectal temperature as an alternate technique to lower their body temperature and maintain a comfortable environment (Sejian et al. 2021b). During heat stress, if all the physiological mechanisms fail to maintain the goat's body temperature, rectal temperature increases (Marai and Haeeb 2010). According to Kadzere et al. (2002), an increase in rectal temperature even by 1 °C is considered to impair the welfare of livestock. Conversely, in a study conducted by Coloma-García et al. (2020), a decrease in rectal temperature by 2 °C was recorded in goats exposed to cold stress. This might hamper the welfare of goats, as the energy needed to maintain homeostasis is deviated towards thermoregulation.

4.3 Pulse Rate and Heart Rate

The pulse rate and heart rate contribute to maintaining homeostasis of circulation along with the general metabolic status (Sejian et al. 2010). When goats are exposed to high ambient temperatures for prolonged periods of time, their pulse rate increases. Therefore, pulse rate is affected by hot and humid environments. Shilja et al. (2016) reported significantly higher pulse rates in heat-stressed Osmanabadi goats and this reflects compromised welfare. Heart rate increases under hot and humid conditions, and this in turn increases blood flow from the core body to the periphery (skin) to enhance heat dissipation by sensible (conduction, convection, and radiation) and insensible (dissipation of water from the skin, sweating) mechanisms. Further, in a high-altitude environment, Tucker et al. (1976) observed reduced myocardial function in goats due to chronic hypobaric hypoxia, which normally occurs at high altitude. According to the researchers, depression of myocardial function is reversible by chronic, but not by acute, relief of hypoxia.

4.4 Skin Temperature

In mammals, skin is an important organ playing a significant role for heat exchange between the body and the surrounding environment (Ribeiro et al. 2018). The skin temperature of goats is affected by blood flow to skin surface areas. The increased skin blood flow through cutaneous vasodilation facilitates heat dissipation from the body to the surrounding environment. A high skin temperature in response to heat stress alters the blood flow, redistributing the blood, further increasing blood flow to the skin surfaces. Therefore, exposure of goats to high ambient temperature and humidity increases heat losses from the skin to the surrounding environment, thus maintaining the homeothermy.

5 Global Climate Change Impact on Goat Welfare

Anthropogenic climate change is posing one of the most dangerous threats that humankind has ever faced. Reports also predict that developing countries are under greater risk of being adversely affected by the changing climatic scenario due to their higher dependence on natural resources, poor financial stability, and limited institutional capacity-building programmes (Sejian 2013). Moreover, the traditional farming systems practised in most of the developing countries rely on rain-fed agriculture with few technological advancements to adapt to erratically changing climatic conditions (Feleke et al. 2016). However, in response to global warming, the impact of high ambient temperatures, once limited to tropical countries, has now extended also to northern latitudes. For example, the number of days with THI > 72 has been increasing in the northern United States, Canada, and Europe (Polsky and von Keyserlingk 2017).

Small ruminants, especially goats, are considered as the most promising climate resilient species for their extreme resistance to heat stress and drought, the minimal investment required, their high feed conversion efficiency, disease resistance, and so on (Sejian et al. 2021a). Though these species are considered very rustic, their ability to thrive in harsh environmental conditions, under poor and/or inadequate management and feeding practices, is often overrated (Sevi et al. 2009; Sejian et al. 2021b). From a general welfare and production point of view, animals need to be free from hunger, thirst, and disease (Lu and Miller 2019); however, these are among the main variables to be significantly impaired due to climate change. Climate change directly (altered temperature, humidity, photoperiod, etc.) and/or indirectly (feed, fodder, and water scarcity, shrinking pasture and disease outbreak) impairs the health status and welfare of goats, with added impact on their productive and reproductive potential (Sejian et al. 2021b).

There have been rising concerns towards addressing the impact of climate change, especially heat stress, on livestock production. Though most of these studies target larger ruminants, there are studies focusing on goats too, which revealed that heat stress leads to declined growth (12%; Pragna et al. 2018), milk (3–10%; Salama et al. 2014), and meat (4%; Hashem et al. 2013) production in goats. A significant decline in milk yield and milk composition (total solids, fat, protein, and casein) was reported in Murciano-Granadina lactating dairy goats when exposed to heat stress in comparison to their control group (Contreras-Jodar et al. 2018). In addition to this, heat-stressed goats have also been reported to have altered milk coagulation properties. Further, heat stress severely impacts the goat dairy and cheese making industry. Reduced feed intake, poor fodder/pasture quality, inadequate water, and direct effect of thermal stress on mammary gland are the primary reasons related to climate change that may induce milk production losses (Contreras-Jodar et al. 2018). Reproductive performance of goats is also significantly impaired by climatic stressors, resulting in altered oestrus cycles, decreased oestrus detection with subsequent prolongation of the age at first insemination/successful insemination, reduced pregnancy rates due to aggravation of early embryonic fatalities, reduced birth weight of kids, and so on (Darcan and Silanikove 2018).

Another severe consequence of climate change posing a threat to goat welfare is its impact on the health status of animals and increased disease outbreaks, which may also have negative economic consequences. It is an established concept that the spread, distribution, and occurrence of many livestock infections/diseases are climate dependent (Marino et al. 2016). Variations in the humidity and temperature highly influence the life cycle and geographical distribution of a number of vectors and microbes. Rift Valley fever in East Africa and bluetongue in Europe are two of the well-documented evidence of vector-borne diseases in goats having high association with climate change (Lancelot et al. 2008). Such disease outbreaks are not only a concern for the animal welfare and from the public health point of view but they also result in declined production and huge economic losses.

In geographic areas where reduction in biomass availability and yield occurs, climate change imposes several other fodder and pasture associated impacts that can indirectly influence goat welfare and production. Some of these include reduced nutrient quality of pasture and feed resources due to poor soil profile and droughts; shift of herbaceous pasturelands to shrubs and trees; conversion of pastureland to cropland; increased occurrence of weeds and pest infestation thereby reducing fodder production (Uddin and Kebreab 2020).

One aspect of climate change on livestock production that is often less emphasized is the deleterious effect on pastoral livestock diversity. Though locally adapted breeds are considered to be more resilient to climatic alteration than exotic ones, the former could be under greater morbidity and mortality risks (Hoffmann 2010b). This is primarily due to the fact that exotic breeds, mainly reared in high-output intensive systems, are provided with better housing and management (Fig. 1) that relatively protects them better from the climatic adversities and provide higher standard of livestock well-being (Sevi et al. 2009) when compared to the local breeds in extensive grazing systems. Additionally, the exponentially rising food demand and climatic adversities will pressurize the low input and extensive production systems

Fig. 1 Goat farming in Niger. This area is characterized by a hot climate, with very high temperatures year-round, and two different seasons: a dry season from October–May, and a brief rainy season. These goats usually do not make use of shelters and are farmed using a transhumant system, as a strategy to optimize resource use (Photo: Silvana Mattiello)

for improved production efficiency and reduction in their environmental footprints. This would further lead to decline in preference of locally adaptive breeds for livestock farming which may have severe consequences in future sustainable farming. As mentioned earlier, there is a scarcity of studies focusing on assessing the welfare of goats amidst the changing climatic scenario. Therefore, given the possibility for potential changes in future production conditions and livestock production objectives, it is necessary to secure the animal genetic diversity (Hoffmann 2010a).

6 Welfare Issues of Goats in Different Tropical and Sub-tropical Regions

Traditionally, goats are reared under extensive grazing systems and occupy marginal lands, mountainous regions, arid and semi-arid areas (Simões et al. 2021) (Fig. 1). Goats can survive in different environmental conditions from sub-tropical arid and humid areas, and Mediterranean climates (e.g., North Africa and Southern Europe; Silanikove 2000) to continental climates (e.g., Northern Europe and North America; Battini et al. 2016). The welfare of goats, as well as the welfare of all farm animals, has been gaining importance during the current climate change scenario since animals are forced to divert their energy away from the productive process to an adaptive process, in order to cope with the adverse environment. The general concern about welfare issues of farm animals is aimed at intensive systems, therefore lots of research focused on these systems, whereas issues related to extensive systems have been neglected (Temple and Manteca 2020). This may be due to the belief that extensive systems provide many benefits in terms of animal welfare. Therefore, identifying the main welfare problems likely to be encountered in extensive goat farming systems is of paramount importance. Goats reared in extensive systems may encounter a range of challenges to their welfare, including nutritional and walking stress, limited water supply, climatic extremes, parasitic infestations, vector-borne diseases, lameness, and inappropriate management (Turner and Dwyer 2007; Sejian et al. 2021b). For example, it has been reported that under extensive conditions, animals spend 43% more energy at pasture than when kept indoors, and energy expenditure is even higher in arid regions due to the long distance they have to travel looking for food sources (Lachica and Aguilera 2003).

Adverse environmental conditions, particularly heat stress conditions prevailing in tropical and sub-tropical countries, negatively influence growth, milk and meat production and compromise the immune response of goats, thereby significantly reducing their welfare under extensive farming systems and during transport. Under these unfavourable environmental conditions, goats drastically adjust their physiological and metabolic functions to cope with the adverse environment. For example, in case of very hot environments, goats respond by enhancing their heat dissipation, through decrease in heart rate and increase in respiration rate, sweating and panting with/without salivation (Darcan et al. 2007). Increased water intake (Olsson et al. 1995) and decreased dry matter intake (Silanikove 2000) are frequently observed changes in goats under extreme heat stress conditions.

Heat stress is one of the major environmental factors which severely affects animal welfare and leads to economic losses to the farmers. In developed countries with a tropical climate, the awareness about animal welfare is higher and several mitigation strategies have been developed by following the international animal welfare standards. This leads to the implementation of appropriate management practices. Unfortunately, in developing tropical countries like Africa, livestock farmers are often unaware of, and unfamiliar with, animal welfare. Also, the literature about the animal welfare concepts in these countries is scarcely available (Fraser 2008). Ndou et al. (2011) reported that a low priority has been given to animal welfare in developing African countries, because goat farming is linked to traditional customs and beliefs, unskilled animal handling, and below standard handling facilities. Various animal welfare issues observed in Kenya include abandonment, overloading, overcrowding, over walking, physical injury, starvation, undernutrition, confinement, unsuitable transportation and poor slaughter facilities, inhumane treatment, and handling at slaughter (Mogoa et al. 2005). The NGOs in African countries are promoting animal welfare; however, these activities are restricted to developed/urban areas and are unreachable from remote areas (Herrero et al. 2009).

In tropical countries, when a shelter is present, one of the top methods to prevent goats from getting exposed to heat stress is through improving the shelter design

Fig. 2 Housing model depicting the elevated perforated flooring for proper channelling of faeces and urine, to maintain the floor dry as possible, especially in high rainfall and high humid areas to maintain the normal behaviour of goats (Source: CCRAAS, ICAR-NIANP)

and management. Ventilation is particularly important to ensure thermal comfort in hot environments: in hot and humid conditions, the recommended ventilation space is 100% of floor area, i.e., the longer sides of the shelter should be kept totally open, whereas in hot and dry weather conditions the recommended ventilation space is 70% of floor area (Misra et al. 2010). Furthermore, the orientation of the shed should favour free air movement to remove moisture and ammonia odour inside the shelter (Ambazamkandi et al. 2015). A high roof promotes airflow and allows sun to enter more readily, potentially increasing radiant heat load but is also more susceptible to damage from heavy winds. A roof vent can contribute to adequate ventilation. The roof offers shelter from various environments. There are different types of roof style such as shed, gable, or modified gable style. Simple thatched roofs are common in rural areas. Thatched roofs require a steeper pitch than iron sheets. In regions with abundant rainfall, a steeper slope is also advantageous. Depending on the production system, material availability, and environment, roofs may be constructed of iron sheet, grass/bushes, wood, stone/brick, or soil (Ambazamkandi et al. 2015). Paddy straw was the most common material used for roofing in rural West Bengal, India, in a study conducted by Nandi et al. (2011). Other common roofing materials included earthen tiles and tin sheets. The white painting of the roof surface (outer side) and black in the inner side should be done to curtail radiation. As to the floor, it is of utmost importance to maintain it as dry as possible, especially in high rainfall and highly humid areas (Fig. 2).

Fig. 3 Cashmere goat farming in the Himalayan region (Photo: Silvana Mattiello)

7 Welfare Issues of Goats at High Altitudes

High altitude is characterized by low atmospheric oxygen pressure, cold weather, bright sunshine, cold breeze, seasonal snowfall as well as rainfall. The intra-day and inter-day weather fluctuation in high altitude is extremely high and this requires a great thermoregulation effort to maintain the goats' homeostasis. Goats might adapt to these challenges by modifying their behavioural and cardinal physiological parameters.

Welfare issues in goats at high altitude arise mainly in relation to environmental threats, feeding, and healthcare aspects. Additionally, for cashmere goats (Fig. 3), management aspects, especially in relation to cashmere harvesting, are also very relevant for animal welfare.

7.1 Welfare Issues Due to Environmental Threats

At high altitudes (>3000 m a.s.l.), goats are mainly reared in extensive systems as well as tethering systems due to low flat land available for constructing proper goat sheds. In both farming systems, during the daytime goats are allowed to graze and during the night they are housed indoors. However, shelters are not always adequate (e.g. lack proper floor, roof, windows, and mangers) and do not provide enough protection from the cold and windy weather. So, ventilation, resting surface, and space availability can be severely compromised. The flooring is usually uneven and made of locally available *kutcha* materials, which results in injury and pain especially in the housing system followed in semi-arid tropical environments. During severe cold stress conditions, especially during night hours, 10 cm thick bedding using locally available resources or raised wooden slats at 15 cm height from the floor are recommended to protect kids and lactating does from cold stress (Maurya et al. 2013).

When goats are out on pasture, they are particularly exposed to low environmental temperatures and precipitations (rain and snowfall). Exposure to strong winds should also be seriously taken into account: in fact, wind speed is one of the most important climatic factors which affect thermal comfort by worsening the impact of the environmental temperature in cold environments, causing wind chill (Bickert and Mattiello 2016). The risk of cold stress may be further enhanced by precipitation, which increases heat dissipation (Webster et al. 1974). Some particularly intense meteorological events, called "zud", may also occur in some areas in Asia, causing dramatic conditions that lead to the death of many animals (Batima et al. 2008).

For the above-mentioned reasons, in outdoor conditions it is important to provide natural (e.g. trees, if available in the area) or artificial shelters (e.g. walls oriented perpendicular to the direction of prevailing winds) from the wind, and roof shelters to protect animals from rain and snowfall (Bickert and Mattiello 2016).

7.2 Welfare Issues in Feeding

Feed scarcity can be considered as a major problem in high-altitude regions. These regions are characterized by rough terrain and, as the altitude increases, vegetation density decreases. Soil at high altitude is deficient of crucial micro and macro nutrients due to continuous removal by erosion of surface soil layers during precipitation. The edible leaves and stems of fodder trees and shrubs available for browsing and grazing may be deficient in some minerals such as Mg, Na, Cu, Co, and therefore goats are more exposed to metabolic diseases. Also, tannin content in most of the available forages is high, hence digestibility is compromised. High altitudes are also characterized by extremely low temperatures and abundant snowfall during the winter, hence vegetation cover during this season is limited and feed and fodder availability becomes scarce. Therefore, animals have to travel long distances searching for food, thus increasing their energy requirements (e.g. up to 10× in Himalayan and mountainous/hill regions; Shinde 2010). Zud refers to a winter-spring phenomenon triggered by severe climatic events, leading to a significant loss of animal life primarily due to starvation or the inability to graze. Particularly in regions such as semi-deserts and deserts in Mongolia and Central Asia, the "zud" phenomenon results in a substantial mortality rate among livestock, especially during exceptionally harsh winters. Different forms of zud are identified, each associated with specific climatic conditions—winter zud may be induced by an impenetrable ice crust, while summer zud can arise due to drought (Batima et al. 2008).

Along with the mentioned feeding issues, water scarcity is emerging to be one of the major welfare concerns in high-altitude regions. The capacity of glaciers and seasonal snow cover to store water changes as the temperature increases. Glaciers give more water to rivers when their mass balances are negative. Negative mass balances, on the other hand, lower glacier volume and area, reducing total melt water from glaciers (Ragettli et al. 2016). Fluctuations in river flow might affect water availability for livestock dwelling in high-altitude regions. Apart from a few open springs, water is mainly obtained through ground wells, and herders either manually or mechanically pump the water into troughs. In the summer, goats are watered once or twice a day, and in the winter, this is reduced to every other day (Zaramarama 2019). Dehydration might therefore be a serious welfare problem in these environments.

7.3 Welfare Issues in Healthcare

High altitude is characterized by low oxygen pressure and cold winds. So, goats have to modify their physiological mechanisms to maintain their homeostasis. This includes reduced respiratory rates to conserve the body's heat, concentrated and increased red blood cells to transport the available oxygen, and increased haematocrit. High altitude in temperate climates is also characterized by a very high number of rainy days. As precipitations are more abundant at high altitude, parasitic infestation is also common (Schai-Braun et al. 2019). This compromises the growth and welfare of infested goats. Healthcare facilities are often scarce at high altitudes. Geographically, high altitudes, especially in temperate regions of Himalayas, are characterized by rough and steep terrain. So, even though goats have evolved over the years to adapt to this kind of terrain, they are still susceptible to fall and to incur lameness problems, which may cause pain and may hamper their freedom to express their natural behaviour.

7.4 Welfare Issues in Cashmere Production

World renowned cashmere fibre is produced from cashmere goats. It has been estimated that there are 700 million cashmere goats concentrated in the areas of China, Mongolia, India, Australia, Turkey, Afghanistan, Iraq, and Iran. Cashmere fibre is of high economic value to the textile industry because of its characteristics: it is lighter, softer, and stronger than sheep wool and has a higher insulating power. Cashmere goats are usually reared in the high and cold regions of the world (Rafat 2020). Therefore, as well as other goats in such challenging environments, they often face water scarcity, feed scarcity, increased moving distance in search of pasture, and constraining movements in the hilly landscape (Liesegang et al. 2010). Furthermore, they are submitted to an additional welfare challenge during the cashmere fibre harvesting process.

Welfare issues during the harvesting procedure concern handling and combing. Poor handling during harvesting causes elevated fear and stress. The combs used during the harvesting process have sharp teeth at the end; therefore, the process can be painful and might lead to injuries. The harvesting process may take a long time, especially if it is performed during a non-moulting period, and this may further increase the stress for the goats. Therefore, according to the best practice guidelines issued by certification bodies, such as the International Cooperation Committee on Animal Welfare (ICCAW and SFA 2022) and the Sustainable Fibre Alliance (SFA 2020), the fibre should be harvested when the goats go through natural shedding and particularly when temperature is warm and stable. In fact, cashmere fibre protects goats from low temperatures and its removal may be detrimental to animal welfare if performed in unfavourable climatic conditions.

8 Welfare Issues of Goat Farming in Urban and Peri-Urban Contexts

Cities now house around 55% of the world's population, with that percentage predicted to rise to 66% by 2050. These changes have been more marked in African and Asian countries (Innes and Casabianca 2021). Among the top ten urbanizing countries, seven are located in African countries and the Asian countries have an urban growth rate of 2.5% per annum. In South Asian countries, the urban population has been continuously following an increasing trend (Ellis and Roberts 2015). In India, the rural population has declined from 72% to 69% during the period 2000 to 2011. Thus, the major goat owning continents, Africa and Asia, are experiencing a demographic transition phase.

This demographic transition opens interesting market opportunities for goat farming and drives farmers towards urban and peri-urban areas in countries such as Nigeria, Burkina Faso, Mali, Uganda, Pakistan, India, and Sri Lanka, where goats are reared for meat, milk and also for the purpose of social status, pleasure, and capital asset (Dossa et al. 2015). Unfortunately, there are several challenges for animal welfare in urban and semiurban contexts. In these contexts, goats' feeding includes diverse feeds ranging from hay and concentrate to vegetable market and restaurant wastes, which might not always meet goats' nutritional requirements. For example, these feeding practices may lead to ruminal acidosis, a commonly reported metabolic disease in peri-urban goats (Padmaja and Praveena 2011). The abundant presence of garbage also represents a threat for goats' welfare in these contexts, as animals may ingest plastic or other inadequate materials, which can provoke serious problems, for example blocking rumen motility (Figs. 4 and 5). Moreover, these farming systems may have to compete with other sectors for access to the limited available resources and therefore face challenges for providing adequate housing, access to grazing and drinking water.

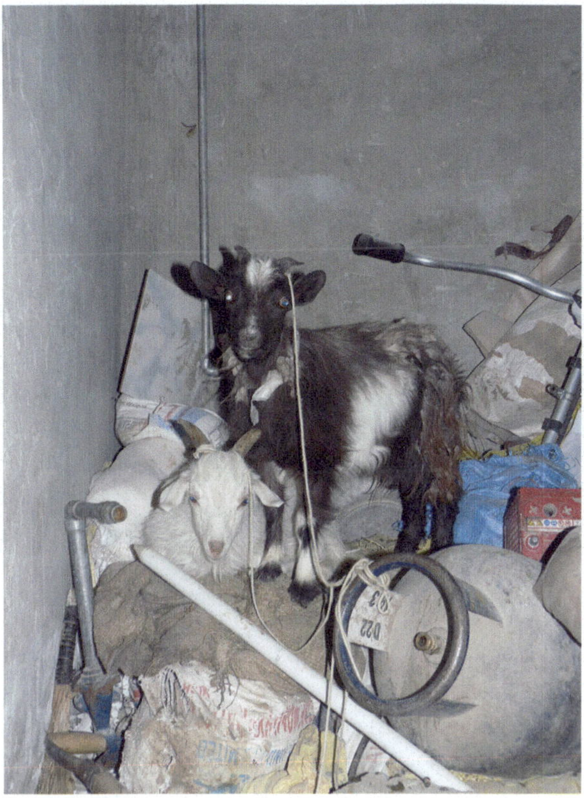

Fig. 4 Cashmere goats in an urban context in the Himalayan region (Photo: Silvana Mattiello)

9 Genetic Approaches to Ensure Appropriate Welfare in Goats Reared in Adverse Environments

Besides the implementation of management strategies that can divert the energy used for thermoregulatory purposes to maintain optimum homeostasis in goats raised under adverse environments, a genetic approach will help to improve goats' coping abilities.

Augmentation of multiple strategies to combat heat susceptibility is a major need for the successful adaptation of goats under the current climate change scenario. No single method can suffice for the above-mentioned need. Marker-assisted selection (MAS) is an indirect selection process through which the trait of interest is selected based on the marker linked to it. MAS is more advantageous over conventional methods for traits with low heritability (e.g. fertility), sex-limited traits (e.g. milk yield) and for the traits in which phenotype is not the better indicator of genotype (e.g. disease resistance) (Hiendleder et al. 2005). A combination of traditional methods and genetic markers is considered as a reliable and promising approach for

Fig. 5 Goats in an urban context in the Sahel region (Photo: Silvana Mattiello)

improving welfare traits in goats and enabling them to cope with adverse environments.

Quantitative Trait Locus (QTL) mapping of the genes that are associated with the thermo-tolerance ability of goats can greatly help in improving the resilient capacity of the animals. The approach can further aid in identification of QTLs linked to thermo-tolerance and categorizing heat tolerant and susceptible breeds. In addition to this, the identified QTLs can be used for marker-assisted introgression (MAI) of the genes into the heat susceptible breeds through crossbreeding programmes. Furthermore, since this method does not require progeny testing, it allows farmers to reduce the generation interval and can help in improving the thermo-tolerance in a short time.

Genomic selection is regarded as an even better version of MAS, in which the disadvantages faced by MAS, like low association between markers and QTLs (Boichard et al. 2016) for few traits, are taken care of. Genomic selection can be described as a form of MAS in which genetic markers in the entire genome are used, such that almost all QTLs are in linkage disequilibrium with one genetic marker at least. Better prediction of breeding value is attained with genomic selection (Ibtisham et al. 2017). In addition to improving genetic gain at a quicker rate, genomic selection also ensures more reliability by increasing the accuracy of estimated breeding value (EBV). Hence, genomic selection will be a better tool to utilize in breeding programmes for improving the thermo-tolerance capacity of goats by taking into account the breeding value of individuals.

In a recent report by Zidi et al. (2014), several significant single nucleotide polymorphisms (SNPs) associated with susceptibility/tolerance to heat stress in Spanish

Florida dairy goats were identified using genome wide association study (GWAS). In another study led by Kim et al. (2016), signatures of natural selection for adaptation to hot arid environments in Bakri sheep and goats were assessed. The authors identified selection signatures spanning candidate genes that were directly or indirectly associated with traits like thermo-tolerance (FGF2, GNAI3, PLCB1), energy and digestive metabolism (ALDH1A3, MYH, TRHDE), body size and development (BMP2, BMP4, GJA3, GJB2) and also immune response associated genes (GRIA1, IL2, IL7, IL21, IL1R1). Such studies aid in providing the basis of ruminant evolution and functional genomics of species living in a similar environment thereby enhancing the selection for climate resilient animals.

The welfare of animals relates strongly to disease resistance and disease tolerance ability. Conventional methods of disease management like antibiotics and vaccination are failing. Development of antibiotic resistance and vaccine development against a wide range of diseases and multi-aetiology diseases has its own disadvantages (Gogoi et al. 2021). Hence, development of disease-resistant breeds of goat is of utmost importance. Quantification of genetic variation observed between the animals for disease resistance and mapping the linked QTL or gene responsible for such variation and utilizing this evidence significantly in the breeding programme could be a way forward towards disease-resistant goat breed development. Since genes underlying this trait are polygenic in nature, as discussed earlier, genomic selection or MAS will be a prompt approach. SNPs of various goat diseases have been identified and utilized in breeding programmes. But as per available literature, more focus is towards sheep, and much more attention is needed for genetic selection of disease-resistant goat breeds.

If the genetic potential of goats is explored or enriched in terms of disease resistance, a major relief can be witnessed in terms of welfare as well as production. A few examples for identified genetic variability among the goat species for different diseases can be useful for understanding the concept of variability of genes linked to disease resistance. A major constraint in goat production is gastrointestinal nematodes. DR beta 1 (DRB) is the highly polymorphic locus linked with nematode resistance in goats. Upregulation of DRB is noticed in goats more susceptible for nematode infestation than the resistant ones. It is evident that TLR7 is supposed to have a vital role in producing resistance to goats against Peste des petits ruminants (PPR), a deadly viral disorder. Polymorphic loci have been identified in goat populations in PrP genes, which are associated with susceptibility to scrapie disease (Pal and Chakravarty 2020). However, the biggest challenge associated with selection for resistance against a disease is that it may indirectly select for another trait associated with susceptibility to another disease(s). Detailed underlying mechanisms remain unclear. Comparative transcriptome analysis can be a reliable approach to understand the deep molecular mechanism underlying disease resistance traits of goats (Aboshady et al. 2019). Recent biotechnological advances have led to the development of genetic engineering tools that can manipulate the animal's genome. Adopting such techniques that alter the genetic make-up of an animal by either introducing disease resistance-conferring genes and/or deleting genes associated with susceptibility, could also be explored.

In addition to the popular methods of attaining genetic improvement like cross-breeding, selection within breeds, and breed substitution, there is yet another option for direct genetic modification like gene editing techniques using CRISPR, which is important in the modern era. Furthermore, QTLs associated with naturally polled breeds of goats should be explored and can be used in breeding programmes for developing hornless breeds of goats. This strategy is believed to make a huge difference in selection from a welfare point of view by avoiding much injury.

10 Conclusions and Future Perspectives

In the coming years, sustaining goat production will become more significant, especially as goats are predicted to be the farm animals of the future, given the changing environment scenario. The socioeconomic importance of goats demands more research to address the challenges of goat farming, with proper consideration for welfare aspects. Along with production, efficient extension methods need to be developed to update farmers on ethical ways of managing goats from birth to death. Further, the stakeholder's knowledge and perception of climate change and its impact on animal welfare widely vary across the globe. This demands immediate attention from policymakers to bridge the gap between climate change research and stakeholders' knowledge, perception, and impact of climate change. More studies on goats' behaviour in a variety of adverse environments, and on the breed-specific adaptation mechanisms, are required to fully comprehend the intricate details of their welfare. Furthermore, genetic strategies should be implemented to offer a permanent solution through marker-assisted breeding selection. Such an approach could revolutionize goat breeding to improve the livelihood of poor and marginal farmers and to improve goats' welfare.

References

Aboshady HM, Mandonnet N, Stear MJ, Arquet R, Bederina M, Sarry J, Tosser-Klopp G, Klopp C, Johansson AM, Jonas E, Bambou JC (2019) Transcriptome variation in response to gastro-intestinal nematode infection in goats. PLoS One 14(6):e0218719

Aleena J, Sejian V, Bagath M, Krishnan G, Beena V, Bhatta R (2018) Resilience of three indigenous goat breeds to heat stress based on phenotypic traits and PBMC HSP70 expression. Int J Biometeorol 62(11):1995–2005

Al-Haidary AA (2004) Physiological responses of Naimey sheep to heat stress challenge under semi-arid environments. Int J Biometeorol 6(2):307–309

Ambazamkandi P, Thyagarajan G, Sambasivan S, Davis J, Shanmugam S, Joseph BA (2015) Shelter design for different livestock from a climate change perspective. In: Sejian V, Gaughan J, Baumgard L, Prasad C (eds) Climate change impact on livestock: adaptation and mitigation. Springer India, pp 399–424

Batima P, Natsagdorj L, Batnasan N (2008) Vulnerability of Mongolia's pastoralists to climate extremes and changes. In: Climate change and vulnerability, vol 2. Earthscan, London, pp 67–87

Battini M, Barbieri S, Vieira A, Stilwell G, Mattiello S (2016) Results of testing the prototype of the AWIN welfare assessment protocol for dairy goats in 30 intensive farms in Northern Italy. Ital J Anim Sci 15(2):283–293
Bernabucci U, Lacetera N, Baumgard LH, Rhoads RP, Ronchi B, Nardone A (2010) Metabolic and hormonal acclimation to heat stress in domesticated ruminants. Animal 4(7):1167–1183
Bickert WG, Mattiello S (2016) Stress in dairy animals: cold stress: management considerations. In: Reference module in food science. Elsevier
Boichard D, Ducrocq V, Croiseau P, Fritz S (2016) Genomic selection in domestic animals: principles, applications and perspectives. C R Biol 339(7–8):274–277
Carvajal MA, Alaniz AJ, Gutiérrez-Gómez C, Vergara PM, Sejian V, Bozinovic F (2021) Increasing importance of heat stress for cattle farming under future global climate scenarios. Sci Total Environ 801:149661
Coloma-García W, Mehaba N, Such X, Caja G, Salama AAK (2020) Effects of cold exposure on some physiological, productive, and metabolic variables in lactating dairy goats. Animals 10(12):2383
Contreras-Jodar A, Salama AA, Hamzaoui S, Vailati-Riboni M, Caja G, Loor JJ (2018) Effects of chronic heat stress on lactational performance and the transcriptomic profile of blood cells in lactating dairy goats. J Dairy Res 85(4):423–430
da Silva WE, Leite JHGM, de Sousa JER, Costa WP, da Silva WST, Guilhermino MM, Asensio LAB, Façanha DAE (2017) Daily rhythmicity of the thermoregulatory responses of locally adapted Brazilian sheep in a semiarid environment. Int J Biometeorol 61(7):1221–1231
Darcan NK, Silanikove N (2018) The advantages of goats for future adaptation to climate change: a conceptual overview. Small Rumin Res 163:34–38
Darcan NK, Cedden F, Cankaya S (2007) Spraying effects on some physiological and behavioural traits of goats in a subtropical climate. Ital J Anim Sci 7(1):77–85
Dossa LH, Sangaré M, Buerkert A, Schlecht E (2015) Production objectives and breeding practices of urban goat and sheep keepers in West Africa: regional analysis and implications for the development of supportive breeding programs. SpringerPlus 4(1):281
Ellis P, Roberts M (2015) Leveraging urbanization in South Asia: managing spatial transformation for prosperity and livability. World Bank Publications
FAO (2020) Five practical actions towards resilient, low-carbon livestock systems: in brief. FAO. https://www.fao.org/publications/card/en/c/CB2007EN/
Feleke FB, Berhe M, Gebru G, Hoag D (2016) Determinants of adaptation choices to climate change by sheep and goat farmers in northern Ethiopia: the case of southern and Central Tigray, Ethiopia. SpringerPlus 5(1):1692
Fraser D (2008) Toward a global perspective on farm animal welfare. Appl Anim Behav Sci 113(4):330–339
Gogoi A, Das B, Chabukdhara P, Phookan A, Phangchopi D (2021) Livestock breeding for disease resistance: a perspective review. Agric Rev
Hamzaoui S, Salama AAK, Albanell E, Such X, Caja G (2013) Physiological responses and lactational performances of late-lactation dairy goats under heat stress conditions. J Dairy Sci 96(10):6355–6365
Hashem M, Hossain M, Rana M, Hossain M, Islam M, Saha N (2013) Effect of heat stress on blood parameter, carcass and meat quality of black Bengal goat. Bangladesh J Anim Sci 42(1):57–61
Herrero M, Thornton PK, Gerber P, Reid RS (2009) Livestock, livelihoods and the environment: understanding the trade-offs. Curr Opin Environ Sustain 1(2):111–120
Hiendleder S, Bauersachs S, Boulesteix A, Blum H, Arnold GJ, Fröhlich T, Wolf E (2005) La genómica funcional: Herramientas para mejorar la sanidad y el bienestar del ganado: -EN- Functional genomics: tools for improving farm animal health and welfare -FR- La génomique fonctionnelle: outils pour améliorer la santé et le bien-être des animaux d'élevage -ES-. Revue Scientifique et Technique de l'OIE 24(1):355–377
Hoffmann I (2010a) Impacts of climate change on livestock genetic diversity and selection. In: Proceedings of the world conference on genetics applied to livestock production, Leipzig.

http://www.wcgalp.org/system/files/proceedings/2010/impacts-climate-change-livestock-geneticselection-and-diversity.pdf. Accessed 9 March 2020
Hoffmann I (2010b) Climate change and the characterization, breeding and conservation of animal genetic resources. Anim Genet 41:32–46
Ibtisham F, Zhang L, Xiao M, An L, Ramzan MB, Nawab A et al (2017) Genomic selection and its application in animal breeding. Thai J Vet Med 47(3):301–310
ICCAW and SFA (2022) International Agricultural Cooperation International Cooperation Committee on Animal Welfare and Sustainable Fibre Alliance. http://www.iccaw.org.cn/uploads/soft/200817/1-200QG15328.pdf. Accessed 16 January 2022
Innes CR, Casabianca B (2021) Creating housing markets in emerging market economies. http://hdl.handle.net/10986/36569
Kadzere CT, Murphy MR, Silanikove N, Maltz E (2002) Heat stress in lactating dairy cows: a review. Livest Prod Sci 77(1):59–91
Kim ES, Elbeltagy AR, Aboul-Naga AM, Rischkowsky B, Sayre B, Mwacharo JM, Rothschild MF (2016) Multiple genomic signatures of selection in goats and sheep indigenous to a hot arid environment. Heredity 116(3):255–264
Kumar S, Pant KP (2003) Development perspective of goat rearing in India: status, issues and strategies. Indian J Agric Econ 58(4):752
Kumar D, Yadav B, Choudhury S, Kumari P, Madan AK, Singh SP, Rout PK, Ramchandran N, Yadav S (2018) Evaluation of adaptability to different seasons in goat breeds of semi-arid region in India through differential expression pattern of heat shock protein genes. Biol Rhythm Res 49(3):466–478
Lachica M, Aguilera JF (2003) Estimation of energy needs in the free-ranging goat with particular reference to the assessment of its energy expenditure by the 13C-bicarbonate method. Small Rumin Res 49(3):303–318
Lancelot R, Rocque S, Chevalier V (2008) Bluetongue and rift valley fever in livestock: a climate change perspective with a special reference to Europe, the Middle East and Africa. Cambridge University Press, pp 87–89
Lebbie SHB (2004) Goats under household conditions. Small Rumin Res 51(2):131–136
Liesegang A, Kaulfers C, Lauff K, Schmid S, Kreuzer M, Wanner M, Leiber F (2010) Influence of high altitude pasture on bone metabolism in goats and sheep. Bone 46:S74
Lu CD, Miller BA (2019) Current status, challenges and prospects for dairy goat production in the Americas. Asian Australas J Anim Sci 32(8):1244–1255
Marai IFM, Haeeb AAM (2010) Buffalo's biological functions as affected by heat stress—a review. Livest Sci 127(2–3):89–109
Marino R, Atzori AS, D'Andrea M, Iovane G, Trabalza-Marinucci M, Rinaldi L (2016) Climate change: production performance, health issues, greenhouse gas emissions and mitigation strategies in sheep and goat farming. Small Rumin Res 135:50–59
Maurya VP, Sejian V, Naqvi SMK (2013) Effect of cold stress on growth, physiological responses, blood metabolites and hormonal profile of native Malpura lambs under hot semi-arid tropics of India. Indian J Anim Sci 83:370–373
Misra RP, Singh SP, Ramachandran N (2010) Shelter management for sheep and goats for mitigating climatic stress. In: Karim SA, Joshi A, Sankhyan SK, Shinde AK, Shakyawar DB, Naqvi SMK, Tripathi BN (eds) Climatic stress & stress management: sheep and goat production. Satish Serial Publishing House, Delhi, India, pp 207–230. ISBN No. 81-89304-81-X
Mogoa EG, Wabacha JK, Mbithi PM, Kiama SG (2005) An overview of animal welfare issues in Kenya. Kenya Vet 29:48–52
Nair MRR, Sejian V, Silpa MV, Fonsêca VFC, de Melo Costa CC, Devaraj C, Krishnan G, Bagath M, Nameer PO, Bhatta R (2021) Goat as the ideal climate-resilient animal model in tropical environment: revisiting advantages over other livestock species. Int J Biometeorol 65(12):2229–2240
Nandi D, Roy S, Bera S, Kesh S, Samanta A (2011) The rearing system of black Bengal goat and their farmers in West Bengal, India. Vet World 254

Nardone A, Ronchi B, Lacetera N, Ranieri MS, Bernabucci U (2010) Effects of climate changes on animal production and sustainability of livestock systems. Livest Sci 130(1–3):57–69
Ndou SP, Muchenje V, Chimonyo M (2011) Assessment and implications of animal welfare in beef production systems in developing countries. Afr J Biotechnol 10(1049):64
Olsson K, Josäter-Hermelin M, Hossaini-Hilali J, Hydbring E, Dahlborn K (1995) Heat stress causes excessive drinking in fed and food deprived pregnant goats. Comp Biochem Physiol A Physiol 110(4):309–317
Padmaja K, Praveena G (2011) Rumen acidosis in goats. Intas Polivet 12(2):318–319
Pal A, Chakravarty AK (2020) Disease resistance for different livestock species. In: Genetics and breeding for disease resistance of livestock. Elsevier, pp 271–296
Polsky L, von Keyserlingk MAG (2017) Invited review: effects of heat stress on dairy cattle welfare. J Dairy Sci 100(11):8645–8657
Pragna P, Sejian V, Bagath M, Krishnan G, Archana PR, Soren NM, Beena V, Bhatta R (2018) Comparative assessment of growth performance of three different indigenous goat breeds exposed to summer heat stress. J Anim Physiol Anim Nutr 102(4):825–836
Rafat SA (2020) New look on an ancient fiber of cashmere. In: Goats (capra): from ancient to modern. IntechOpen
Ragettli S, Immerzeel WW, Pellicciotti F (2016) Contrasting climate change impact on river flows from high-altitude catchments in the Himalayan and Andes Mountains. Proc Natl Acad Sci 113(33):9222–9227
Rahimi J, Mutua JY, Notenbaert AMO, Marshall K, Butterbach-Bahl K (2021) Heat stress will detrimentally impact future livestock production in East Africa. Nat Food 2(2):88–96
Ribeiro MN, Ribeiro NL, Bozzi R, Costa RG (2018) Physiological and biochemical blood variables of goats subjected to heat stress—a review. J Appl Anim Res 46(1):1036–1041
Salama AAK, Caja G, Hamzaoui S, Badaoui B, Castro-Costa A, Façanha DAE, Guilhermino MM, Bozzi R (2014) Different levels of response to heat stress in dairy goats. Small Rumin Res 121(1):73–79
Sarmento W, Biel M, Berger J (2019) Seeking snow and breathing hard—behavioral tactics in high elevation mammals to combat warming temperatures. PLoS One 14(12):e0225456
Schai-Braun SC, Posautz A, Alves PC, Hackländer K (2019) Gastrointestinal parasite infestation in the alpine mountain hare (*Lepus timidus varronis*): are abiotic environmental factors such as elevation, temperature and precipitation affecting prevalence of parasite species? Int J Parasitol Parasites Wildl 9:202–208
Sejian V (2013) Climate change: impact on production and reproduction, adaptation mechanisms and mitigation strategies in small ruminants: a review. Indian J Small Rumin 19(1):1–21
Sejian V, Maurya VP, Naqvi SMK (2010) Adaptability and growth of Malpura ewes subjected to thermal and nutritional stress. Trop Anim Health Prod 42(8):1763–1770
Sejian V, Silpa MV, Reshma Nair MR, Devaraj C, Devapriya A, Ramachandran N, Awachat VB, Shashank CG, Fonsêca VFC, Bhatta R (2021a) Goat as the ideal future climate resilient animal model. In: Sejian V, Chauhan SS, Devaraj C, Malik PK, Bhatta R (eds) Climate change and livestock production: recent advances and future perspectives. Springer Singapore, pp 279–292
Sejian V, Silpa MV, Reshma Nair MR, Devaraj C, Krishnan G, Bagath M, Chauhan SS, Suganthi RU, Fonseca VFC, König S, Gaughan JB, Dunshea FR, Bhatta R (2021b) Heat stress and goat welfare: adaptation and production considerations. Animals 11(4):1021
Sevi A, Casamassima D, Pulina G, Pazzona A (2009) Factors of welfare reduction in dairy sheep and goats. Ital J Anim Sci 8(sup1):81–101
Sharma R, Singh PK, Maitra A, Pandey AK, Mukesh M, Singh SR, Singh B (2013) Molecular characterization, body parameters and management practices of Purnea cattle. Indian J Anim Sci 83(5):536–541
Shilja S, Sejian V, Bagath M, Mech A, David CG, Kurien EK, Varma G, Bhatta R (2016) Adaptive capability as indicated by behavioral and physiological responses, plasma HSP70 level, and PBMC HSP70 mRNA expression in Osmanabadi goats subjected to combined (heat and nutritional) stressors. Int J Biometeorol 60(9):1311–1323

Shinde AK (2010) Stress factors for small ruminants under field existence and strategies for their amelioration. In: Karim SA, Joshi A, Sankhyan SK, Shinde AK, Shakyawar DB, Naqvi SMK, Tripathi BN (eds) Climatic stress & stress management: sheep and goat production. Satish Serial Publishing House, Delhi, India, pp 231–247. ISBN No. 81-89304-81-X

Silanikove N (2000) Effects of heat stress on the welfare of extensively managed domestic ruminants. Livest Prod Sci 67(1–2):1–18

Simões J, Abecia JA, Cannas A, Delgadillo JA, Lacasta D, Voigt K, Chemineau P (2021) Review: managing sheep and goats for sustainable high yield production. Animal 15:100293

Sustainable Fibre Analysis Impact Report (2020) https://sustainablefibre.org/wp-content/uploads/2020/07/SFAImpact-Report_Jan-2020.pdf

Temple D, Manteca X (2020) Animal welfare in extensive production systems is still an area of concern. Front Sustain Food Syst 4:545902

Tucker CE, James WE, Berry MA, Johnstone CJ, Grover RF (1976) Depressed myocardial function in the goat at high altitude. J Appl Physiol 41(3):356–361

Turner S, Dwyer C (2007) Welfare assessment in extensive animal production systems: challenges and opportunities. Anim Welf 16(2):189–192

Uddin ME, Kebreab E (2020) Review: impact of food and climate change on pastoral industries. Front Sustain Food Syst 4:543403

Webster AJF, Monteith JL, Mount LE (1974) Heat loss from cattle with particular emphasis on the effects of cold. In: Heat loss from animals and man assessment and control. Elsevier, pp 205–231

Zaramarama (2019, May 22) Animal welfare and cashmere—a personal perspective. Goats and Gers. https://goatsandgers.wordpress.com/2019/05/22/animal-welfare-and-cashmere-production/. Accessed 02 Feb 2023

Zidi A, Abo-Shady H, Molina A, Menéndez-Buxadera A, Sánchez-Rodríguez M, Díaz C, Carabaño MJ, Serradilla JM (2014, Aug) Genome wide association for heat stress tolerance/susceptibility in Florida dairy goats. In: 10th World Congress on Genetics Applied to Livestock Production, pp 17–22

Enhancing Animal Welfare Through Disease Prevention in Goats

George Stilwell and Ana Lopes Vieira

Abstract

In this chapter, *disease* is defined as any physical or mental condition where the normal function of a local tissue, an organ, a system, or the whole body is compromised. Preventing the occurrence of diseases is crucial: the main diseases and how they impact the welfare of goats of all ages are here summarized. Diseases frequently result in pain, discomfort, sleep disturbances, stress, anxiety, fear, undernutrition, and the inability to engage in natural behaviours. However, it is also true that certain factors can increase the likelihood of developing a disease. The main objective of this chapter is to describe how to recognize the different diseases at a very early stage, how to identify the main risk factors involved, and how to prevent their occurrence. Goats, being prey species, have evolved stoicism, masking signs of illness and pain to avoid predation. However, this behaviour can delay detection and treatment by inexperienced stockpersons. Detecting subtle changes in behaviour is important for early recognition of sick animals. Very little will be said about treatment as this is not intended to be a medical/clinical chapter.

Supplementary Information The online version contains supplementary material available at https://doi.org/10.1007/978-3-031-62182-6_8. The videos can be accessed individually by clicking the DOI link in the accompanying figure caption or by scanning this link with the SN More Media App.

G. Stilwell (✉)
Animal Behaviour and Welfare Lab, CIISA, and Associate Laboratory for Animal and Veterinary Sciences (AL4AnimalS), Faculty of Veterinary Medicine, University of Lisbon, Lisbon, Portugal
e-mail: stilwell@fmv.ulisboa.pt

A. L. Vieira
CEG-IST, Instituto Superior Técnico, Universidade de Lisboa, Lisbon, Portugal

S. Mattiello, M. Battini (eds.), *The Welfare of Goats*, Animal Welfare 25,
https://doi.org/10.1007/978-3-031-62182-6_8

Keywords

Goat · Disease · Pain · Prevention · Metaphylaxis · Herd health

1 Why Do Diseases Cause Poor Welfare?

Good Health is one of the four Principles used by the AWIN protocol to assess the welfare of goats (AWIN 2015). The absence of disease is also one of the Five Freedoms globally recognized as the foundation of animal welfare assessment (Farm Animal Welfare Council 1979), encompassing both mental and physical well-being. From a veterinary point of view, we could be tempted to say that the absence of disease is the most important freedom, as the presence of disease generally compromises all the other four freedoms: disease will lead to pain and discomfort, resting problems, stress/anxiety and fear, undernutrition or malnutrition, inability to express natural behaviours, etc. Although this can be seen as a (vet) biased approach to welfare it seems to be the logical basis for a chapter named Welfare and Disease.

It should be recalled that signs of weakness or shortcomings, that would be expected in animals with disease/pain, may be very subtle in ruminants (Delano et al. 2002; Häger et al. 2017). Prey species, such as ruminants, developed the evolutionary advantage of stoicism, meaning that those that could mask the signs of illness and pain would not be the first targeted by predators (Stafford 2014). Yet, goats are unique in some ways—for example, goats are more vocal animals than sheep or cattle (Goldberg 2018). Goats will express, sometimes very loudly, signs of acute stress or pain, but will endure and disguise signs of prolonged disease, discomfort, frailties, or chronic pain, especially when kept in a group (Kata et al. 2017). Although this has evolutionary advantages for the wild animal, it may delay detection and treatment of sick goats by less experienced stockpersons.

Malaise in goats is most often assessed by observing and interpreting changes from normal behaviour. Stockpersons should have training in detecting these subtle alert signs, as success in treatment or disease outbreak control may depend on early recognition of sick animals. Classic behavioural indicators of pain in goats include: isolation, altered demeanour, apathy, fatigue, subdued behaviour, peculiar positioning or prolonged time lying down, altered gait, decreased appetite, reduced rumination, grinding of teeth, cheek and facial muscle contracture, low ears, open nostrils and rapid and shallow breathing (Smith and Sherman 2009). Although goats may be very expressive when subject to acute pain (e.g. injection or disbudding), bleating occurs very rarely in cases of chronic pain or illness in goats. When it occurs it can be considered as a poor prognosis sign. A Numeric Rating Scale for the assessment of pain after surgery in goats (Staffieri et al. 2009) and a facial expression scale for sheep (McLennan et al. 2016) have been developed, but none has been tested and validated for sick goats on farms. However, empirical evidence and some preliminary studies on changes in face expression in kids after disbudding (Ajuda et al. 2020) (Fig. 1) indicate that these methods also apply to goats on farms.

Fig. 1 Expression of discomfort in a goat kid

Fig. 2 Pain can be assessed by recording changes in behaviour after use of analgesic. For example, goats may become more relaxed after treatment (Photo: Silvana Mattiello)

Another way to recognize clinically masked pain is by assessing behaviour changes after administering an analgesic (Fig. 2) or after performing a nerve block. This has been used in studies on pain management in arthritis, pregnancy toxaemia or after disbudding or other surgeries (Smith and Sherman 2009; Zamir et al. 2009; Okafor et al. 2015; Adams 2017; Smith et al. 2021).

There are many ways disease can impact welfare. The main path is the predictable association with pain, undernutrition, and altered social behaviour. Prolonged pain often results from inflammation being caused by infection or trauma. When this happens an "inflammatory soup" activates silent nociceptors and increase C-fibres' responsiveness to stimuli. Additionally, infection agents and inflammation metabolites act centrally causing generalized pain, swelling/oedema will cause local and regional compression of tissues, bowel distention or ischaemia will cause visceral pain, among many other interrelating factors. Inflammation will eventually lead to secondary peripheral hyperalgesia, by which adjacent areas to a traumatized or diseased organ will also be a source of pain (Fig. 3) (Whay et al. 1998; Viñuela-Fernández et al. 2007). Furthermore, central hyperalgesia may result from persistent C-fibres and silent fibres-sourced stimuli reaching the central nervous system. Moreover, behaviour changes originated by illness (e.g. prolonged lying down allied with muscle atrophy or limb oedema) will often lead to generalized musculoskeletal pain (Smith and Sherman 2009). The knowledge that animals, even when being treated (for example, for lameness), may remain very sensitive to the pain caused by an otherwise innocuous mechanical stimulus, is very important (Whay et al. 1998; Herzberg et al. 2020). People responsible for handling or moving farm animals should be aware of this possibility, so that they understand the significance of strange behaviours or reluctance to follow instructions such as walking along a gravelly path (Whay et al. 1998; Anderson and Muir 2005; Reppert et al. 2020) or signs of generalized pain in animals with mastitis (Medrano-Galarza et al. 2012).

Illness, with or without the presence of pain, will reduce appetite, leading to loss of body condition and weakness that results from energy depletion, dehydration and other specific nutritional deficits, and from increased catabolism (Brown and Bradford 2021). Pro-inflammatory cytokines will cause the liver to produce acute-phase proteins, stimulate the mobilization of amino-acids from muscle, and induce cachexia and wasting syndromes, especially in prolonged diseases (Kuhla 2020).

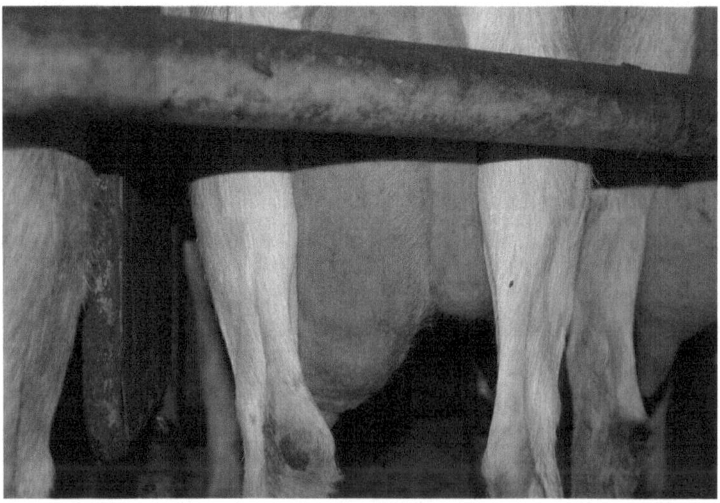

Fig. 3 In goats with mastitis the non-affected half may also be painful due to secondary hyperalgesia

Thus, the association between disease and malnutrition is twofold: physiological—hindered brain regulation of appetite and satiety and increased catabolism; and behavioural—reluctance to walk to the feed trough or to browse, because of pain. Weakness and pain may thwart goats from walking long distances with the flock (Fig. 4), climbing to hard to access areas or browsing branches high above the ground when kept in extensive systems, especially in more arid regions. Several metabolic diseases may occur as a complication of this starvation state (Smith and Sherman 2009), as further discussed below.

Another way disease will affect overall goat welfare is by inhibiting social behaviour. Sick animals will avoid confrontation and seek less and be less targeted by initiators of affiliative behaviours. On the other hand, not being able to tackle or to avoid agonistic behaviours will lead to fear, stress, and loss of hierarchy status. Synchrony with the flock, especially the inability to accompany or follow other animals when browsing, may also be affected, causing additional stress, anxiety, and frustration (Smith and Sherman 2009).

Because typically young goat kids have very low energy reserves, disease in these animals may be particularly challenging. Hypophagia will quickly lead to hypoglycaemia, hypothermia, and immune depression. Likewise, any disease or disorder that prevents kids from following the dam (e.g. septic arthritis) or suckling (e.g. orf) will have rapid and usually severe consequences. Identification of sick kids should be done as early as possible, and treatment of the more severe cases should encompass parenteral or forced feeding and fluid therapy.

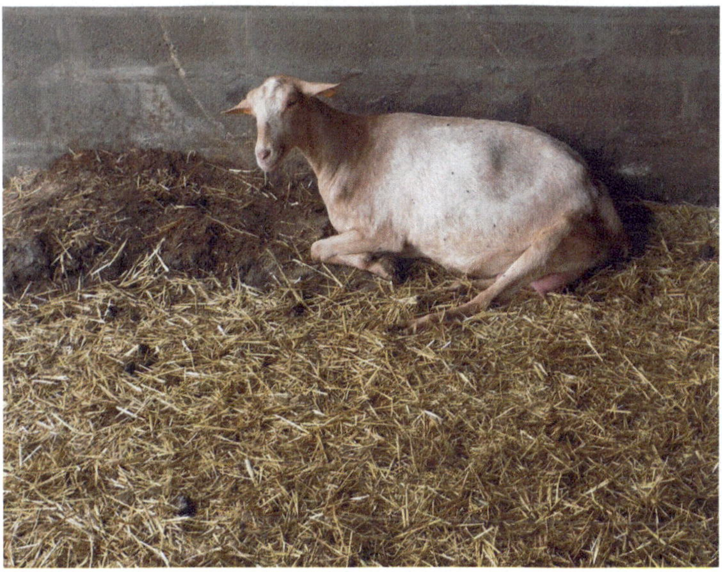

Fig. 4 A sick goat that refused to go out to pasture with the flock, aggravating negative energy balance and predisposing to metabolic diseases

Finally, disease of goats may affect indirectly the welfare of kids, especially the newborns, as milk from sick goats may be scarce or less nutritious. Also, goats in pain (e.g. those with mastitis) may not allow kids to suck. Kids from goats affected by disease may go hungry, show higher susceptibility to disease, and even starve to death. Treatment of goats with suckling kids should always consider the possibility of having to supplement the young ones.

In summary, most diseases cause poor welfare in multiple and diverse ways. Early detection and competent treatment are keys in reducing the impact of disease on welfare. There is compelling evidence that the eradication of some infectious diseases (e.g. Caprine arthritis encephalitis or paratuberculosis) correlates with better overall welfare (Muri et al. 2016).

2 Poor Welfare and Disease

The relationship between welfare and disease can occur in one other direction—poor welfare leading to disease (Hempstead et al. 2021). Factors impacting welfare usually lead to greater susceptibility to disease because of reduced or impaired immune response to pathogens. Some of the most common and well-studied examples are respiratory disease following transport and comingling; or increased incidence of disease in submissive animals in overstocked conditions (Fig. 5) (Greenwood and Shutt 1992; Linklater and Smith 1997; Kannan et al. 2000; Lacasta et al. 2008; Smith and Sherman 2009). The reasons for this correlation are multiple and complex. Chronic release of cortisol (as a stress hormone) is often blamed as a

Fig. 5 Overstocking in poorly ventilated and illuminated barns increases the susceptibility to respiratory disease in young goats

crucial factor (Greenwood and Shutt 1992), but other elements, such as prolonged undernutrition or thirst, can also reduce the immune capacity to fight commensal microorganisms.

So, in conclusion, disease is certainly a cause of poor welfare but poor welfare may also be the trigger for disease.

3 Reducing Disease and Reducing the Welfare Impact of Diseases

There are many ways farmers can use to prevent diseases and reduce their impact on welfare. These will be specified with the description of the different diseases, but, in general, the most common welfare hazards can be grouped as:

- Nutrition source—Although there is the idea (myth?) that goats will endure well while eating sticks or paper, it is evident that a strong and efficient immune system will not work in animals with energy and other nutrients' deficits. This is especially the case in lactating dairy goats, in does carrying multiple foetuses, and in new-born kids. In animals kept in extensive systems, supplementation must be ensured when pasture is poor or scarce. In intensive systems, diet must be adequate as there is sometimes a tendency to look at goats as "small cows". For example, the use of maize silage may lead to severe ruminal acidosis as goats are very skilled in choosing the grains. In some dairy herds, the problem may be obesity, as goats will accumulate abdominal fat with a negative effect on rumen fill (Fig. 6) (Lima et al. 2012a). The inclusion of a specialized nutritionist is strongly recommended for large dairy goat farms. For more detailed information about nutrition and welfare, please see Chap. 4 (Housing and feeding).

- Housing source—Several flaws are seen in installations that may lead to higher disease prevalence. For example, in more traditional and rural settings, goats are most times kept at night (and in extreme weather conditions) in ill-ventilated small pens with very poor light and hygiene. Sometimes water is not permanently available or is provided in old and dirty troughs. Also, potentially traumatic objects (e.g. barbed wire, shovels, axes, etc.) are sometimes kept in the same area. In the intensive systems, the feed trough is sometimes placed too low for goats and the environment is bare, leading to frustration and stress, and eventually to reduced immune competence. More detailed information about housing factors which may affect goats' health can be found in Chap. 4 (Housing and feeding).
- Management source—This includes the adoption of a well thought and designed vaccination programme against the main infectious diseases. Once more, vaccination schedule should be species-specific, as rules that apply to cows or even sheep may not be adequate for goats. For example, vaccination against clostridium should be more frequent than that advocated for sheep. Deworming and

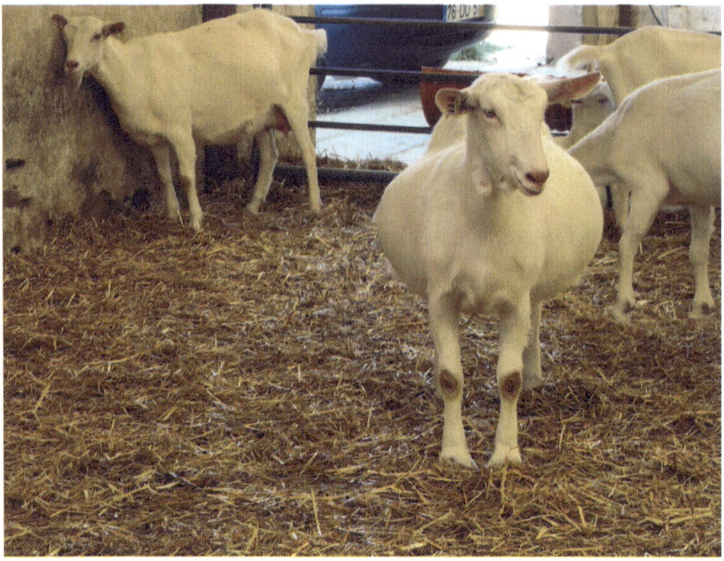

Fig. 6 Overfed goats accumulate fat in the abdominal cavity increasing the risk for pregnancy toxaemia

other anti-parasite treatments (including against ectoparasites) should be modelled to each farm's requisites, based on the risk evidence. Anti-parasite resistance prevalence is growing mostly because of its excessive use (Antunes et al. 2022). Goats in intensive systems may not need deworming, at least not in adult animals.

– Goats in very intensive systems may have overgrown and deformed claws (Ajuda et al. 2019). This may reduce mobility and access to the feed trough, increasing the risk of pregnancy toxaemia. Thus, regular and competent claw trimming is an essential procedure to reduce disease incidence and increase overall welfare (Fig. 7).

Also included in management is the need for competent and trained supervision of animals of all ages. As was said above, goats may disguise signs of disease so that only an expert eye can identify animals with problems. Training in goat behaviour and welfare for all staff working closely with the herd should be sought. Additional information on the impact of management practices and stockperson competence on goat welfare can be retrieved from Chap. 5 (Management).

Although this is not a book on disease treatment in goats, it is important to underline here that there are huge physiologic differences between ruminant species, and so extrapolation to goats from clinical studies' results on cattle or sheep should be done very cautiously. As much as possible, treatment selection (including dosage) should be based on strong scientific evidence and not on ideas or experience in other species.

Fig. 7 Claw trimming is an essential procedure to ensure the welfare of dairy goats in intensive systems

4 Welfare Impact of the Most Common Diseases of Goats

The most common and the most important diseases in terms of impact on welfare will be discussed. Emphasis will be given to early signs, welfare impact, and prevention measures. The objective is to provide information on the most common diseases so that rapid alert and adequate intervention are provided at an early stage of the disease, in order to minimize the negative impact on goats' welfare.

4.1 Metabolic Diseases

Dairy goats are especially vulnerable to metabolic diseases when their nutritional requirements are not met during late pregnancy and early lactation. Under certain conditions, morbidity and mortality are very high. The welfare impact of the clinical disease is usually huge, characterized by extreme suffering of both goat and foetus/kid.

4.1.1 Pregnancy Toxaemia

There are several paths to get to clinical pregnancy toxaemia. The common ground is the negative energy balance that may occur in goats at the end of twin or triplet gestation, as multiple foetuses require and consume more energy than the dam is able to ingest and absorb. Both types of pregnancy toxaemia (overeating and undernutrition) usually occur in the last 6 weeks of gestation.

Obese pregnancy toxaemia occurs because intra-abdominal fat stores and a uterus filled with two or three foetuses occupy most of the abdomen, so that the rumen capacity is immensely reduced (Lima et al. 2012a). Leptin, high in obese animals, may also play a role by further reducing the goats' appetite (Smith and Sherman 2009). On the other hand, undernutrition pregnancy toxaemia or starvation ketosis develops when not enough food or very poor and coarse feedstuff is provided to end-of-gestation does (Fig. 8). Although there is no extra energy requirement until the last 2 months of pregnancy, it is established that if Net Energy (NE) supplementation in late pregnancy is under 500 kcal/d, goats with multiple gestation will be more susceptible to pregnancy toxaemia (Smith and Sherman 2009). Even if food is not scarce, pregnancy toxaemia can occur in more submissive animals due to lack of access to feed, namely because of insufficient feeder space, continuous exposure to extreme weather conditions, overstocking, etc.

A similar but much less frequent disorder can occur soon after parturition (lactational ketosis) with negative energy balance being caused by inability to compensate for energy depletion in milk production.

Thus, pregnancy toxaemia is usually a flock problem as it results mainly from farm dietary or management circumstances (Lima et al. 2012b). The exception may be secondary pregnancy toxaemia, by which an individual goat will reduce food intake because of another disease or condition (e.g. lameness, poor teeth condition).

The most evident signs of the clinical disease are: progressive loss of appetite, reluctance to get up and move, increasing signs of depression, neurologic signs (e.g. stargazing, nystagmus, ataxia, tremors), apparent blindness, teeth grinding, bloat, swollen distal limbs, lateral recumbence, and coma (Fig. 9). Low glucose and elevated ketone bodies and urea can be found in blood and ketone bodies will be easily detected in urine (Lima et al. 2012a, 2016; Rodolfo et al. 2014). Close supervision

Fig. 8 Pregnant goats kept in very arid areas without supplementation are more prone to develop pregnancy toxaemia

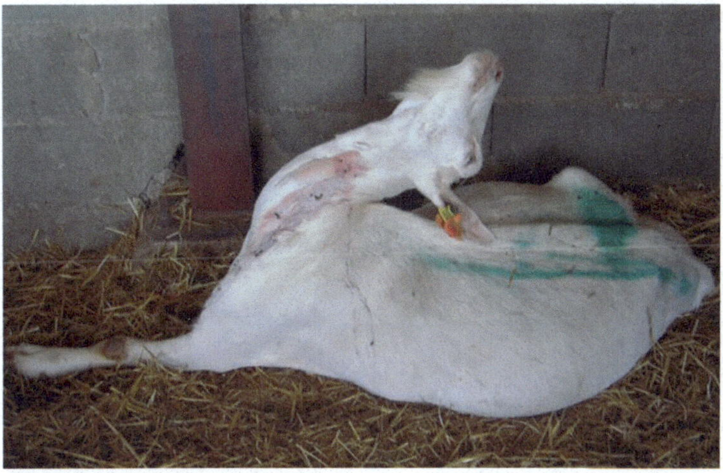

Fig. 9 Neurological signs (stargazing) in goat with pregnancy toxaemia (Photo: Miguel Saraiva Lima)

of animals at risk (last 6 weeks of pregnancy) is crucial to detect what can be the start of an outbreak and to deliver early treatment to affected goats.

Affected animals not treated immediately will suffer greatly until death occurs. In goats with very poor prognosis, removal of the foetuses may be necessary (C-section or parturition induction) (Lima et al. 2012a, b). This may bring about additional welfare issues—the welfare of premature kids and post-surgery recovery, as the survival rate of the dam is below 60%. Both dam and kids will be exposed to extreme stress and should be the object of intensive care (comfortable bedding, protection from cold, hand-feeding of colostrum to the kids…). Euthanasia of severely affected goats (Fig. 10) and very premature and unresponsive kids should be considered on welfare grounds.

Prevention measures include: well programmed and designed nutrition plan, easy access to food, energy supplementation in case of poor pastures, early treatment of any disease in pregnant does (for example, lameness), protection from inclement weather as this may reduce food availability while increasing energy expenditure, etc.… In large farms, pregnant goats should be grouped, according to: body condition score; stage of pregnancy; and number of foetuses. This distribution should be re-evaluated every fortnight (Smith and Sherman 2009; Brozos et al. 2011).

4.1.2 Hypocalcaemia and Hypomagnesaemia

These two less common metabolic diseases occur shortly before or up to 2 weeks after parturition, but in some ways can be confused with pregnancy toxaemia.

Hypocalcaemia is characterized by isolation and progressive paralysis, stiff gait with muscle tremors, recumbence, cold ears, bloat and finally lack of conscience. Types of pasture (e.g. cereals), plants rich in oxalates and stress, are predisposing factors.

Hypomagnesaemia occurs primarily in goats grazing young, rapidly growing spring/autumn pastures (low in magnesium). Adult goats have a limited ability to

Fig. 10 Pregnancy toxaemia can cause extreme suffering. Prompt treatment is essential. Euthanasia is sometimes recommended on welfare grounds

summon magnesium from body reserves, so are dependent on daily intake. Although clinical cases may be rare, these are characterized by teeth grinding, tetany, seizures, recumbency and may eventually lead to death. Prevention is achievable by supplementing with magnesium (lick blocks), avoiding pastures with very young grass, or reducing potassium intake before parturition (Dalley 2002).

4.1.3 Ruminal Acidosis

The main cause of ruminal acidosis is the excessive eating of starch-rich diets (grain or concentrate) or reduced ingestion of long fibre that increases the need to ruminate (Braun et al. 1992; Giger-Reverdin 2018). Because goats can easily select particles in the feed, this condition can arise from feeding maize silage as goats will look for the grain. Two degrees can develop: subclinical acidosis and clinical acidosis. The first is characterized by anorexia, reduced rumen motility with a filled rumen, mild bloat, diarrhoea, milk drop, teeth grinding, and mild signs of abdominal pain. It is very correlated to the sorting behaviour characteristics of goats and so may affect only part of the flock (Giger-Reverdin 2018). It is more common in dairy goats in intensive systems. Recovery occurs spontaneously, especially if the diet is corrected, and so the welfare impact is mild (Rapetti and Bava 2008).

In contrast, acute ruminal/systemic acidosis follows rapid consumption of large amounts of grain/concentrate. This may happen when total mixed diets are poorly formulated and grain is easily selected by animals. Rapidly fermented sugars and starches lead to the production of lactic acid (and other acids) which draws huge amounts of fluids into the rumen. Rumen contractions are absent and the rumen is filled with fluid and gas, or sometimes foam (Fig. 11), causing severe abdomen enlargement.

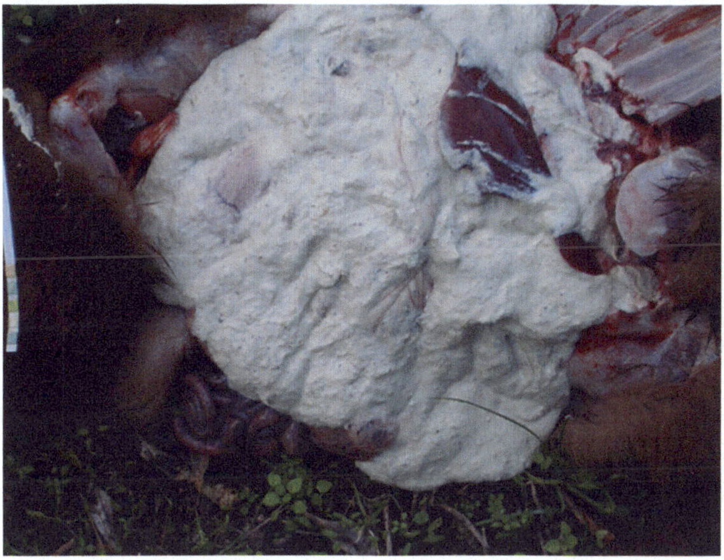

Fig. 11 Foam in rumen of young goat that died from acute ruminal acidosis

Other symptoms may include constipation followed by diarrhoea, bloat, colic, muscle tremors, and teeth grinding. This will be followed by splashy sounds when the abdomen is shaken, abdominal distension and signs of dehydration (sunken eyes, skin tenting, cold ears), neurological signs and eventually hypovolaemic shock (Ermilio and Smith 2011). Polioencephalomalacia (with ataxia, blindness, and other neurologic signs) is a very common complication of ruminal lactic acidosis as biotin-producing bacteria will die in low pH (Chigerwe and Aleman 2016). Vocalization represents a very poor prognosis. Death usually occurs if treatment is not immediate (Smith and Sherman 2009; Chigerwe and Aleman 2016).

4.1.4 Urolithiasis

Urolithiasis is the formation of crystals and calculi in the urinary tract (Fig. 12). The urethra of male goats is convoluted and the diameter is thin; this facilitates the entrapment of crystals at the sigmoid flexure or in the urethral process of small ruminants. Obstructive urolithiasis, most frequently seen in fattening kids in intensive systems, castrated males, or overfed show or pet bucks exposed to a high grain/low roughage diet (Fig. 12), results from the obstruction of the urinary tract (i.e. urethra) by these calculi (Sickinger and Windhorst 2022). Females are seldom affected.

The first signs of obstruction are restlessness, tail twitching, and anxiety. Animals then start to evidently strain to urinate. Untrained stockpersons will assume that the animal is constipated. Urine with blood may be seen dripping from the prepuce and crystals (like sand) may be palpated attached to the preputial hairs. In extreme cases, rupture of the bladder/urethra may occur with an accumulation of fluid seen in the ventral abdomen. The welfare of animals with obstructive urolithiasis is deeply compromised due to pain, discomfort when urinating and, probably, metabolic/electrolyte alterations related to kidney insufficiency.

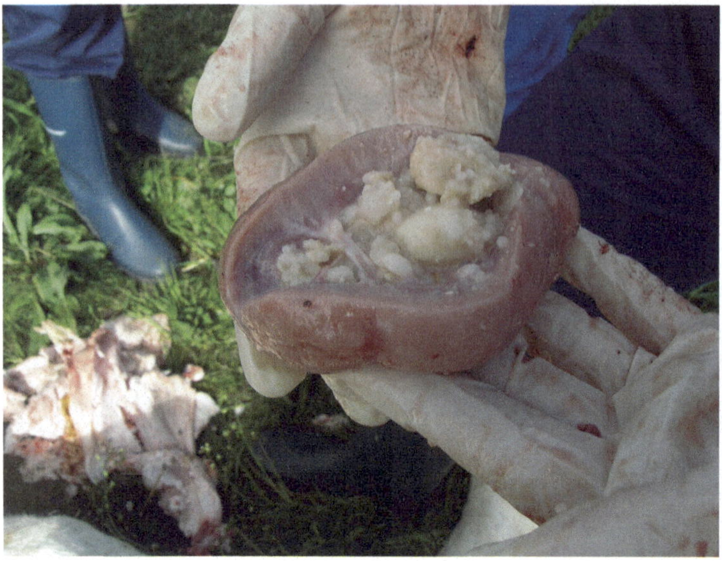

Fig. 12 An extreme case of urolithiasis in a pet buck. Both kidneys were full of calculi (Photo: Miguel Saraiva Lima)

There are three main points for the prevention of urolithiasis:

(a) Increasing urine output by facilitating and promoting water drinking. Access to a clean, reliable, and adequate water supply must be assured. Planning water provision in barns or outdoors should always consider that goats have a very strong hierarchy, which means that dominant animals may keep others away from water troughs. Drinking places should be enough and access should be easy
(b) Acidifying the urine. Ammonium chloride (2.25%) will reduce the pH of goats' urine to 6.5 (Mavangira et al. 2010)
(c) Reducing eventual excess of minerals in the feed and controlling the calcium/phosphorus ratio (Jones et al. 2017)

4.2 Infectious Diseases: Bacteria

The bacterial diseases that have the highest impact on welfare (potentially affecting large numbers and causing severe malaise and suffering) are briefly described.

4.2.1 Clostridiosis

Changes in environment, host resistance, or gut milieu may initiate rapid multiplication, tissue invasion, and release of powerful exotoxins by pathogenic clostridia bacteria. Intake of large amounts of starch, sugar, or soluble protein, sudden changes in diets or access to fresh and lust pastures are all considered risk factors in the development of these diseases. These toxins may damage and destroy vital organs that are distant from where the bacteria multiply.

Several clostridia cause severe disease, being responsible for intense suffering and almost inexorably leading to death. The most common are known as enterotoxaemias and are caused by various types of *Clostridium perfringens* (Uzal and Songer 2008). In goats, most enterotoxaemia cases seem to be caused by *C. perfringens* type D. Enterotoxaemia is clinically expressed in three distinct forms (Blackwell et al. 1991; Smith and Sherman 2009; Simpson et al. 2018): (i) peracute, that is more common in kids, may cause death with almost no previous clinical sign. However, most cases will show profound depression, signs of severe abdominal pain (colic) that includes very loud bleating and watery, bloody, and mucous diarrhoea, and generalized haemorrhages (Fig. 13); (ii) acute, with similar but less severe and abrupt signs and soft faeces that will eventually lead to diarrhoea. Death usually occurs after a 4–5 days' course; (iii) chronic, in which mature animals apparently recover from mild clinical signs only to relapse back to disease after a few days. This can continue for weeks, during which goats show dullness, anorexia, low milk yield, weight loss, and poor-quality hair coat. Death, euthanasia, or culling usually follow. Abomasitis may also occur due to clostridial infection (Simpson et al. 2018).

Other clostridia species cause disease when getting access to the animal organism through wounds or by ingesting large quantities of spores. Diseases such as tetanus, botulism, malignant oedema, blackleg, etc., are caused by different clostridia toxins, resulting in neurologic or muscle disorders. This clostridiosis is not as common in goats as in sheep, probably because as browsing animals, goats are less exposed to the spores that are usually on the soil. Poorly maintained contaminated feed (e.g. haylage, silage) is also a common source of spores (Smith and Sherman 2009).

Vaccination is the best way to reduce the impact of clostridiosis (Lacasta et al. 2015). Commercial vaccines (toxoids) include the most common toxins. However, several setbacks are frequent in goats: failure of vaccination to protect from clinical disease; the need for more frequent re-vaccination (every 4 months) compared with sheep; not including all possible toxins. Prevention measures also include

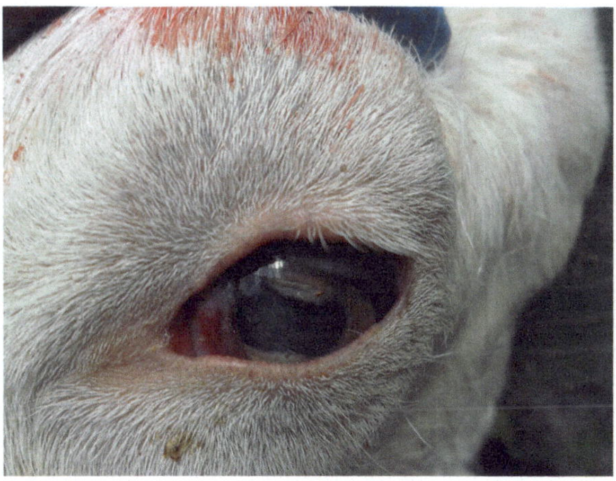

Fig. 13 Eye haemorrhage in young doe with peracute clostrial disease

circumventing sudden changes in diets or excess carbohydrate-rich feeds that may trigger rapid bacteria multiplication (Simpson et al. 2018); for grazing/browsing animals, turnout to thriving pasture should be gradual; also, immunosuppressant factors such as stress should be avoided.

4.2.2 Foot Scald, Footrot, and Foot Abscesses

Although these three conditions are considered different foot diseases, they are difficult to distinguish in the field. Foot scald results from the infection by *Dichelobacter (Bacteroides) nodosus;* foot rot is a more severe and deeper infection of the interdigital space with virulent strains of *D. nodosus* and *Fusobacterium necrophorum* (Depiazzi et al. 1991); and foot abscesses which usually result from the previous diseases when deep tissue are also infected with *Trueperella pyogenes*, and an abscess is formed.

Goats seem to be less prone to interdigital infections compared with some sheep breeds (Ghimire et al. 1999). However, in certain conditions (e.g. wet pasture/bedding and warm temperatures), prevalence can be very high. Other environmental and management factors, such as overgrown claws, overstocking, introducing infected animals to a susceptible flock, etc., may predispose goats to these diseases. Although usually less severe than in sheep, lameness can still be marked when there is underrunning of the hoof horn. In more severe cases, and when the four limbs are affected, goats may walk on their knees or show reluctance to get up and walk. Redness, swelling, pain at palpation, teeth grinding, reluctance to rise or walk, or gait alterations are common signs of the three conditions. Pus may be visible at the coronet. A foul smell is more characteristic of foot rot (Smith and Sherman 2009).

Prevention of these infectious diseases of the feet includes keeping animals in clean and dry bedding, frequent pasture rotation, periodic foot-baths, culling chronic cases and, in very special cases, vaccination (Egerton et al. 2002).

However, probably the most worrying foot problem in goats is the overgrowth and deformation of the claws in intensively kept dairy goats (Fig. 14) (Ajuda et al.

Fig. 14 (**a**, **b**) Severely overgrown and deformed hooves in a dairy goat, due to reduced wear (Video 1 Lame goat) (▶ https://doi.org/10.1007/000-c4p)

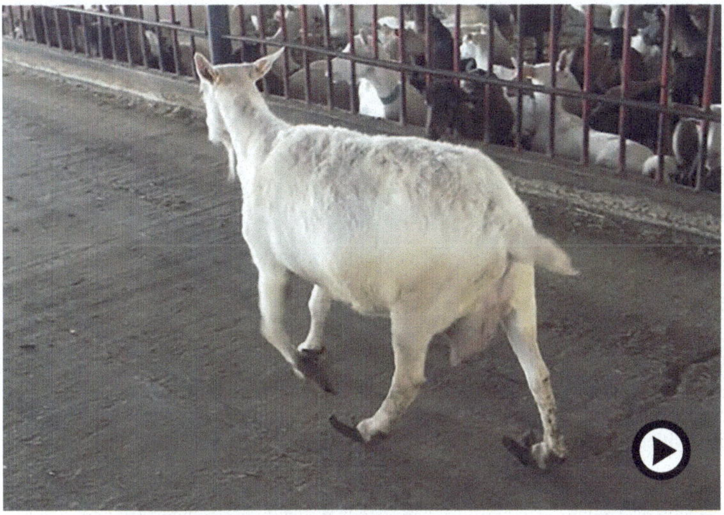

Fig. 14 (continued)

2019). This condition results from lack of wear in these animals kept in pens with deep straw bedding. It may cause severe lameness, although it may not be clear in all cases if this lameness results from mechanical malfunctions or from pain (Ajuda et al. 2013). According to our personal experience, severely deformed claws are associated with chronic pain and chronic lameness, even after curative trimming (Ajuda et al. 2019). This issue is further addressed in Chap. 6 when talking about claw-trimming procedures.

4.2.3 Caseous Lymphadenitis (CLA)

This is one of the various chronic and insidious diseases that affect domestic goats (and sheep). It is caused by *Corynebacterium pseudotuberculosis*. The bacteria are very resistant in the environment and gain access to the animals' body through wounds in the mucous membranes or the skin and eventually colonize regional lymph nodes, producing abscesses. The infected external lymph nodes are usually very evident (5–15 cm in diameter) (Figs. 15 and 16), but infection can go unnoticed if only internal lymph nodes become enlarged. However, these can cause severe disorders due to the compression of nerves or other organs (Fig. 17).

The most commonly infected lymph nodes are the parotid, pre-scapular, and retromammary. The pus in these abscesses is thick (pasty) and yellowish or greenish. For some time, goats will continue to eat and produce milk as healthy animals. Evidence of poor welfare in goats with external abscesses has been described: affected goats show a poorer body condition and higher heart rate, white blood cell count, and total protein concentration than goats without abscesses. Furthermore, goats with abscesses spend significantly less time feeding (Mattiello et al. 2018). Pneumonia, vestibular disease (Fig. 17), mastitis, and meningitis caused by *C. pseudotuberculosis* may occur, but are rare (Stilwell, personal experience, 2022).

Fig. 15 Goat with evident parotid abscess

Fig. 16 Goat with evident prefemoral abscess (Photo: Silvana Mattiello)

If a farm is free from CLA, no measure is too much to keep the microorganism away—very strict biosecurity rules and quarantine of any animal brought into the farm are crucial (Williamson 2001). However, prevention of new infections in farms with an endemic status is not easy. Culling or quarantining infected animals until it is possible to drain and disinfect the abscesses is feasible only in small well-run farms. Avoid abscesses leaking pus onto the goats' living area and pastures at all costs. Only open and clean out abscesses in isolated and easily cleaned areas to which healthy goats cannot get. The use of self-produced bacterins is a possibility,

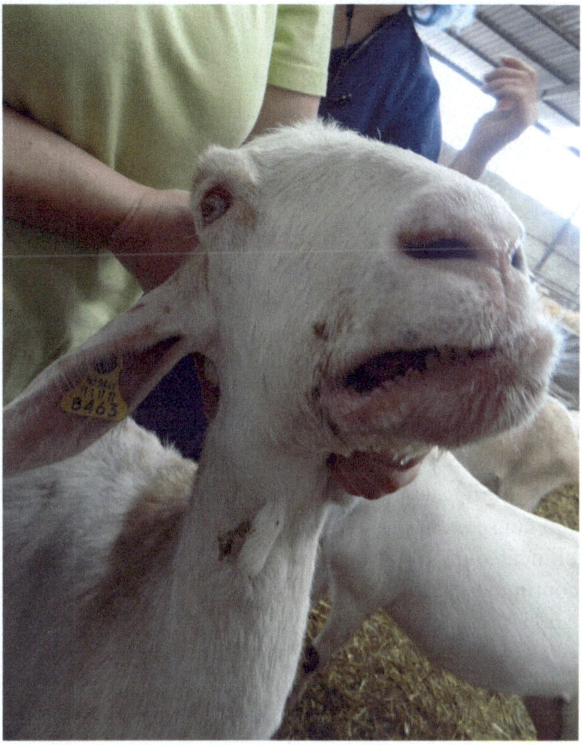

Fig. 17 Facial paralysis due to CLA abscess in brain

but eradication will not be achieved before many years of a vaccination programme (de Pinho et al. 2021).

4.2.4 Abortion Causing Agents: Brucellosis, Chlamydiosis, Leptospirosis, Salmonellosis, Q Fever, and Others

Although the impact on goats' welfare can be minimal, these diseases may cause suffering for the foetus and sometimes for young kids. They may also lead to foetal-membranes retention and toxic metritis in does.

All these bacteria *(Brucella melitensis, B. abortus, Chlamydophila abortus, Leptospira spp, Salmonella* spp., and *Coxiella burnetii*) may cause abortion in goats and sheep, and all are considered zoonotic (Braun et al. 2002; Smith and Sherman 2009; Roest et al. 2012). In many cases, abortion may be the only clinical sign, and laboratory assistance is needed for aetiologic diagnosis. Brucellosis is eradicated from many countries and has been replaced by others as the main culprits in abortion outbreaks. Chlamydial abortion in goats typically occurs in the last 2 months of pregnancy, especially in the last 2 weeks. Leptospirosis will also cause septicaemia and neurologic signs, usually with haemoglobinuria and sudden death, in young goats. Salmonellosis is a cause of abortion but also causes septicaemia, pneumonia, and fibrinous enteritis in young stock, thus reducing goats' welfare. Q Fever is caused by the obligate intracellular rickettsial organism *Coxiella burnetii*. Outbreaks in the first gestation are common and then abortion becomes less prevalent in the next pregnancies. Abortion occurs at the end of gestation time and infected goats will act as carriers (Roest et al. 2012; Van den Brom et al. 2015).

Other possible agents for abortion in goats are intracellular protozoan parasites such as *T. gondii* and *N. caninum*. Domestic and wild felids or canines are the main carriers and responsible for shedding infecting oocytes. The disease rarely causes clinical signs in the goat and the incidence of infections is unknown in most countries (Basso et al. 2022).

Once any of these organisms is identified as the cause of abortion, a control and eradication programme should be established by a veterinarian and extra human health prevention measures should be implemented (especially for pregnant women).

4.2.5 Paratuberculosis

Mycobacterium avium subsp. *paratuberculosis* is responsible for this chronic disease, which affects the digestive apparatus and causes a reduction in the absorption of nutrients and a consequent dramatic weight loss. Although many infected goats will not show clinical signs of paratuberculosis, it is a disease to consider in goats losing weight continuously (Windsor 2015). Contrary to the clinical signs in cattle, goats may not show persistent and watery diarrhoea. This means that paratuberculosis is probably a common but underdiagnosed cause of progressive weight loss in adult goats. As clinical signs (weight loss, soft stool, decreased milk production, slight anaemia, intermandibular oedema) are not characteristic, laboratory diagnosis including isolation of relevant biomarkers is necessary to differentiate the disease from other chronic debilitating pathologies (Buczinski et al. 2019; Fiorentina et al. 2021). Infected asymptomatic animals are very frequent even if clinical cases are only seldom evident.

Prevention is essential. In a naïve farm, very strict biosecurity rules must be applied, as the import of a *M. paratuberculosis* carrier will rapidly lead to an infected herd. Infection of individuals very often occurs by ingestion of colostrum/milk from infected does. This means that management around parturition is a crucial part of any control/eradication programme (Juste and Perez 2011). Immediate separation of the kid, high standards for hygiene in maternities, disinfection of these facilities between parturitions, and using pasteurized colostrum or colostrum replacement (e.g. from dairy cows) are some of the main prevention measures (Windsor 2015). Other measures for prevention of infection of individual animals in an infected farm are: vaccination that will reduce excretion of bacteria from carriers and thus the occurrence of new infections; identifying, isolating, and culling infected goats (Juste and Perez 2011).

The zoonotic potential of the disease (suspected to be associated with Crohn's Disease) is not clear.

4.2.6 Mastitis

Clinical mastitis is relatively rare in goats, compared with cows. Mastitis in goats is mainly caused by bacteria, but also by virus (e.g. Caprine Arthritis Encephalitis

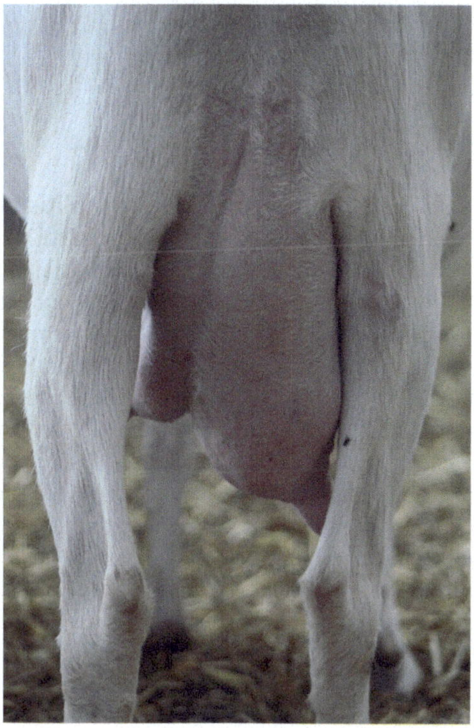

Fig. 18 Chronic mastitis can lead to the atrophy of the half affected

Virus—CAEV). *Staphylococcus* spp. are the most common isolates from mastitis cases and severity differs greatly. For example, *Staphylococcus aureus* may cause both hyperacute clinical mastitis (gangrenous mastitis) and subclinical mastitis (Mishra et al. 2018). Contagious agalactia by *Mycoplasma* spp. may be an important cause of clinical and subclinical mastitis in endemic areas (Contreras et al. 2007).

The signs of acute clinical mastitis in goats are similar to those in other ruminant species—redness, swelling causing asymmetry of udder halves (Fig. 18), pain response even to gentle palpation and changes in milk appearance (watery milk, changes in colour, presence of clots, etc.). Hyperacute cases will show areas of cold purplish skin with reduced sensitivity to palpation. Milk will rapidly change to a reddish or brownish fluid. Pain, discomfort, and toxaemia are the main paths by which mastitis affects the welfare of goats. It can also indirectly impact the welfare of sucking kids by reducing the quantity and quality of the milk ingested. This may lead to permanent hunger, reduced immunity and, in the worse cases, death from starvation.

The use of somatic cell count (SCC) in goats is not recommended for the identification of subclinical mastitis, as this species may present very high SCC in uninfected udders due to a high apocrine component in milk secretion and the possibility that a large number of non-infectious factors will cause a rise in milk cells (Paape et al. 2001). This means that culture is the best way for diagnosing subclinical mastitis at the moment.

Prevention of mastitis depends on good hygiene both in milking as in housing conditions (e.g. dry and clean bedding) and culling chronic cases (Mattiello 2016).

4.3 Infectious Diseases: Virus

4.3.1 Caprine Arthritis Encephalitis Virus

Caprine Arthritis Encephalitis Virus (CAEV) causes lifelong infection, resulting in incurable inflammation in multiple organs, namely joints, brain, lungs, and mammary gland. Goats can act as carriers without showing any clinical signs for years (Michiels et al. 2018). Clinical signs of infection, mainly arthritis, are frequently subtle and may take years to develop (Smith and Sherman 2009). Goats affected by CAEV will show swollen joints (most frequently the carpus joint) (Fig. 19), stiffness, and reluctance to move and will frequently walk on their knees. These signs are usually associated with weight loss due to reduced visits to the feed trough and probably chronic discomfort. Neurologic forms (encephalitis characterized by depression, progressive paresis of the limbs, blindness, head tremor, head tilt, circling and facial muscle paralysis) occur mainly in 1- to 6-month-old kids. Even this form may be very slow, evolving over a period of weeks. The respiratory syndrome is less common in goats than in sheeps' lentivirus infection (Maedi-Visna).

Transmission is mainly by colostrum, but horizontal (e.g. airborne) and iatrogenic transmission through blood-contaminated needles has been proven (East et al. 1993).

Although the impact on milk yield may be negligible (Leitner et al. 2010), this disease is often included in eradication programmes established at farm or at

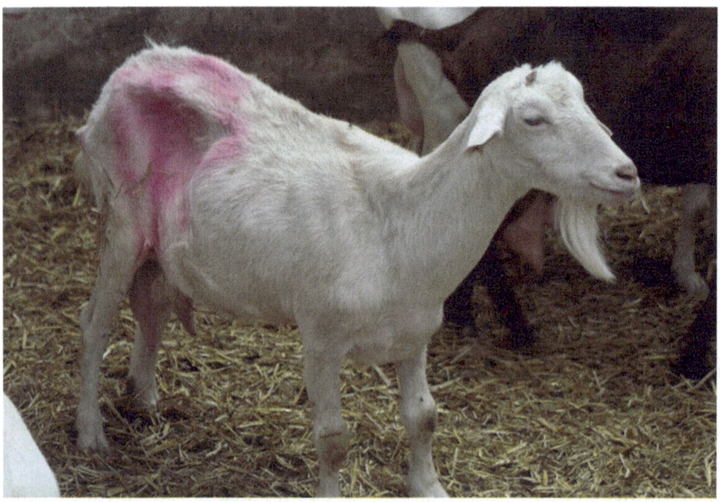

Fig. 19 Adult goat with signs of CAEV—loss of body condition, stiff walking, and swollen carpal joints

country level, because it is highly transmissible and causes occasional outbreaks of clinical disease, and because it can affect very negatively the welfare of goats (Peterhans et al. 2004; Muri et al. 2016). Preventing the spread of CAEV infection includes the removal of kids from their dams before suckling, pasteurization of colostrum (Leitner et al. 2010), and the segregation and culling of seropositive animals (Turin et al. 2005). In some farms, cow colostrum has been used to avoid using contaminated goat colostrum (note: bovine colostrum is unlikely to protect the kid against some caprine pathogens). If eradication programmes are in place, it is highly recommendable to avoid using common needles in vaccination.

4.3.2 Contagious Ecthyma

Contagious ecthyma (Orf) lesions on the teat end predispose to mastitis by affecting the physiological barrier of the teat canal and by harbouring infectious agents (often orf lesions are secondarily infected with staphylococci). Thus, the hygiene at milking is very much compromised in goats with orf lesions. Infection usually results from mouth-to-teat contact by infected kids when suckling, but the hands and teat cups at milking can also be the source of infection (Hosamani et al. 2009).

The clinical signs of infection are the presence of very characteristic lesions on the mouth, nose, ears, and face of kids (Figs. 20 and 21) or on the udder of lactating goats. The lesions are proliferative, looking very much like warts or papillomas. Mastitis in goats and hunger, by not being able to suck, or because of dysphagia (stomatitis), pain/discomfort, and weight loss in kids, are the main welfare consequences of this very contagious viral disease. Prevention of outbreaks depends on quick identification and isolation of affected cases, good hygiene during milking

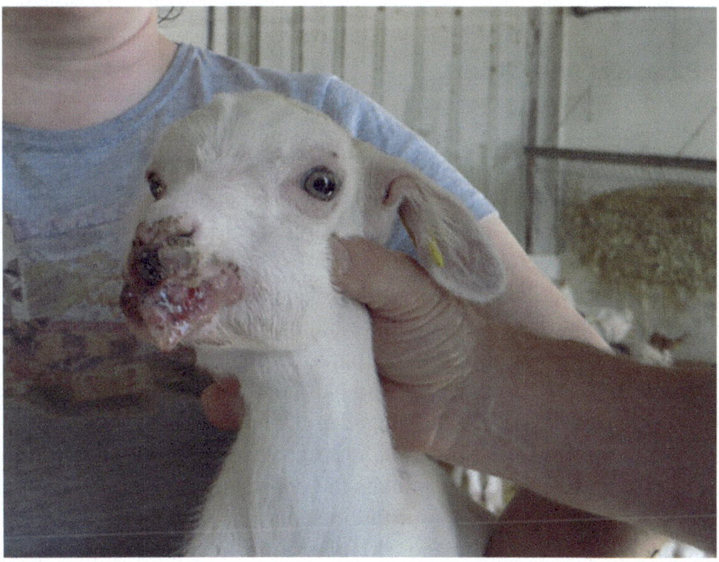

Fig. 20 Kid with very widespread contagious ecthyma—very severely affected kids may die because suckling is impossible

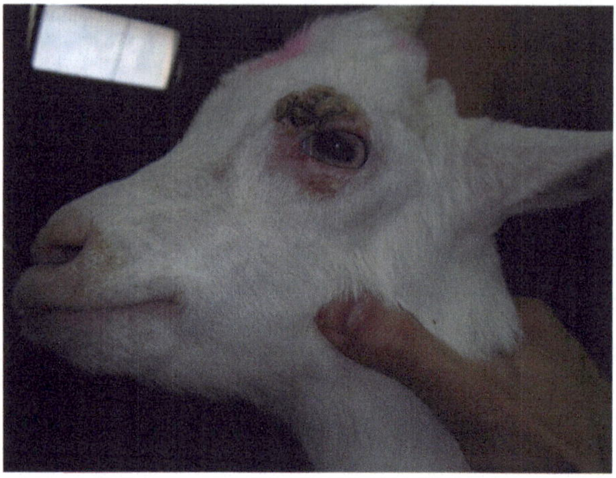

Fig. 21 Although orf lesions are more often seen around the mouth they can develop in other parts of the head and in teats

procedures (use of gloves, disinfection of equipment, etc.), and avoiding sucking of healthy goats by infected kids. Orf-free farms should maintain a closed herd or at least impose a strict quarantine on imported animals. All-in-all-out measures may be needed in case of outbreaks in kid-raising pens, as scabs falling from lesions may be a source of infection for weeks or months (McKeever and Reid 1986). Possibility of vaccination should be discussed with the farm veterinarian, as different rules apply in different countries.

4.4 Parasites and Others

Many kind of parasites affect goats. These are usually divided in two groups: external and internal (mainly gastro-intestinal, GI) parasites.

In extensive or semi-extensive systems, ecto- and endo-parasites are a frequent source of very poor health and welfare. Several species of GI worms, ticks, lice, mites, and flies may infect goats and their relative importance depends on environmental and climatic variables. Nematodiasis and other GI parasites are a bigger problem at pasture because they are more likely to complete their life cycle. The syndrome produced by the different GI parasites may be indistinguishable clinically. The main signs are: weight loss, bad hair condition, diarrhoea, anaemia, submandibular oedema, for GI parasites (Fig. 22) (Domke et al. 2011).

In heavy external parasites' infection, it is expected to see hair loss (through the parasite direct activity or because animals are constantly rubbing the body against objects or self-biting), anaemia, pruritus, discomfort, restlessness, reduced milk production, and ill-thrift. A close and careful examination of the skin (Fig. 23) will reveal arthropods on the skin or attached to the hair. In some types of mange, collecting hair and hair follicles may be necessary for the diagnosis.

a b

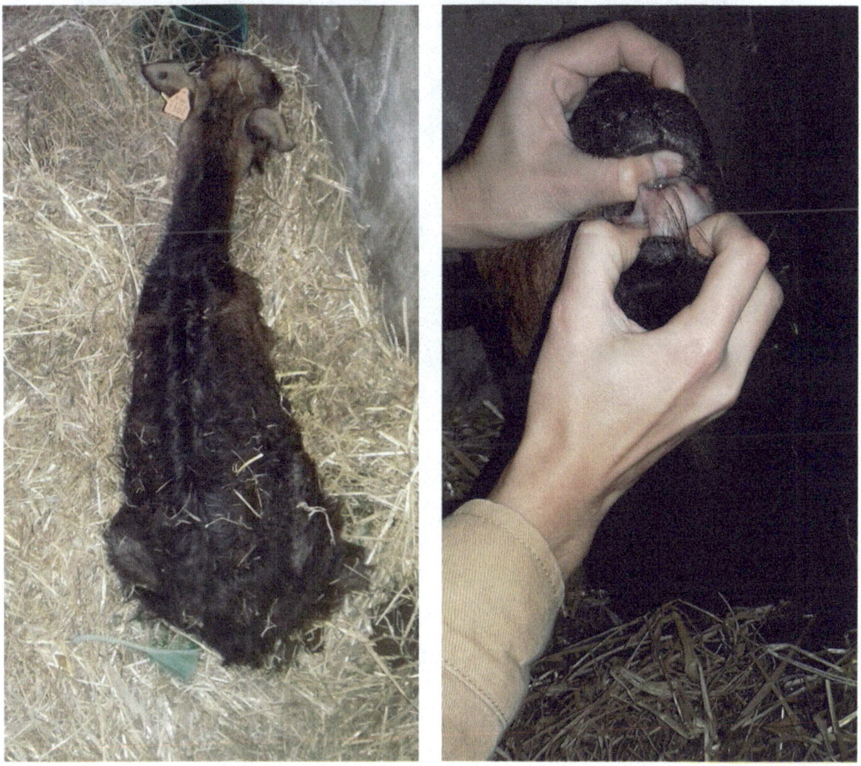

Fig. 22 (**a**) Thin goat retrieved from pasture due to heavy infestation with GI helminth (*Bunostomum* sp.). (**b**) Thin goat with very white oral mucosae was evident

In intensive systems, GI parasites are much less common or at least do not usually cause clinical/welfare problems. This means that, at a time when parasite resistance to several deworming products is increasing, it is highly recommended to only use them when severe infection is identified by lab testing and, even then, mainly on the young stock.

Coccidiosis is an exception as it occurs indistinguishable in intensive and extensive systems. Outbreaks of coccidiosis occur mainly in groups of young goats. The acute form is usually easily diagnosed by clinical signs—diarrhoea, dysentery, dehydration, pale mucosae, weight loss and sometimes sudden death. Some kids that survive clinical coccidiosis will often continue to show ill-thrift, loose faeces, and depression, perhaps because permanent damage to the GI mucosae causes a malabsorption syndrome. Likewise, external parasites may be abundant in both pastured and permanently housed goats, especially when husbandry conditions are poor.

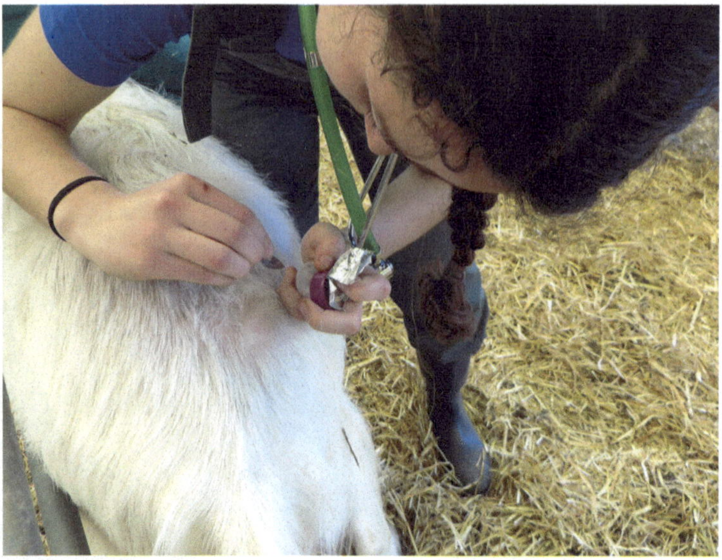

Fig. 23 Collecting hair and skin in a suspected mange infected goat

5 The Gold Principles for Good Herd Health Management

Because many important diseases are very insidious but also very contagious, it is crucial that goat farms remain as closed as possible. The introduction of carriers of diseases such as CAEV, Q Fever, paratuberculosis, tuberculosis, or caseous lymphadenitis is a huge risk and may jeopardize the health and welfare of many animals and compromise the success of commercial farms. Even farms that run very extensive systems should be aware that contact with other herds or pastures can be a risk. Likewise, the practice of taking goats to fairs, exhibitions, or contests should be discouraged. If import of new animals or contacts with "foreign" animals are unavoidable, firm isolation and quarantine procedures should be pursued.

A high immune status is an obligatory requisite for a high herd health status. An efficient, effective, and quick acting immune response will not be possible if nutritional requirements are not met. Overall hunger or energy, protein or minerals (calcium, magnesium, selenium, iodine, cobalt, copper, etc.) deficiencies are many times at the base of outbreaks of devastating diseases. It should be noted that heavy parasite (internal or external) burden is an equivalent way to get to these dangerous imbalances, even if food is plentiful. Nutritional and feeding good practices are detailed in Chap. 4.

In the same way, stress is a very relevant and common risk factor for disease. Stressed animals will eat less and may have high levels of cortisol that will increase susceptibility to infectious agents. Additionally, stressed animals will answer weakly to active immunization. In these cases, farmers may relax in their prevention measures, trusting that goats are protected by vaccination, only to find (too late) that they are still susceptible to infection. Stress is prevented by good stockmanship, good installations, and good husbandry in general.

Two particular problems arise when checking goats in extensive systems: sick animals may be detected late; and, clinical history is often absent or very imprecise. Additionally, it may be difficult to confirm the type of plants eaten during the previous days, the possibility of contact with feral, wild animals or other livestock or the exposure to predators.

Finally, a word on the need to have competent and close veterinary assistance. More than being there to treat and save sick animals, veterinarians are a crucial element in setting rational and science-based vaccination and deworming programmes, and in implementing diagnostic tools including use of lab tests and use of new technologies. Also, in these times when the use of antimicrobials should be limited to treating bacterial diseases, only a veterinarian familiar with the characteristics and particularities of the farm can eventually implement evidence-based metaphylactic antimicrobial protocols that will save many lives without compromising the fight against antimicrobial resistance. Thus, veterinarians should be closely involved in herd health planning and in the evaluation of disease outbreaks, in order to rationalize the use of antimicrobials. Only these professionals can decide if metaphylactic or prophylactic use of antimicrobials is acceptable in view of the risk factors or lack of alternatives.

In summary, good nutrition, good housing, and good human–animal relationship, together with well thought and well-designed prevention programmes, will result in high levels of welfare.

References

Adams J (2017) Assessment and management of pain in small ruminants. Livestock 22(6):324–328

Ajuda IG, Vieira AC, Almeida FF, Stilwell G (2013, August 11–14) Lameness on dairy goats: what is the level of pain experienced and how can we evaluate it? Preliminary results. In: Proceedings: international conference on lameness in ruminants. Bristol, UK

Ajuda I, Battini M, Stilwell G (2019) The role of claw deformation and claw size on goat lameness. Vet Anim Sci 8:100080

Ajuda I, Battini M, Mattiello S, Arcuri C, Stilwell G (2020) Evaluation of pain mitigation strategies in goat kids after cautery disbudding. Animals 10:277

Anderson DE, Muir WW (2005) Pain management in ruminants. Vet Clin North Am Food Anim Pract 21(1):19–31

Antunes MI, Lima MS, Stilwell G, Romeiras MI, Fragoso L, Madeira de Carvalho LM (2022) Anthelmintic efficacy in sheep and goats under different management and deworming systems in the region of Lisbon and Tagus Valley, Portugal. Pathogens 11(12):1457

AWIN (2015) AWIN welfare assessment protocol for goats

Basso W, Holenweger F, Schares G, Müller N, Campero LM, Ardüser F, Moore-Jones G, Frey CF, Zanolari P (2022) Toxoplasma gondii and Neospora caninum infections in sheep and goats in Switzerland: seroprevalence and occurrence in aborted foetuses. Food Waterborne Parasitol 28:e00176

Blackwell TE, Butler DG, Prescott JF (1991) Differences in signs and lesions in sheep and goats with enterotoxemia induced by intraduodenal infusion of Clostridium perfringens type D. Am J Vet Res 52(7):1147–1152

Braun U, Rihs T, Schefer U (1992) Ruminal lactic acidosis in sheep and goats. Vet Rec 130(16):343–349
Braun U, Stehle C, Ehrensperger F (2002) Clinical findings and treatment of listeriosis in 67 sheep and goats. Vet Rec 150(2):38–42
Brown WE, Bradford BJ (2021) Invited review: mechanisms of hypophagia during disease. J Dairy Sci 104(9):9418–9436
Brozos C, Mavrogianni VS, Fthenakis GC (2011) Treatment and control of peri-parturient metabolic diseases: pregnancy toxemia, hypocalcemia, hypomagnesemia. Vet Clin North Am Food Anim Pract 27(1):105–113
Buczinski S, Arsenault J, Kostoulas P, Corbière F, Fecteau G, Dendukuri N (2019) Accuracy of paratuberculosis diagnostic tests in small ruminants: protocol for a systematic review and meta-analysis. Anim Health Res Rev 20(1):98–102
Chigerwe M, Aleman M (2016) Seizure disorders in goats and sheep. J Vet Intern Med 30(5):1752–1757
Contreras A, Sierra D, Sánchez A, Corrales JC, Marco JC, Paape MJ, Gonzalo C (2007) Mastitis in small ruminants. Small Rumin Res 68:145–153
Dalley DE (2002) Diseases of dairy animals, noninfectious - grass tetany. In: Roginski H (ed) Encyclopedia of dairy sciences. Elsevier
de Pinho RB, de Oliveira Silva MT, Bezerra FSB, Borsuk S (2021) Vaccines for caseous lymphadenitis: up-to-date and forward-looking strategies. Appl Microbiol Biotechnol 105(6):2287–2296
Delano ML, Mischler SA, Underwood WJ (2002) Biology and diseases of ruminants: sheep, goats, and cattle. In: Laboratory animal medicine, pp 519–614
Depiazzi LJ, Richards RB, Henderson J, Rood JI, Palmer M, Penhale WJ (1991) Characterisation of virulent and benign strains of Bacteroides nodosus. Vet Microbiol 26(1–2):151–160
Domke AV, Chartier C, Gjerde B, Leine N, Vatn S, Osterås O, Stuen S (2011) Worm control practice against gastro-intestinal parasites in Norwegian sheep and goat flocks. Acta Vet Scand 53(1):29
East NE, Rowe JD, Dahlberg JE, Theilen GH, Pederson NC (1993) Modes of transmission of caprine arthritis-encephalitis virus infection. Small Rumin Res 10:251–262
Egerton JR, Ghimire SC, Dhungyel OP, Shrestha HK, Joshi HD, Joshi BR, Abbott KA, Kristo C (2002) Eradication of virulent footrot from sheep and goats in an endemic area of Nepal and an evaluation of specific vaccination. Vet Rec 151(10):290–295
Ermilio EM, Smith MC (2011) Treatment of emergency conditions in sheep and goats. Vet Clin North Am Food Anim Pract 27(1):33–45
Farm Animal Welfare Council (1979) Five freedoms. https://webarchive.nationalarchives.gov.uk/20121010012427/http://www.fawc.org.uk/freedms.html
Fiorentina P, Martino C, Mancini Y, De Iorio MG, Williams JL, Minozzi G (2021) Using omics approaches in the discovery of biomarkers for early diagnosis of Johne's disease in sheep and goats. Animals 11(7):1912
Ghimire SC, Egerton JR, Dhungyel OP (1999) Transmission of virulent footrot between sheep and goats. Aust Vet J 77(7):450–453
Giger-Reverdin S (2018) Recent advances in the understanding of subacute ruminal acidosis (SARA) in goats, with focus on the link to feeding behaviour. Small Rumin Res 163:24–28
Goldberg ME (2018) Pain recognition and scales for livestock patients. J Dairy Vet Anim Res 7(5):236–239
Greenwood PL, Shutt DA (1992) Salivary and plasma cortisol as an index of stress in goats. Aust Vet J 69(7):161–163
Häger C, Biernot S, Buettner M, Glage S, Keubler LM, Held N, Bleich EM, Otto K, Müller CW, Decker S, Talbot SR, Bleich A (2017) The sheep grimace scale as an indicator of post-operative distress and pain in laboratory sheep. PLoS One 12(4):e0175839
Hempstead MN, Lindquist TM, Shearer JK, Shearer LC, Plummer PJ (2021) Health and welfare survey of 30 dairy goat farms in the Midwestern United States. Animals 11(7):2007
Herzberg D, Strobel P, Ramirez-Reveco A, Werner M, Bustamante H (2020) Chronic inflammatory lameness increases cytokine concentration in the spinal cord of dairy cows. Front Vet Sci 7:125

Hosamani M, Scagliarini A, Bhanuprakash V, McInnes CJ, Singh RK (2009) Orf: an update on current research and future perspectives. Expert Rev Anti Infect Ther 7(7):879–893
Jones ML, Gibbons PM, Roussel AJ, Dominguez BJ (2017) Mineral composition of Uroliths obtained from sheep and goats with obstructive Urolithiasis. J Vet Intern Med 31(4):1202–1208
Juste RA, Perez V (2011) Control of paratuberculosis in sheep and goats. Vet Clin North Am Food Anim Pract 27(1):127–138
Kannan G, Terrill TH, Kouakou B, Gazal OS, Gelaye S, Amoah EA, Samaké S (2000) Transportation of goats: effects on physiological stress responses and live weight loss. J Anim Sci 78(6):1450–1457
Kata C, Rowland S, Goldberg ME (2017) Pain recognition in companion species, horses and livestock. In: Goldberg ME (ed) Pain management for veterinary technicians and nurses. Wiley
Kuhla B (2020) Review: pro-inflammatory cytokines and hypothalamic inflammation: implications for insufficient feed intake of transition dairy cows. Animal 14(S1):s65–s77
Lacasta D, Ferrer LM, Ramos JJ, González JM, De las Heras, M. (2008) Influence of climatic factors on the development of pneumonia in lambs. Small Rumin Res 80(1–3):28–32
Lacasta D, Ferrer LM, Ramos JJ, González JM, Ortín A, Fthenakis GC (2015) Vaccination schedules in small ruminant farms. Vet Microbiol 181(1–2):34–46
Leitner G, Krifucks O, Weisblit L, Lavi Y, Bernstein S, Merin U (2010) The effect of caprine arthritis encephalitis virus infection on production in goats. Vet J 183(3):328–331
Lima MS, Pascoal RA, Stilwell GT (2012a) Glycaemia as a sign of the viability of the foetuses in the last days of gestation in dairy goats with pregnancy toxaemia. Ir Vet J 65(1):1
Lima MS, Pascoal RA, Stilwell GT, Hjerpe CA (2012b) Clinical findings, blood chemistry values, and epidemiologic data from dairy goats with pregnancy toxemia. Bov Pract 46(2):102–110
Lima MS, Silveira JM, Carolino N, Lamas LP, Pascoal RA, Hjerpe CA (2016) Usefulness of clinical observations and blood chemistry values for predicting clinical outcomes in dairy goats with pregnancy toxaemia. Ir Vet J 69:16
Linklater KA, Smith MC (1997) Respiratory diseases. In: Diseases and disorders of the sheep and goat. Mosby-Wolfe
Mattiello S (2016) Husbandry practices and animal health. In: Tsakalidou E, Papadimitriou K (eds) Non-bovine milk and milk products. Elsevier
Mattiello S, Battini M, Mantova E, Noè L, Grosso L, Barbieri S (2018) Evidence of poor welfare in goats with external abscesses. Large Anim Rev 24:113–118
Mavangira V, Cornish JM, Angelos JA (2010) Effect of ammonium chloride supplementation on urine pH and urinary fractional excretion of electrolytes in goats. J Am Vet Med Assoc 237(11):1299–1304
McKeever DJ, Reid HW (1986) Survival of orf virus under British winter conditions. Vet Rec 118(22):613–614
McLennan KM, Rebelo CJB, Corke MJ, Holmes MA, Leach MC, Constantino-Casas F (2016) Development of a facial expression scale using footrot and mastitis as models of pain in sheep. Appl Anim Behav Sci 176:19–26
Medrano-Galarza C, Gibbons J, Wagner S, de Passillé AM, Rushen J (2012) Behavioral changes in dairy cows with mastitis. J Dairy Sci 95(12):6994–7002
Michiels R, Van Mael E, Quinet C, Welby S, Cay AB, De Regge N (2018) Seroprevalence and risk factors related to small ruminant lentivirus infections in Belgian sheep and goats. Prev Vet Med 151:13–20
Mishra AK, Sharma N, Singh DD, Gururaj K, Abhishek, Kumar V, Sharma DK (2018) Prevalence and bacterial aetiology of subclinical mastitis in goats reared in organized farms. Vet World 11(1):20–24
Muri K, Leine N, Valle PS (2016) Welfare effects of a disease eradication programme for dairy goats. Animal 10(2):333–341
Okafor ROS, Remi-Adewunmi BD, Fadason ST, Ayo JO, Muhammed SM (2015) Evaluation of the effect of piroxicam and ascorbic acid combination on live weight and rectal temperature of savannah brown goats during post-operative pain management. J Appl Anim Res 43(4):439–444

Paape MJ, Poutrel B, Contreras A, Marco JC, Capuco AV (2001) Milk somatic cells and lactation in small ruminants. J Dairy Sci 84:E237–E244

Peterhans E, Greenland T, Badiola J, Harkiss G, Bertoni G, Amorena B, Eliaszewicz M, Juste RA, Krassnig R, Lafont JP, Lenihan P, Pétursson G, Pritchard G, Thorley J, Vitu C, Mornex JF, Pépin M (2004) Routes of transmission and consequences of small ruminant lentiviruses (SRLVs) infection and eradication schemes. Vet Res 35(3):257–274

Rapetti L, Bava L (2008) Feeding management of dairy goats in intensive systems. In: Cannas A, Pulina G (eds) Dairy goats feeding and nutrition. CABI International

Reppert EJ, Kleinhenz MD, Viscardi A, Montgomery SR, Crane AR, Coetzee JF (2020) Development and evaluation of two different lameness models in meat goats, a pilot study. Transl Anim Sci 4(4):193

Rodolfo SJC, Augusto AJB, Mendonça CL, Cleyton CCD, Alonso SFP, Jobson CFP, Elizabeth LHF, Pierre SC (2014) Biochemical, electrolytic and hormonal findings in goats affected with pregnancy toxemia. Pesqui Vet Bras 33(10):1174–1182

Roest HJ, van Gelderen B, Dinkla A, Frangoulidis D, van Zijderveld F, Rebel J, van Keulen L (2012) Q fever in pregnant goats: pathogenesis and excretion of Coxiella burnetii. PLoS One 7(11):e48949

Sickinger M, Windhorst A (2022) A systematic review on urolithiasis in small ruminants according to nutrition-dependent prevalence and outcome after surgery. Vet World 15(3):809–817

Simpson KM, Callan RJ, Van Metre DC (2018) Clostridial Abomasitis and enteritis in ruminants. Vet Clin North Am Food Anim Pract 34(1):155–184

Smith MC, Sherman DM (2009) Goat medicine. Wiley-Blackwell. ISBN: 978-0-781-79643-9

Smith JS, Schleining J, Plummer P (2021) Pain management in small ruminants and camelids: applications and strategies. Vet Clin North Am Food Anim Pract 37(1):17–31

Staffieri F, Driessen B, Lacitignola L, Crovace A (2009) A comparison of subarachnoid buprenorphine or xylazine as an adjunct to lidocaine for analgesia in goats. Vet Anaesth Analg 36(5):502–511

Stafford KJ (2014) Recognition and assessment of pain in ruminants. In: Egger CM, Love L, Doherty T (eds) Pain management in veterinary practice. Wiley-Blackwell. ISBN: 978-0-813-81224-3

Turin L, Pisoni G, Giannino ML, Antonini M, Rosati S, Ruffo G, Moroni P (2005) Correlation between milk parameters in CAEV seropositive and negative primiparous goats during an eradication program in Italian farm. Small Rumin Res 57:73–79

Uzal FA, Songer JG (2008) Diagnosis of Clostridium perfringens intestinal infections in sheep and goats. J Vet Diagn Investig 20(3):253–265

Van den Brom R, van Engelen E, Roest HI, van der Hoek W, Vellema P (2015) Coxiella burnetii infections in sheep or goats: an opinionated review. Vet Microbiol 181(1–2):119–129

Viñuela-Fernández I, Jones E, Welsh EM, Fleetwood-Walker SM (2007) Pain mechanisms and their implication for the management of pain in farm and companion animals. Vet J 174(2):227–239

Whay HR, Waterman AE, Webster AJ, O'Brien JK (1998) The influence of lesion type on the duration of hyperalgesia associated with hindlimb lameness in dairy cattle. Vet J 156(1):23–29

Williamson LH (2001) Caseous lymphadenitis in small ruminants. Vet Clin North Am Food Anim Pract 17(2):359–371, vii

Windsor PA (2015) Paratuberculosis in sheep and goats. Vet Microbiol 181(1–2):161–169

Zamir S, Rozov A, Gootwine E (2009) Treatment of pregnancy toxaemia in sheep with flunixin meglumine. Vet Rec 165(9):265–266

Assessing the Welfare of Goats

Monica Battini, Stefania Celozzi, and Silvana Mattiello

Abstract

The assessment of goat welfare must necessarily be based on science, using valid, reliable, and feasible animal-based indicators. In recent years, interest in goats and their welfare has increased significantly, culminating in the publication of the European AWIN welfare assessment protocol in 2015. Although this protocol is widely used, there are still many gaps to be filled for a comprehensive goat welfare assessment. Most studies address the welfare of dairy goats, while virtually no work specifically exists on meat or fibre goats. Only a few indicators have been developed to assess the welfare of bucks and kids, while some studies are available for extensively farmed goats. Feral and wild goats are rarely considered in animal welfare studies. Most validated indicators are useful for measuring the health status of goats, while few are suitable for assessing emotions. There is a great expectation towards Precision Livestock Farming technologies and how it will improve and simplify welfare assessment. While this is partly true, we must not forget that relying completely on technology brings with it risks that can be detrimental to animals.

Keywords

Animal-based indicators · Welfare assessment · Certification schemes · Precision Livestock Farming · Five domains

Supplementary Information The online version contains supplementary material available at https://doi.org/10.1007/978-3-031-62182-6_9. The videos can be accessed individually by clicking the DOI link in the accompanying figure caption or by scanning this link with the SN More Media App.

M. Battini (✉) · S. Celozzi · S. Mattiello
Department of Agricultural and Environmental Sciences—Production, Landscape, Agroenergy, University of Milan, Milan, Italy
e-mail: monica.battini@unimi.it

S. Mattiello, M. Battini (eds.), *The Welfare of Goats*, Animal Welfare 25,
https://doi.org/10.1007/978-3-031-62182-6_9

1 Animal Welfare Assessment Framework

Animal welfare is a complex and multifaceted issue with a scientific, ethical, legal, religious, and cultural dimension. Furthermore, the global interpretation of the animal welfare concept intertwines national and international public policies and has strong economic implications (e.g. commercial trade policy). The responsibility of caring for animal welfare must be shared between governments, citizens, school and universities, veterinarians, scientists, and the people who directly own and use animals.

Animals, including goats, may be kept for production of food, fibre, and other products, for scientific and educational purposes, as working or companion animals, and may be transported and traded internationally. The World Organisation for Animal Health (formerly known as OIE, now WOAH) recognizes all these purposes as legitimate in recognition of the sentience of animals but fosters ethical responsibilities to ensure humane use of animals. In 2017, all the WOAH Member Countries adopted the WOAH Global Animal Welfare Strategy with the objective of achieving: "A world where the welfare of animals is respected, promoted and advanced, in ways that complement the pursuit of animal health, human well-being, socio-economic development and environmental sustainability" (OIE 2018).

This Global Animal Welfare Strategy, while based on science, does not assess animal welfare, but it provides guidelines for proper animal management. The purposes of the assessment are manifold. Animal welfare can be assessed for ethical reasons, for certification purposes to satisfy the market's demand, or for highlighting problems on the farm and consequently advising solutions. In this last case, the assessment outcome should not only provide information on the animals' welfare state but it should also help to highlight the causes responsible for the poor welfare status, and farmers might be encouraged to intervene to solve problems. Finally, welfare can be assessed also for scientific purposes: for example, the comparison between welfare in different housing or management systems can help identifying better solutions for the animals. Animal welfare assessment involves examining the animals to ensure good health and welfare as well as monitoring for any signs of pain, distress, or suffering. This requires a multidimensional approach with specific measures that may be useful to interpret different aspects of animal welfare (Mason and Mendl 1993) and corresponds to a multi-criteria evaluation, with all relevant indicators that do not compensate each other (Blokhuis et al. 2010). The aim of an assessment is to determine the actual welfare of animals, including both their physical and mental state and to this aim the European Food Safety Authority (EFSA) suggests a preferable use of animal-based measures to assess animals' responses to external inputs and their ability to adapt to a context (EFSA 2012). Further, a set of measures and not a single measure is needed to provide a comprehensive assessment of welfare. Non-animal-based measures (i.e. resource- and management-based measures) are often considered easier to record than animal-based measures and usually are consistent over time. However, they cannot really measure the welfare state of an animal, but rather provide information about possible risks for its welfare in case the individual is not able to adapt to its surrounding environment. Therefore, according to EFSA (2012), these measures should be considered as risk factors or, in case they promote positive animal welfare, as benefits (e.g. access to pasture or to palatable food), while animal-based measures reflect the actual welfare state achieved by the animals. Although many animal-based measures are simple and easy to use even under commercial conditions, in some cases they may require further analysis in a laboratory, e.g., metabolic profiling, may be time-consuming to collect, e.g., changes in behavioural rhythms, or may require an extensive training, e.g., Qualitative Behaviour Assessment can be used to identify animals whose welfare is poor, but there is a potential further application of these measures for early prediction of the deterioration of welfare in some animals (e.g. the increase of the somatic cell counts can predict a mastitis outbreak).

Animal-based measures are divided into observations and measures that can be collected on animals during their life (e.g. body condition score, agonistic behaviours) or after the slaughter or killing of the animals (e.g. lesions to internal organs), and data-driven information obtained by farm records (e.g. mortality rate, milk production). To obtain an effective and appropriate welfare assessment, measures must be valid, reliable, and feasible (Battini et al. 2014). Validity is the relation between a variable and what it is supposed to measure or predict and might be situation dependent: sometimes measures that are valid for adult animals are not applicable to young animals or measures used in intensive farming are not effective for extensive husbandry systems. This implies to refine or at least to check if validity criterion is still confirmed in different contexts (EFSA 2012). Reliability is the extent to which a measurement is repeatable and consistent. There are several types of reliability: the inter-observer reliability that is the agreement between different observers during a simultaneous observation; the intra-observer reliability defined as the agreement between successive observations of the same individual or group by a single observer; the test-retest reliability that refers to the agreement between observations performed on the same individual or group on at least two different occasions. A specific case is the consistency over time of animal-based measures, where results must represent a long-term condition and should not be too sensitive to changes, provided that the circumstances have not changed significantly, for example due to seasonal changes (Can et al. 2017). The last attribute is the feasibility that is the practical chance of using the measure for the welfare assessment. Feasibility is strictly dependent on the purposes of the assessment: for on-farm inspections, time, cost, and personnel are major constraints. After consultation, stakeholders of the goat sector industry suggested that a maximum acceptable time for on-farm welfare assessment could be set at 2 h per farm. Measures should not require to be further processed after collection (e.g. for laboratory analysis), effectively ruling out, e.g., blood sampling and hair collection from on-farm welfare assessments and should not require more than one person to be collected (Battini et al. 2014). For different purposes, e.g., scientific use, such constraints may not be relevant.

Based on the literature available, many different animal-based measures can be potentially used to assess animal welfare. The most appropriate measure for a particular situation depends on several different issues, e.g. the purpose of the assessment, the skills of the observer, the conditions under which it is to be gathered.

2 Available Animal-Based Indicators for Goat Welfare Assessment on Farm

The interest towards goats and their welfare is recent (Caroprese et al. 2009) and so far, scientific research has always regarded small ruminants as a single entity, with sheep receiving much more attention than goats. Most of the current research is mainly focused on goat farming and especially dairy goat farming, due to the intensification of husbandry and the consequent increase of welfare problems in such systems (Minnig et al. 2021). That said, it is not excluded that some measures may also be suitable for other categories or, for example, for goats kept in sanctuaries or petting zoos, but some studies are needed to confirm this.

Animal-based measures were commonly grouped as physiological, pathological, productive, and behavioural measures, but this was not an official classification. More recently, the Welfare Quality® project proposed a classification for welfare assessment schemes based on four Principles, identifying the main areas of animal needs (Good Feeding, Good Housing, Good Health, Good Behaviour), split into 12 independent Criteria, each of which corresponds to a key welfare question (Keeling 2009). This classification was then adapted for goats in the AWIN project (Table 1).

Further, the Five Domain Model proposed by Mellor and Beausoleil (2015) suggested four domains related to internal states and external circumstances (i.e. Nutrition, Environment, Health, and Behaviour), and a fifth domain, i.e., Mental State, which is a final component showing positive or negative affective engagement resulting from the sum of other domains. This last approach seems the most appropriate to present a list of relevant animal-based measures.

2.1 Nutrition

This section describes measures that can be used to assess signs of hunger, thirst, and malnutrition.

Table 1 Welfare principles and criteria proposed during the Welfare Quality® project. Indicators are those included in the AWIN welfare assessment protocol for goats

Welfare principles	Welfare criteria
Good feeding	Appropriate nutrition
	Absence of prolonged thirst
Good housing	Comfort around resting
	Thermal comfort
	Ease of movement
Good health	Absence of injuries
	Absence of disease
	Absence of pain and pain induced by management procedures
Appropriate behaviour	Expression of social behaviour
	Expression of other behaviours
	Good human–animal relationship
	Positive emotional state

The assessment of body condition in goats is particularly challenging without specific training and previous experience because these animals generally have important visceral and internal fat deposits rather than subcutaneous fat. Body condition can be expressed using a Body Condition Score (BCS). Several BCS systems have been developed for goats, either merely based on visual inspection or requiring palpation (or both) (Battini et al. 2014). A practical 3-point visual scale ("very thin", "normal", "fat") was developed by Vieira et al. (2015) (Fig. 1).

The observer locates behind the single standing animal and checks the general condition and the rump region, evaluating the presence of prominent or not discernible hip and pin bones and if the line that connects the hipbone to the thurl is concave or convex. This method is reported to be highly feasible but at present, it has been validated only for Saanen, Alpine and their crossbred goats. Also, Anzuino et al. (2010) chose a 3-point scale method ("obviously thin", "normal", "obviously fat"); however, this method requires palpation and is therefore more time-consuming due to the need for restraining the animals. More detailed scales requiring palpation were developed and used by several authors (McGregor and Butler 2008; Muri et al. 2013; Santucci et al. 1991). However, simplified scales are more suitable for on-farm welfare assessment schemes where the purpose of BCS evaluation is to find detrimental situations with animals extremely lean or fat as a quick screening before further in-depth analysis. Visual methods are less stressful for animals, but may lose some accuracy (Minnig et al. 2021). BCS is prone to variation during the year, but this is expected if affected by changes in lactation stage or seasons (Can et al. 2017; McGregor and Butler 2008). So far, BCS has been applied only to goats for milk and fibre production in intensive and extensive situations.

Poor body conditions in goats may be associated to mineral deficiencies or surplus, and this can be reflected in the hair coat condition, which can therefore be

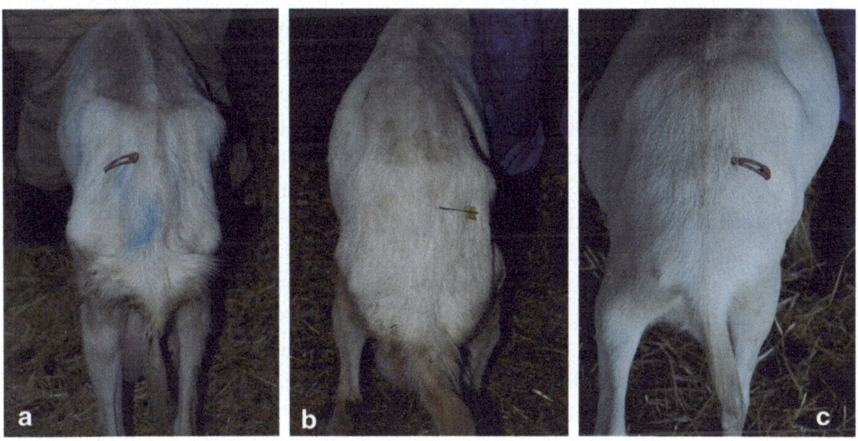

Fig. 1 Example of very thin (**a**), normal (**b**), and very fat (**c**) goats according to the 3-point scale scoring developed to visually assess the body condition from a distance, avoiding palpation (Photos: Ana Vieira)

considered as a further valid and feasible measure to be visually assessed to evaluate hunger and malnutrition (Battini et al. 2015a) (see Sect. 2.3).

Goats require fresh and clean water always available. This can be a critical issue mainly in extensive husbandry systems, also due to the global warming that is leading to a continuous increase in temperature with more hot days and drought periods (Sejian et al. 2021). Literature is scarce on measures to assess if goats suffer from prolonged thirst or are dehydrated. Changes in duration or frequency of drinking behaviour can be an alert of compromised welfare in dairy goats (Al-Ramamneh et al. 2012; Darcan et al. 2008). The AWIN protocol for welfare assessment of goats in intensive farming systems suggests the observation of the number of goats queueing at the drinker to assess whether goats are free from thirst (AWIN 2015a). However, this indicator can be applied only in intensive farming systems, and no alternative indicator has been identified for extensive or semi-extensive systems yet (Battini et al. 2021) Water intake, urine and faecal outputs, plasma volume, and haemoconcentration were used by Jaber et al. (2019), Rahardja et al. (2011) to assess the effect of restricted water availability in meat goats, but this required laboratory analysis, implying a low feasibility due to time and cost constraints.

2.2 Environment

This section presents measures that evaluate how goats live in their environment, in terms of physical and thermal comfort.

The assessment of hair coat cleanliness or dirtiness is relevant for goats, as these animals dislike lying on damp surfaces. However, in indoor conditions wet bedding is unexpected due to absorbing properties of straw and usually dry faeces of goats. Different methods can be used to assess cleanliness in dairy goats, such as a 3-point scale from optimal to poor (Tiezzi et al. 2019) or a detailed scale for many body regions (Anzuino et al. 2010). However, the assessment of cleanliness can be affected by hair coat colour, with white-coloured breeds looking dirtier than dark-coloured breeds (Battini et al. 2016b, c). This last issue might be overcome by an accurate individual assessment and more time spent on the evaluation.

Goats prefer hard surfaces as they evolved in rocky areas and hidden places for resting (Zobel et al. 2019). They also prefer lying alone or in pairs and this behaviour is influenced by space availability (Fig. 2).

Behavioural observations can be performed to assess preferences and choices of goats in the environment (e.g. time spent lying, resting synchronization, body postures, resting by the wall, lying alone), but direct observations or video recordings are usually time-consuming and unpractical (Andersen and Bøe 2007; Celozzi et al. 2021). As a more feasible alternative, some of these behaviours, such as lying time, may be evaluated taking advantage of newly developed Precision Livestock Farming techniques, for example fitting accelerometers to the rear goats' legs (Maurmann et al. 2021; Zobel et al. 2015).

Fig. 2 Goats generally prefer resting against walls, seeking limited or no contact with other goats. However, overcrowded pens can adversely affect their welfare (Photo: Monica Battini)

Inadequate temperatures, high humidity, wind, and rain are the main factors affecting thermal comfort in goats (Bøe and Ehrlenbruch 2013). Information available so far mainly covers the needs for cosmopolitan breeds of dairy goats housed indoors and suggests an optimal temperature of 10–18 °C, relative humidity between 60 and 80%, and wind speed of 0.5 m/s, with a Temperature Humidity Index (THI) that ranges from 55 to 70 (Battini et al. 2016a). Both cold and heat stress can be visually assessed using two different 3-point scales ("no signs", "hair horripilation", "shivering" for cold stress (Fig. 3); "normal respiration", "elevated respiration", "panting" (Video 1) for heat stress (Battini et al. 2016a)) or by collapsing the system in a binary scoring with presence/absence of any sign of thermal stress (AWIN 2015a) (Fig. 4).

Some physiological and blood parameters are also valid indicators of thermal stress. In goats suffering from heat stress, water intake, rumination rate, rectal temperature, pulse and respiration rate, skin temperature (Darcan et al. 2008), glucose, total cholesterol, urea, and cortisol (Sejian and Srivastava 2010) are likely to increase, whereas total protein and aldosterone tend to decrease (Sejian and Srivastava 2010). However, it is questionable whether these measures are specific of thermal stress and even feasible in different contexts.

Fig. 3 Goats may experience cold-related discomfort, exhibiting signs like hair horripilation and shivering in severe instances. Following a period of acclimation to lower temperatures, they typically develop a thicker coat (Photo: Monica Battini)

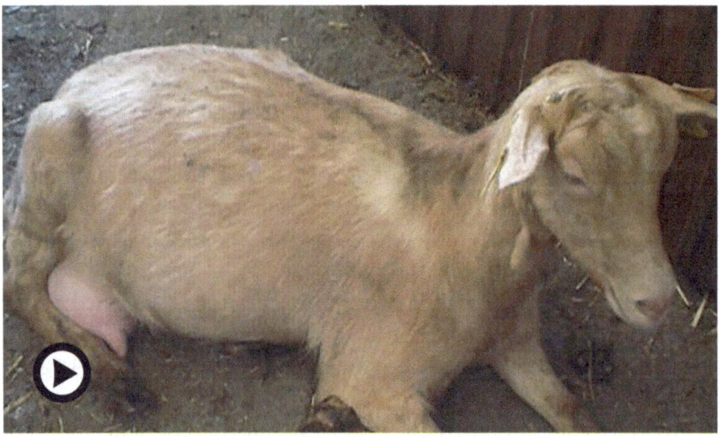

Fig. 4 Goat panting with closed mouth (Video by Monica Battini) (Video 1) (▶ https://doi.org/10.1007/000-c4s)

2.3 Health

This section discusses the measures of health, including lesions and diseases, and pain or pain related to management procedures. Research on health and disease aspects has always been much more extensive than research on other aspects of welfare, hence a wide range of health indicators is currently available, and only the most relevant measures of goat welfare will be presented in this section. This section does not aim to be completely exhaustive given the breadth of the topic.

Lameness is one of the major welfare issues in dairy cows, but information on goats is scarce. It is known that lameness induces pain and may alter normal behaviours, for example limiting the possibility of moving and of getting access to food. Several lameness scales are reported for goats, but the main issue is where and when to assess the presence of this problem. In fact, this could be more difficult in goats housed indoors, because they usually walk on soft bedding that may reduce the evidence of irregular gait and mild lameness. However, moving the animals to concrete flooring is not always feasible. These are the reasons behind some authors' choice (Battini et al. 2016b, c; Can et al. 2016) to evaluate only severely lame animals (checking for the presence of some specific traits, i.e. abnormal gait, head nodding, spine curvature, kneeling in the pen), with the assessor moving inside the pen. Muri et al. (2013) reported difficulties when assessing lameness from outside the pen on a 4-point scale system, due to low visibility in crowded pens. A 4-point scale system was used by Anzuino et al. (2010) to assess the animals both in the pen and in the milking parlour, but this method was judged time-consuming and therefore unfeasible (Fig. 5).

However, the evaluation of lameness in two locations confirmed that the assessment in the pen underestimates the prevalence of lame animals. Under semi-extensive conditions, Battini et al. (2021) evaluated severe lameness in dairy goats after the morning milking, while animals were on the way to pasture. Presence of severe lameness was also evaluated in fenced extensive and semi-extensive meat goat farms, observing animals moving in fenced areas (Leite et al. 2020a).

Fig. 5 Example of four goats from no lame to severely lame (Video by Inês Ajuda and Ana Vieira) (Video 2) eliminare (▶ https://doi.org/10.1007/000-c4r)

Lameness is frequently associated to deformed and overgrown claws that is a major problem in goats, especially in intensive farming systems, due to lack of claw wear due to soft bedding, absence of access to pasture, and insufficient trimming routine (Ajuda et al. 2019). Claw overgrowth can be visually assessed when goats are restricted at the feed rack or in the milking parlour, observing only the rear feet and scoring the overgrowth as acceptable or not (Battini et al. 2016b, c), or as moderate or severe (Anzuino et al. 2010). Alternatively, goats can be restrained in their home pen, and claw overgrowth can be evaluated lifting the feet to score each claw on a 4-point scale system (mild, severe, extreme, deformed; (Muri et al. 2013)); however, this system is less feasible and more stressful to animals than other methods.

Integument alterations (neck, body, legs, ear tears due to tagging) and discharges (nasal, ocular, vulvar) are reported by several authors as measures of injuries and disease (Anzuino et al. 2010; Battini et al. 2016b, c; Can et al. 2016; Muri et al. 2013) (Fig. 6).

However, there is high inconsistency in definitions to compare results, and some authors reported low prevalence and scarce inter-observer reliability. Lesions in goats housed indoors are seldom observed and are mainly represented by hairless areas caused by inadequate structures. In extensively reared goats, the recording of lesions could be more relevant due to the natural environment and the increased risk for animals being injured, but it can also be more challenging, due to the need to restrain free-ranging animals.

The presence of external abscesses (mainly caused by *Corynebacterium pseudotuberculosis*) also seems to be associated to altered health conditions and behavioural changes and can therefore be used as an animal-based indicator in welfare assessment protocols (Mattiello et al. 2018). It can be easily assessed by visual inspection mainly around lymph node areas.

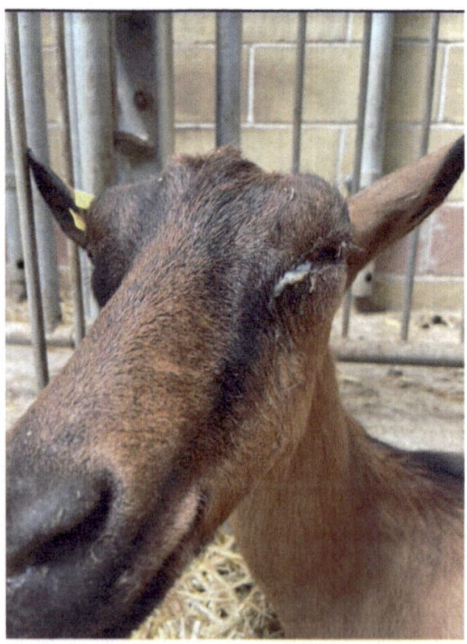

Fig. 6 Goat with unilateral ocular discharge (Photo: Monica Battini)

Mastitis is a frequent infectious disease in dairy animals. During clinical mastitis, the udder appears swollen, warm, and reddened, but mastitis can also be subclinical which is not easily identifiable and may require palpation. Untreated subclinical mastitis could lead to subsequent fibrosis and gland atrophy, associated to pain and discomfort. The presence of an asymmetric udder in which one half is at least 25% longer than the other is a sign of past intra-mammary infection and its visual assessment is reported to be a reliable and feasible welfare measure (Can et al. 2016).

As mentioned above, hair coat condition is also related to the animal's health state. Healthy goats have shiny hair, with a good sheen, homogeneous and well adherent to the body, whereas matted, rough, or scurfy hair is frequently associated to chronic diseases (e.g. abnormal lung sounds) (Battini et al. 2015a) (Fig. 7).

Cortisol (collected from plasma, saliva, serum, urine, faeces) concentration can be used as indicator of stress during painful (e.g. disbudding, oral drenching, subcutaneous vaccination) or stressful (e.g. handling, restraining) procedures, but its collection has a low feasibility in many contexts (Alvarez and Gutiérrez 2010; Hempstead et al. 2018b; Kruger et al. 2016; Toinon et al. 2023). Hair cortisol is also a valid measure of chronic stress, but feasibility is still low due to expensive laboratory analysis (Battini et al. 2015a).

Vocalizations during disbudding can be a sign of pain; in addition, behavioural alterations (e.g. increased time spent lying, lethargy, restlessness, high frequency of kicks) may be long-lasting signs of discomfort after the procedure is ended (Ajuda et al. 2020; Alvarez et al. 2015; Hempstead et al. 2017). Head shaking, head scratching, and head rubbing may also indicate pain and/or discomfort in goat kids subjected to disbudding (Hempstead et al. 2018a, 2020). In Switzerland, farmers' ability to disbud properly goat kids using anaesthesia was evaluated grading from 1 to 4 the frequency of occurrence of specific behaviours, i.e. vocalization,

Fig. 7 Poor coat condition is described as matted, rough, or scurfy hair (Photo: Monica Battini)

limb-movement (paddling, kicking, pull-up limb), and head lifting. The minimum score corresponds to the absence of such behaviours, whereas in the maximum score the behaviours occur more than 7 times (Wagmann et al. 2018). The study reported that 35.8% of the animals showed moderate to strong behavioural reactions during the disbudding procedure performed by the farmers. Hence, anaesthesia and analgesia of goat kids resulted inadequate. A pain visual analogue scale was used by Abril-Sánchez et al. (2018) including the number, intensity, and duration of vocalizations, the contractions and movements provoked by electroejaculation, and the intensity and number of facial contractions, orbital tightening, abnormal ear position and abnormal nostril and philtrum shape, similarly to the facial grimace scale developed by McLennan et al. (2016) for sheep.

2.4 Behaviour

Animals should be able to express their natural behaviour. If they cannot behave normally for several different reasons, such as a disease or an inadequate surrounding environment or a bad management, one of their first attempts to cope with a negative situation will be a modification of their behaviour. This section deals with behaviours that may be indicative of poor welfare, either because they are abnormal, or because they are performed with anomalous frequencies or duration, or because they are performed in inappropriate contexts. On the other side, behaviours that may be indicative of positive welfare are also taken into account.

Activity-budget is usually collected to gather information about the allocation of time of activities during daytime. This is particularly relevant because changes in frequency or time of behaviours may be indicative that a modification in how animals perceive the environment or feel is occurring. Diurnal time-budgets were first studied on feral goats in Scotland and New Zealand (Shi et al. 2003; Stronge et al. 1997) where it was found that daily activities may be affected by season, feed availability, and daylight, but how time is allocated to behaviours also differs depending on sex and age of the animal. These studies are remarkable to gain insights into the welfare of feral goats, even if the general interest in this is low, because of the relatively small economic implications. Furthermore, since environment and latitude can alter the behaviour of these animals, studies in other areas of the world would also be needed. Studies carried out on feral goats can clarify how farmed goats behave and what changes in their behaviour may be a sign of a welfare problem. For example, changes in activity-budget were found in dairy goats due to heat stress (decline in walking, increase in standing still; (de Albuquerque Paulo and Lopes 2014)), presence of a disease (decrease of feeding, increase of chewing; (Mattiello et al. 2018)), space availability (less feeding activity when animal density increases; (Vas and Andersen 2015)), and pregnancy stage (less time feeding and more on resting during the last period of gestation; (Vas and Andersen 2015)). Since goats' natural behaviour is usually highly synchronized, synchronization of behaviours (e.g. feeding/grazing, resting) may be a promising indicator of positive welfare as it responds to a behavioural need. An attempt to use this indicator was made in extensive conditions but the authors state that more information is needed to know the right moment to perform the observations depending on different management routines (e.g. milking time, gathering of animals at night) (Battini et al. 2021). The scan and instantaneous sampling method is preferable for recording activity-budget and synchronization of behaviours; however, duration of scan intervals should be carefully evaluated to obtain reliable information without compromising the feasibility.

Goats are herd-living and gregarious animals as well as synchronized, hence the presence of animals isolated from the group can be a sign of poor health or reduced welfare (Battini et al. 2015b) (Figs. 8 and 9).

Even if female goats are more aggressive than other ungulates and agonistic interactions regularly occur (Fournier and Festa-Bianchet 1995), the recording of changes in the frequency of occurrence of agonistic behaviour was found to be a valid indicator of competition in farm conditions (Aschwanden et al. 2009; Ehrlenbruch et al. 2010; Fernández et al. 2007; Hillmann et al. 2014). Aggression is usually frequent around feeding time, in cases of lack of resources (e.g. insufficient number of feeding places or resting space) or due to routine changes (e.g. regrouping). Some authors reported scarce feasibility of this indicator when included in on-farm assessment protocols, as observations usually require prolonged periods of time to obtain reliable information (Battini et al. 2014; Nordmann et al. 2011).

Fig. 8 The presence of goats exhibiting signs of physical or mental isolation should be carefully evaluated (Photo: Monica Battini)

Fig. 9 A goat may be considered isolated if it remains separated from the rest of the group, shows disinterest in activities, or maintains a listless posture (Video by Monica Battini) (Video 3) (▶ https://doi.org/10.1007/000-c4q)

However, if the aim is to specifically record aggressive behaviour, this indicator can be successfully used, keeping in mind the importance of scheduling the observations with the best timing. Among agonistic behaviours (e.g. threats, frontal clashes, chases, butts, displacements), the most feasible to record is displacement (physical removal of another goat from the feeding place), as it is frequently associated to a specific moment (e.g. feeding time (Neave et al. 2018)).

Avoidance and approach reactions of goats towards humans can be measured to gather information about their perception of people (e.g., farmers, owners, keepers) and the quality of the human–animal relationship (HAR) (Celozzi et al. 2022). First tests were developed in late 80s, e.g., (Lyons 1989; Lyons and Price 1987), but most tests and measures are more recent and improved compared to the previous ones. The latency to the first contact test is a feasible measure applied by different authors with minimum variations: 5–15 min is the time for the observer to remain motionless with the back to the wall waiting for contact with one or more goats (Battini et al. 2016b, c; Jackson and Hackett 2007; Leite et al. 2020b; Mersmann et al. 2016). Battini et al. (2016b, c) warn that this test is valid to assess the HAR of dairy goats if the buck is not in the same pen, as its presence may change the behaviour of females. The latency to the first contact test is a good option in commercial and indoor husbandry systems, where goats are not used to humans moving around in the pen (Video 4). However, this test has been validated only for dairy goats, and it is not known if it can be applied to meat goats (Fig. 10).

Some authors applied an avoidance distance test with the assessor approaching individual animals in the pen when in group with peers from a distance of 200 or 300 cm (Battini et al. 2016b, c; Mattiello et al. 2010). However, the feasibility of this test is low, as many other authors reported the impossibility to apply the test following the same procedure due to the strong avoidance reaction of goats that started running around as a whole flock when the assessor entered the pen (Mersmann

Fig. 10 Latency to the First Contact Test applied in a dairy goat farm (Video by Monica Battini) (Video 4) (▶ https://doi.org/10.1007/000-c4t)

et al. 2016; Muri et al. 2013). Mersmann et al. (2016) also suggested an avoidance distance test performed at the feeding rack with the assessor first walking along the trough and recording the number of goats remaining in the rack after their passage and then measuring the individual reaction of goats when approaching frontally. Both tests are valid but can be limited by the design of the structures (e.g. narrow feed alley). Tests and measures described so far can be used in on-farm and experimental contexts on single or grouped animals. In extensive conditions, these tests cannot be applied as they require the presence of a pen or at least a corral. Some authors applied a Familiar Human Approach Test used for extensively managed sheep (Richmond et al. 2017) to measure the flight distance induced by the stockperson approach in meat and dairy goat does (Battini et al. 2021; Leite et al. 2020a). The feasibility of this test seems high, but its validity in goats still needs to be confirmed.

Play behaviour is also suggested as a welfare measure in goats possibly highlighting the presence of positive welfare. This indicator has been used, for example, to compare the welfare of goat kids in enriched environments compared to kids in unstructured pens (Tölü et al. 2016). In the enriched environment, kids played more, suggesting that different pen structures may offer different stimuli to the animals and the presence of enrichments may elicit a positive behaviour, such as play. However, behaviours such as run and jump, that are usually included in play behaviour repertoire, occurred more in artificially-reared kids than in dam-reared kids (Toinon et al. 2021). Further, artificially-reared kids were also found to vocalize and rear more than dam-reared kids and appeared to be more stressed due to maternal deprivation. Given that results, the use of play behaviour needs to be carefully evaluated.

2.5 Mental State

The evaluation of animals' emotions is a relatively new and challenging field of research in animal welfare science, and few measures have been developed so far in this respect.

The Qualitative Behaviour Assessment (QBA) is a well-recognized measure to evaluate emotions in animals. QBA is a holistic approach that is based on the ability of humans to judge postures, expression of behaviours of goats, by scoring descriptors on a visual analogue scale. A first attempt was applied by Muri et al. (2013), but a fixed list of terms was later developed and validated by Grosso et al. (2016) evaluating the emotions of herds of goats at pasture and indoors. This method is suitable to be used in different contexts, but inter-observer reliability can be critical and requires an extensive training (Battini et al. 2018). Miller et al. (2018) applied QBA to assess the habituation of feral rangeland goats to an intensive farming system, using the free choice profile method. This method is highly reliable, but usually less suitable for on-farm evaluation because a high number of assessors is needed and because the phase in which assessors generate their own terms is time-consuming.

Vocalizations are a promising indicator of emotions. Research on this indicator is still growing and so far, we know that goats produce two main types of bleats: open-mouth bleats when in pain or in a negative situation (e.g. isolation (Siebert et al. 2011)), as well as in positive situations (e.g. food anticipation), and closed-mouth bleats with still unclear meaning (Lenhardt 1977). More recently, Briefer et al. (2015) found that the fluctuation in fundamental frequency is less pronounced during positive emotions (e.g. anticipation of food reward) than in negative emotions (e.g. food related frustration or isolation from conspecifics). Furthermore, fast rate, high amplitude, and high frequency have been recorded in situations of high arousal (Briefer et al. 2015). Abril-Sánchez et al. (2018) found that goat bucks vocalize more during painful procedures. However, vocalizations are still far from being a feasible measure of welfare, and more research in this direction is required. Precision livestock farming techniques might be of help in the future to assist in the automated recognition of animals' emotions based on their vocalizations.

Facial expressions may also provide information on animals' emotions. For example, ear postures are used to communicate and express emotions. The duration of ears positioned forward is higher in goats expecting palatable food, indicating a general level of activity across emotional situations, or attention to the reward (Baciadonna et al. 2020). Further research is needed to add information about the valence of ear posture.

3 Protocols to Assess the Welfare of Goats on Farm

In the last decades, research on goat welfare has made significant progress with the development and testing of assessment protocols in Northern (Anzuino et al. 2010; Muri et al. 2013), Southwestern (Can et al. 2016), South (Battini et al. 2016b, c, 2021; Tiezzi et al. 2019), Southeastern Europe (Nenadović et al. 2021), and in North (Hempstead et al. 2021), Central (Silva Salas et al. 2021), and South America

(Larrondo et al. 2021; Leite et al. 2017, 2020a, b). Protocols include a set of selected animal-based indicators useful to assess a wide range of issues related to welfare. Protocols may also include resource- and management-based indicators and are targeted to a specific context (e.g. intensive systems, animals at pasture).

The forerunner work was performed in the United Kingdom by Anzuino et al. (2010). When this general overview of farmed dairy goat health and welfare in 24 farms located in England and Wales was published, the authors pointed out that only few studies were already available on goats and mostly assessed specific issues (e.g. udder health, lameness, diet, and feeding behaviour). The so-called standard protocol of direct animal observations published by Anzuino et al. (2010) was constructed from a review of the existing literature by consulting the Goat Veterinary Society and by extrapolating principles of welfare assessment for other species. The application of this protocol can be made by a single observer who has to perform on-farm observations over the course of 24 h, including two milking routines. For group assessment, every pen of goats is observed for a period of 10 min recording obviously sick or dull animals, excessive scratching and rubbing, oral behaviour directed at the bars and walls of pen structures or at the teats, dyspnoea, coughing, kneeling in the pen and at the trough. Then the observer enters the pen to estimate the prevalence of lameness. For individual assessment, observations are performed during milking using a systematic sample of goats from the entire herd, with every second to sixth goat sampled. A small, handheld cosmetic mirror must be used to examine the teats and the ventral surface of the udder for dirt and lesions with no need for handling. Lameness is checked again on animals at the release after milking. When it was published, this protocol provided useful insights on welfare and health conditions of dairy goats, but its use in commercial situation has some constraints that reduce its feasibility: the first issue is that the assessment is time-consuming, if one observer is engaged for 24 h; the second issue is that the use of handheld objects (e.g. mirror) during the assessment is not practical.

The protocol developed and tested by Muri and colleagues on 30 Norwegian dairy farms in 2013 added more information on the welfare of goats and the role of the stockperson (Muri et al. 2013). It was also the first protocol that included an attempt of QBA for goats. After checking the inter-observer reliability, the protocol was tested by three observers on each farm and on the same day, starting an hour after completion of the morning feeding. The authors suggest that the application of the protocol starts with a 5-min period to habituate the animals to the presence of the observer, then the QBA and the recording of coughs, lame and obviously sick or dull animals are performed on the whole group for 20 min. Then, the observer scores the quality of human–goat interaction by evaluating the farmers' behavioural style and the goats' behavioural responses when the farmer is asked to mark individual animals. Gait score is also evaluated on a sample of individually marked animals. After this, the observer enters the pen to measure the fear level of goats to unfamiliar humans assessing the animals' reaction to an attempt to touch them on the chin. Then, on individual animals, the observer records: discharges from eyes and nostrils, mild or severe skin lesions, damage to the ears partially or totally ripped out due to ear tags, enlarged lymph nodes, Body Condition Score, chest girth, udder conformation, asymmetry, impetigo, teat lesions, swellings on joints, callus formation, claw overgrowth, cleanliness. As to the feasibility of this protocol, the authors pointed out that, even if the assessment can be performed during one working day by a single observer, the mean time taken to apply the whole protocol can reach up to 6 h 30 min, suggesting that further work is required to make the protocol more feasible for routine animal welfare assessments conducted by farm advisory services or by the authorities.

So far only the European Animal Welfare Indicators (AWIN) project, funded in 2011 by the EC within the seventh Framework Programme, officially published a protocol for the assessment of dairy goat welfare in intensive and semi-intensive husbandry systems (AWIN 2015a; Battini et al. 2015b). The AWIN project aimed to develop welfare assessment protocols for sheep, goats, horses, donkeys, and turkeys. All the developed protocols, except that for turkeys, include a two-level approach to increase feasibility and acceptability among stakeholders. The first level consists of a quick screening, based on the use of robust, feasible, and, in some cases, multi-criteria (i.e. one indicator can cover more than one welfare criterion) animal-based indicators, which can be readily applied and require no or minimal handling of the animals. Depending on the outcome of the first-level welfare assessment, a second level, consisting of a more comprehensive and in-depth assessment, may be recommended. After the on-farm testing of the protocol, the authors reported that the average time needed by one assessor for the first-level assessment of one or two pens simultaneously was 90 min; for the second level assessment, 30–45 s per individual goat were required, provided that animals were restrained preferably at the feed rack. The final AWIN welfare assessment protocol for goats includes 18 measures: six are collected on individual animals, ten at group level, one measure is collected both at individual and at group level depending on the level of evaluation, one (bedding) is a resource-based measures, as no valid animal-based measure was identified for the evaluation of comfort during resting (Fig. 11).

For the development of a practical welfare assessment scheme, it is necessary to develop a sampling strategy to make the protocol feasible in terms of time and costs yet maintaining its validity and reliability. In the AWIN welfare assessment protocol for goats, the sample of goats to be individually evaluated is determined on the total number of lactating goats housed in a farm according to the following table (Table 2).

The results of the testing process of the AWIN welfare assessment protocol in Portugal and Italy were published at the end of the project (Battini et al. 2016b, c; Can et al. 2016). In both Countries, 30 dairy farms classified depending on herd size (from small to very large) were assessed. Before being included in the final protocol, measures were tested for inter-observer reliability between pairs of observers in both Countries (Vieira et al. 2018). The inter-observer reliability for group-level measures was substantial, whereas for individual-level measures it ranged from fair-good to excellent in some cases. Considerations about the consistency over time of the outcome of the assessment were also considered for defining the final protocol: in fact, Can et al. (2017) highlighted that some measures were prone to variation during the year. This was the case, for example, of the hair coat condition, which the authors suggested not to evaluate during the moulting season.

The first comprehensive evaluation of goat welfare in Chile was performed in 2019 by Larrondo et al. (2021) using the AWIN welfare assessment protocol for

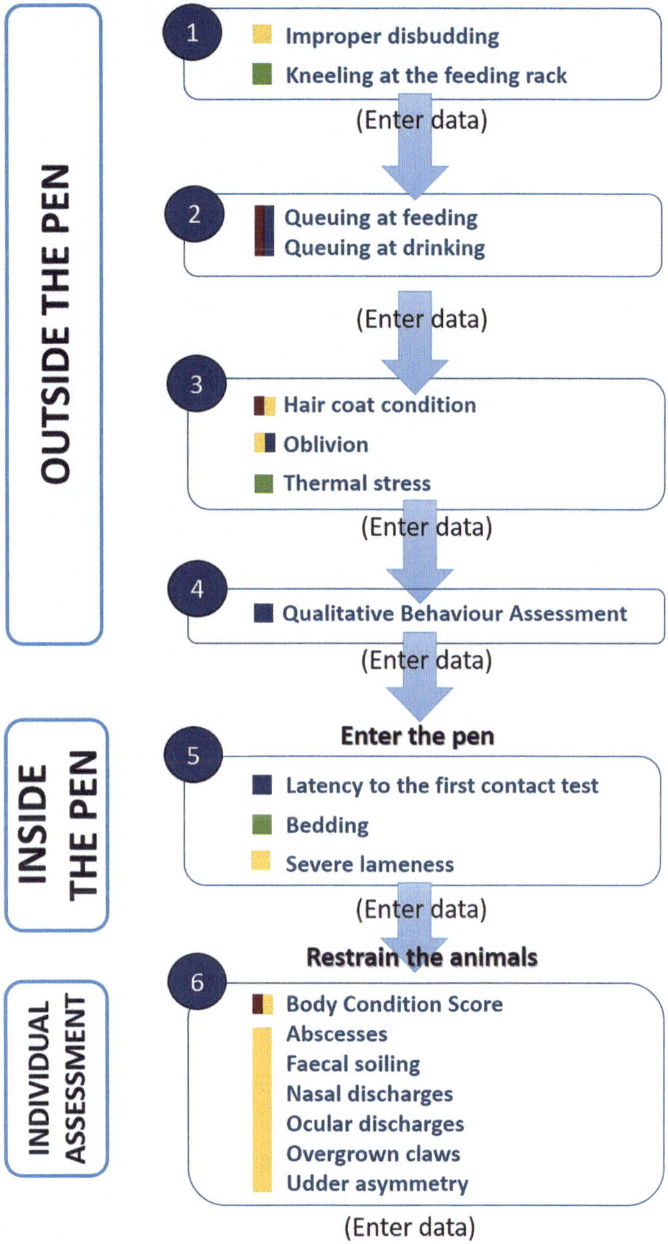

Fig. 11 Flowchart for data collection of the AWIN welfare assessment protocol for goats (AWIN 2015a)

goats in 22 goat farms in semi-intensive husbandry systems. The AWIN protocol was integrally adopted apart from QBA, due to insufficient availability of time and staff to collect this indicator. The assessment was performed on dairy goats; however, half of the goats evaluated were Creole breed, a dual-purpose breed for milk

Table 2 Sample size of goats to be evaluated for the individual welfare assessment. Suggested and minimum sample are calculated assuming a 50% prevalence and 10% accuracy, but suggested sample considers IC 95%, whereas minimum sample IC 90%

Farm size—number of lactating goats	Suggested sample	Minimum sample
<15	All animals	All animals
15–19	13	13
20–24	17	16
25–29	20	19
30–34	23	21
35–39	26	24
40–44	29	26
45–49	31	28
50–59	33	29
60–69	37	32
70–79	41	35
80–89	44	37
90–99	47	39
100–124	49	41
125–149	55	44
150–174	59	47
175–199	63	49
200–224	65	51
225–249	68	53
250–299	70	54
300–349	73	56
350–399	76	57
400–449	78	57
450–499	80	58
500–599	81	59
600–699	83	60
700–799	85	61
800–899	86	62
900–999	87	63
1000–1099	88	63
1100–1299	89	64
1300–1499	90	65
1500–1699	91	65
1700–1799	91	66
>1800	92	66

and meat production. This opens the possibility of effectively applying the AWIN protocol to meat goats and not only to dairy goats.

A group of researchers in Brazil applied a modified version of the AWIN welfare assessment protocol for dairy goats to lactating, non-lactating, and meat goats in intensive, semi-intensive, and extensive husbandry systems (Leite et al. 2017, 2020a b). A similar approach was adopted in Italy for the welfare assessment of dairy goats in semi-extensive farming systems (Battini et al. 2021) and in Serbia for extensively managed goats (Nenadović et al. 2021). Some measures from the original AWIN protocol were considered not applicable to some specific situations (e.g. queuing at the feeding rack was excluded in extensive systems due to the absence of a feed trough) and were therefore removed from the protocol. Other measures were adapted from the AWIN welfare assessment protocol for sheep (e.g. Familiar Human Approach Test for the evaluation of human–animal relationship in extensive systems instead of the Latency to the First Contact Test that requires that goats are housed in a pen) (AWIN 2015b). The application of the AWIN welfare assessment protocol for goats worked properly for the evaluation of non-lactating goats in intensive and semi-intensive husbandry systems, whereas for the assessment of meat goats, it was suggested integrating the AWIN protocol with more indicators, such as aggressive behaviours, stereotypies, coughs, and vulvar discharge (Leite et al. 2017). More criticisms were encountered in the application of the AWIN protocol in more extensive situations. For example, Battini et al. (2021) reported that the application of the protocol under semi-extensive conditions took more time (approximately, 4 h per farm for group measures and 3 min per goat at individual level) than the application in intensive farming conditions. Time was found to be critical (from 1 to 3 h) also for the assessment of meat goat farms in Brazil, especially in view of its application on large herds. In extensive conditions, the use of binoculars was needed to observe animals from a distance, but this reduced the feasibility and the accuracy of the assessment. Thick vegetation and fog are further issues to be considered when planning the assessment, especially on mountain pastures (Nenadović et al. 2021).

To assess the welfare of dairy goats in intensive and semi-intensive husbandry systems in Mexico, researchers applied a protocol consisting of two sections (Silva Salas et al. 2021). The first section derives from the Animal Needs Index (ANI-35L) developed by Bartussek et al. (2000) and comprises four evaluation sheets (locomotion, flooring, environment, management) and additional resource-based indicators; the second section includes a set of validated animal-based measures, focused on physical conditions and clinical signs of disease, derived from the AWIN welfare assessment protocol for goats. Many changes were applied to the original AWIN protocol, e.g. Body Conditions Score was evaluated on a 5-point scale and FAMACHA (i.e. a diagnostic card that matches eyelid colour to anaemia level) was included. The authors concluded that the European protocol could be effectively applied to a semi-desert context of farming, as that evaluated in Mexico, even though several changes were required to better adapt it to this environment.

Until 2021, no published reports of large-scale welfare assessment for dairy goats on farms in the Midwestern United States were available (Hempstead et al. 2021). Hempstead et al. (2021) evaluated 30 dairy goat farms using a protocol developed by modifying already published protocols. The protocol included 13 group-level measures and 9 individual-level measures and required no animal handling. The authors planned the evaluation starting from the milking routine for individual assessment, followed by group assessment in the pen, but also vice versa. This strategy allowed the visit of multiple farms on the same day. The complete application of this protocol lasted approximately 3–5 h. Among many indicators, this protocol includes measures related to ear pathology that has received scarce attention in other protocols; however, it is lacking information on human–animal relationship and emotional state. Table 3 shows the results of welfare assessments of goats in studies conducted to date in different farming systems and productive categories (Table 3).

Table 3 Welfare problems in different husbandry systems and in goats with different production attitudes

	Dairy goats in intensive systems					Dairy goats in semi-intensive systems		Dairy goats in extensive systems		Meat goats in semi-intensive systems	Meat goats in extensive systems
Animal-based welfare indicators	Anzuino et al. (2010)	Battini et al. (2016)	Can et al. (2016)	Hempstead et al. (2021)	Silva Salas et al. (2021)	Larrondo et al. (2021)	Silva Salas et al. (2021)	Battini et al. (2021)	Nenadović et al. (2021)	Leite et al. (2020a, b)	Leite et al. (2020a, b)
Body condition score (% of animals)											
Very thin	3.4	13	4.9	4	0	16.59	33.4	32.1	26.74	18.54	14.5
Thin					16.9		8.1			38.41	35.23
Normal body condition					61	78.25	30.7	67.9		33.11	37.3
Fat					22		27.8			5.96	10.88
Very fat	2.7	6.2	17.4	3.9	0	5.16	0	0		3.67	2.07
Cleanliness (dirty animals) (% of animals)											
Cleanliness				14.9	11		0				
Udder cleanliness		4.3	2.4								
Hind quarters cleanliness		32.7	18						31.4		
Lower legs cleanliness		34.5	11.1								
Dirty teats	12.7										
Very dirty teats	1.8										
Dirty udder	16.3										
Very dirty udder	1.4										
Dirty limbs	36.4										
Very dirty limbs	6.9										
Dirty body	7.3										
Dirty head	1.7										

Lesions (% of animals)											
Hind quarter lesions		3.8	10.4								
Lower legs lesions	6.2	2.6	8								
Body lesions	19.9	11.5	15.2	8.9	3		1				
Neck lesions	14.2	4	28.4								
Periorbital skin lesion	6.3										
Head lesions		35.5	25.4								
Nasal skin lesion	9.2										
Udder skin lesion	7.6										
Severe udder skin lesion	1										
Any teat abnormality	7.6										
Teat skin wound	2.4										
Teat skin inflamed	2.5										
Teat irregular swelling	2.2										
Teat end lesion	0.6										
Torn ears	6.2			2.1							
Missing ear tags				2.4							
Integument alterations					0		0				
Improper disbudding	6.4	12.7	10.8	5.5	9	5.83		.			
Full horns	1.5			1.4							
Joint swelling	0.7										

(continued)

Table 3 (continued)

	Dairy goats in intensive systems					Dairy goats in semi-intensive systems		Dairy goats in extensive systems		Meat goats in semi-intensive systems	Meat goats in extensive systems
Animal-based welfare indicators	Anzuino et al. (2010)	Battini et al. (2016)	Can et al. (2016)	Hempstead et al. (2021)	Silva Salas et al. (2021)	Larrondo et al. (2021)	Silva Salas et al. (2021)	Battini et al. (2021)	Nenadović et al. (2021)	Leite et al. (2020a, b)	Leite et al. (2020a, b)
Gait alterations (% of animals)											
Minor lameness	12.9									1.68	0.37
Lame	3.1				0		0			0	1.13
Severe lameness	3.2	3.05	2.1	1.2		0.22		0.59		0	0
Kneeling in the pen_Outside the pen		0.15	0.4	0.3							
Kneeling in the pen_Inside the pen		0.18									
Overgrown claws	79.8	55.5	35.5	51.4	38	10.99	1	0			
Mild overgrown claws	47.8			33.8							
Severe overgrown claws	32			17.7							
Slight knee calluses	23.8	84	84.2	55.7							
Moderate knee calluses	56.8										
Severe knee calluses	18.3	8.9	19.6	25.2							
Discharges (% of animals)											
Vulvar discharge	5	0.4	0.5							0	1.03
Ocular discharge	6	0.9	9	2.9	6	1.35	1	0		15.89	8.29
Nasal discharge	0.6	5.7	4.5	6.9	3	3.14	1	0		7.28	4.14

Udder conformation (% of animals)											
Udder asymmetry		3.8	5.9	3.3		2.47		3.82	18.6		
Moderate udder asymmetry	15.8										
Severe udder asymmetry	6.2										
Any udder abnormality	33.8										
Udder irregularity	3.3										
Accessory	1.3										
Moderately pendulous udder	5										
Very pendulous udder	0.3										
External abscesses (% of animals)											
Abscesses					4	3.81	0	11.9	19.78	16.55	9.84
Udder abscesses		3.9	2.2								
Hind quarter abscesses		0.9	1								
Body abscesses		8.4	9.9								
Neck abscesses		7.6	7.6								
Head abscesses		5.4	13.7								
Lymphnode swelling	0.4										
Thermal stress (% of animals)											
Thermal stress						0				0	4.16
Panting score		0.38	4	3.7							

(continued)

Table 3 (continued)

	Dairy goats in intensive systems					Dairy goats in semi-intensive systems		Dairy goats in extensive systems		Meat goats in semi-intensive systems	Meat goats in extensive systems
Animal-based welfare indicators	Anzuino et al. (2010)	Battini et al. (2016)	Can et al. (2016)	Hempstead et al. (2021)	Silva Salas et al. (2021)	Larrondo et al. (2021)	Silva Salas et al. (2021)	Battini et al. (2021)	Nenadović et al. (2021)	Leite et al. (2020a, b)	Leite et al. (2020a, b)
Shivering score		3.87	4.8	0.1							
Human–animal relationship (% of animals)											
Latency to first contact test[a]		53.04	125.9								
Avoidance distance test											
Contact		2.27	1								
Acceptance		7.02	1.3								
Other health indicators (% of animals)											
Hair coat condition		24.13	22.9	8.3		12.78		2.67	62.79	8.4	12.12
Faecal soiling	9.8	15.3	2.6	3.5		11.66		0		3.31	3.1
Respiration quality										0.66	0
Subclinical mastitis					6		5				
Prolapse					0		0				
Any ear pathology				8							
Frostbitten ears				3.3							
Infected ear tags				0.1							
Anaemia					4		12				
Diarrhoea					0		0				
Behaviours (% of animals)											
Queuing at feeding		7.23	22.8	6.8		0.9					
Queuing at drinking		1.35	3.6	2		1.12					

Kneeling at the feeding rack_outside pen		2.55	0	0.7		0					
Kneeling at the feeding rack_inside the pen		2.29									
Oblivion		0.06	0.5			0				4.2	0.37
Abnormal lying		0.36	0.4								

[a]Time expressed as seconds elapsed before the goat shows the first contact towards the assessor

4 Animal-Based Indicators for Welfare Assessment During Slaughter and Transport

In 2021, the European Food Safety Authority (EFSA) published a scientific opinion on the welfare of sheep and goats at slaughter (EFSA AHAW Panel et al. 2021). This opinion deals with the process of slaughtering from the arrival of animals until their death. The panel of experts identified several animal-based indicators, some already listed in this chapter (e.g. panting and shivering score for thermal stress on arrival and at lairage (Battini et al. 2016a), lameness as a sign of pain at unloading from the truck (AWIN 2015a)), whereas others were included to cover specific compromissions of welfare likely to be encountered during the whole slaughter. Exhaustion (animal recumbent on the floor or reluctant to move, albeit conscious) and tachypnoea (excessive respiration rate, e.g. >60 bpm (Reddy et al. 2019)) can be used to assess fatigue on arrival or during lairage; vocalizations (Goldberg 2018), attempts to escape and turning back (modified after AWIN (2015a)) are suggested indicators to assess fear; aggression at the water troughs and increased water intake are signs of prolonged thirst, as well as aggressive behaviour is a sign of social stress (Tölü and Savaş 2007). No indicator is available to measure prolonged hunger or restriction of movement, even if space allowance can be considered a good proxy. Any sign of consciousness (i.e. breathing, muscle tone, spontaneous blinking, corneal and palpebral reflex, and vocalizations) after stunning is considered a prerequisite to experience pain and therefore it should be carefully evaluated.

In 2022, EFSA also published a scientific opinion on the welfare of small ruminants during transport (EFSA AHAW Panel et al. 2022). Many animal-based indicators were identified for sheep, but not for goats. Similar welfare concerns as for sheep were identified, however no specific welfare indicators during transport is suggested for goats.

5 Certification Schemes for High-Standard Products

Certification schemes for animal products are processes that guarantee compliance with precise standards. Not all the certifications deal with animal welfare, some of them focus on food quality or fibre quality, some others inform the consumer or the final user about the husbandry systems and practices on farm. In this section, we will focus on certifications that also include animal welfare in their schemes.

Certifications are typically issued by non-profit organizations, retailers, supermarkets (Waitrose & Partners 1999), or animal welfare audit organizations and the monitoring of compliance with welfare standards is generally entrusted to independent third-party audits (Ohioline, 2017). Scientific committees ensure that the standards required by certification schemes are always up to date from a scientific and technological progress perspective. Farmers' participation to certification programmes is usually on a voluntary basis and requires incurring costs, but certified farmers may benefit of premiums and better consumers' perception of their products. Logos or labels on certified products contribute to the farm image and to drive the consumer's choices. Certification schemes that include welfare standards for farmed goats are quite recent and date back to the early 2000s. However, most of these certifications are not specifically designed for goats, but reference to their welfare is included in schemes dealing with their products (e.g. milk, meat, fibre). Only few certifications are exclusively dedicated to goats, with most of them including requirements on improving fibre quality over time (i.e. cashmere and mohair, with five certifications schemes (The Good Cashmere Standard 2019; Sustainable Fibre Alliance 2020; Sustainable cashmere 2020; ICCAW 2013; ICCAW 2015; Responsible Mohair Standard 2022)) and, in one case, meat production (INTEROVIC 2022). Among the animal welfare certifications for goat products, those concerning fibre play a role of primary importance due to the economic relevance of the end-product; most of these certifications are private and linked to large manufacturing groups and are geographically concentrated in the areas of fibre production. This explains why, to the authors' knowledge, there are at least four certification schemes for cashmere goat production from Mongolia and China and at least one for mohair fibre collected from Angora goats. In 2020, Mongolia adopted the standard MNS 6891:2020 for Responsible Nomads Code of Practices (MONTSAME, 2022) for sustainable nomadic livestock production. This standard includes six key indicators that herders must comply with, one of which is about ensuring the animal welfare.

Certification schemes for goat milk and meat production are more widespread (e.g. USA, United Kingdom, Germany, Austria). Spain has the only certification scheme specifically designed for meat goats: the animal welfare INTEROVIC was developed by the Interprofessional Organization of Sheep and Goat Meat and applies to the primary production of goat meat, to the transport and slaughter of animals, including stunning procedures, and to the industrial processing phase of meat and leather products. This organization will shortly release a certification for dairy goats in intensive and extensive husbandry systems derived from the AWIN welfare assessment protocol for dairy goats. Most certification schemes are designed around the concept of the Five Freedoms of animal welfare, originated in response to a 1965 UK Government report on livestock husbandry, and formalized in 1979 by the UK Farm Animal Welfare Council (Farm Animal Welfare Council 2009). Some schemes are grounded on the use of science-based assessment protocols set up on animal-based indicators, whereas others are based mainly on the respect of minimum standards of resources and management, often referring to best practice guidelines. Science-based standards are, for example, Red Tractor Certified Standards 2021, Animal Welfare Approved (Animal Welfare Institute 2015; A Greener World 2018; Certified Humane® (2014), Global Animal Partnership (2019), Tierschutz Kontrolliert (2022) that include animal-based indicators (e.g. expression of species-specific behaviour like feeding and social behaviours), whereas American Grassfed Certified (2010), Naturland (2022) focus on resource-based indicators (e.g. access to pasture or specific requirement for unfettered pasture, with exceptions only due to inclement weather). American Humane Certified™ (American Humane 2021a, b) and Certified Humane® (2014) are mainly focused on how humans interact with goats, with the latter specifically focused on compassionate handling protocols supporting goats' ability to engage in their natural behaviours.

Certification schemes for fibre production are essentially based on the respect for the Five Freedoms (or Five Domains Models for the new Sustainable Fibre Alliance standard) of animal welfare and pose special attention to animal management, particularly to the way in which fibre is collected. For example, they require that fibre collection is carried out without causing pain to the animals and considering the animals' thermoregulation needs, and therefore fibre collection should be performed during the moulting season, when days become warmer, and animals do not suffer from cold stress. Moreover, stringent provisions apply to the restraint of animals, which can be a potential cause of injury and thus of pain. For this purpose, the goats must be restrained for the shortest possible period, by using ropes that are not too tight. Furthermore, during restraint, three limbs of the animal must not be tied at the same time and the operator must not step on any part of the goat's body. The Responsible Nomads certification also includes some resource-based indicators (e.g. providing access to pasture, water, and shelter) and requires a constant monitoring of goat health in close collaboration with veterinary services (MONTSAME 2022). Certification schemes also address issues related to the transport and humane slaughter of animals.

Many of the certifications for cashmere production also include environmental requirements (The Good Cashmere Standard 2019; Sustainable Fibre Alliance 2020; Responsible Mohair Standard 2022). This is related to the fact that the intensification of husbandry is becoming one of the major causes of landscape degradation and desertification in some areas. Due to the volatile price of cashmere, the cashmere goat population in Mongolia has quadrupled in less than 20 years. This uncontrolled intensification, together with the reduction of rain events, has led to a progressive depletion of pastures. This situation may have detrimental impacts on the welfare of animals, whose diet may be insufficient or unbalanced due to the limited feed resource availability.

In conclusion, the scarce diffusion of welfare certification schemes for the goat species is likely to be explained by the modest economic interest for its products. Where this interest is greater, such as for the production of high-quality yarns for the fashion industry, several certification schemes are in place, although some of them possibly need to be updated based on the most recent scientific research, and some of them are therefore currently under revision. In fact, although there have been considerable scientific advances in the assessment of goat welfare, these have mostly concerned dairy goats in intensive and semi-intensive husbandry systems (AWIN 2015a), whereas the scientific knowledge available for fibre and meat producing goats is now somewhat dated. Furthermore, it is important to highlight that several countries, such as Australia and New Zealand, have put in place codes of practice for goat breeding and management, which could lay the foundations for future certification schemes (The Australian

Livestock Processing Industry Animal Welfare Certification System (AAWCS) 2020; AMIC 2018; Ministry for Primary Industries 2013).

Certification schemes listed and described in this section are the result of a careful and detailed research mainly based on Internet search. To the best of authors' knowledge information are truthful and verifiable on available sources. It is not our aim to endorse any public or private organizations above-mentioned in this section.

6 The Use of Precision Livestock Farming Technologies in the Assessment of Goat Welfare

Recent years have seen an exponential increase of research dedicated to Precision Livestock Farming (PLF) technologies in animal husbandry. Compared to other species, small ruminants (sheep and goats) pay the price of being economically less relevant than, for example, cattle for which many technologies have been developed and some of them are already commercially available. Confirming the interest in this topic, the European Commission recently funded two projects (TechCare in 2020 and Sm@rt Small Ruminant Technologies Platform in 2021) with the objective of both integrating and sharing knowledge on innovative digital solutions for small ruminants.

These projects are still ongoing, but other research and researchers are active in the identification of technologies suitable for goats (Caja et al. 2020). However, most of the research is focused on tools that can improve the efficiency of goat production or support farmers in their decision-making process. Only few technologies have been investigated to have positive impact on goat welfare (Tuyttens et al. 2022). For example, pain and stress might be quantified by using thermal imaging cameras (e.g. eye temperature in meat goats (Bartolome et al. 2019)); lying and comfort behaviours can be evaluated using accelerometers fitted to the rear goats' legs (Maurmann et al. 2021; Zobel et al. 2015); gait alteration can be more precisely monitored by using a pressure-sensing walkway (Rifkin et al. 2019); feed intake and feeding behaviour can be assessed by applying signal-processing algorithms to automatically identify and classify sound-producing jaw movements (Halachmi et al. 2019); abnormal behaviours can be recorded by an automatic tracking system based on surveillance videos (Su et al. 2021); vocalizations can be automatically recorded and used as early warning for farmers (Ntalampiras et al. 2023).

In outdoor conditions, drone cameras may assist the welfare assessment of extensively managed goats, as well as feral goats. However, insufficient information is available about the possible disturbance effect that drones may elicit in animals' activities (Vayssade et al. 2019). Furthermore, PLF has the potential to help managing goats in extensive systems, for example using virtual fencing systems, although the impact of such systems on animal welfare needs to be further investigated (Aaser et al. 2022)

7 Future Perspectives

Despite the ever-increasing interest in goat welfare, there are still some gaps to be filled for proper assessment. First, most measures and protocols are developed for adult dairy goats on intensive farms, although some attempts have been made for meat or extensively managed goats. These attempts are usually successful in terms of feasibility, but validity and reliability have not always been checked. Many authors applied or partly applied the AWIN welfare assessment protocol for goats in different situations other than lactating goats in intensive farming systems. However, the removal or replacement of some measures does not guarantee that the whole protocol is still valid. Furthermore, some measures, such as QBA, are frequently removed from the protocol because of the lack of time or impossibility to provide extensive training; however, in doing that a relevant part of the assessment is lost. Despite the heavy body of literature on cognitive abilities of goats (Zobel and Nawroth 2020), the assessment of emotions in this species is still underrepresented. Some studies investigated the valence and arousal of vocalizations (Laurijs et al. 2021), but there is much more work to do on facial expressions and body postures.

Valid measures to assess the welfare of bucks and kids are still insufficient. Some measures developed for females might also be suitable for males, but some necessarily need to be tested in terms of validity, reliability, and feasibility. Measures of pain due to castration are currently lacking, and scarce information is available about the effect on behaviour of keeping bucks isolated in an individual pen.

The welfare of feral goats, as well as that of goats kept for recreational and educational purposes (e.g. zoos, petting farms, schools), has been completely ignored so far. Again, it is possible that some measures developed for lactating goats can be effectively applied to other categories or contexts, but specifically designed measures might be required under some circumstances. It is also important to point out that although some measures may be applicable, it seems that there is no "culture'" of assessing the welfare of goats kept for recreational and educational purposes. This is probably due to an underestimation of the problem. It is desirable that in the future, few and easy-to-collect indicators could be available for staff members, keepers, and volunteers that work with goats in contexts other than farms.

Feasibility is a major constraint for the acceptability and use of welfare assessment protocols. However, with continuous technical developments, automatic recordings, and advancements in PLF techniques, it is likely that the collection of several currently impracticable animal-based measures will become less expensive and more feasible in the future (Wathes et al. 2008). The use of PLF devices can also help assessing the welfare of individual goats that usually is not considered in welfare assessment protocols. However, as stated by Tuyttens et al. (2022), this boom in PLF technologies is due to hopes and claims that are still unproven to have a beneficial effect on the welfare of animals. This technological revolution in farms cannot be stopped, but we should be aware of potential threats that can directly (e.g. inaccurate predictions, harmful effects of exposure/wearing of hardware components) or indirectly (e.g. farmers spending less quality time with animals or becoming over-reliant on PLF devices, stockpeople loosing experience in animal husbandry) affect the animals.

In conclusion, although many progresses have been done and many are on the way, a lot of research is still needed to allow an accurate welfare assessment in all goat categories (e.g. does, bucks, kids) and in all possible situations. This is very important, given that goats can be found in a wide variety of contexts, ranging from intensive or extensive commercial farms to zoos, sanctuaries, and other confined contexts, as well as in free-ranging animals. If, on one side, considerable skills and tools have been developed, especially for intensively farmed goats, there is still a lack of information for the other contexts, particularly for wild and feral goats. Furthermore, the availability of valid and reliable indicators of positive welfare is still very limited in goats, and research in this field should be strongly encouraged to guarantee that animals can have a high quality of life.

A final comment is that given the diverse environments and husbandry systems in which goats are raised, assessing their welfare through localized studies or specific contexts is likely to yield results that are unique to that group of animals. Consequently, findings may not readily apply to or be replicable in other individuals. To foster broader and more robust results essential for advancing the care and management of goats, a worldwide network of researchers and laboratories has embarked on a mission to conduct reproducible research in the realm of goat behaviour and welfare science. Inspired by previous initiatives in human psychological research, such as ManyLabs and ManyBabies, and similar undertakings in non-human species like ManyBirds and ManyDogs, the ManyGoats network, detailed at www.themanygoatsproject.com, initiated its collaborative efforts in 2021.

Appendix

See Tables 4 and 5.

Table 4 List of welfare certifications retrieved from the Internet for meat and milk goats. Only certifications available in English are included in the list

Name of certifier	Year of first publishing	Certification	Main aspects of welfare addressed	Certification target	Certification issued by	Audit		Certification duration	Certification logo
						Auditors	Frequency		
Humane Choice True Free Range	–	Approved/rejected	Free range reared animals Other requirements for handling, transport, and slaughter	Australian and New Zealand Meat goat farmers	NASAA Organic (National Association for Sustainable Agriculture Australia)	NASAA Organic (National Association for Sustainable Agriculture Australia)	–	–	
American Grassfed Certified	2009	Approved/rejected	Animals born and reared in the same farm and raised on pasture. Feeding based on grass and pasture. No use of hormones and use of antibiotics only in case of sick animals	Goat farmers	American Grassfed Association	Independent third-party audit	At least every 15 months	Annual	
Red Tractor Certified Standards	2000	Approved/rejected	Herd health, animal handling	Dairy goat farmers	SAI Global (independent certification body)	SAI Global (independent certification body)	Every 18 months plus any spot visits	Annual	
Waitrose & Partners	2018	Approved/rejected	Five freedoms, positive behaviour, QBA	Goat farmers	Waitrose & Partners	Waitrose & Partners	–	Annual	

Naturland	2010	Approved/ rejected	Organic farming; particular importance attached to animals reared in free-range areas and access to pasture	Goat farmers	Naturland (non-profit organic association)	Independent third-party audit	At least once per year	Annual	Naturland
Animal Welfare Approved (optional addition "Grassfed Certified")	–	Approved/ rejected	Guarantees that the animal is raised outside (pasture or range) for their entire lives. Enable the natural behaviour of goats, through the promotion of their social interactions, comfort, and physical and psychological well-being	Goat farmers	A Greener World (non-profit organization)	A Greener World staff or agents	One visit per year at minimum	Annual	CERTIFIED GRASSFED BY AGW
American Humane Certified™	2017	Approved/ rejected	Five Freedoms (Farm Animal Welfare Council); humane treatment of farm animals	License supplier; goat farmers; goat supplier	American humane farm programme	Third-party audit	–	–	American Humane CERTIFIED™

(continued)

Table 4 (continued)

Name of certifier	Year of first publishing	Certification	Main aspects of welfare addressed	Certification target	Certification issued by	Audit		Certification duration	Certification logo
						Auditors	Frequency		
Certified Humane	1998	Approved/ rejected	Goats raised in humane conditions (with compassion and kindness), free from abnormal stress and allowed to express their natural behaviours	Goat farmers, haulers, slaughterhouse operators	Humane Farm Animal Care DBA Certified Humane® (non-profit organization)	Third-party farms inspectors (veterinarians and people who have an advanced degree in Animal Science)	Annual	Annual	
Global Animal Partnership (G.A.P.) (for goats reared for meat)	2016	Four levels	Recognition of feed behavioural needs (grazing, browsing, climb when feeding)	Meat goat farmers	Global Animal Partnership (non-profit charitable organization; G.A.P.)	Authorized third-party certification companies	Every 15 months	Between 14 and 18 months, necessary to evaluate the entire production cycle	
Tierschutz Kontrolliert ("Animal welfare-controlled" seal of approval)	2012	Two levels (gold and silver)	Husbandry conditions, transport, and slaughter The gold level offers the animal the chance to meet their natural needs. Animal based-indicators	Goat farmers	Four Paws (International animal welfare organization)	State-approved inspection bodies	At least once a year	Annual	

Australian Livestock Processing Industry Animal Welfare	2005	Approved/ rejected	Handling of goats at the meat processing establishments; stunning methods	Goat meat processing establishments	Australian Meat Industry Council (AMIC)	It depends on the audit body chosen by the meat processing establishment	It depends on the audit body chosen by the meat processing facility	Annual	
Animal Welfare INTEROVIC Spain (BAIE)	–	Approved/ rejected	Five Freedoms (Farm Animal Welfare Council); other requirements are referred to: environment, biosecurity, housing conditions, food, water, health.	Primary production centres, slaughtering centres, food industry, leather industry; Certification bodies	Organización Interprofesional del Ovino y Caprino de Carne	Seven different certification bodies	At least one audit every three years, in the hottest months, that is, the most unfavourable ones	Triennial	

Table 5 List of welfare certifications retrieved from the Internet for fibre goats. Only certifications available in English are included in the list

Name of certifier	Year of first publishing	Certification	Aspects of welfare addressed	Certification target	Certification issued by	Audit		Certification duration	Certification logo
						Auditor	Frequency		
The Good Cashmere Standard®	2019	Two types of certifications: GCS Certificate (for cashmere producers)	Five Freedoms (Farm Animal Welfare Council);	Cashmere producers (i.e. buying stations/ dehairing stations); Cashmere herders subjected to a self-assessment questionnaire and applied to AbTF for initial approval	Supported by Trade Foundation (AbTF; non-profit organization)	ELEVATE (an independent audit company)	–	Annual	THE GOOD CASHMERE STANDARD AbTF
		GCS Registration Certificate (for spinning mills and yarn traders)	Rules for combing and shearing, handling and transport and euthanasia and slaughter						

SFA (Sustainable Fiber Alliance) Certified—the Cashmere Standard	2015 (year of creation)	Three levels[a]:	Five freedoms 1. Feeding and drinking water	Cashmere herders	Sustainable Fiber Alliance (SFA; international NGO)	Conformity Assessment Body (CAB)	Annual	Annual	
	2017 (year of implementation)	Bronze (compliance with minimum criteria)	2. Farming environment	Cashmere processors					
	2023 (release of the new standard)	Silver (compliance with minimum criteria and some requirements)	3. Husbandry management						
		Gold (compliance with all the criteria)	4. Health						
			5. Fibre harvesting						
			6. Transport and farm transfer						
			7. Humane slaughter						
			The new standard will be based on the Five Domains Model						

(continued)

Table 5 (continued)

Name of certifier	Year of first publishing	Certification	Aspects of welfare addressed	Certification target	Certification issued by	Audit		Certification duration	Certification logo
						Auditor	Frequency		
Farm Animal Welfare Requirements Cashmere Goat	2019	Three levels[a]:	Five freedoms	Cashmere herders	China Association for the Promotion of International Agricultural Cooperation International Cooperation Committee on Animal Welfare (ICCAW; no profit organization)	–	Annual	Annual	ICCAW 中国农业国际合作促进会 动物福利国际合作委员会
		Bronze (compliance with minimum criteria)	Husbandry	Cashmere processors					
		Silver (compliance with minimum criteria and some requirements)	Feeding and drinking						
		Gold (compliance with all the criteria)	Farming environment						
			Management						
			Health						
			Harvesting fibre (sharing and combing)						
			Transportation and farm transfer						
			Humane slaughtering						
			Traceability and recording						

AVSF - Sustainable Cashmere	2016	Approved/ rejected	Animal feeding management	Mongolian cashmere goat farmers (Sustainable Cashmere Union – Mongolia (SCU) cooperative)	AVSF (Agronomes & Vétérinaires sans Frontières)	Sustainable Cashmere Certification Committee (S3C)	Multi-level control mechanisms: 1. Herders (self-evaluation) 2. Cooperative (internal controls)	–	
			Animal genetic management				3. S3C (external controls)		
			Animal health management						
			Global herding management						
Responsible Mohair Standard Certified	Standard published on March 17, 2020	Approved/ rejected	Five Freedoms (Farm Animal Welfare Council)	The entire production chain, from the mohair farmers, to the garment manufacturer or brand	Textile exchange (global non-profit organization)	Different certification bodies	Annual	Annual	

[a]Compliance with criteria and requirements of Animal Husbandry and Cashmere Harvesting Code of Practice

References

Aaser MF, Staahltoft SK, Korsgaard AH, Trige-esbensen A, Kristian A, Alstrup O, Sonne C, Pertoldi C, Bruhn D, Frikke J, Linder AC (2022) Is virtual fencing an effective way of enclosing cattle? Personality, herd behaviour and welfare. Animals 12:842

Abril-Sánchez S, Crosignani N, Freitas-De-Melo A, Terrazas A, Damián JP, Beracochea F, Silveira P, Ungerfeld R (2018) Sedation or anaesthesia decrease the stress response to electroejaculation and improve the quality of the collected semen in goat bucks. Animal 12(12):2598–2608

Ajuda IDGG, Battini M, Stilwell GT (2019) The role of claw deformation and claw size on goat lameness. Vet Anim Sci 8

Ajuda I, Battini M, Mattiello S, Arcuri C, Stilwell G (2020) Evaluation of pain mitigation strategies in goat kids after cautery disbudding. Animals 10(2)

Al-Ramamneh D, Riek A, Gerken M (2012) Effect of water restriction on drinking behaviour and water intake in German black-head mutton sheep and Boer goats. Animal 6(1):173–178

Alvarez L, Gutiérrez J (2010) A first description of the physiological and behavioural responses to disbudding in goat kids. Anim Welf 19(1):55–59

Alvarez L, De Luna JB, Gamboa D, Reyes M, Sánchez A, Terrazas A, Rojas S, Galindo F (2015) Cortisol and pain-related behavior in disbudded goat kids with and without cornual nerve block. Physiol Behav 138:58–61

Andersen IL, Bøe KE (2007) Resting pattern and social interactions in goats—the impact of size and organisation of lying space. Appl Anim Behav Sci 108(1–2):89–103

Anzuino K, Bell NJ, Bazeley KJ, Nicol CJ (2010) Assessment of welfare on 24 commercial UK dairy goat farms based on direct observations. Vet Rec 167(20):774–780

Aschwanden J, Gygax L, Wechsler B, Keil NM (2009) Loose housing of small goat groups: influence of visual cover and elevated levels on feeding, resting and agonistic behaviour. Appl Anim Behav Sci 119(3–4):171–179

AWIN (2015a) AWIN welfare assessment protocol for goats

AWIN (2015b) AWIN welfare assessment protocol for sheep

Baciadonna L, Briefer EF, McElligott AG (2020) Investigation of reward quality-related behaviour as a tool to assess emotions. Appl Anim Behav Sci 225(November 2019):104968

Bartolome E, Azcona F, Maria CA, Perdomo-González D, Ribes-Pons J, Terán E (2019) Testing eye temperature assessed with infrared thermography to evaluate stress in meat goats raised in a semi-intensive farming system: a pilot study. Arch Anim Breed 62(1):199–204

Bartussek H, Leeb C, Held SDE (2000) Animal needs index for cattle. In: Federal Research Institute for Agriculture in Alpine Regions BAL Gumpenstein, A 8952 Irdning, of the Federal Ministry of Agriculture and Forestry, Environment and Water management. http://www.bartussek.at/pdf/anicattle.pdf

Battini M, Vieira A, Barbieri S, Ajuda I, Stilwell G, Mattiello S (2014) Invited review: animal-based indicators for on-farm welfare assessment for dairy goats. J Dairy Sci 97(11):6625–6648

Battini M, Peric T, Ajuda I, Vieira A, Grosso L, Barbieri S, Stilwell G, Prandi A, Comin A, Tubaro F, Mattiello S (2015a) Hair coat condition: a valid and reliable indicator for on-farm welfare assessment in adult dairy goats. Small Rumin Res 123(2–3):197–203

Battini M, Stilwell G, Vieira A, Barbieri S, Canali E, Mattiello S (2015b) On-farm welfare assessment protocol for adult dairy goats in intensive production systems. Animals 5(4)

Battini M, Barbieri S, Fioni L, Mattiello S (2016a) Feasibility and validity of animal-based indicators for on-farm welfare assessment of thermal stress in dairy goats. Int J Biometeorol 60(2):289–296

Battini M, Barbieri S, Vieira A, Stilwell G, Mattiello S (2016b) Results of testing the prototype of the AWIN welfare assessment protocol for dairy goats in 30 intensive farms in northern Italy. Ital J Anim Sci 15(2):283–293

Battini M, Barbieri S, Waiblinger S, Mattiello S (2016c) Validity and feasibility of human-animal relationship tests for on-farm welfare assessment in dairy goats. Appl Anim Behav Sci 178:32–39

Battini M, Barbieri S, Vieira A, Can E, Stilwell G, Mattiello S (2018) The use of qualitative behaviour assessment for the on-farm welfare assessment of dairy goats. Animals 8(7):123

Battini M, Renna M, Giammarino M, Battaglini L, Mattiello S (2021) Feasibility and reliability of the AWIN welfare assessment protocol for dairy goats in semi-extensive farming conditions. Front Vet Sci 8(October):1–14

Blokhuis HJ, Veissier I, Miele M, Jones B (2010) The welfare quality® project and beyond: safeguarding farm animal well-being. Acta Agric Scand A Anim Sci 60(3):129–140

Bøe KE, Ehrlenbruch R (2013) Thermoregulatory behavior of dairy goats at low temperatures and the use of outdoor yards. Can J Anim Sci 93(1):35–41

Briefer EF, Tettamanti F, McElligott AG (2015) Emotions in goats: mapping physiological, behavioural and vocal profiles. Anim Behav 99

Caja G, Castro-Costa A, Salama AAK, Oliver J, Baratta M, Ferrer C, Knight CH (2020) Sensing solutions for improving the performance, health and wellbeing of small ruminants. J Dairy Res 87(S1):34–46

Can E, Vieira A, Battini M, Mattiello S, Stilwell G (2016) On-farm welfare assessment of dairy goat farms using animal-based indicators: the example of 30 commercial farms in Portugal. Acta Agric Scand A Anim Sci 66(1):43–55

Can E, Vieira A, Battini M, Mattiello S, Stilwell G (2017) Consistency over time of animal-based welfare indicators as a further step for developing a welfare assessment monitoring scheme: the case of the animal welfare indicators protocol for dairy goats. J Dairy Sci 100

Caroprese M, Casamassima D, Rassu SPG, Napolitano F, Sevi A (2009) Monitoring the on-farm welfare of sheep and goats. Ital J Anim Sci 8(SUPPL. 1):343–354

Celozzi S, Mattiello S, Battini M, Zucali M, Pirovano L, Albano C, Brasca M, Bava L (2021) Effect of a reduced amount of straw bedding on goats' comfort and hygienic characteristics of milk and straw. Ital J Anim Sci 20(1):995–1001

Celozzi S, Battini M, Prato-Previde E, Mattiello S (2022) Humans and goats: improving knowledge for a better relationship. Animals 12(6):1–33

Darcan N, Cedden F, Cankaya S (2008) Spraying effects on some physiological and behavioural traits of goats in a subtropical climate. Ital J Anim Sci 7(1):77–85

de Albuquerque Paulo JL, Lopes F (2014) Daily activity patterns of Saanen goats in the semi-arid northeast of Brazil. Rev Bras Zootec 43(9):464–470

EFSA (2012) Statement on the use of animal-based measures to assess the welfare of animals. EFSA J 10(6):1–29

EFSA AHAW Panel, Nielsen SS, Alvarez J, Bicout DJ, Calistri P, Canali E, Drewe JA, Garin-Bastuji B, Gonzales Rojas JL, Gortázar Schmidt C, Herskin M, Miranda Chueca MÁ, Padalino B, Pasquali P, Roberts HC, Spoolder H, Stahl K, Velarde A, Viltrop A et al (2021) Welfare of sheep and goats at slaughter. EFSA J 19(11)

EFSA AHAW Panel, Nielsen SS, Alvarez J, Bicout DJ, Calistri P, Canali E, Drewe JA, Garin-Bastuji B, Gonzales Rojas JL, Gortázar Schmidt C, Michel V, Miranda Chueca MÁ, Padalino B, Pasquali P, Roberts HC, Spoolder H, Stahl K, Velarde A, Viltrop A et al (2022) Welfare of small ruminants during transport. EFSA J 20(9):1–101

Ehrlenbruch R, Jørgensen GHM, Andersen IL, Bøe KE (2010) Provision of additional walls in the resting area-the effects on resting behaviour and social interactions in goats. Appl Anim Behav Sci 122(1):35–40

Farm Animal Welfare Council (2009) Farm animal welfare in Great Britain: Past, present and future

Fernández MA, Alvarez L, Zarco L (2007) Regrouping in lactating goats increases aggression and decreases milk production. Small Rumin Res 70(2–3):228–232

Fournier F, Festa-Bianchet M (1995) Social dominance in adult female mountain goats. Anim Behav 49(6):1449–1459

Goldberg ME (2018) Pain recognition and scales for livestock patients. J Dairy Vet Anim Res 7(5):236–239

Grosso L, Battini M, Wemelsfelder F, Barbieri S, Minero M, Dalla Costa E, Mattiello S (2016) On-farm qualitative behaviour assessment of dairy goats in different housing conditions. Appl Anim Behav Sci 180:51–57

Halachmi I, Guarino M, Bewley J, Pastell M (2019) Smart animal agriculture: application of real-time sensors to improve animal Well-being and production. Annu Rev Anim Biosci 7:403–425
Hempstead MN, Waas JR, Stewart M, Cave VM, Sutherland MA (2017) Behavioural response of dairy goat kids to cautery disbudding. Appl Anim Behav Sci 194(March):42–47
Hempstead MN, Waas JR, Stewart M, Cave VM, Sutherland MA (2018a) Evaluation of alternatives to cautery disbudding of dairy goat kids using behavioural measures of post-treatment pain. Appl Anim Behav Sci 206(April):32–38
Hempstead MN, Waas JR, Stewart M, Dowling SK, Cave VM, Lowe GL, Sutherland MA (2018b) Effect of isoflurane alone or in combination with meloxicam on the behavior and physiology of goat kids following cautery disbudding. J Dairy Sci 101(4):3193–3204
Hempstead MN, Lindquist TM, Shearer JK, Shearer LC, Sutherland MA, Plummer PJ (2020) Acute cortisol and behavior of dairy goat kids administered local anesthesia, topical anesthesia or systemic analgesia prior to cautery disbudding. Physiol Behav 222(May):112942
Hempstead MN, Lindquist TM, Shearer JK, Shearer LC, Cave VM, Plummer PJ (2021) Welfare assessment of 30 dairy goat farms in the Midwestern United States. Front Vet Sci 8(April):1–12
Hillmann E, Hilfiker S, Keil NM (2014) Effects of restraint with or without blinds at the feed barrier on feeding and agonistic behaviour in horned and hornless goats. Appl Anim Behav Sci 157:72–80
Jaber LS, Duvaux-Ponter C, Hamadeh SK, Giger-Reverdin S (2019) Mild heat stress and short water restriction treatment in lactating alpine and Saanen goats. Small Rumin Res 175(October 2018):46–51
Jackson KMA, Hackett D (2007) A note: the effects of human handling on heart girth, behaviour and milk quality in dairy goats. Appl Anim Behav Sci 108(3–4):332–336
Keeling L (2009) An overview of the development of the welfare quality assessment systems. Wageningen Academic Publishers
Kruger LP, Nedambale TL, Scholtz MM, Webb EC (2016) The effect of environmental factors and husbandry practices on stress in goats. Small Rumin Res 141:1–4
Larrondo C, Leiva J, de la Cruz-Cruz L (2021) Dairy goat welfare in semi-intensive production systems and drought conditions. Anim Welf 30(4):371–379
Laurijs KA, Briefer EF, Reimert I, Webb LE (2021) Vocalisations in farm animals: a step towards positive welfare assessment. Appl Anim Behav Sci 236:105264
Leite LO, Stamm FO, Garcia RCM (2017) Indicators to assess goat welfare on-farm in the semi-arid region of Brazilian northeast. Ciência Rural 47(9):1–8
Leite LO, Stamm FO, Corrêa Maceno MA, Camarinha Filho AJ, Garcia RCM (2020a) On-farm welfare assessment in meat goat does raised in semi-intensive and extensive systems in semi-arid regions of ceará, northeast, Brazil. Ciencia Rural 50(10):1–13
Leite LO, Stamm FO, Souza RA, Camarinha Filho JA, Garcia RCM (2020b) On-farm welfare assessment in dairy goats in the Brazilian northeast. Arq Bras Med Vet Zootec 72(6):2308–2320
Lenhardt ML (1977) Vocal contour cues in maternal recognition of goat kids. Appl Anim Ethol 3:211–219
Lyons DM (1989) Individual differences in temperament of domestic dairy goats and the inhibition of milk ejection. Appl Anim Behav Sci 22:269–282
Lyons DM, Price EO (1987) Relationships between heart rates and behaviour of goats in encounters with people. Appl Anim Behav Sci 18:363–369
Mason G, Mendl M (1993) Why is there no simple way of measuring animal welfare? Anim Welf 2(4):301–319
Mattiello S, Battini M, Andreoli E, Minero M, Barbieri S, Canali E (2010) Avoidance distance test in goats: a comparison with its application in cows. Small Rumin Res 91(2–3):215–218
Mattiello S, Battini M, Mantova E, Noè L, Grosso L, Barbieri S (2018) Evidence of poor welfare in goats with external abscesses. Large Anim Rev 24:113–118
Maurmann I, Greiner BAE, von Korn S, Bernau M (2021) Lying behaviour in dairy goats: effects of a new automated feeding system assessed by accelerometer technology. Animals 11(8)

McGregor BA, Butler KL (2008) Relationship of body condition score, live weight, stocking rate and grazing system to the mortality of Angora goats from hypothermia and their use in the assessment of welfare risks. Aust Vet J 86(1–2):12–17

McLennan KM, Rebelo CJB, Corke MJ, Holmes MA, Leach MC, Constantino-Casas F (2016) Development of a facial expression scale using footrot and mastitis as models of pain in sheep. Appl Anim Behav Sci 176:19–26

Mellor DJ, Beausoleil NJ (2015) Extending the "five domains" model for animal welfare assessment to incorporate positive welfare states. Anim Welf 24(3):241–253

Mersmann D, Schmied-Wagner C, Nordmann E, Graml C, Waiblinger S (2016) Influences on the avoidance and approach behaviour of dairy goats towards an unfamiliar human - an on-farm study. Appl Anim Behav Sci 179:60–73

Miller DW, Fleming PA, Barnes AL, Wickham SL, Collins T, Stockman CA (2018) Behavioural assessment of the habituation of feral rangeland goats to an intensive farming system. Appl Anim Behav Sci 199(February 2017):1–8

Minnig A, Zufferey R, Thomann B, Zwygart S, Keil N, Schüpbach-Regula G, Miserez R, Stucki D, Zanolari P (2021) Animal-based indicators for on-farm welfare assessment in goats. Animals 11(11):1–22

Muri K, Stubsjøen SM, Valle PS (2013) Development and testing of an on-farm welfare assessment protocol for dairy goats. Anim Welf 22(3):385–400

Neave HW, von Keyserlingk MAG, Weary DM, Zobel G (2018) Feed intake and behavior of dairy goats when offered an elevated feed bunk. J Dairy Sci 101(4):3303–3310

Nenadović K, Ilić T, Jovanović N, Bugarski D, Vučinić M (2021) Welfare of native goat breeds in Serbia—emphasis on parasitological infections. Front Vet Sci 8(August):1–13

Nordmann E, Keil NM, Schmied-Wagner C, Graml C, Langbein J, Aschwanden J, von Hof J, Maschat K, Palme R, Waiblinger S (2011) Feed barrier design affects behaviour and physiology in goats. Appl Anim Behav Sci 133(1–2):40–53

Ntalampiras S, Ludovico LA, Presti G, Vena MV, Fantini D, Ogel T, Celozzi S, Battini M, Mattiello S (2023) An integrated system for the acoustic monitoring of goat farms. Eco Inform 75(March):102043

OIE (2018) Introduction to the recommendations for animal welfare. In: Terrestrial animal health code. OIE, pp 1–4

Rahardja DP, Toleng AL, Lestari VS (2011) Thermoregulation and water balance in fat-tailed sheep and Kacang goat under sunlight exposure and water restriction in a hot and dry area. Animal 5(10):1587–1593

Reddy PR, Kumar BR, Prasad CS, Venkataseshiah C, Hyder I (2019) Erythrocyte fragility based assessment of true thermal resilience in tropical small ruminants. Biol Rhythm Res 53(2):234–245

Richmond SE, Wemelsfelder F, de Heredia IB, Ruiz R, Canali E, Dwyer CM (2017) Evaluation of animal-based indicators to be used in a welfare assessment protocol for sheep. Front Vet Sci 4(December):1–13

Rifkin RE, Grzeskowiak RM, Mulon PY, Adair HS, Biris AS, Dhar M, Anderson DE (2019) Use of a pressure-sensing walkway system for biometric assessment of gait characteristics in goats. PLoS One 14(10):1–14

Santucci P, Branca A, Napoleone M, Bouche R, Aumont G, Poisot F, Alexandre G (1991) Body condition scoring of goats in extensive conditions. In: Morand-Fehr (ed) Goat nutrition. Wageningen Academic Publishers, pp 240–255

Sejian V, Srivastava RS (2010) Effects of melatonin on adrenal cortical functions of Indian goats under thermal stress. Vet Med Int 2010

Sejian V, Silpa MV, Reshma Nair MR, Devaraj C, Krishnan G, Bagath M, Chauhan SS, Suganthi RU, Fonseca VFC, König S, Gaughan JB, Dunshea FR, Bhatta R (2021) Heat stress and goat welfare: adaptation and production considerations. Animals 11(4):1–24

Shi J, Dunbar RIM, Buckland D, Miller D (2003) Daytime activity budgets of feral goats (Capra hircus) on the Isle of Rum: influence of season, age, and sex. Can J Zool 81(5):803–815

Siebert K, Langbein J, Schön PC, Tuchscherer A, Puppe B (2011) Degree of social isolation affects behavioural and vocal response patterns in dwarf goats (Capra hircus). Appl Anim Behav Sci 131(1–2):53–62

Silva Salas MÁ, Mondragón-Ancelmo J, del Rosario Jiménez Badillo M, Rodríguez Licea G, Napolitano F (2021) Assessing dairy goat welfare in Mexican intensive or semi-intensive farming conditions. J Dairy Sci

Stronge DC, Fordham RA, Minot EO (1997) The foraging ecology of feral goats Capra hircus in the Mahoenui giant weta reserve, southern king country, New Zealand. N Z J Ecol 21(1):81–88

Su Q, Tang J, Zhai J, Sun Y, He D (2021) Automatic tracking of the dairy goat in the surveillance video. Comput Electron Agric 187(March):106254

Tiezzi F, Tomassone L, Mancin G, Cornale P, Tarantola M (2019) The assessment of housing conditions, management, animal-based welfare, and its association with productive and reproductive traits. Animals 9(11)

Toinon C, Waiblinger S, Rault JL (2021) Maternal deprivation affects goat kids' stress coping behaviour. Physiol Behav 239(March):113494

Toinon C, Waiblinger S, Palme R, Rault JL (2023) Long-term effects of early maternal deprivation on goat social behaviour. Animal 17(5)

Tölü C, Savaş T (2007) A brief report on intra-species aggressive biting in a goat herd. Appl Anim Behav Sci 102(1–2):124–129

Tölü C, Göktürk S, Savaş T (2016) Effects of weaning and spatial enrichment on behavior of Turkish Saanen goat kids. Asian Australas J Anim Sci 29(6):879–886

Tuyttens FAM, Molento CFM, Benaissa S (2022) Twelve threats of precision livestock farming (PLF) for animal welfare. Front Vet Sci 9(May):1–12

Vas J, Andersen IL (2015) Density-dependent spacing behaviour and activity budget in pregnant, domestic goats (Capra hircus). PLoS One 10(12):1–17

Vayssade JA, Arquet R, Bonneau M (2019) Automatic activity tracking of goats using drone camera. Comput Electron Agric 162(August 2018):767–772

Vieira A, Brandão S, Monteiro A, Ajuda I, Stilwell G (2015) Development and validation of a visual body condition scoring system for dairy goats with picture-based training. J Dairy Sci 98(9):6597–6608

Vieira A, Battini M, Can E, Mattiello S, Stilwell G (2018) Inter-observer reliability of animal-based welfare indicators included in the animal welfare indicators welfare assessment protocol for dairy goats. Animal

Wagmann N, Spadavecchia C, Morath-Huss U, Schüpbach-Regula G, Zanolari P (2018) Evaluation of anaesthesia and analgesia quality during disbudding of goat kids by certified Swiss farmers. BMC Vet Res 14(1):1–8

Wathes CM, Kristensen HH, Aerts JM, Berckmans D (2008) Is precision livestock farming an engineer's daydream or nightmare, an animal's friend or foe, and a farmer's panacea or pitfall? Comput Electron Agric 64(1):2–10

Zobel G, Nawroth C (2020) Current state of knowledge on the cognitive capacities of goats and its potential to inform species-specific enrichment. Small Rumin Res 192(December 2019):106208

Zobel G, Weary DM, Leslie K, Chapinal N, von Keyserlingk MAG (2015) Technical note: validation of data loggers for recording lying behavior in dairy goats. J Dairy Sci 98(2)

Zobel G, Neave HW, Webster J (2019) Understanding natural behavior to improve dairy goat (Capra hircus) management systems. Transl Anim Sci 3(1):212–224

Website

A Greener World (2018). https://agreenerworld.org/certifications/animal-welfare-approved/standards/goat-standards/. Accessed 15 May 2022

American Grassfed (2010). https://www.americangrassfed.org/. Accessed 30 May 2022

American Humane (2021a). https://www.americanhumane.org/humane-heartland/. Accessed 26 May 2022
American Humane (2021b) Dairy-goats-audit-tool. https://www.americanhumane.org/app/uploads/2021/08/Dairy-Goats-Audit-Tool.pdf. Accessed 26 May 2022
AMIC (2018) AMIC-Ed-3-Industry-Animal-Welfare-Standard_Final-effective-1-Jan-2022. https://amic.org.au/wp-content/uploads/2021/10/AMIC-Ed-3-Industry-Animal-Welfare-Standard_Final-effective-1-Jan-2022.pdf. Accessed 27 May 2022
Animal Welfare Institute (2015). https://awionline.org/content/consumers-guide-food-labels-and-animal-welfare. Accessed 29 May 2022
Certified humane (2014). https://certifiedhumane.org/our-standards/. Accessed 17 May 2022
Global Animal Partnership (2019). https://globalanimalpartnership.org/standards/goat/. Accessed 14 May 2022
https://www.imvoconvenanten.nl/-/media/imvo/files/kleding/2019-dierenwelzijn-factsheets.pdf?la=en&hash=9B9771E5FE0683BF888633F854746C53. Accessed 23 May 2022
ICCAW (2013). http://www.iccaw.org.cn. Accessed 13 May 2022
ICCAW (2015). http://www.3ecashmere.com/?category_id=5. Accessed 13 May 2022
Ministry for Primary Industries (2013) Code of welfare goats. https://www.mpi.govt.nz/dmsdocument/46033-Code-of-Welfare-Goats. Accessed 31 May 2022
MONTSAME (2022). https://www.montsame.mn/en/read/300066. Accessed 31 May 2022
NATURLAND (2022). https://www.naturland.de/en/naturland/what-we-stand-for/animal-welfare/sheep-and-goats.html. Accessed 18 May 2022
Ohioline (2017). https://ohioline.osu.edu/factsheet/vme-33. Accessed 26May 2022
Red Tractor Certified Standards (2021). https://saiassurance.co.uk/red-tractor-dairy-goats. Accessed 29May 2022
Responsible Mohair Standard (2022). https://textileexchange.org/standards/responsible-mohair/. Accessed 14 May 2022
Sustainable Cashmere (2020). http://sustainablecashmereplatform.com. Accessed 12 May 2022
Sustainable Fibre Alliance (2020). https://sustainablefibre.org/#section_19. Accessed 10 May 2022
The Australian Livestock Processing Industry Animal Welfare Certification System (AAWCS) (2020). https://aawcs.com.au/certification/. Accessed 25 May 2022
The Good Cashmere Standard (2019). https://thegoodcashmerestandard.org/. Accessed 10 May 2022
Tierschutz-kontrolliert (2022). https://www.tierschutz-kontrolliert.org/ueber-uns. Accessed 31 May 2022
Waitrose & Partners (1999). https://www.johnlewispartnership.co.uk/content/dam/cws/pdfs/Juniper/ethics-and-sustainability/Our-approach-to-animal-welfare-2021.pdf. Accessed 20 May 2022

Where Are Goats Going? The Future of Goats for Farming and Other Purposes

Silvana Mattiello, Paola Crepaldi, and Monica Battini

Abstract

Goats are a species of growing interest all over the world, and their role is evolving beyond their mere use for production purposes, now including also educational, therapeutic, recreational, and companion purposes. Most of the goats are located in developing countries, where farming conditions are often challenging for their welfare. At the same time, welfare-friendly production systems, such as extensive and organic systems, are spreading in developed countries in response to consumers' demand, with positive effects on goats' welfare. Nowadays, the species is facing a number of environmental challenges, such as climate change and the return of large predators, for which it is necessary to implement proper management and breeding strategies. New welfare issues are emerging in relation to the increasing presence of goats in zoos, pet farms, schools, sanctuaries, and private settings, as people interacting with goats might not always be sufficiently informed on how to manage these animals. Information and training of people who take care of goats should therefore be strongly encouraged, in order to guarantee that animals receive adequate care.

Keywords

Farming · Zoos · Sanctuaries · Teaching · Animal-assisted therapy · Companion animals

S. Mattiello (✉) · P. Crepaldi · M. Battini
Department of Agricultural and Environmental Sciences - Production, Landscape, Agroenergy, University of Milan, Milan, Italy
e-mail: silvana.mattiello@unimi.it

S. Mattiello, M. Battini (eds.), *The Welfare of Goats*, Animal Welfare 25,
https://doi.org/10.1007/978-3-031-62182-6_10

1 Evolution of Goat Farming Systems

The human–goat relationship has been dramatically evolving during the last millennia, starting from mere hunting for subsistence to intensive farming systems, passing through a long domestication process, initiated about 10,000 years ago, that made goats a widely used and increasingly economically important commercial species. Through the domestication process, many changes occurred in goats, which had to adapt to different housing and management systems (see chapter "Where Do Goats Come From?"). These systems present a wide range of variation, going from very intensive rearing systems for the more selected and cosmopolitan breeds to extensive pasture-based systems where rural and local breeds are usually farmed. Small holdings and family-run farms are also widespread all around the world, especially in rural communities in the mountains or in tropical and subtropical areas. In developed countries, goat breeding was initially started as a niche activity, but it then followed the path of the dairy cattle sector, with intensification of farming systems, characterised by high degree of specialisation and industrial animal production, and impressive increases of production performances (Dubeuf 2021). These farming systems may threaten animal welfare by limiting the expression of goats' natural behaviour, due, for example, to high stocking densities, sexual segregation, early separation of kids from their mothers, frequent regrouping, and manipulation during critical periods, including gestation and weaning (Miranda-de la Lama and Mattiello 2010). However, intensive farming concerns only a small number of animals bred worldwide and represents a small number of breeds. In fact, as seen in chapter "Where Do Goats Come From?", most animals are raised in regions of the world where they serve as subsistence farming systems to survive to hunger and poverty. For example, in Italy, the intensively farmed cosmopolitan Saanen breed accounts only for about 10% of the officially registered goats (Crepaldi, personal communication). Most local breeds in Italy are raised on semi-extensive or extensive farms occupying marginal areas (Cortellari et al. 2021). Extensive or semi-extensive farming systems, as well as small family farms, have been spreading all over the world, especially in Asia and Africa, where goat population numbers have been particularly increasing in the last 50 years (Fig. 1) and now represent about 94% of the total world population (1.09 billion heads; Fig. 2).

In these continents, the attention towards animal welfare is lower than in Europe and North America, in spite of the fact that extensive and semi-extensive farming systems are more exposed to the effects of adverse meteorological events and to seasonal variations, and this may have detrimental effects on resource availability, which in turn may affect body condition, production, and animal welfare (Chávez-Espinoza et al. 2021). Furthermore, these farming systems may expose goats to a higher risk of predation. These issues will certainly deserve more attention in the future, especially considering the high number of goats raised in these conditions.

Regarding current trends in goat farming systems, particular attention should be given to the development and spread of organic farming and of farming based on agroecological principles, which are increasingly growing in developed countries. Organic farming, also known as "ecological" or "biological" farming, is a

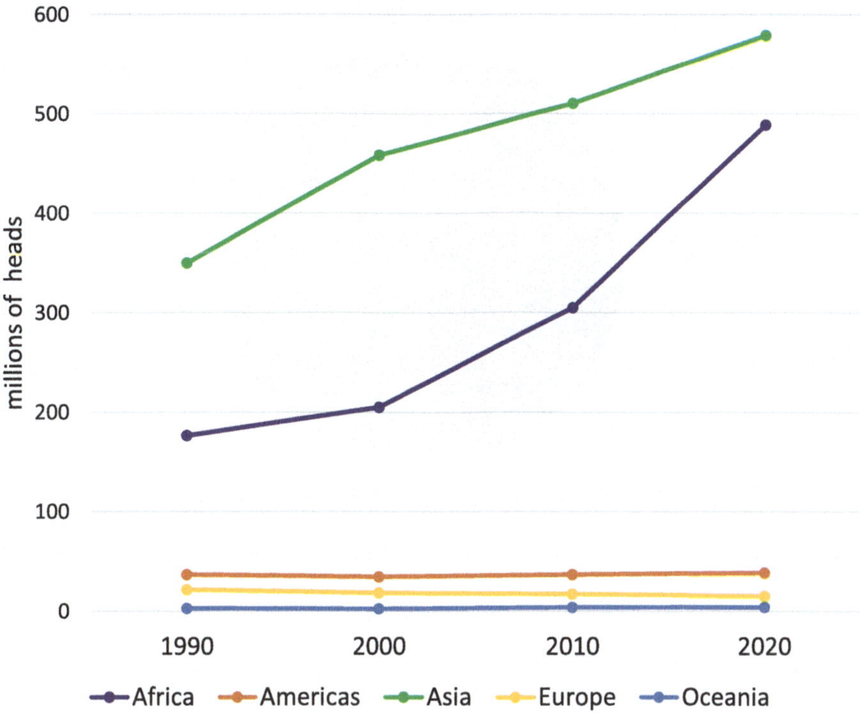

Fig. 1 Trend of the number of goats (millions of heads) in the five continents between 1990 and 2020 (data extracted from https://www.fao.org/faostat/en/#data/QCL)

production system based on natural processes and the use of organic and natural materials, where animal welfare is given a high priority. To this aim, in organic farms goats are raised with a higher attention for their physiological, behavioural, and ecological needs, thus assuring the conditions that allow them to express their genetic potential in terms of growth and development. In order to ensure a high level of adaptation to the farming environment, the adoption of native breeds, well adapted to local growing conditions and resistant to diseases, is strongly encouraged and contributes to maintaining a high welfare level.

In developed countries, the increasing consumers' concern for environmental sustainability and animal welfare, as well as for product safety and quality, has strongly contributed to the development of organic goat farming, which may represent a way to increase the value of products obtained from small goat farms. As an example, the trend of the number of live goats in organic farms in the three main European organic producing countries is shown in Fig. 3.

Dubeuf (2021) considers that goat farming in the world has favourable prospects, thanks to its modernisation and integration into the life of rural populations. However, conditions for this further development should take into account consumers' demand for animal welfare and environmentally sustainable products.

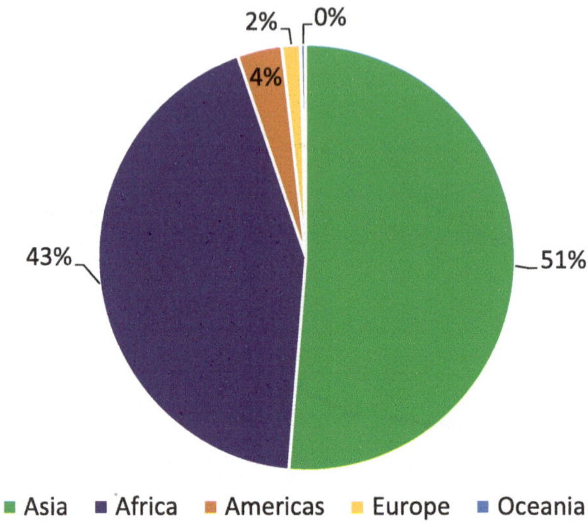

Fig. 2 Percentage number of goat heads in the five continents in 2020 (data extracted from https://www.fao.org/faostat/en/#data/QCL)

2 Future Challenges for Goat Welfare in a Changing World

The main current environmental challenges for goat welfare are represented by the return of predators, which may cause stress and lead to death, and by climate change, which not only induces heat stress and threatens the availability of food resources, but also represents a potential threat for the spread of emerging diseases (Altizer et al. 2013).

2.1 Predation

Predation is an increasingly important matter of concern, because predator populations have been increasing in number in the last decades. Many predator species that were on the edge of extinction have recovered thanks to conservation strategies and hunting control, with obvious benefits for biodiversity conservation and for the welfare of those endangered species. However, in many countries, this exacerbated or led to severe human–wildlife conflicts, especially in areas with extensive or semi-extensive farms of small ruminants. In Europe and North America, wolves (*Canis lupus*) and brown bears (*Ursus arctos*) are the main predator species, and puma (*Puma concolor*) is the most widespread apex predator in the Americas. In Africa,

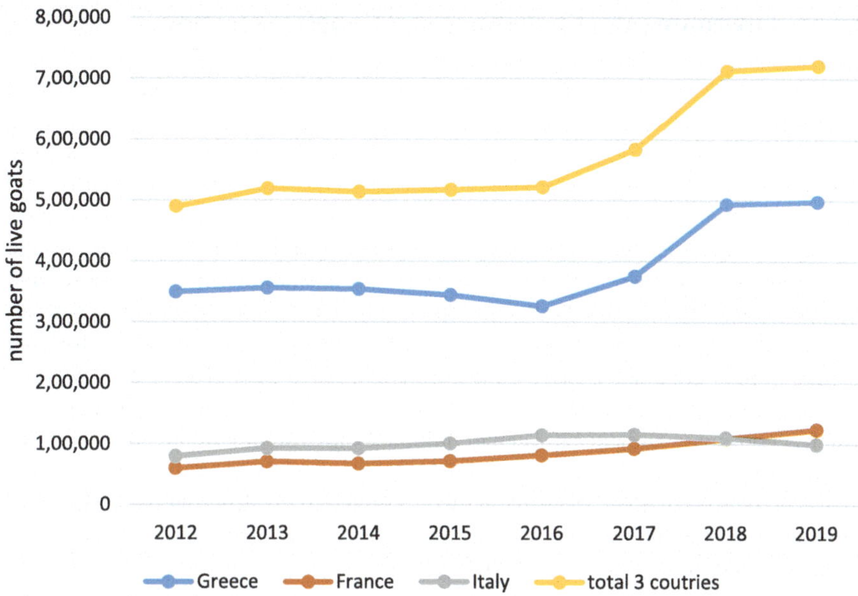

Fig. 3 Trend of the number of live goats in organic farms in the three main European organic goat producing countries (Greece, France, and Italy) from 2012 to 2019 (https://ec.europa.eu/eurostat/databrowser/view/org_lstspec/default/table?lang=en)

important predators are spotted hyenas (*Crocuta crocuta*), African wild dogs (*Lycaon pictus*), lions (*Panthera leo*), leopards (*Panthera pardus*), and golden jackals (*Canis aureus*), whereas in Asia wolves, leopards, golden jackals, jungle cats (*Felis chaus*), and Himalayan black bears (*Ursus thibetanus*) represent the main threats (Baral et al. 2021; Bista et al. 2022).

Serious conservation issues may arise when predators belong to endangered species like the harpy eagle (*Harpia harpyja*) in Amazonia (Miranda et al. 2022), the Pallas's cat (*Otocolobus manul*) in Mongolia, or the snow leopard (*Panthera uncia*) in the mountain ranges of Central and South Asia, which may become victims of illegal killing by farmers when the human–wildlife conflict exacerbates.

Goats often represent one of the main victims of predation events. These events may lead to the death of the goats, which is obviously detrimental for animal welfare. However, a predator's attack may also jeopardise goat welfare by inducing fear and stress, which may compromise growth, milk production, and reproductive performances. These consequences may be interpreted as indicators of poor welfare. Modern farmers should be aware of these problems and should protect their animals from predation events, avoiding unsupervised grazing and adopting appropriate prevention methods, such as guardian dogs, electric fences, or acoustic deterrers.

2.2 Environmental Challenges and Genetic Selection

Selection goals have evolved over time, depending on the breeds and production purposes.

The evolution of selection objectives has shifted from increasing production, which was a need during the period after World War II, to improving quality, and choosing animals, which are resistant to the most common diseases, with morphological characteristics related to a greater functional welfare. Parameters such as the evaluation of locomotion traits and correctness of limbs, the absence of diseases and infections, and the animal's body conditions aimed at avoiding excessive exploitation are currently the basis of goat breeding. For example, genetic indices for goat welfare in Italy are mainly based on the aforementioned traits (Negro et al. 2022).

The most recent challenges for goat breeding are represented by climate issues, which according to Dubeuf (2021) will influence the evolution of the goat sector for the next 30 years. The objectives of animal selection have been evolving through the choice of animals that are resistant and well adapted to the different environmental, climatic, and breeding conditions they face, especially under semi-extensive and extensive conditions (Phocas et al. 2016).

In parallel, the goat sector is facing worldwide challenges stemming from the intensification of livestock systems. The sector needs to respond to the rapidly changing demand for food, in an environmentally and socially sustainable environment, ensuring that the world's poorest people are no longer hungry and adding the consideration of well-being of the living animals (Rauw 2016). As temperatures rise and marginal regions dry up, food availability becomes increasingly problematic, and the identification of animals resistant to hot and dry conditions is now one of the main topics under evaluation by the international scientific community (Simões and Gutiérrez 2018). It is therefore important to better disentangle the genetic complex traits related to physiological processes involved in stress due to climate change.

The rapidly changing demand for food must somehow be satisfied by cultivating an environment that is ecologically and socially sustainable. This sustainability includes appeasing the hunger of the world's poorest people as well as safeguarding the welfare of animals.

With respect to climate change, landscape genomic approaches could help determine which genotypes and genes could make animals and populations more resilient. Animals with resilient genotypes and genes need to be valued. The evolution of breeding techniques and genomic technologies may in the future allow the rapid introgression of genotypes or genetic variants to develop more resilient, less stress-sensitive, and healthier animals, breeds, and populations (Perisse et al. 2021). The selection of disease-resistant genotypes is particularly relevant even because of the recent changes in patterns of trade in animals and animal products, which facilitate the spread of diseases in wider geographic areas (Babiuk et al. 2008).

2.3 Further Future Goals of Genetic Selection

In general, programmes devoted to the maintenance of genetic diversity between and within different breeds and populations could help safeguard the welfare of goats and their future worldwide. Genomic evaluation of the level of homozygosity and heterozygosity of single animals and populations could help to define the best strategy to manage populations in order to avoid excessive levels of inbreeding, which accentuate negative health and behavioural traits due to a lack of fitness in natural and artificial selection. This is particularly relevant for local breeds which count only for a low number of individuals and therefore there is a need to avoid increases in inbreeding, either by using tools to identify genomic inbreeding for a single goat and breed or by resorting to crossbreeding where the situation demands it. This approach requires overcoming the concept of a standardised breed. However, unfortunately, genetic breeding programmes for local breeds are rare; instead, there are programmes to conserve genetic resources. These programmes are more likely to be successful if associated with an income for breeders, for example valuing typical products related to areas dedicated to semi-extensive and extensive farming systems.

A great deal of attention, both from consumers and policy makers, has been devoted in recent decades to the issue of animal welfare (Rauw 2016). Even breeders and their associations recognise that animals free of behavioural, physiological, and immunological problems are the foundation of the public awareness regarding animal welfare and breeding. Most farmers know that animal welfare is their ally because it decreases the frequency and need for health interventions and expensive therapies and helps keeping farms profitable. Animals in harmony with themselves and their environment are the basis of any good breeding system. Among selection criteria, longevity, health, and fertility have been gaining much more importance in recent years (Rauw 2016; Roger 2012). Genetic indices that promote animal welfare, derived from weighted estimates of Body Condition Score (an indicator proved to be linked to adequate physiological condition with good energy balance), limb strength, and locomotion have been recently proposed (Negro et al. 2022).

The use of extended lactation periods to reduce reproductive intensity and distribute milk production over time is also becoming more widespread (Rauw 2016; Salama et al. 2005). Again, genetics and genomics may identify individuals more predisposed to this practice, which appears to have beneficial effects on animal welfare. However, a comprehensive approach to health and welfare is needed to investigate its possible negative effects, for example on udder health (see chapter "Welfare Issues in Goat Farming and Management").

The domestication syndrome, which is mainly related to unintentional selection for tameness, and is responsible for many morphological and physiological modifications, identifies physiological and behavioural traits conducive to reduced stress (Wilkins et al. 2014). In this respect, the behavioural traits of goats need to be better evaluated, using genomic tools to determine which animals are best adapted to different farming and environmental conditions. These animals are supposed to be less sensitive to the stress induced by farming conditions and human contact, and therefore would develop stronger immune systems, which make them less susceptible to infections and parasitic diseases (Gray et al. 1995; Nenadović et al. 2021). Direct selection of resistant animals is also one of the main goals pursued by breeder's associations, especially in cosmopolitan breeds, where it is possible to collect direct and accurate phenotypes for disease resistance.

The inclusion of behaviour-related traits certainly requires a greater investigation of the genetic pathways related to these traits, also because negative emotional and physical experiences might affect not only the subject itself, but also its offspring (Chojnacki et al. 2014). Miranda-de la Lama and Mattiello (2010) reviewed the importance that social behaviour plays in goat welfare. In poultry, Ellen et al. (2014) proposed assessing social interactions that promote the positive effects of cooperation and maternal behaviour and evaluating the negative effects of competition and aggression on animal welfare, productivity, and health, by estimating the social genetic value of an individual relative to its flock mates. It should be mentioned that, in goat breeding, aspects related to social interactions are often indirectly considered in the estimation of "farm" effect in the model used. Both positive and negative aspects are in fact controlled by the farmer, who removes particularly competitive or aggressive animals, forms mating groups according to the characteristics of the animals, and prefers does with good maternal behaviour traits. The collection of detailed phenotypes with respect to these traits and also genomic and epigenetic studies in the presence of high environmental stressors can lead to a better understanding of the genetic basis of social behaviour for the development of better management practices and better models to estimate the genetic and genomic values of animals with respect to these traits.

Farmers, breeders' associations, and researchers are aware of the great challenge posed by determining and developing conditions that promote animal welfare and the sustainability of livestock production. They are also mindful of the need to conserve the traits and characteristics of breeds and populations that have been developed during the ten-thousand-year relationship of goats with humans, which may serve them to face the needs and challenges of the future.

3 Goats as Non-food-producing Animals

Apart from their use as food-producing animals, the role of goats in their relationship with human beings may have different aims. Russell (2017) reports that at various times in Southern Africa hunter-gatherers already used to keep livestock for several reasons which go beyond the provision of food for subsistence. So, these authors wondered why hunters-gatherers kept livestock, while they continued hunting, and suggested that livestock, and particularly goats, played a social function, as they were exchanged as means for establishing and reinforcing social bonds. In this sense, an interesting example is that of a goat that was circulated among five households for different purposes (Ikeya 1993, reported by Russell 2017), which could range from "marriage insurance", payments, sacrifices, magical rites and purification ceremonies to meat feasts and clothing supply. The exchange of goats may render tangible benefits, such as sharing the burden of grazing and watering the animals and circulating prime breeding stock, but also intangible benefits, such as building and sustaining reciprocal ties and gaining prestige and respect. This is also one of the reasons why these communities tend to keep goats rather than slaughtering them, except for ritual sacrifices. In fact, in African communities (e.g. in the Niger Delta of Nigeria), goats play a significant role in burial ceremonies, marriages, religious festivals, and traditional sacrifices (Ikeoji 2010). However, in spite of the fact that for the above-mentioned reasons goats are rarely slaughtered and eaten in these societies (as shown by the scarce amount of livestock remains retrieved in archaeological sites), it seems that they are still considered as "goods", used to provide some kind of service to humans, and that their needs and welfare are not sufficiently taken into account. Furthermore, knowing the importance of establishing a good human–goat relationship for goats' welfare (see chapter "Welfare Issues in Goat Farming and Management"), we can argue that repeated changes of the caretakers might have a detrimental effect in this respect.

In modern developed societies, the role of goats has been changing during recent decades and people are now concerned about the health and welfare of goats, either for food production or non-food production purposes. Talking about non-production purposes, nowadays goats are seen also as animals that can provide intangible social and cultural services to humanity, which go beyond their mere capacity to provide tangible economic services such as milk, meat, and fibre. For example, they are now often considered as companion or pet animals, or even as educational or therapeutic help. This implies that goats are no longer kept only in traditional commercial farms, but they can also be found in a variety of different contexts, such as schools, pet and educational farms, pet zoos, zoos, and sanctuaries, where they play different roles other than that of merely producing food or fibre.

Most of these contexts, except for sanctuaries, are conceived for recreational or educational purposes of people, and therefore they do not position animal welfare as a central priority. For these reasons, most of the research carried out in these contexts aims mainly at quantifying the benefits that humans can obtain from goats, often neglecting the animals' point of view. Therefore, in many of these situations, goats' welfare may be threatened. For example, a recent survey carried out among goat owners who took their animals to a University Veterinary Hospital in Italy revealed that 97.36% of the animals were not kept for production purposes (they belonged to animal welfare associations or to individual owners who kept them as pets) and 37.75% of them showed health problems due to inadequate feeding (Colla 2022).

Petting zoos, as well as petting farms, are facilities where animals can closely interact with humans, and can even be touched by adults and children; however, petting farms differ from petting zoos because, in addition to recreational and educational purposes, petting farms, like city farms in general, also have productive purposes, i.e. they use animals for producing milk and meat for sale (De Malsche and Cornips 2021). The main purpose of these institutions is to facilitate the development of a positive relationship of visitors towards animals by providing them the opportunity of direct interactions. This in turn may promote positive attitudes towards animals in the future, thus hopefully resulting in improved animal welfare. This educational purpose seems to be particularly successful in children. In fact, since the earlier stages of domestication, children have always been attracted by animals, especially by puppies, and enjoy establishing contacts with them, and this is why they actually played a role in the domestication process (Forni 1990). It is therefore not surprising that De Malsche and Cornips (2021) report a strong interest of schools for visiting a goat petting farm. In this respect, teachers declare that they want to allow children to know animals and not to be afraid of them, and parents are often visiting the farm in order to educate their children. De Malsche and Cornips (2021) carried out a study on human–animal interactions in that petting farm, showing that effective communication took place during feeding time between goats and humans, both adults and children, and that, in some cases, the attention of goats towards humans is so strong that they can stop eating their "normal" feed to obtain alternative feed directly from the visitor's hands. Despite the multiple recreational and educational benefits that petting farms and zoos may provide to humans, some considerations are required from the goats' side. First, in order to avoid possible negative consequences on goats' welfare, it is important to provide verbal instructions, such as petting tips and recommendations to leave outside food for human consumption, before allowing visitors to enter the goats' enclosure (Anderson et al. 2002). Furthermore, the effect of human density should be controlled because the close contact with animals may potentially affect the behaviour and welfare of goats housed in this type of exhibit. In fact, an increase of undesirable behaviours (e.g. head tossing, head butting, foot stamping, rearing, nose-blowing, and retreating from visitors), probably aiming at restoring and maintaining critical distances, was observed in pygmy goats in Zoo Atlanta's petting zoo in response to increasing levels of human density in the yard (Anderson et al. 2002, 2004). This negative effect could not be mitigated by the proximity of the zookeeper (Anderson et al. 2004). Contrary to these findings, a study carried out by Farrand et al. (2014) did not find negative effects on the behaviour of goats in a petting zoo, suggesting a limited impact of visitors on goats' welfare. These authors observed that an increasing visitors' density significantly increased the proportion of times that goats entered in contact with them and decreased goats' proximity and contact with conspecifics.

Anderson et al. (2002) also noted that undesirable behaviours exhibited by goats in response to high visitors' densities could be limited by the opportunity for goats to escape from human contact, as observed when a retreat space, consisting of a space isolated from the main enclosure by a wall 1.7 m high made of chicken wire style fencing, was provided into the goats' exhibit. Based on these findings, the authors strongly recommend that the design of the enclosure, and in particular the provision of a retreat space, should be carefully considered when managing a petting zoo in order to improve animal welfare. Unfortunately, this condition is not always met in petting facilities. For example, De Malsche and Cornips (2021) describe the situation of a pet farm where goat kids have no possibility to escape from human presence.

On the other hand, positive contact with humans has been suggested as a possible environmental enrichment under some circumstances (Claxton 2011). However,

Farrand et al. (2014) did not record significant behavioural changes in goats in response to visitors' grooming and therefore suggest that the opportunity to interact with people is not in itself enriching for goats.

Positive relationships can be established between goats and humans, either children or adults, in response to specific education programmes involving daily social contacts, as reported for the zoo of Naples in Italy (Mastellone et al. 2020). This study shows that such long-term socialisation resulted in a more positive attitude of goats towards humans. In particular, highly socialised goats exhibited more interactions with humans than goats with a lower level of socialisation, and tended to perform more human-directed behaviours, especially involving tactile interactions, in order to ask for human help to solve an impossible task, represented by the challenge to open a plastic food container whose lid was locked on a wooden base. This behaviour was interpreted by the authors as a socio-positive behaviour. This suggests that educational activities in zoos implying frequent contacts between goats and humans may be beneficial for goats' welfare, because they contribute to improve the human–animal relationship and thus to reduce possible stress to due handling procedures.

The general temperament of the animals should also be taken into account. For example, sheep seem to be more fearful of humans than goats, and inter-individual differences may occur also within the same species. Therefore, zookeepers will have to observe each individual animal's behaviour to select the more appropriate animals to be managed in a petting zoo or petting farm setting (Anderson et al. 2002).

Other welfare problems for goats in petting farms and zoos, and also in some sanctuaries, can be related to feeding (Figs. 4 and 5). Although specific references to goats are lacking, it is known that the administration of highly digestible and easily available feed (e.g. concentrate feed) might affect ruminants' normal behaviour, reducing the time dedicated to feeding. To encourage animals to express their normal species-specific ethogram, the administration of feeds with high fibre content is therefore recommended in these premises (Górka et al. 2016).

Additionally, visitors, especially school children, can be allowed to deliver grains, grass, or hay to the goats, or to bottle feed kids with milk that can be bought on the farm. This last kind of feeding has been observed to be particularly appreciated by visitors, who prefer to interact with kids rather than with adult animals (De Malsche and Cornips 2021). However, even when feed quality is strictly controlled, this habit might not be optimal for the animals, as it makes impossible to control feed intake, which may exceed recommended levels and therefore lead to metabolic imbalances or diseases. Furthermore, it has been observed that, if animals have already been satiated, they might not be interested in obtaining more food, and this might cause frustration in visitors who are willing to interact with goats. For example, De Malsche and Cornips (2021) observed that sometimes goat kids were not interested in drinking milk from the bottles offered by visitors, and clearly refused them. In this case, it may happen that visitors do not accept this refusal and try to force the animal to accept the milk, which can result in a reduced welfare for the kid, even though the visitor's intention towards the animal is good. This highlights the anthropocentric view of visitors, who are disappointed because they cannot take

Fig. 4 A visitor feeding goats in a goat sanctuary (Photo: Monica Battini)

control of the situation and they cannot meet their expectation to maintain a hierarchic control during the interaction (De Malsche and Cornips 2021). In order to avoid this kind of problems, as well as to prevent the transmission of zoonosis which may occur in petting farms and zoos (reviewed by Stirling et al. 2008), it would be important to provide preliminary instructions to the visitors before allowing them to enter in contact with the animals.

City farms represent a further possibility for humans to interact and learn from animals. City farms, also called educational farms, have been spreading quite recently in developed countries, with the aim to connect the urban and rural world, teaching people of all ages who live in urban environments the inter-relationships that occur among living organisms in natural environments. In Europe, a European Federation of City Farms has been created since 1919, in connection with several other associations around the world, such as city farms within the American 4H farms initiative. Any independent city farm, or city farms working together to form an organisation of city farms, may become members of this network, which therefore acts as a supranational association of national organisations. Only few data are available on this type of usage of goats, although small ruminants are commonly

Fig. 5 A visitor feeding goats in a goat pet farm (Photo: Monica Battini)

found in this kind of farms (Schilling et al. 2012), and they are quite popular especially among children. As an example of activities carried out in city farms, children can be involved in "goat clubs" where each child works in a team and is responsible for his/her own goat and they provide the animals with feed and water, keep their stable clean and brush them. They can even learn to milk the animals or harness them to drive a small cart (European Federation of City Farms 2002). On the one hand, these activities have an educational value for children, but on the other hand we may wonder what the impact of inexperienced personnel on the welfare of goats is. In order to minimise stress effect, children are carefully instructed and they even receive a farm's reference book about goats (European Federation of City Farms 2017), containing guidelines on the management of goats. This contains useful tips on, for example, how to restrain goats, how to milk them, how to take care of their hooves, or which are their spatial and feeding requirements. The purpose of this booklet is therefore commendable, although some instructions may pose a few welfare concerns, as for example the suggestion of tethering goats to a pole to let them graze, when fencing is not available: of course, this practice can severely limit goats' movements, and may therefore be detrimental for their welfare, and even dangerous when in proximity of major roads. The use of goats to drive small carts would also deserve some attention from the welfare point of view, as no evidence is available on its effect on goats' welfare. However, in general, based on health assessment, behavioural observations, and behavioural tests, Schilling et al. (2015) showed no evidence of major stress in goats in German city farms. Last but not least, it is worth underlining that, due to the close human–animal contact, city farms may represent a risk for the transmission of zoonotic agents from animals to humans and *vice versa* (see chapter "Enhancing Animal Welfare Through Disease Prevention in Goats"), and, therefore, optimal hygienic measures should be accounted for, in order to limit these risks (Schilling et al. 2012).

The use of animals as educational tools in urban schools is also becoming more and more common, for example in the USA (Loyd et al. 2021) and in Japan (Koda et al. 2013, 2016), where urban farm to school (FTS) programmes are adopted in order to enhance children's psychological development and learning by promoting children's affection for these animals, respect for living beings, sense of responsibility, and enhanced interpersonal interactional skills. In Japan, most elementary schools keep animals, which are usually represented by small animals such as birds, fish, rabbits, or hamsters (Koda et al. 2016). However, according to several studies (Koda et al. 2013, 2016; Loyd et al. 2021), goats also represent a valuable teaching resource, not only in petting farms and zoos, but also in schools. Besides the above-mentioned considerations that goats have a friendly character and a low fear of humans (Anderson et al. 2002), and that children have a particularly strong attitude for animals' care, it is worth remarking also that, as goats are medium-sized animals, they tend to have more equitable interactions with children than small animals like birds or hamsters, and visual signals may be particularly effective between goats and children because goats' faces are close to those of children, who can better observe goats' facial expressions and easily interact with them (Koda et al. 2016). It is probably because of these many reasons that the use of goats resulted to be quite successful in school educational programmes.

Children who had the opportunity to interact with goats became more familiar with them, and expressed ideas about better coexistence with goats than children who did not interact with these animals (Koda et al. 2013). According to Koda et al. (2016) goats in schools provided the expected benefits to the children, stimulating their interest, cooperation, and empathy, even though keeping goats resulted to be more demanding compared to smaller animals in terms of responsibility and labour. Koda et al. (2013) also noted that girls were more interested in taking care of the animals than boys, which is not surprising if we think about the fundamental role that women played in the process of domestication, reflected in the more positive attitude of women to take care of animals (Forni 1990). Loyd et al. (2021) carried out a survey among parents whose children had experiences with goats at school, in order to evaluate their personal attitude towards goats and how they perceived the effect of keeping goats on their children's educational and outdoor experiences and interests. The results showed that parents' attitude towards goats was positive from the beginning, but it became even more positive after the experience of having goats at school. Parents also believed that goats had a very positive effect on their children as regards interest and enthusiasm for school. In general, in line with what is suggested by Koda et al. (2016), through cooperation in goat-keeping this experience reinforced the relationships between parents, children, school, and the community in general, with overall positive social benefits. Even though Koda et al. (2013) could not observe an increased interest towards natural science and scientific professions in children who had experiences with goats at school, Koda et al. (2016) suggest that goats in schools might promote an interest for scientific studies.

Therefore, taking into account all the above-mentioned considerations, goats in schools seem to be a great opportunity, at least from an anthropocentric point of view. However, in this book, we would like to focus mainly on the goats' point of view and, in this respect, it is necessary to highlight some potential welfare issues. According to Koda et al. (2016), most of the problems may arise above all from the insufficient knowledge of teachers on keeping animals, especially regarding animals' needs in terms of housing, feeding, behavioural requirements, and identification and treatment of illness and injuries. Therefore, when using goats, as well as other animals, for teaching programmes, it is mandatory that teachers, and educational staff in general, receive adequate instruction on animal care and that veterinary support is secured when required.

An adequate training of the personnel who take care of the animals is also required in normal zoos. This is not always easy to achieve, because zookeepers have to deal with a great variety of species and should therefore be able to know the species-specific needs of each species. According to the European Council Directive 1999/22/EC relating to the keeping of wild animals in zoos, a zoo, also called zoological garden, is defined as "a permanent establishment where animals of wild species are kept for exhibition to the public for seven or more days a year". However, despite being composed mainly of wild species, zoo collections might also include domestic species, such as farm animals like goats (European Commission 2015).

The role of zoos has evolved over time. Until the nineteenth-century, zoos were places where animals were raised in cramped cages, with iron bars, without the possibility of living in a natural environment, in poor welfare conditions (Hosey et al. 2013), whereas in modern zoos animals have open spaces that more resemble their natural habitat, with the presence of elements such as caves, shelters, elevated points (e.g. trunks or rocks), trees or waterfalls (Fig. 6), and therefore their role moved from being a simple display of exotic animals to a fundamental commitment to supporting and promoting the conservation of threatened species through breeding programmes, maintaining an excellent state of animal welfare and promoting research and education (Hosey et al. 2013). For these reasons, zoos now take care of providing appropriate housing, feeding, and management of their animals. For these aspects, general recommendations for keeping goats, as described in chapters "Welfare Issues in Goat Farming: Housing and Nutrition" and "Welfare Issues in Goat Farming and Management", could probably supply adequate guidelines to follow. However, an additional aspect of the housing design should be taken into account regarding the possibility of visually or physically interacting with humans. Some negative effects have been observed in zoo animals during the interactions with humans, especially if under uncontrolled circumstances (Learmonth 2020).

Fig. 6 An exhibit for goats in a zoo. This environment provides adequate shelter, shade, and elevated points where goats may climb to satisfy their natural behaviour requirements (Photo: Monica Battini)

This seems to be confirmed also for goats in the above-mentioned study carried out by Anderson et al. (2002) in a petting zoo, which recommends providing animals with a retreat space to withdraw from excessive interactions with humans.

Besides good housing, feeding, and management, the fundamental pillars for achieving high levels of animal welfare in zoos are zookeepers' training, animals' training programmes, and environmental enrichment. Zookeepers must receive adequate training and attend specific courses which should address all animal species, including goats, so that they are aware of species-specific physical and mental requirements. These courses should also be addressed to motivate zoo staff and raise their enthusiasm for their profession (Bacon et al. 2021). Additionally, training programmes for the animals should be designed and carried out to teach them to perform specific behaviours, such as moving towards a given direction, or accepting specific health treatments, or being submitted to procedures such as blood sampling or faecal sampling, with the aim to facilitate the execution of these necessary management operations with no or limited stress for the animals (Mellen and Ellis 1996). Training in zoos is usually carried out using positive reinforcements, such as, in most cases, palatable food or water, which the animals can associate with a conditioned reinforcer, such as a light or a sound (typically a clicker; Pfaller-Sadovsky et al. 2020). It has been observed that, in some species, using a secondary reinforcement may help to accelerate the acquisition of operant learning tasks and delays the time to extinction of the learned task. This technique is often used in zoos to habituate or desensitise animals to be handled or to undergo specific conditions which might otherwise be stressful for the animals. For example, Fernandez and Rosales-Ruiz (2021) successfully applied in a petting zoo in Texas an 8-step shaping programme based on the use of a clicker sound followed by a piece of food any time the goat exhibited the desired behaviour. At the end of the shaping programme, goats easily accepted to wear a halter and to be led via the halter. This was quite useful for a less stressful animal management, as before shaping goats used to engage in undesired behaviours (e.g. jumping and chewing), whereas after the training they became easier to lead or move throughout their exhibit, without being forced. This resulted in an easier task for the zookeepers, but also in an improved welfare for the goats. The association of a primary stimulus (e.g. water) with a secondary stimulus (e.g. audio signals of different tones) successfully trained dwarf goats to discriminate between different shapes and can therefore be effectively used in zoos to train goats (Langbein et al. 2007).

Finally, according to the European Council Directive 1999/22/EC, environmental enrichment programmes should be planned to improve goats' welfare in zoos. The importance of providing complex and cognitively enriching environments that promote natural behaviour and stimulate cognitive abilities is well recognised in zoos. In particular, the presence of rocky, steep, and elevated areas or trunks is important for goats; even caves may be useful to make goats feel safe and protected from their natural threats, such as predators, and to allow individuals to isolate themselves from conspecifics in case of social conflicts (Zobel et al. 2019). A recent study (Araújo do Nascimento et al. 2022) suggested that enrichments such as classical music, plastic bottles with food, suspended tyres, and logs of wood contributed to positively affect the thermoregulatory and behavioural responses of indoor-housed goats, thus improving their welfare, even though the mechanisms underlying the relationship between enrichments and thermal stress are not completely clear. In zoos, successful experiments have been carried out playing classical music to the animals: animals seemed to react to music, which positively engaged them, and visitors also showed appreciation for the music and were amused by the animals' reactions (Creation Museum 2017).

Although zoo animal welfare is now a priority in most institutions worldwide, it is not the same for countries where the access to the relevant information is not granted and where behavioural, cultural, and financial barriers may prevent the application of practices aiming to improve animal welfare (Ward et al. 2020). Therefore, efforts should be made to make this information available in zoos all around the world.

Sanctuaries (Fig. 7) are defined as places where animals live under human protection and their keeping conditions resemble their lives in their natural habitat as much as possible (Pierce and Bekoff 2018). Different from zoos, whose primary aims are conservation, education, and research and that therefore do not have animal welfare as their primary objective, animal sanctuaries are defined as facilities that "rescue and provide shelter and care for animals that have been abused, injured,

Fig. 7 Goats in a sanctuary (Photo: Monica Battini)

abandoned or are otherwise in need, where the welfare of each individual animal is the primary consideration in all sanctuary actions. In addition, the facility should enforce a non-breeding policy and should replace animals only by way of rescue, confiscation, or donation" (Global Federation of Animal Sanctuaries 2023). This implies a considerable difference in the way zoos and sanctuaries see animals. Sanctuaries therefore host formerly abused, neglected, and/or unwanted goats, which may have previously experienced situations in which their welfare was seriously compromised, for example due to inappropriate housing (e.g. tethering) or feeding and untreated injuries. These adverse past events can be remembered by goats and might have negative consequences on their present welfare (Mendl and Paul 2004). However, Briefer and McElligott (2013) suggest that, after several years of good care in a sanctuary, goats can probably recover from these negative conditions, as shown by the fact that these animals, that had previously been submitted to poor welfare conditions, exhibited long-term optimistic bias, probably in response to the release from previous stress.

More than 200 sanctuaries are listed in the Global Federation of Animal Sanctuaries (2023) and, although the majority of them hosts wild animals, 29 of them focus on farm animals, such as the Goodheart Animal Sanctuary (2022) in Worcestershire or the Safe Haven Farm Sanctuary (2023) in New York, both hosting also goats. There are even some sanctuaries which are particularly dedicated to goats. They can be found in different geographic regions: for example, we can find

The Buttercups Sanctuary for Goats (2023) in England, or the Goatlandia Farm Animal Sanctuary (2023) in California, or the Goats of Anarchy (2023) in New Jersey, or the Black Goat Farm and Sanctuary (2020) in Ontario. This provides evidence of a high human sensitivity to the welfare of goats.

Very little information is available on the use of pet goats for therapeutic purposes, and this scarce information again focuses on the humans' point of view, overlooking the goats' point of view. For example, the research carried out by Scholl et al. (2008) in a residential institution for deaf and multiple-disabled adults confirmed that goats, as well as other domestic animals, can play a positive role for disabled people, increasing their interest and empathy towards the animals, and in particular towards goat kids. Contact with goats also seemed to provide additional stimuli and enriched the life of disabled persons, increasing their attentiveness and expression of joy. However, in this study, the attention declared by the authors for goats' welfare was limited to the choice of animals with a gentle temperament and those that had been previously well socialised with people, and to the supervision provided by the staff who took care that the animals were not harmed by the interactions with the disabled persons. This last aspect was guaranteed by the permanent possibility granted to the animals of retreating and avoiding any contact with humans, if they were disturbed by their presence. The authors also declared that goats were chosen for this experiment, because they are particularly socially responsive and gentle animals that have undergone a long process of domestication and have therefore been selected for tameness and improved ability to communicate with humans, which make them particularly suitable for this therapeutic context. These remarks suggest that goats' welfare was to a certain extent considered, but further studies would be required in this kind of interaction context to observe not only human behaviour but also goat behaviour to guarantee that animal welfare is respected.

Finally, among the many purposes of keeping goats, it is worth citing the use of goats for controlling invasive or overgrown vegetation. In fact, thanks to their feeding habits, goats represent a valid alternative approach to weed management, although at present they are probably underexploited for this purpose (Hart 2001). Cashmere goats are particularly adapted to this aim, and they are commercially exploited in North America (e.g. https://goatseatweeds.com/press/cashmere-goats-join-in-the-noxious-weed-battle/) and Europe (e.g. http://www.chianticashmere.com/en/Cashmere-goats/The-cashmere-goat/) to improve the quality of pastures and abandoned farmlands that other herbivores are not able to value. However, specific stocking rate criteria and production systems need to be developed to support a welfare-friendly use of goats for weed and bush control in different environments (Hart 2001). Even in urban and semi-urban areas, goats may be hired for vegetation control and, in the USA, these services are emerging to provide landscaping services to the communities (Loyd et al. 2021). Despite the growing popularity that goats are acquiring thanks to these services, one might wonder if their management by inexperienced persons is always suitable to guarantee goats' welfare.

4 Conclusions

Goats are facing a wide range of social, cultural, and environmental changes (e.g. changes of the societal attitude towards animals, climate change, the return of large predators), which may impact their welfare at various levels. Thanks to their ability to adapt to different environments and to their high resilience to climate change, goats are a species of growing interest not only in developed countries but also in developing countries, where most of the goats are concentrated and where farming conditions are often challenging for their welfare. In this respect, it may be worthwhile considering that in developing countries, where the conditions of poverty and hunger induce local populations to consider goats only as a means of subsistence, the improvement of the living conditions of such populations could help to increase their attention towards animal welfare. On the other hand, welfare-friendly production systems, such as extensive and organic systems, are spreading in developed countries in response to consumers' demand, with positive effects on goats' welfare. At the same time, goats are no longer considered only as a production species, but they are kept also for several other purposes, for example as companion animals or for therapeutic, recreational, and teaching purposes. Even though goats in zoos and other human interaction contexts will probably have a low numerical importance in the future, these situations have become more frequent and will probably gain further significance. This poses new welfare issues, as people interacting with goats might not always be sufficiently informed on how to manage these animals. Information and training of people who take care of goats should therefore be strongly encouraged, in order to guarantee that animals receive adequate care (in terms of feeding, shelter, health management, breeding, and proper handling) and that caregivers understand that goats can exhibit either interest or disinterest in interacting with humans, but it is always important to respect their wills and to allow them to express their preferences without forcing them. Furthermore, since keeping goats for non-productive purposes is a relatively recent phenomenon, scarce information is currently available on their welfare under these conditions, and most of the research carried out in these contexts is centred on humans and on the benefits that humans can receive from goats, while attention towards the welfare of goats is rarely taken into account and certainly deserves more consideration.

References

Altizer S, Ostfeld RS, Johnson PTJ, Kutz S, Harvell DC (2013) Climate change and infectious diseases: from evidence to a predictive framework. Science 341:514–519

Anderson US, Benne M, Bloomsmith MA, Maple TL (2002) Retreat space and human visitor density moderate undesirable behavior in petting zoo animals. J Appl Anim Welf Sci 5(2):125–137

Anderson US, Maple TL, Bloomsmith MA (2004) A close keeper-nonhuman animal distance does not reduce undesirable behavior in contact yard goats and sheep. J Appl Anim Welf Sci 7(1):59–69

Araújo do Nascimento AP, Martins Castro MS, de Sousa Oliveira D, Freitas Silveira RM, Ortiz Vega WH, Medeiros Nobre ME, Andrioli A, Maria de Vasconcelos A (2022) Environmental

enrichment in dairy goats in a semi-arid region: thermoregulatory and behavioral responses. J Therm Biol 106(May):103248
Babiuk S, Bowden TR, Boyle DB, Wallace DB, Kitching RP (2008) Capripoxviruses: an emerging worldwide threat to sheep, goats and cattle. Transbound Emerg Dis 55(7):263–272
Bacon H, Vigors B, Shaw DJ, Waran N, Dwyer CM, Bell C (2021) Zookeepers–The most important animal in the zoo? J Appl Anim Welf Sci 00(00):1–13
Baral K, Sharma HP, Kunwar R, Morley C, Aryal A, Rimal B, Ji W (2021) Human wildlife conflict and impacts on livelihood: a study in community forestry system in mid-hills of Nepal. Sustainability (Switzerland) 13(23)
Bista A, Chanchani P, Subedi N, Bajracharya SB (2022) The peri-urban leopards of Kathmandu: assessing determinants of presence and predation on domestic animals. Oryx 56(1):91–100
Black Goat Farm and Sanctuary (2020) Black Goat Farm and Sanctuary. https://www.blackgoatsanctuary.com/. Accessed 21 Sep 2023
Briefer EF, McElligott AG (2013) Rescued goats at a sanctuary display positive mood after former neglect. Appl Anim Behav Sci 146(1–4):45–55
Buttercups Sanctuary for Goats (2023) Buttercups Sanctuary for Goats. https://www.buttercups.org.uk/. Accessed 21 Sep 2023
Chávez-Espinoza M, Cantú-Silva I, González-Rodríguez H, Montañez-Valdez OD (2021) Sistemas de producción de pequeños rumiantes en México y su efecto en la sostenibilidad productiva. Revista MVZ Cordoba 27(1):1–14
Chojnacki RM, Vas J, Andersen IL (2014) The effects of prenatal stocking densities on the fear responses and sociality of goat (Capra hircus) kids. PLoS One 9(4)
Claxton AM (2011) The potential of the human-animal relationship as an environmental enrichment for the welfare of zoo-housed animals. Appl Anim Behav Sci 133(1–2):1–10
Colla F (2022) Indagine sulle modalità di management degli ovicaprini allevati a scopi non zootecnici: costruzione di un questionario informativo. University of Milan
Cortellari M, Bionda A, Negro A, Frattini S, Mastrangelo S, Somenzi E, Lasagna E, Sarti FM, Ciani E, Ciampolini R, Marletta D, Liotta L, Ajmone Marsan P, Pilla F, Colli L, Talenti A, Crepaldi P (2021) Runs of homozygosity in the Italian goat breeds: impact of management practices in low-input systems. Genet Sel Evol 53(1)
Creation Museum (2017) Goats Love Music Too! https://creationmuseum.org/blog/2017/07/13/goats-love-music-too/. Accessed 21 Sep 2023
De Malsche F, Cornips L (2021) Examining interspecies interactions in light of discourse analytic theory: a case study on the genre of human-goat communication at a petting farm. Lang Commun 79:53–70
Dubeuf J-P (2021) Future prospects on the goat activities for the coming decades in the context of a world in transition. In: Goat science – Environment, health and economy. IntechOpen, pp 1–14
Ellen ED, Bas Rodenburg T, Albers GAA, Elizabeth Bolhuis J, Camerlink I, Duijvesteijn N, Knol EF, Muir WM, Peeters K, Reimert I, Sell-Kubiak E, van Arendonk JAM, Visscher J, Bijma P (2014) The prospects of selection for social genetic effects to improve welfare and productivity in livestock. Front Genet 5(Nov):1–14
European Commission (2015) EU Zoos directive good practices document. Publications Office of the European Union
European Federation of City Farms (2002) Driving the goat-cart. Teamwork in a goat club. Good Practice Data base EFCF. https://cityfarms.org/files/attachments/.360/Driving_the_goat_cart.pdf. Accessed 21 Sep 2023
European Federation of City Farms (2017) The Goat book. A guidebook for Goat Clubs. https://cityfarms.org/files/attachments/.1078/Goat_book_for_a_goat_club.pdf. Accessed 21 Sep 2023
Farrand A, Hosey G, Buchanan-Smith HM (2014) The visitor effect in petting zoo-housed animals: aversive or enriching? Appl Anim Behav Sci 151:117–127
Fernandez EJ, Rosales-Ruiz J (2021) A comparison of fixed-time food schedules and shaping involving a clicker for halter behavior in a petting zoo goat. Psychol Rec 71(3):487–491
Forni G (1990) Gli albori dell'agricoltura. Redi

Global Federation of Animal Sanctuaries (2023) Helping sanctuaries help animals. https://www.sanctuaryfederation.org/. Accessed 21 Sep 2023

Goatlandia Farm Animal Sanctuary (2023) Goatlandia Farm Animal Sanctuary. https://www.goatlandia.org/. Accessed 21 Sep 2023

Goats of Anarchy (2023) Goats of Anarchy. https://www.goatsofanarchy.org/. Accessed 21 Sep 2023

Goodheart Animal Sanctuary (2022) Goodheart Animal Sanctuary. https://www.goodheartanimalsanctuaries.com/. Accessed 21 Sep 2023

Górka P, Przybyło M, Kański J, Kloska A (2016) Effect of pelleted cereal-based feed used in the diet on feed intake, eating behaviour, rumination and nutrient digestibility in antelope sitatunga (*Tragelaphus spekii*). J Anim Feed Sci 25:125–133

Gray G, Woolaston R, Eaton B (1995) Breeding for resistance to infectious diseases of small ruminants in Europe (ACIAR mono). Australian Centre for International Agricultural Research

Hart SP (2001) Recent perspectives in using goats for vegetation management in the USA. J Dairy Sci 84:E170–E176

Hosey G, Melfi V, Pankhurt S (2013) Zoo animals behaviour, management, and welfare. Oxford University Press

Ikeoji C (2010) Socio-economic and cultural functions of goats in Niger Delta of Nigeria: implication for education of farmers. Ghana Journal of Agricultural Science 42(1–2):75–81

Koda N, Hirose T, Watanabe G (2013) Relationships between caregiving to domestic goats and gender and interest in science. Innovative Teaching 2:5

Koda N, Kutsumi S, Hirose T, Watanabe G (2016) Educational possibilities of keeping goats in elementary schools in Japan. Front Vet Sci 3(DEC):1–8

Langbein J, Siebert K, Nuernberg G, Manteuffel G (2007) The impact of acoustical secondary reinforcement during shape discrimination learning of dwarf goats (*Capra hircus*). Appl Anim Behav Sci 103(1–2):35–44

Learmonth MJ (2020) Human–animal interactions in Zoos: what can compassionate conservation, conservation welfare and duty of care tell us about the ethics of interacting, and avoiding unintended consequences? Animals 10(11):1–14

Loyd DD, King EG, Thompson JJ (2021) Goats in schools: parental attitudes and perceived benefits. Anthrozoös 34(1):139–155

Mastellone V, Scandurra A, D'aniello B, Nawroth C, Saggese F, Silvestre P, Lombardi P (2020) Long-term socialization with humans affects human-directed behavior in goats. Animals 10(4):1–9

Mellen JD, Ellis S (1996) Animal learning and husbandry training. In: Kleiman D, Allen M, Thompso K, Lumpki S (eds) Wild mammals in captivity: principles and techniques. University of Chicago Press, pp 88–99

Mendl M, Paul ES (2004) Consciousness, emotion and animal welfare: insights from cognitive science. Anim Welf 13:17–25

Miranda EBP, Peres CA, Downs CT (2022) Landowner perceptions of livestock predation: implications for persecution of an Amazonian apex predator. Anim Conserv 25(1):110–124

Miranda-de la Lama GC, Mattiello S (2010) The importance of social behaviour for goat welfare in livestock farming. Small Rumin Res 90(1–3):1–10

Negro A, Cortellari M, Bionda A, Grande S, Cesarani A, Carta A, Macciotta N, Biffani S, Crepaldi P (2022) Empowerment of the Italian small ruminant sector: new traits and tools toward a sustainable management. In: Veerkamp RF, de Haas Y (eds) Proceedings of the 12th World Congress on Genetics Applied to Livestock Production (WCGALP). Wageningen Academic Publishers, pp 373–375

Nenadović K, Ilić T, Jovanović N, Bugarski D, Vučinić M (2021) Welfare of native goat breeds in Serbia—emphasis on parasitological infections. Front Vet Sci 8(August):1–13

Perisse IV, Fan Z, Singina GN, White KL, Polejaeva IA (2021) Improvements in gene editing technology boost its applications in livestock 11(January):1–21

Pfaller-Sadovsky N, Hurtado-Parrado C, Cardillo D, Medina LG, Friedman SG (2020) What's in a click? The efficacy of conditioned reinforcement in applied animal training: a systematic review and meta-analysis. Animals 10(10):1–31

Phocas F, Belloc C, Bidanel J, Delaby L, Dourmad JY, Dumont B, Ezanno P, Fortun-Lamothe L, Foucras G, Frappat B, González-García E, Hazard D, Larzul C, Lubac S, Mignon-Grasteau S, Moreno CR, Tixier-Boichard M, Brochard M (2016) Review: towards the agroecological management of ruminants, pigs and poultry through the development of sustainable breeding programmes: I-selection goals and criteria. Animal 10(11):1749–1759

Pierce J, Bekoff M (2018) A postzoo future: why welfare fails animals in zoos. J Appl Anim Welf Sci 21:43–48

Rauw WM (2016) Improving animal welfare through genetic selection. In: Rauw WM (ed) Frontiers in genetics. Frontiers Media SA

Roger PA (2012) Welfare issues in the reproductive management of small ruminants. Anim Reprod Sci 130(3–4):141–146

Russell T (2017) 'Where goats connect people': cultural diffusion of livestock not food production amongst southern African hunter-gatherers during the Later Stone Age. J Soc Archaeol 17(2):115–137

Safe Haven Farm Sanctuary (2023) Safe Haven Farm Sanctuary. https://safehavenfarmsanctuary.org/. Accessed 21 Sep 2023

Salama AAK, Caja G, Such X, Casals R, Albanell E (2005) Effect of Pregnancy and Extended Lactation on Milk Production in Dairy Goats Milked Once Daily. J. Dairy Sci. 88:3894–3904

Schilling AK, Hotzel H, Methner U, Sprague LD, Schmoock G, El-Adawy H, Ehricht R, Wöhr AC, Erhard M, Geue L (2012) Zoonotic agents in small ruminants kept on city farms in Southern Germany. Appl Environ Microbiol 78(11):3785–3793

Schilling AK, Reese S, Palme R, Erhard M, Wöhr AC (2015) Stress assessment in small ruminants kept on city farms in southern Germany. J Appl Anim Welf Sci 18:119–132

Scholl S, Grall G, Petzl V, Röthler M, Slotta-Bachmayr L, Kotrschal K (2008) Behavioural effects of goats on disabled persons. Int J Therapeutic Commun 29(3):297–309

Simões J, Gutiérrez C (2018) Sustainable goat production in adverse environments. In: Sustainable goat production in adverse environments, vol 1. Springer

Stirling J, Griffith M, Dooley JSG, Goldsmith CE, Loughrey A, Lowery CJ, McClurg R, McCorry K, McDowell D, McMahon A, Millar BC, Rao J, Rooney PJ, Snelling WJ, Matsuda M, Moore JE (2008) Zoonoses associated with petting farms and open zoos. Vector Borne Zoonotic Dis 8(1):85–92

Ward SJ, Williams E, Groves G, Marsh S, Morgan D (2020) Using zoo welfare assessments to identify common issues in developing country zoos. Animals 10(11):1–15

Wilkins AS, Wrangham RW, Tecumseh Fitch W (2014) The "domestication syndrome" in mammals: a unified explanation based on neural crest cell behavior and genetics. Genetics 197(3):795–808

Zobel G, Neave HW, Webster J (2019) Understanding natural behavior to improve dairy goat (Capra hircus) management systems. Transl Anim Sci 3(1):212–224

Index

S. Mattiello, M. Battini (eds.), *The Welfare of Goats*, Animal Welfare 25, https://doi.org/10.1007/978-3-031-62182-6

The manufacturer's authorised representative in the EU is Springer Nature Customer Service Centre GmbH, Europaplatz 3, 69115 Heidelberg, Germany. If you have any concerns regarding our products, please contact ProductSafety@springernature.com

Printed and bound by CPI Group (UK) Ltd, Croydon, CR0 4YY
21/09/2026
02220015-0011